Antimicrobial Resistance in Aquaculture and Aquatic Environments

Preetham Elumalai • Sreeja Lakshmi
Editors

Antimicrobial Resistance in Aquaculture and Aquatic Environments

Springer

Editors
Preetham Elumalai
Department of Marine Biology
Microbiology and Biochemistry
School of Marine Sciences
Cochin University of Science
and Technology
Cochin, Kerala, India

Sreeja Lakshmi
IVVN (International Veterinary Vaccinology
Network) Fellow
University of Edinburgh
United Kingdom

ISBN 978-981-97-7319-0 ISBN 978-981-97-7320-6 (eBook)
https://doi.org/10.1007/978-981-97-7320-6

This Springer imprint is published by the registered company Springer Nature Singapore Pte Ltd.
The registered company address is: 152 Beach Road, #21-01/04 Gateway East, Singapore 189721, Singapore

I dedicate this book to my parents, my guiding stars, who made me capable to achieve my goals and succeed in life.

Preetham Elumalai

Foreword by Prof. M. Junaid Bushiri

Research in the last couple of years has been actively involved in expanding the knowledge on the importance of antimicrobial resistance in aquaculture as a growing concern that can threaten both animals as well as humans. The current book *Antimicrobial Resistance in Aquaculture and Aquatic Environments* narrates antimicrobial resistance (AMR) from the basic level. Aquaculture is one of the highly demanding fields for livelihood and economic stability at the global level. The management of health and sustainability of the industry is of supreme importance. Intensification of fresh water fish farming and the introduction of new species of fish to the farm coupled with the frequent use of antibiotics to fight bacterial infections in aquatic animals are some of the common causes of antimicrobial resistance. Persistence of antibiotics increases the existence of disease-resistant bacteria which spread to the ecosystem causing severe health threats to human beings through the food chain. It is of at most importance to popularize the knowledge of the severe consequences of AMR and finding fruitful applicable strategies to tackle the issue.

The book is articulated in such a way that it covers all the important aspects of AMR starting from the historical perspective on how aquaculture contributes to the AMR global pandemic. Various chapters discuss the AMR in freshwater, marine and farm aquatic organisms. The book takes us through the screening tools to monitor AMR and the pros and cons of the diagnostic tools, and shares valuable knowledge on phage therapy, prebiotics, probiotics and bacteriocins as antimicrobial agents, biosecurity measures and vaccines as effective tools to combat AMR in aquaculture.

It is a pleasure to write a foreword to this resourceful compilation of chapters enriched with the diverse aspects of antimicrobial resistance in aquatic system. I sincerely appreciate the efforts of the authors and editors in articulating this book. I am certain that this book will provide updated information on AMR to a wide range of educationists from academics to research.

Prof. Junaid Bushiri
Vice Chancellor
Cochin University of Science and Technology
Cochin, Kerala, India

Foreword by Prof. Myron Christodoulides

Resistance occurs when organisms capable of causing disease do not respond to treatment with drugs. The increasing levels of resistance of microbial pathogens to treatment with antiviral, antibacterial, antifungal and anti-ecto/endoparasite compounds represents probably the greatest threat to human and animal health this century.

For example, bacterial antimicrobial resistance (AMR) accounted for 1.27 million global human deaths in 2019 and contributed to 4.95 million deaths. O'Neill's authoritative review on AMR published in 2014 estimated that by 2050, AMR would account for 10 million human deaths annually. Though these numbers may be statistically insignificant for a predicted global population of 10 billion by 2050, the estimated cost to the world's economy is estimated to be over 100 trillion USD in lost output. Tackling AMR requires a 'one health' approach to human and veterinary medicine, coupled with an understanding of the contributory roles played by the environment and increasingly by climate change. We now appreciate the importance of good stewardship in the use of antimicrobials for treating human infections, but indiscriminate use of antimicrobials in veterinary medicine is often under-appreciated and needs to be addressed. It has been estimated that ~100,000–200,000 tons of antibiotics are used globally, which is having a direct impact on human and animal health and on food safety. The use and misuse of antimicrobials in animals is driven by the need for food security, both for individual nations and for the global population interconnected by multilateral trading. Notably, aquaculture is seen as a critical source of global food security, as it provides 70% of the world's fish supply.

Now, Drs. Preetham Elumalai and Sreeja Lakshmi have edited an important collection of high-level papers that make a significant contribution to the literature to our understanding of how antimicrobial resistance impacts on aquaculture and aquatic environments. The book begins with chapters that explain what AMR is and the historical perspective and contribution of aquaculture to AMR globally and an examination of AMR in both freshwater and marine environments. Antibiotics have been used since the 1950s to control bacterial infections in fish and as growth promotors, and their use is highly variable amongst countries. It is also clear that antimicrobials/antibiotics are introduced into freshwater via different contaminating routes, usually in discharges from households, farm leaching, the pharmaceutical industry, hospitals and wastewater treatment plants. Aquaculture farming, industrial pollution and discharge present in freshwater all contribute to contaminating the marine environment with antimicrobials. There is a chapter that outlines antibiotic use and the presence of AMR genes in emerging multidrug resistant pathogens and the prevalence of infectious diseases in farmed aquatic organisms, including economically important trout and salmon industries. The zoonotic potential of AMR from aquaculture is also an important issue addressed. Several chapters are included that discuss methods used to screen and monitor AMR in wastewater facilities, recreation facilities and other aquatic environments, and the pros and cons of diagnostics methods for AMR surveillance. The need to identify AMR bacteria in aquaculture feed samples, to monitor changes in the prevalence of AMR over time, and track AMR spread between different species is presented.

The book ends with several chapters that address alternative methods to potentially reduce and/or replace antimicrobial use in aquaculture. Under the umbrella of biosecurity, these include phage therapy, the use of bacteriocin and predatory bacteria, the application of probiotics, prebiotics and synbiotics, and finally the development and introduction of vaccines against the major disease-causing pathogens. All of these are 'works-in-progress' and the successful introduction of *Piscirickettsia salmonis* vaccines in Norway point the way; since their introduction in the late 1980s and early 1990s, these vaccines have reduced significantly the use of antibiotics for farmed Atlantic salmon.

The very final chapter brings the various themes together in the context of a global approach for tackling aquaculture AMR through agreed international action plans. Briefly, countries prepare their National Action Plans (NAPs) under the guidance of the Food and Agriculture Organization (FAO) of the United Nations, the United Nations Environment Programme (UNEP), the World Health Organization (WHO) and the World Organisation for Animal Health (WOAH). Central to these NAPs is a mission to raise awareness of AMR in aquaculture amongst other uses.

In summary, this book will be useful to a wide audience, including academic microbiologists, health professionals, AMR experts and veterinarian researchers in aquaculture, and more broadly when considering the provenance of antimicrobials present in the environment. The chapters included bring the field up-to-date and will stand as a reference for future research in AMR and aquaculture. The editors and commissioned authors are to be congratulated for producing a very fine book.

Prof. Myron Christodoulides
Faculty of Medicine
University of Southampton
United Kingdom

Foreword by Dr. David Verner-Jeffreys

Antimicrobial resistance, the so-called silent pandemic, is one of the biggest challenges facing humankind. Unless measures are put in place to check the emergence and spread of antibiotic-resistant bacteria, modern medicine will cease to function effectively. Routine treatments, such as hip replacements and biopsies, will become highly hazardous, and otherwise inconsequential infections, resulting from minor scratches and grazes, will threaten patients' lives. We have become complacent in taking for granted these wonder drugs. However, they are vital for both human medicine and for ensuring reliable production of terrestrial and aquatic animals, thereby representing key tools for ensuring food security. AMR is a true One Health challenge, as resistant bacteria and the resistance genes that make them less susceptible to antibiotics can readily be spread between humans and animals, including fish and shellfish.

Drs. Preetham Elumalai and Sreeja Lakshmi have edited a very important and timely book that brings together a collection of authoritative papers that collectively summarize what is known about the potential role and impact of AMR in aquaculture. This very important food sector, critical for livelihoods and food security of hundreds of millions of people globally, is potentially threatened by the emergence and spread of AMR. The information provided will be of great interest and use to researchers, farmers, veterinarians, policy makers, and public health experts as well as others with an interest in this important area.

The initial chapters provide a solid grounding in what is known at present about antimicrobial resistance, how it developed and spread, and the issues it causes in aquaculture, as well as more generally in the aquatic environment. There are then a series of chapters that look at particular aspects and risks associated with aquaculture production, the broader aquatic environment, such as potential impact on risks associated with food borne zoonoses, and those associated with environmental discharge of antibiotics and resistant bacteria. Advice on how to diagnose drug-resistant infections and how to monitor AMR in aquaculture, including feed, and more broadly in aquatic systems, is also provided.

Finally, there are some important chapters that investigate key intervention strategies, such as vaccination and biosecurity, as well as promising alternative treatments, such as the use of bacteriophage therapy and probiotics. There is also a concluding chapter that provides guidance as to how AMR in aquaculture should be addressed globally through development of One Health-based national action plans.

All the chapters have been written by recognized international and Indian experts and represent authoritative overviews of the state of knowledge of the important and emerging area. The editors and authors are to be commended for bringing this important collection of papers together to all those working in the field of AMR in aquaculture and the wider aquatic environment.

David Verner-Jeffreys
WorldFish Headquarters (Malaysia)
Bayan Lepas, Penang, Malaysia

Preface

Antimicrobial resistance (AMR) is contemplated as one of the greatest threats to food security and global public health. AMR exemplifies the condition in which disease causing microorganisms become resistant to the drugs used to treat them. Following the extensive dissemination of AMR, food production sectors, especially in low- and middle-income countries, are found to be at high risk. Aquaculture is the fastest growing food sector worldwide and the intensification of aquaculture is vulnerable to disease outbreaks. Lack of management practices to maintain biosecurity, disease outbreaks, and the increased use of antimicrobial drugs are often found as the risk factors of AMR. Improper usage of antibiotics damages the natural environment and changes the microbial population in the livestock and aquaculture. Therefore it carries urgent attention to mitigate AMR risk through efficient regulation and a surveillance system that affirms food security and global public health.

Our book, *Antimicrobial Resistance in Aquaculture and Aquatic Environments*, provides an overview on the importance of the awareness of antimicrobial resistance, causes, and consequences, and practices to be introduced towards the mitigation of AMR for healthy aquaculture environment. We really appreciate the authors for the contribution of their expertise and skilful renditions. We anticipate the book will provide a comprehensive illustration of grounds and determents of AMR.

Kochi, Kerala, India

Preetham Elumalai
Sreeja Lakshmi

Contents

Chapter 1
Understanding AMR

Rodolphe E. Gozlan

Abstract Most rural households depend on agriculture for their livelihoods, and livestock are often an integral part of their production systems. Achieving food security in the face of population growth, climate change, resource degradation, water scarcity, poverty, food loss, and conflict requires a multidimensional approach. Reducing the pressure of food production on the environment while feeding an ever-growing human population is one of the greatest challenges facing humanity. Aquaculture in particular is increasingly seen as a critical source of food security, providing 70% of the world's fish supply. Sustainability will be a key focus for aquaculture production in the future. There will be an increased emphasis on reducing environmental impacts, improving resource efficiency, and minimizing the use of antibiotics and chemicals. Unfortunately, this growth in aquaculture and the associated use of antibiotics has also brought with it a growing threat of antimicrobial resistance (AMR). Here, I try to unravel the drivers behind the rise of antibiotic resistance and the way forward to deal with the threat of antibiotic resistance, the role of environmental drivers in the rise of antibiotic resistance, and in particular the threat of antibiotic resistance in aquaculture as aquaculture is at the crossroads between global warming and antibiotic resistance. Beyond aquaculture, I also look at the threat of antibiotic resistance to the ecosystem with some perspectives on how to reduce the risks of antibiotic resistance. The complexity of managing and reducing the impact of the emergence of antibiotic-resistant bacteria on human, animal, and environmental health (OneHealth) is a major public health problem that requires urgent global action. The time to act is now, and we can be part of the solution.

Keywords Food security · Sustainable aquaculture · Climate change · Poverty · Production

R. E. Gozlan (✉)
ISEM, Université de Montpellier, CNRS, Montpellier, France
e-mail: rudy.gozlan@ird.fr

P. Elumalai, S. Lakshmi (eds.), *Antimicrobial Resistance in Aquaculture and Aquatic Environments*, https://doi.org/10.1007/978-981-97-7320-6_1

1.1 Overall Global Context

The world's population is expected to reach 9 billion by 2050, and some 2.6 billion people in the developing world are estimated to be living on less than $2 a day, of whom some 1.4 billion are 'extremely' poor, living on less than $1.25 a day (Our World in Data 2023). Nearly three-quarters of the extremely poor—1 billion people—live in rural areas. Most rural households depend on agriculture for their livelihoods, and livestock are often an integral part of their production systems. Projected consumption of animal protein in 2030 per km^2 indicates consumption of between 500 and 5000 kg/km^2 in most urban areas of the world. This does not include fish and seafood (FAO 2011). Achieving food security in the face of population growth, climate change, resource degradation, water scarcity, poverty, food loss, and conflict necessitates a multi-dimensional approach. Unsustainable land use practices, deforestation, soil erosion, and water scarcity degrade natural resources that are essential for food production. Soil degradation reduces fertility, water pollution harms ecosystems, and deforestation diminishes biodiversity and disrupts ecological balance. Also, significant amounts of food are lost or wasted throughout the supply chain, from production to consumption. Post-harvest losses, inadequate storage facilities, inefficient transportation, and consumer behaviour contribute to food waste. Conflict, political instability, and displacement also disrupt farming activities, leading to food insecurity. Displaced populations often face limited access to land, resources, and markets, exacerbating their vulnerability to food shortages. Integrating sustainable agricultural practices, investing in rural development, strengthening resilience to climate change, improving water management, reducing food waste, and fostering inclusive governance are key strategies to overcome these challenges. Collaboration among governments, international organizations, researchers, farmers, and consumers is crucial to ensure sustainable and resilient food systems that can provide sufficient and nutritious food for all.

1.2 The Role of Sustainable Food Production

The goal of sustainable development is to reduce the pressure of food production on the environment, while feeding an ever-growing human population is one of the major challenges facing humanity. Implementing sustainable land and water management practices, promoting reforestation, and adopting precision agriculture techniques can help conserve resources and ensure long-term food security. Minimizing food loss and waste through improved infrastructure, better storage and transportation systems, consumer awareness campaigns, and collaboration among stakeholders can reduce pressure on food production and enhance food security. However, the magnitude of the environmental impacts of food production, largely related to land use, has motivated the evaluation of the environmental and health benefits of dietary changes, typically away from meat towards other sources, including fish and seafood. For example, demand for seafood is expected to increase significantly in the coming decades (OECD/FAO 2021). Aquaculture in particular is increasingly seen as a critical source of food security, providing 70% of the world's fish supply. Looking at the projections for fish production in 2030, capture production data level off at just over 90 million tonnes, similar to today, while aquaculture production is expected to increase by over 20% from the current 70 to 90 million tonnes (OECD/FAO 2021).

1.3 The Future of Aquaculture in Insuring Sustainable Food Security

Sustainability will be a key focus in the future of aquaculture production. There will be an increased emphasis on reducing environmental impacts, improving resource efficiency, and minimizing the use of antibiotics and chemicals. Innovative practices such as integrated multi-trophic aquaculture (IMTA), where multiple species are farmed together to create a more balanced ecosystem, and recirculating aquaculture systems (RAS), which minimize water usage and waste, will gain prominence. Advancements in technology will revolutionize aquaculture production. Automation, robotics, and artificial intelligence (AI) will play significant roles in optimizing production processes, improving feed management, monitoring water quality, and enhancing disease detection and prevention. Remote sensing, big data analytics, and genetic technologies will aid in more precise and efficient breeding programs and disease control. With increasing pressure on coastal areas and concerns over environmental impacts, there will be a shift towards offshore aquaculture. Offshore farms, located in deeper waters, can mitigate some environmental concerns and provide opportunities for sustainable expansion. Additionally, high-tech systems such as floating or submersible cages, offshore aquaponics, and land-based closed-containment systems will gain popularity for their potential to minimize environmental impacts and improve production efficiency. Feed availability and sustainability are critical considerations in aquaculture. As the demand for fishmeal and fish oil, traditionally used in aquaculture feeds, continues to increase, alternative and sustainable feed sources will become more prevalent. This includes the development of feed formulations using plant-based ingredients, algae, insects, and single-cell proteins. These alternative feeds can reduce the pressure on wild fish stocks, lower the environmental footprint of aquaculture, and enhance the nutritional profile of farmed seafood. Aquaculture will play a significant role in integrated food systems, where various agricultural practices, such as crop farming and livestock production, are combined to create synergies. Aquaponics, for instance, integrates fish farming with hydroponics, enabling nutrient recycling and maximizing resource use efficiency. Such integrated systems can enhance food production, increase sustainability, and provide opportunities for diversified and resilient farming practices. Consumer demand for sustainable and responsibly produced seafood is increasing. Traceability systems and certification schemes will continue to gain importance, ensuring transparency in the supply chain and enabling consumers to make informed choices. Blockchain technology, which allows for secure and immutable record-keeping, will likely play a role in enhancing traceability and ensuring the integrity of product information including, for example, any antibiotic treatment (e.g. period of treatment, type of antibiotics).

1.4 The Impact of Global Warming on Aquaculture Diseases

Meanwhile, climate change poses significant challenges to aquaculture. Rising water temperatures, ocean acidification, and extreme weather events impact both the health of farmed organisms and the infrastructure supporting aquaculture operations. Global warming is expected to increase the mean water temperature of the disease outbreak season (Karvonen et al. 2010). Such an increase in water temperature has been shown to inevitably lead to an increase in disease risk in aquatic systems and particularly in farmed aquatic species (Reverter et al. 2021; Combe et al. 2023). Similar observations have been made in the ocean, with, for example, an increased risk of *Vibrio* emergence at higher latitudes in response to ocean warming (Baker-Austin et al. 2013; Vezzulli et al. 2016). Such an increase in animal disease will inevitably lead to an increase in prophylactic (prevention) and metaphylaxis (control) treatments to maintain production and ensure food security in the years to come.

1.5 The Rise of Antibiotic Resistance

In 1942, Anne Miller became the first US patient to receive penicillin at New Haven Hospital. Since then, the use of antibiotics has been transferred to animal production, including aquaculture. Unfortunately, this growth in aquaculture and the associated use of antibiotics has also brought with it a growing threat of antimicrobial resistance (AMR). AMR occurs when microorganisms, such as bacteria, become resistant to the drugs used to treat them. Antibiotics are generally used to treat infections in humans, but they are also widely used in the production of food-producing animals. Their use in this context includes treating infections, preventing common diseases, and promoting growth. However, any use of antibiotics carries the risk of encouraging the selection of resistant bacterial strains. In the case of aquaculture fish or shrimp farms, where low doses of antibiotics are regularly administered to stimulate growth and prevent disease, this use creates ongoing selective pressure on resistant bacterial strains (Ferri et al. 2022; Hossain et al. 2022; Kampouris et al. 2022). As a result, new resistant bacterial strains of animal origin are emerging, some of which can colonize and infect humans (Dutil et al. 2010). This resistance can be caused by a variety of factors, including overuse of antibiotics, poor sanitation and hygiene, and the use of substandard drugs. AMR can also be caused by the transmission of resistant bacteria from one organism to another, either through direct contact or through the environment. As bacterial resistance increases, the effectiveness of drugs used to treat infections decreases, leading to an increased risk of infection, increased mortality and even the spread of drug-resistant infections to humans, making it a serious global health concern (Howard et al. 2014). The emergence of resistant pathogens in medical settings is fuelled by the use of antibiotics and the transmission of bacteria among patients, medical staff, and visitors, ultimately diminishing the efficacy of these drugs (Baker et al. 2021; Foreman et al. 2018; Zelmer et al. 2022). Not only does antibiotic use in outpatient populations contribute to resistance in the community, but also in several countries, the availability of over-the-counter antibiotics also leads to inappropriate use and resistance, including instances where a full course of antibiotics is not completed (Laxminarayan et al. 2013). Given the potential for AMR to cause serious damage to both the environment and human health, it is imperative that we address this growing threat in aquaculture. The Centers for Disease Control and Prevention (CDC 2019) in the United States states 'Stop talking about a coming post-antibiotic era—it is already here. You and I are living in a time when some wonder drugs no longer work wonders and families are being torn apart by a microscopic enemy. The time for action is now, and we can be part of the solution'.

1.6 The Way Forward to Deal with the Threat of Antibiotic Resistance

Edith Heard, the scientist who heads the European Molecular Biology Laboratory, warned in 2021: 'If we don't do something, today's antibiotics won't be able to fight infections. This will be the next pandemic'. The CDC has also warned stakeholders and regulators to 'stop playing the blame game'. Every person, every industry, every country can influence the development of antibiotic resistance. We all have a role to play and should be held accountable to make meaningful progress against this threat (CDC 2021). We also need to stop relying solely on new antibiotics that are slow to reach the market and that, sadly, will one day become ineffective against these bacteria. We need to adopt aggressive strategies that keep the bugs away and prevent infections in the first place. We also need to believe that antibiotic resistance is a problem 'over there' in someone else's hospital, state or country—not in our own backyard (CDC 2021). In the United States, it is estimated that up to 50% of antibiotics used in the medical sector are unnecessary or prescribed with incorrect dosage or duration (CDC 2019), exacerbating the problem by promoting additional, and preventable, selection for resistance. Antibiotic resistance has been found in every state in the United States and in every country in the world. There is no place safe from antibiotic resistance, but everyone can do something about it. Do what you can, from washing your hands to improving the use of antibiotics. Despite international efforts to tackle the problem of AMR, global levels of AMR continue to rise. This highlights what is commonly referred to as the 'action gap'. Recent reports from the World Bank and the Global AMR Leaders Group show that the solution to reducing this action gap between the urgent need to reduce AMR and the actual reduction in the environment and production systems, particularly in low- and middle-income countries (LMIC), lies in integrating AMR into broader sustainable development policies. These reports call for a major effort in AMR implementation research to improve our ability to identify and characterize the important links with more global sustainable development strategies.

1.7 The Role of Environmental Drivers on the Rise of Antibiotic Resistance

However, there are aggravating factors for the rise of AMR. In particular, temperature and population density, where, in the United States, *Escherichia coli* antibiotic resistance to all antibiotics was found to be associated with a 30-year average minimum temperature and local population density (MacFadden et al. 2018). This means that an increase of 10 °C across regions leads to a 3% increase in antibiotic resistance for common pathogens. Minimum temperature is widely utilized as a climatic parameter to assess the survival capabilities of species in diverse environments. Its significance becomes particularly evident when examining two crucial aspects: (1) regions that can support the persistence of environmental and/or colonizing bacteria and (2) regions with an increased propensity for the growth and transmission of resistant strains. While further investigation is necessary to fully comprehend the underlying mechanisms driving the observed associations between resistance and temperature, several hypotheses have been put forth. Firstly, temperature may play a facilitating role in horizontal gene transfer, which involves the exchange or acquisition of resistance genes. Secondly, temperature exerts a substantial influence on bacterial growth rates, thus serving as a pivotal factor in the carriage and transmission of resistant strains between humans and animals. The economic level of countries also has an indirect effect on the level of antibiotic resistance. For example, in LMICs, the proportion of antimicrobial compounds with resistance greater than 50% (P50) has increased significantly over time (more than doubling) in chicken, pig, and cattle production from 2000 to 2018 (Van Boeckel et al. 2019), and new hotspots of resistance to multiple drugs have been found in all LMICs across all continents.

1.8 The Threat of Antibiotic Resistance in Aquaculture

The threat of AMR is not limited to human health. Food production systems in a variety of environments are generally recognized as being particularly at risk from the emergence and spread of AMR, particularly in LMICs, where traditional extensive farming practices are gradually evolving towards intensive production systems. This threat can also have serious consequences for aquatic organisms such as fish, shellfish, and other aquatic animals. AMR can lead to increased risk of infection, increased mortality, and reduced growth and reproduction. In addition, AMR can spread through the environment, leading to the spread of drug-resistant bacteria to other organisms. The threat of AMR is particularly worrying for aquaculture, where the use of antimicrobial drugs is becoming increasingly common. Aquaculture is the fastest growing food sector worldwide, with Asia leading the way in global production. The expansion of this industry has been greatly facilitated in recent years by the rapid transition of existing aquaculture farms to intensive production systems, mainly in LMICs in Asia. Antibiotic use in these farms is high and is expected to rise as disease levels intensify and increase (Schar et al. 2020; Lulijwa et al. 2020). This is due to the high stocking densities in aquaculture, which can lead to an increased risk of disease, and the lack of effective disease control measures. As a result, antibiotics are often used to prevent and treat disease in animal production, leading to a greater potential for the development of AMR (Fig. 1.1). Also, the use of antibiotic treatments in aquaculture varies greatly depending on the type of production, from rural (e.g. rice–fish farming), extensive (e.g. pond farming), intensive (e.g. offshore fish cages), or hyper-intensive (e.g. raceways), reflecting the increasing level of control conditions, costs and incomes, and animal density (Ottinger et al. 2016). The use of antibiotics, which is largely unregulated, is also an issue. For example, in Europe, the use of antibiotics for growth promotion or prophylaxis (prevention in healthy animals before clinical signs) is prohibited, but this is not the case in LMICs (Sarter et al. 2019). Furthermore, some of the antibiotics used are the same antibiotics that are critical for human health (amoxicillin, ampicillin,

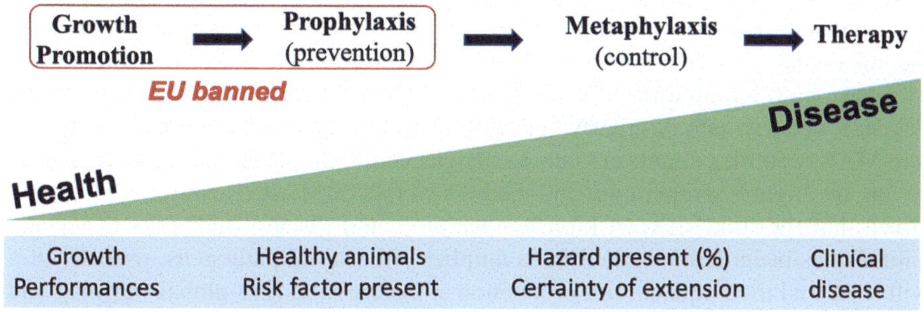

Fig. 1.1 Schematic representation of the link between aquaculture production, disease risk, and antibiotic use

erythromycin, streptomycin, gentamycin, neomycin, oxolinic acid, enrofloxacin, flumequine) or highly important (e.g. chloramphenicol, florfenicol). Aquaculture is also often considered a hotspot for AMR due to its position as a collector of antibiotics from wastewater (urban, hospital, agricultural, etc.), surface run-off from land production and the use of antibiotics in aquaculture itself. The international community is well aware of the unique risks that aquaculture environments pose to AMR on a global scale (Bondad-Reantaso et al. 2020; Thornber et al. 2020). However, when looking at the global situation, only three of the five major aquaculture-producing countries have national AMR action plans that specifically address the aquaculture sector (WHO 2023). The lack of effective management strategies for AMR in aquaculture means that the existing national action plans focus mainly on surveillance and control, rather than directly addressing the underlying socio-economic factors.

1.9 A Crossroad Between Global Warming, Aquaculture, and Antibiotic Resistance

A recent study (Reverter et al. 2020) calculated the multi-antibiotic resistance (MAR) index specifically for individual bacterial isolates (i.e. strains or species) where possible or for groups of isolates (same bacterial genus) as the ratio between the number of resistant bacterial isolates and the number of total combinations tested (number of antibiotics x number of isolates tested). They found that the incidence of MAR in aquaculture was highest in Asia (China, Indonesia, Japan, India, Iran), Africa (Saudi Arabia, Nigeria, Zambia) and Latin America (Mexico, Uruguay) (Reverter et al. 2020). It also highlighted that among a number of socio-economic (i.e. gross domestic product (GDP) per capita, population density, sanitation, malnutrition, and water supply), clinical (i.e. clinical MAR, antibiotic use), and environmental (i.e. animal import, aquaculture, animal production, animal import, animal density, and pesticide use) factors, only temperature and clinical MAR were correlated with aquaculture MAR. This is an important finding as it highlights the pathways underlying AMR. It also shows that the correlation between clinical MAR and aquaculture MAR identifies a link based on economic status, where high-income countries have a low clinical and aquaculture MAR and low-income countries have both a high clinical and aquaculture MAR with countries such as India, Pakistan, and Vietnam. Similarly, the correlation between temperature and aquaculture MAR identifies countries such as Bangladesh, India, Pakistan, and Vietnam as having the highest temperature and the highest risk of MAR (Reverter et al. 2020). Given that the link between high temperatures and disease outbreaks in aquatic animals has been demonstrated for a number of bacterial pathogens, most LMICs will soon find their aquaculture production at the crossroads of global warming and AMR. show the complexity associated with the issue of AMR in aquaculture environments, in particular the interweaving of present and future environmental aspects as well as socio-economic aspects. In fact, the issue of antibiotics is often used as a justification for 'quick-fix' solutions to complex social and economic situations (Broom et al. 2021; Willis and Chandler 2019; Hinchliffe et al. 2018). In order to implement effective AMR strategies, it is therefore necessary to go beyond the simple monitoring and management of antibiotic use at local (farm) level and look more closely at the implementation of proactive and preventive approaches. This is a general problem for AMR strategies, as international agreements on AMR remain essentially focused on research, surveillance, and stewardship rather than on more upstream preventive measures as advocated in the 2015 Global Action Plan on AMR (Tejpar et al. 2022). The issue of AMR should also be integrated into broader sustainable development strategies and thus make use of the funding, resources, and activities associated with them (Fig. 1.2).

1.10 Beyond Aquaculture the Threat of Antibiotic Resistance to the Ecosystem

The threat of AMR is not limited to aquaculture, but encompasses the wider farm environment. We have these AMR phenomena just about everywhere in freshwater and marine environments, where antibiotics are used to treat diseases in wild fish and crustaceans that combine with other sources from urban effluents or agricultural farms. The consequences of this wider diffusion within the ecosystem can be serious, as the spread of drug-resistant bacteria can lead to endemic contamination of water and soil, and even a reduction in aquatic biodiversity. It can also lead to a significant increase in potentially dangerous organisms, such as pathogenic bacteria, and a loss of beneficial organisms, such as those that play a role in the nutrient cycle. It is therefore important to understand that the nature of aquaculture production, which involves open systems connected to the wider ecosystem, makes the risk of AMR in freshwater and marine environments of particular concern (Bondad-Reantaso et al. 2020; Thornber et al. 2020). These environments are home to a wide range of species, many of which are already threatened or endangered, so the problem of AMR also concerns issues linked to the conservation of biodiversity. The spread of resistant bacteria can have serious consequences for these rare or vulnerable species, leading to increased mortality and reduced populations. In addition, the spread of resistant bacteria can lead to the spread of drug-resistant infections in humans, who could become infected through contact with these environments during leisure activities such as swimming or fishing. This makes AMR in aquaculture a serious global health problem that goes far beyond aquaculture production.

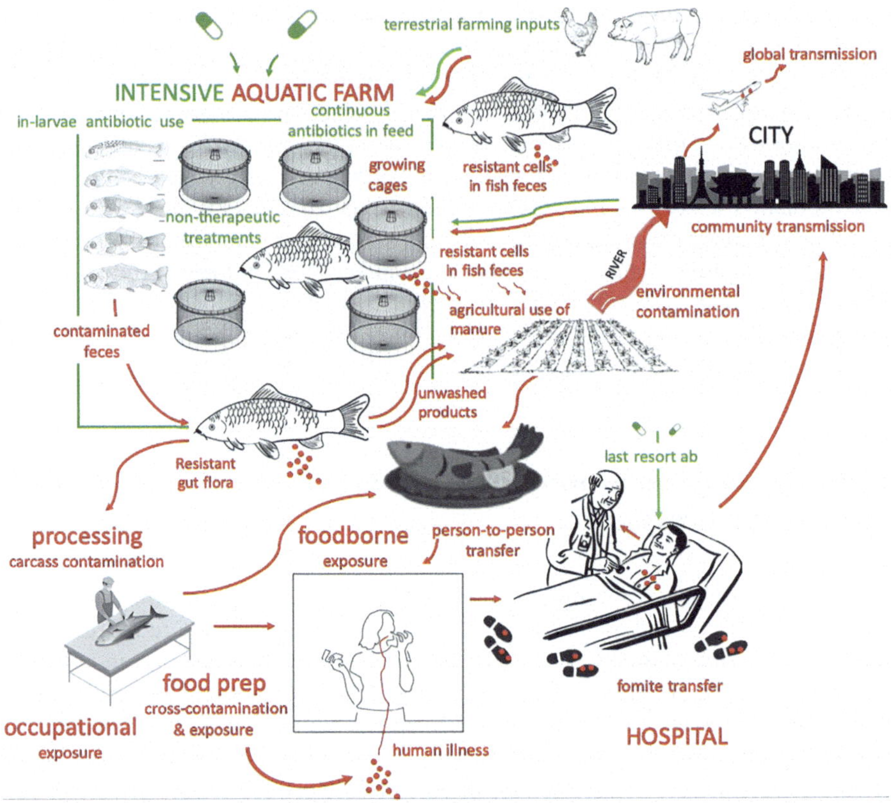

Fig. 1.2 There are numerous pathways linking antibiotic use (green) and antibiotic resistance (red) in the aquaculture and human health sectors. Most antibiotics currently sold in developing countries support fish and shrimp production and are used prophylactically as growth promoters and/or to treat, control, and prevent disease. There is also significant overlap between the spectrum of antibiotics used in animals and humans. Aquaculture is typically a hotspot for antibiotic resistance, as not only does this sector use antibiotics, but aquacultures also collect surplus antibiotics from land-based farms (e.g. pigs, poultry industry) or urban run-off. They receive either antibiotics from these external sources or bacteria that are already resistant to certain antibiotics. Systematic exposure to antibiotics selects for resistant bacteria that can persist on fish/shrimp or in waste during rendering. Humans are exposed to and can be infected by resistant bacteria in aquaculture and fish/shrimp processing plants, as well as by preparing and consuming contaminated fish/shrimp. Waste containing resistant bacteria can infiltrate surrounding waters. Waste containing resistant bacteria can enter nearby aquatic and terrestrial ecosystems and can also be used as fertilizer in crops, which are then consumed by humans. These food-borne bacteria can cause antibiotic-resistant infections in humans. Certain types of resistant pathogens can persist in the human population and cause further infections in the community or return to animal food production and be re-amplified to resist other classes of drugs. With the increase in global connectivity through transport, the risk of the spread of resistant bacterial strains is no longer a local problem but is becoming a global one. (Modified from Koch et al. 2017)

1.11 Some Perspectives to Reduce the Risks of Antibiotic Resistance

As we have seen throughout this chapter, the threat of AMR is indeed a serious global health problem, which is cross-sectoral and goes beyond animal or human health alone, but has ramifications for the overall health of aquatic and terrestrial ecosystems (OneHealth; see). It is therefore essential that on aquaculture farms, we take rapid and practical steps to deal with this growing threat as effectively as possible. This requires a combination of strategies to combat AMR with a multifaceted approach that includes improving the appropriate use of antimicrobials in humans, animals, and agriculture. Some strategies have already been proposed to combat AMR such as the following:

1. Promoting the responsible use of antimicrobial agents in aquaculture, such as developing and implementing antimicrobial use regulations and developing alternative methods to prevent and treat bacterial infections
2. Improving surveillance and monitoring of AMR in aquaculture, e.g. by establishing national and international surveillance systems, improving laboratory capacity for antimicrobial susceptibility testing, and promoting the exchange of data on AMR in aquaculture
3. Promoting good aquaculture practices, such as improving water quality, reducing stocking densities and promoting the use of vaccines and other alternative disease prevention methods
4. Raising public awareness of AMR in aquaculture, such as promoting responsible seafood consumption, raising awareness of the risks associated with the misuse of antimicrobial agents, and supporting research and development of alternative methods of disease prevention and treatment

In order to achieve these objectives, there are already potential avenues that could lead to greater sustainability in production, and thus to less recourse to antibiotics, by relying essentially on the resilience of production systems, on strengthening the immune systems of cultivated organisms and finally on better control of infectious risks. I propose a number of pragmatic avenues to explore, such as the following:

1. In some areas, aquaculture could easily be combined with forest planting to ensure food, energy, and water resilience, while limiting the impact of global warming by improving ecosystem services through fertilization, water reuse, carbon storage, and biomass bioenergy production (e.g. Shifflett et al. 2016). This approach shows that coupled agroecosystems could produce efficient food and bioenergy yields and at the same time promote water reuse while complying with water quality regulations and thus reduce disease risk emergence. This is an approach that needs to be explored further, breaking down the silos of thought that sometimes exist between different disciplinary fields and even different ecosystem compartments.

2. There are, of course, the more conventional sanitary tools, with future prospects for vaccines for industrial fish farming (e.g. Brudeseth et al. 2013). Until recently, this approach has been limited to rich countries, but as many traditional fish farms in LMIC are moving to more intensive modes of production, such vaccine approaches could also appear in new production areas. High-density production requires effective vaccines to control various diseases, with a significant positive impact on reducing the use of antibiotics. This was, for example, the case in Norway, which saw a rapid drop in antibiotic consumption following the introduction of vaccines for the production of Atlantic salmon (*Salmo salar*) in the late 1980s and early 1990s.
3. Beyond vaccination, the use of plant extracts in aquaculture as an alternative to antibiotics (e.g. Reverter et al. 2021) could become a booming sector. Indeed, this natural approach to health has become very popular in recent years, in the treatment of a number of human diseases, and consumers are increasingly concerned about production methods, particularly the use of antibiotics in livestock farming. Plant-enriched diets can significantly improve the growth, immunity, and disease survival of animals. Plant supplements are also versatile alternatives that can benefit both small-scale breeders and large-scale producers.
4. In the context of alternatives to vaccination and in line with the use of plants, the use of traditional pharmacopoeia in small-scale freshwater fish farming as metaphylactic treatments offers real advantages for maintaining food security in relatively isolated areas (e.g. Caruso et al. 2013). The use of plants, as well as their variety and number, has been shown to depend not only on the species cultivated, production systems, and production areas but also on social characteristics such as the professional experience of fish farmers and their ethnic origin.
5. Finally, new avenues of research are opening up, notably with the use of probiotics as a means of controlling diseases in aquaculture (e.g. Hoseinifar et al. 2018). Probiotics, prebiotics, synbiotics, and medicinal plants are among the most promising dietary supplements for the control or treatment of bacterial, viral, and parasitic diseases in aquaculture. There are several ways in which probiotics could protect against disease, either by boosting the immune response, competing for binding sites, or producing antibacterial substances and competing for nutrients.

1.12 Conclusion

This chapter sets out the basic issues surrounding antibiotic resistance, with a focus on the aquaculture sector. The complexity of managing and reducing the impact of the emergence of antibiotic-resistant bacteria on human, animal, and environmental health (OneHealth) is well understood. However, AMR is a major public health problem that requires urgent global action. In the animal production sector and in aquaculture in particular, the excessive and abusive use of antimicrobial agents can contribute to the emergence and spread of AMR in aquatic environments, as well as to an indirect effect of global warming, which will lead to an increase in the number and frequency of bacterial infections in the production chain. Combating AMR in aquaculture therefore requires a global approach that includes reducing the use of antimicrobial agents, promoting good aquaculture practices, improving the surveillance and control of AMR, and raising public awareness of the problem, as well as other natural approaches to limiting the use of antibiotics and reducing the risk of bacterial infection. Taking measures to combat AMR in aquaculture could ensure the sustainable development of aquaculture, safeguard public health, and protect aquatic ecosystems.

References

Baker RE, Mahmud AS, Miller IF, Rajeev M, Rasambainarivo F, Rice BL, Takahashi S, Tatem AJ, Wagner CE, Wang L-F, Wesolowski A, Metcalf CJE (2021) Infectious disease in an era of global change. Nat Rev Microbiol 20:193–205

Baker-Austin C, Trinanes JA, Taylor NGH, Hartnell NR, Siitonen A, Martinez-Urtaza J (2013) Emerging *Vibrio* risk at high latitudes in response to ocean warming. Nat Clim Chang 3:73–77

Bondad-Reantaso MG, Lavilla-Pitogo CR, Karunasagar I, Arthur JR, Hao B, Irde E, Garrido GE, Peñarubia OR (2020) Outputs and activities of FAO project FMM/RAS/298/MUL on antimicrobial resistance in fisheries and summary of FAO's recent work on antimicrobial resistance in aquaculture. FAO, Rome

Broom A, Kenny K, Prainsack B, Broom J (2021) Antimicrobial resistance as a problem of values? Views from three continents. Crit Public Health 31:451–463

Brudeseth BE, Wiulsrød R, Fredriksen BN, Lindmo K, Løkling K-E, Bordevik M, Steine N, Klevan A, Gravningen K (2013) Status and future perspectives of vaccines for industrialised fin-fish farming. Fish Shellfish Immunol 35(6):1759–1768

Caruso D, Lusiastuti AM, Taukhid, Slembrouck J, Komarudin O, Legendre M (2013) Traditional pharmacopeia in small scale freshwater fish farms in West Java, Indonesia: an ethnoveterinary approach. Aquaculture 416–417:334–345

CDC (2019) Antibiotic resistance threats in the United States, 2019. U.S. Department of Health and Human Services, CDC, Atlanta, GA

CDC (2021) Combating antimicrobial resistance, a global threat. https://www.cdc.gov/drugresistance/index.html

Combe M, Reverter M, Caruso D, Pepey E, Gozlan RE (2023) Impact of global warming on the severity of viral diseases: a potentially alarming threat to sustainable aquaculture worldwide. Microorganisms 11(4):1049

Dutil L, Irwin R, Finley R, Ng LK, Avery B, Boerlin P, Bourgault A-M, Cole L, Daignault D, Desruisseau A, Demczuk W, Hoang L, Horsman GB, Ismail J, Jamieson F, Maki A, Pacagnella A, Pillai DR (2010) Ceftiofur resistance in Salmonella enterica serovar Heidelberg from chicken meat and humans, Canada. Emerg Infect Dis 16:48–54

FAO (2011) World livestock 2011—livestock in food security. FAO, Rome

Ferri G, Lauteri C, Vergara A (2022) Antibiotic resistance in the finfish aquaculture industry: a review. Antibiotics 11(11):1574

Foreman KJ, Marquez N, Dolgert A, Fukutaki K, Fullman N, McGaughey M et al (2018) Forecasting life expectancy, years of life lost, and all-cause and cause-specific mortality for 250 causes of death: reference and alternative scenarios for 2016–40 for 195 countries and territories. Lancet 392:2052–2090

Hinchliffe S, Butcher A, Rahman MM (2018) The AMR problem: demanding economies, biological margins, and co-producing alternative strategies. Palgrave Commun 4:142

Hoseinifar SH, Sun YZ, Wang A, Zhou Z (2018) Probiotics as means of diseases control in aquaculture, a review of current knowledge and future perspectives. Front Microbiol 9:2429

Hossain A, Habibullah-Al-Mamun M, Nagano I, Masunaga S, Kitazawa D, Matsuda H (2022) Antibiotics, antibiotic-resistant bacteria, and resistance genes in aquaculture: risks, current concern, and future thinking. Environ Sci Pollut Res 29:1–22

Howard SJ, Hopwood S, Davies SC (2014) Antimicrobial resistance: a global challenge. Sci Transl Med 6(236):236ed10

Kampouris ID, Klümper U, Kramer L, Sorum H, Wedekind H, Berendonk TU (2022) Dissemination of antibiotic resistance in antibiotic-free recirculating aquaculture systems. J Hazard Mater Adv 8:100201

Karvonen A, Rintamäki P, Jokela J, Valtonen ET (2010) Increasing water temperature and disease risks in aquatic systems: climate change increases the risk of some, but not all, diseases. Int J Parasitol 40(13):1483–1488

Koch BJ, Hungate BA, Price LB (2017) Food-animal production and the spread of antibiotic resistance: the role of ecology. Front Ecol 15(6):309–318

Laxminarayan R, Duse A, Wattal C, Zaidi AKM, Wertheim HFL, Sumpradit N, Vlieghe E, Hara GL, Gould IM, Goossens H, Greko C, So AD, Bigdeli M, Tomson G, Woodhouse W, Ombaka E, Peralta AQ, Qamar FN, Mir F, Kariuki S, Bhutta ZA, Coates A, Bergstrom R, Wright GD, Brown ED, Cars O (2013) Antibiotic resistance—the need for global solutions. Lancet Infect Dis 13:1057–1098

Lulijwa R, Rupia EJ, Alfaro AC (2020) Antibiotic use in aquaculture, policies and regulation, health and environmental risks: a review of the top 15 major producers. Rev Aquac 12(2):640–663

MacFadden DR, McGough SF, Fisman D, Santillana M, Brownstein JS (2018) Antibiotic resistance increases with local temperature. Nat Clim Chang 8(6):510–514

OECD/FAO (2021) OECD-FAO agricultural outlook 2021-2030. OECD Publishing, Paris. https://doi.org/10.1787/19428846-en

Ottinger M, Clauss K, Kuenzer C (2016) Aquaculture: relevance, distribution, impacts and spatial assessments—a review. Ocean Coast Manag 119:244–266

Our World in Data (2023). https://ourworldindata.org/poverty

Reverter M, Sarter S, Caruso D, Avarre JC, Combe M, Pepey E, Pouyaud L, Vega-Heredía S, de Verdal H, Gozlan RE (2020) Aquaculture at the crossroads of global warming and antimicrobial resistance. Nat Commun 11:1870

Reverter M, Tapissier-Bontemps N, Sarter S, Sasal PP, Caruso D (2021) Moving towards more sustainable aquaculture practices: a meta-analysis on the potential of plant-enriched diets to improve fish growth, immunity and disease resistance. Rev Aquac 13(1):537–555

Sarter S, Reverter M, Gozlan RE, Avarre J-C, Caruso D (2019) Antimicrobial resistance in aquaculture: challenges and alternatives in the context of global warming. In: Ecological intensification: a new paragon for sustainable aquaculture international conference, 28–30 Oct 2019, Bogor, Indonesia

Schar D, Klein EY, Laxminarayan R, Gilbert M, Van Boeckel TP (2020) Global trends in antimicrobial use in aquaculture. Sci Rep 10:1–9

Shifflett SD, Culbreth A, Hazel D, Daniels H, Nichols EG (2016) Coupling aquaculture with forest plantations for food, energy, and water resiliency. Sci Total Environ 571:1262–1270

Tejpar S, Van Katwyk RS, Wilson L, Hoffman SJ (2022) Taking stock of global commitments on antimicrobial resistance. BMJ Glob Health 7:e008159

Thornber K, Verner-Jeffreys DW, Hinchliffe S, Rahman MM, Bass D, Tyler C (2020) Evaluating antimicrobial resistance in the global shrimp industry. Rev Aquac 12:966–986

Van Boeckel TP, Pires J, Silvester R, Zhao C, Song J, Criscuolo NG, Gilbert M, Bonhoeffer S, Laxminarayan R (2019) Global trends in antimicrobial resistance in animals in low- and middle-income countries. Science 365(6459):eaaw1944

Vezzulli L, Grande C, Reid PC, Hélaouët P, Edwards M, Höfle MG, Brettar I, Colwell RR, Pruzzo C (2016) Climate influence on Vibrio and associated human diseases during the past half-century in the coastal North Atlantic. PNAS 113(34):E5062–E5E50

Willis LD, Chandler C (2019) Quick fix for care, productivity, hygiene and inequality: reframing the entrenched problem of antibiotic overuse. BMJ Glob Health 4:e001590

World Health Organization (2023) Library of AMR national action plans. https://www.who.int/teams/surveillance-prevention-control-AMR/national-action-plan-monitoring-evaluation/library-of-national-action-plans. Accessed 18 Jun 2023

Zelmer AR, Nelson R, Richter K, Atkins GJ (2022) Can intracellular *Staphylococcus aureus* in osteomyelitis be treated using current antibiotics? A systematic review and narrative synthesis. Bone Res 10(1):53

Chapter 2
Historical Perspective and Contribution of Aquaculture to the AMR Global Pandemic

Miriam Reverter, Sarahi Vega-Heredia, and Philip J. Warburton

Abstract The global rise in the emergence and spread of pathogens (bacteria, viruses, fungi, and parasites) that have developed or acquired new resistance mechanisms to drugs, commonly referred to as antimicrobial resistance (AMR) is alarming. The emergence and spread of AMR, which is mostly attributed to the use (often overuse and misuse) of antimicrobial substances such as antibiotics, is not a new phenomenon. In fact, AMR was discovered several decades ago. However, exponential increases in global antimicrobial use have led to a rise in the emergence and spread of AMR, which is nowadays considered as one of the top ten global public health threats by the World Health Organization (WHO). Antimicrobials, and in particular antibiotics, were first used in aquaculture practices in the 1950s, as a source to control bacterial infections. However, as in other food-producing sectors, the use of non-therapeutic antimicrobials in aquaculture (e.g. as prophylaxis treatments, to promote animal growth or as preservative to prevent fish and shellfish spoilage) becomes widespread soon after. The pervasive use of antimicrobials in aquaculture has facilitated the emergence and dissemination of antibiotic resistance worldwide, (1) by directly selecting for resistant pathogens within the aquaculture facilities, (2) by disseminating antimicrobial resistant microorganisms and genes through high levels of international trade of reared aquatic species, and (3) by allowing antimicrobials to enter freshwater and marine ecosystems, where they can select and transfer AMR genes to the wild populations and environment (i.e. water, sediment). In fact, aquatic environments, and in particular aquaculture inland facilities, are often considered as hotspots of AMR, since they may not only contribute to AMR emergence through antimicrobial use, but may accumulate and disseminate

M. Reverter (✉)
School of Biological and Marine Sciences, University of Plymouth, Plymouth, UK
e-mail: miriam.reverter@plymouth.ac.uk

S. Vega-Heredia
UABC School of Marine Sciences, Universidad Autónoma de Baja California, Ensenada, Mexico

P. J. Warburton
School of Biomedical Sciences, University of Plymouth, Plymouth, UK

P. Elumalai, S. Lakshmi (eds.), *Antimicrobial Resistance in Aquaculture and Aquatic Environments*, https://doi.org/10.1007/978-981-97-7320-6_2

antimicrobial resistant microorganisms and genes coming from other AMR-contaminated sources (e.g. urban and agricultural effluents), amplifying AMR selection pressure. The aquaculture contribution to the global AMR pandemic has now been recognized by the WHO and other international organizations such as the Food and Agriculture Organization of the United Nations (FAO), and several efforts have been launched over the past decade to regulate antimicrobial use in aquaculture. Although structured and systematic surveillance efforts of AMR in aquaculture are still lacking, increasing scientific literature is revealing geographic aquaculture AMR hotspots, which may help in developing local targeted surveillance measures.

Keywords AMR selection · AMR transfer · AMR dissemination · Aquatic environments · Aquatic animals

2.1 History of Antibiotics and Antimicrobial Resistance (AMR)

Antibiotics are one of the key medical interventions of the twentieth century, revolutionizing the treatment of infectious diseases and enabling life-saving medical procedures to become not only possible but also routine. While penicillin is typically thought of as the first antibiotic to be discovered and used to fight infections, remedies with apparent antibiotic effects have been used to treat illnesses dating back centuries, long before the discovery of microbes and their role in causing infection. There are several references to the use of moulds in folk medicine cures (Wainwright et al. 1992) and evidence that ancient Egyptians appear to have used a variety of compounds to treat several potentially infectious diseases (Metwaly et al. 2021). Furthermore, Harrison et al. (2015) replicated a tenth century Anglo-Saxon remedy and demonstrated its ability to kill methicillin-resistant *Staphylococcus aureus* (MRSA) in a mouse chronic wound model.

Modern-day antibiotics are natural, semi-synthetic, or synthetic compounds that use either a narrow- or broad-spectrum mechanism of activity to either kill bacteria (bactericidal) or inhibit their ability to grow (bacteriostatic). Sir Alexander Fleming's discovery of penicillin is usually accredited as the discovery of the first antibiotic. However, Bartolomeo Gosio's discovery of mycophenolic acid from *Penicillium glaucum* (renamed *Penicillium brevicompactum*) inhibiting *Bacillus anthracis* comes several decades earlier (Gosio 1896), as does the report on arsphenamine, commonly called Salvarsan or compound 606, a synthetic organoarsenic antibacterial agent introduced in the 1910s as a treatment for syphilis (Ehrlich and Bertheim 1912). Given these reports were not published in English, this likely led to the lack of wider reporting.

The well-known discovery of penicillin (penicillin G) in 1929 (Fleming 1929) was a fortuitous discovery resulting from a *Penicillium notatum* (renamed *P. chrysogenum*) contamination that showed inhibition of *Staphylococcus aureus* on an agar plate. Penicillin G acts by preventing the formation of the bacterial wall, rendering the cell osmotically unstable and leads to lysis. The first test of penicillin in 1941 on a patient with a sepsis infection resulted in an initial drastic improvement in their condition; however supplies ran out before the patient could be fully treated. Subsequent trials with adequate supplies resulted in successful treatment of other patients, confirming penicillin as an effective treatment (Abraham et al. 1941). The discovery of several more compounds followed, which included tyrothricin (Dubos 1939), streptomycin (Jones et al. 1944) with efficacy against *Mycobacterium tuberculosis*, chloramphenicol (Ehrlich et al. 1947), aureomycin (renamed chlortetracycline) (Duggar 1948), and cephalosporin C (Brotzu 1948; Newton and Abraham 1955) with successive analogues of this compound proving to be very effective. The identification of so many new compounds triggered the 'Golden Age' of antibiotic discovery, a period between the late 1940s and 1960s that accounted for the discovery of approximately half of the compounds commonly used today (Davies 2006). This includes classes of antibiotics such as the streptogramins (Charney et al. 1953), erythromycins (Flynn et al. 1954), and glycopeptides in the late 1950s (Griffith 1981), along with the synthesized compounds trimethoprim (Roth et al. 1962), an antifolate compound, and the quinolone precursor, nalidixic acid (Lesher et al. 1962). While the oxazolidinones were originally discovered in the 1970s, their potential as an antibiotic was only realized almost 30 years later with the approval of linezolid for clinical use in 2000 (Ford et al. 2001). Since the 1970s, any new compound entering the clinic has been a derivative or modified version of an existing drug class with no new antibiotic class being found over that 40-year period. However, the implementation of novel screening methods has subsequently led to the discovery of teixobactin in 2015 (Ling et al. 2015) and marks the first new class of antibiotic in decades, indicating that there are still new compounds to be discovered.

When antibiotics were first introduced, it was suggested that the development of bacterial resistance to these compounds was unlikely due to low mutation rates. However, Fleming warned of the dangers of penicillin misuse and that exposure to subinhibitory concentrations led to antibiotic resistance (Fleming 1945). While resistance through mutation has been observed, antibiotic resistance is inevitable as microorganisms that produce antibiotics must possess the mechanisms to be able to protect themselves from their own toxic compounds. Mechanisms of resistance include drug inactivation, reduced uptake or active efflux of the antibiotic, and altered target site or modification of an essential biosynthetic pathway that render the antibiotic ineffective (Reygaert 2018). Resistance can be either intrinsic, meaning the bacterial cell does not possess the antibiotic target site and is unaffected, or acquired through alteration in their genetics that confer resistance; this can occur either through mutation, or the acquisition of resistance genes via transduction (phage-mediated transfer), transformation (update of free DNA), or conjugation (acquisition through mobile genetic elements such as a plasmid).

Initial observations of resistance to penicillin were first reported in the 1940s, around the time it was being trialled for use, with *Escherichia coli* (Abraham and Chain 1940) and *Staphylococcus aureus* strains (Rammelkamp and Maxon 1942) found to capable of resisting penicillin; with the discovery of Salvarsan-predating penicillin, resistance to this compound was also reported earlier. Within 20 years, over 80% of hospital staphylococci isolates analyzed were found to the resistant to penicillin, while over 65% of those from the community were also resistant (Chambers 2001). The development of methicillin, a second-generation penicillin capable of resisting beta-lactamase enzymes that inactivate the antibiotic by breaking its structure, became available in 1959. However, its introduction was quickly followed by reports of MRSA from 1961 onwards (Jessen et al. 1969), and a similar pattern of resistance is seen for other classes of antibiotics following their release.

Antibiotics have been a ubiquitous and successful cure for infectious diseases; however, increasing resistance is rendering existing drugs less effective, and fewer new compounds are being brought to market to replace them. It is estimated that around 25% of antibiotics are inappropriately administered in the clinic (Ray et al. 2019), but this is compounded by their continued prophylactic use in agriculture and aquaculture in many countries (Schar et al. 2020; Tiseo et al. 2020), antibiotics as growth promoters were banned in the United Kingdom and the EU since 2006. As a result of overuse and misuse, antibiotic resistance continues to increase and is now severely impacting patient treatment with an estimated 1.27 million deaths worldwide in 2019 as a direct result of antibiotic-resistant infections (Murray et al. 2022). With predictions of 10 million annual deaths worldwide by 2050 because of antibiotic resistant infections (O'Neill 2016), there needs to be considerable emphasis on mobilizing global initiatives to tackle antimicrobial resistance (AMR).

2.2 Historical Use of Antimicrobials in Aquaculture

Antimicrobials, and in particular antibiotics such as tetracyclines, were first used in aquaculture practices in the 1950s, as a source to control bacterial infections (Kirchhelle 2018). However, as in other food-producing sectors, the use of non-therapeutic antimicrobials in aquaculture (e.g. as prophylaxis treatments, to promote animal growth or as preservative to prevent fish and shellfish spoilage) becomes widespread soon after (Cabello 2006; Kirchhelle 2018). Antibiotics use in aquaculture increased especially around the 1970s following bacterial and fungal outbreaks, which triggered the widespread use of antibiotics like oxytetracycline, amoxicillin, and sulfadimethoxine–ormetoprim (Romet 30) (Kirchhelle 2018). Antibiotics risks, especially in regard to AMR emergence, lead to regulations on the use of antibiotics in animal husbandry. The European Economic Community (ECC), Switzerland, and Britain were among the first to restrict the use of medically relevant antibiotics, and the non-therapeutic use of antibiotics in Europe was banned in 2001 (2001/82/EC) (Kirchhelle 2018). Improved management and the use of alternative treatments such as vaccinations allowed Norwegian salmon aquaculture to decrease their antimicrobial consumption by 90% between 1987 and 1993 (from 50 to 5 tonnes) (Kirchhelle 2018). However, the use of antibiotics is highly variable among countries and type of cultured organisms. For example, antibiotic use in salmon aquaculture (*Salmo* spp.) ranges from ~0.02 to 0.39 g antibiotic/tonne of produced salmon in Scotland and Norway to ~660 g/tonne in Chile (Millanao et al. 2011; Watts et al. 2017).

In 2017, the global antimicrobial consumption in aquaculture was estimated at 10,259 tonnes, with Asia (mostly China, India, Indonesia, and Vietnam) accounting for most of the antimicrobials consumed (~94%) (Schar et al. 2020). Regulations regarding antibiotic use in aquaculture vary among Asian countries, but the major aquaculture producers (China, Indonesia, Vietnam, and India, which accounted for ~88% of the global aquaculture production in 2020) all have national regulations that regulate the use of veterinary drugs and set standards for antibiotic residues and withdrawal periods (FAO 2022). However, a review in 2017 by the World Health Organization South East Asia Region (SEAR) found that many SEAR countries did not enforce regulations to monitor antibiotic use in aquaculture (Goutard et al. 2017). Of special concern is the use of antibiotics classed by the WHO as Critically and Highly Important Antibiotics (CIA and HIA), since they are considered essential as last-resort treatment for multidrug-resistant infection in humans (Collignon et al. 2016). A recent article found that 35% of antibiotics used in global aquaculture were classified as CIA and 61% as VIA, highlighting the potential global health risks from aquaculture-driven AMR (Schar et al. 2020).

High levels of antibiotic resistance have been found across countries, especially in Southeast Asia, Central and South America, and Africa (Reverter et al. 2020; Schar et al. 2021). Recent research found that 28 countries over 40 have presented high multiantibiotic resistant (MAR) levels, with countries such as Zambia, Mexico, Tunisia, Saudi Arabia, and Iran presenting the highest MAR (Reverter et al. 2020). Another study identified aquaculture MAR hotspots in Turkey, India, China, and Vietnam (Schar et al. 2021), highlighting the need to reinforce antibiotic regulations in these countries.

In recent years, the implementation of more stringent regulations and the raise of awareness of antibiotic risks has triggered an increasing focus on finding alternatives to antibiotics in aquaculture. These include the use of single- and multi-pathogen vaccines, probiotics, prebiotics such as the use of immunostimulant plants, and improved management practices to increase animal welfare (Brudeseth et al. 2013; Tarkhani et al. 2020; Reverter et al. 2021). Although these new avenues cast light in the future of aquaculture, more efforts are needed to fully understand and mitigate the current role of aquaculture practices on AMR emergence and dissemination.

2.3 Contribution of Aquaculture to Global AMR

2.3.1 Direct and Indirect Contributions of Aquaculture to Emergence, Accumulation, and Spread of AMR

Environments in which aquaculture takes place such inland water bodies, coastal areas, and offshore cages have been shown to act as hotspots for AMR emergence and dissemination (Cabello et al. 2016; Graham et al. 2019; Reverter et al. 2020). These environments not only contribute to AMR emergence through the direct use of antimicrobials but are also often highly connected with surrounding ecosystems, facilitating indirect accumulation, dissemination, and amplification of AMR processes (e.g. microbial mutation, recombination, and horizontal gene transfer) (Fig. 2.1) (Hatosy and Martiny 2015; Graham et al. 2019).

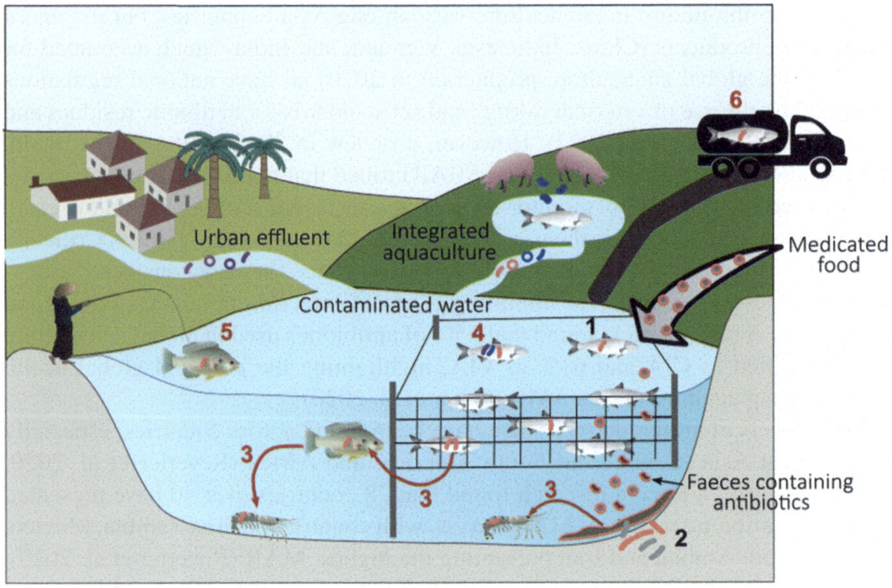

Fig. 2.1 Schema of main processes leading to AMR emergence (black numbers) and dissemination (red numbers) in aquaculture. (1) Emergence of AMR in reared organisms as a result of antibiotic administration. (2) Emergence of AMR in surrounding environment (e.g. water, sediment) as a result of antibiotic accumulation (in uneaten food and faeces from reared organisms). (3) Dissemination of AMR from reared to wild organisms. (4) Amplification of AMR emergence and dissemination through exposure of aquaculture systems and reared organisms to contaminated sources such as urban effluents or residues from other production systems (e.g. integrated livestock–aquaculture farms). (5) Dissemination of AMR through the food chain. (6) Dissemination of AMR through national and international trade of reared organisms containing antibiotic residues and AMR microbes

2.3.1.1 Aquaculture Practices and AMR Emergence

AMR Emergence in Reared Aquatic Animals

The best studied mechanism by which aquaculture contributes to global AMR is by the administration of antimicrobials to reared animals (Fig. 2.1). Antimicrobials are most often delivered to reared animals (e.g. fish or shellfish) via medicated feed (often ad libitum), but bathing (i.e. immersion of animals in water containing diluted antimicrobials) is also used (Cabello et al. 2013). Contrary to most terrestrial-farming techniques, antimicrobial treatments in aquaculture are always performed at the cohort level (i.e., all fish present in a tank or system are given antimicrobials, regardless of their individual illness status), generally following disease diagnosis of few individuals (Cabello et al. 2013). Using such untargeted, large-scale use of antimicrobials in aquaculture contributes by itself significantly to the emergence of global AMR. However, the problem of AMR in aquaculture is multi-pronged, and understanding the different socio-economic and environmental aspects is paramount to design successful mitigation programs (Collignon et al. 2018; Thornber et al. 2022).

Most of the world's aquaculture (>90%) occurs in low- and middle-income countries (LMICs), where aquaculture practices are little regulated, creating vast opportunities for exchanges between human, animal, and environmental bacteria, and emergence of AMR (Thornber et al. 2020; FAO 2022). Misinformation on suitable practices to prevent and treat disease outbreaks (e.g. biosecurity and quarantine procedures) and appropriate therapeutical use of antimicrobials (e.g., suitability of antimicrobials depending on etiological agent, appropriate dosing, and minimum length of treatment) is common in LMICs (Nadimpalli et al. 2018; Ström et al. 2019; Thornber et al. 2022). Accessibility to regulated high-quality antibiotics is often limited in LMICs, with many fish farmers acquiring antibiotics 'over the counter' and administering them to reared aquatic animals according to own preferred practices (Phu et al. 2015; Laxminarayan et al. 2016; Henriksson et al. 2018; Ström et al. 2019). Although the magnitude of this problem (i.e. use of unregulated antibiotics) remains unknown, recent research has highlighted that banned antibiotics important for human medicine (e.g. CIA, HIA) (Collignon et al. 2016) are still widely used in aquaculture practices in several countries (Lulijwa et al. 2020; Schar et al. 2020, 2021; Luu et al. 2021).

Research has also shown that routine antibiotic use (e.g. as growth promoter or prophylaxis) is still prevalent in some countries, which exacerbates the problem of indiscriminate antibiotic use and AMR emergence (Henriksson et al. 2018; Schar et al. 2020; Luu et al. 2021). Administering sub-inhibitory antibiotic doses over sustained periods of time provides the perfect conditions for the selection of antimicrobial resistant strains (Aarestrup and Carstensen 1998; Gullberg et al. 2011; Lau et al. 2022). Low antibiotic concentrations increase horizontal gene transfer (HGT) by activating DNA recombinases and hence mediating the spread of antimicrobial resistant genes (ARGs) (Gullberg et al. 2011; Andersson and Hughes 2014). Chronic antibiotic administration in fish has also been shown to suppress innate immunity (Guardiola et al. 2012; Limbu et al. 2018) and reduce gut microbial diversity (Navarrete et al. 2008; Rosado et al. 2019), which could potentially lead to enhanced disease susceptibility and need of further antimicrobials. Stressful rearing conditions such as crowding and poor environmental conditions (e.g. oxygen, nutrients, pollutant exposure) have all been shown to compromise animal's immunity and microbial homeostasis, increasing the risk of disease outbreaks, leading to higher use of antimicrobials (Bly et al. 1997; Boutin et al. 2013).

Finally, disease-causing etiological agents are commonly not identified (or misdiagnosed) in rural aquaculture settings, leading to the (mis)use of broad-spectrum antibiotics to treat a variety of diseases including non-bacterial (e.g. antifungal and antiviral) infections (Henriksson et al. 2018). It is worth noting, however, that identification of the etiological agents of aquatic animal diseases is a major global challenge of the industry both in LMICs and high-income countries (HICs). The limited research in aquatic diseases in comparison with terrestrial pathologies, together with the high diversity of animals reared, and the complexity of the infections (infections are often multi-pathogen), results in overall misunderstanding of many aquatic infections and consequent use of broad-spectrum antimicrobials (Bayliss et al. 2017; Sanguanrut et al. 2018).

AMR Emergence in Aquaculture Environments

A second major problem of aquaculture practices is that large amounts of antibiotics administered to aquatic animals leak and accumulate in the surrounding environment (e.g. water, sediment), altering the microbiota and favouring the emergence and transmission of AMR and ARGs (Fig. 2.1) (McPhearson et al. 1991; Muziasari et al. 2016; Rico et al. 2017). Some estimates suggest that up to 80% of the antimicrobials administered to reared aquatic animals are deposited in the surrounding environment (Cabello et al. 2016). Medicated uneaten food as well as excretions (stools, urine) containing unabsorbed antimicrobials dissipate and accumulate in water and sediment. Although some antibiotics undergo rapid degradation in the water (hydrolysis or photodegradation) (Babić et al. 2013), many adsorb on dissolved organic matter and sediment, remaining stable for prolonged time periods (Yuan et al. 2019). Experimental studies showed that a wide range of antibiotics such as oxytetracycline, flumequine, florfenicol, sulfadiazine, and trimethoprim persist over 6 months in marine sediments (Samuelsen et al. 1992; Hektoen et al. 1995) and may persist longer (>12 months) in stagnant anoxic conditions, such as those often found in aquaculture farms (Björklund et al. 1990). Antibiotics in aquaculture farm sediments are often found at sub-inhibitory concentrations (Lalumera et al. 2004; Xiong et al. 2015; Yuan et al. 2019), which are known to favour AMR emergence and dissemination (Gullberg et al. 2011). Recent studies have revealed that composition and functionality of bacterial communities in antibiotic-containing sediments is altered and persists even after antibiotic removal, favouring the proliferation of ARGs hosts, which contribute to ARG propagation (Hornick and Buschmann 2018; Deng et al. 2020; Xu et al. 2022; Feng et al. 2023). Increases in AMR-associated mobile genetic elements (MGEs) have also been observed in sediments of aquaculture farms, increasing the risk of further AMR dissemination (Chen et al. 2022).

2.3.1.2 Dissemination of AMR Derived from Aquaculture Practices

Aquaculture farms are highly connected with surrounding environments, and therefore AMR microbes and ARGs acquired in aquaculture settings can easily be disseminated into the environment and reach humans (Fig. 2.1). Aquaculture effluents have been shown to contain high quantities of antibiotics and ARGs, acting as a source of antibiotics in environmental water (Jang et al. 2018; Pham et al. 2018; Jo et al. 2021). Aquaculture effluents are dispersed over large distances affecting ecosystems that can be several kilometres away (Serrano-Grijalva et al. 2011; Herbeck and Unger 2013). Through this journey, diluted antibiotics can exert selection pressures leading to further AMR emergence and ARGs can be passed via HGTs through the trophic chain (EFSA Panel on Biological Hazards (BIOHAZ) et al. 2021). Antibiotics and ARGs in the water effluents can also sink or be biologically transferred to seabed or river sediments, where they accumulate, further creating opportunities for AMR emergence, dissemination, and amplification (Andrieu et al. 2015). Although this problem is especially prevalent in areas with low biosecurity where aquaculture wastewater is directly (i.e., without treatment) discarded into environmental/local water sources (Opiyo et al. 2018), removal of antibiotics, AMR bacteria, and ARGs from wastewater remains challenging (Hiller et al. 2019; Pallares-Vega et al. 2019). Traditional wastewater treatments are unable to remove antibiotics completely, and although a multitude of novel wastewater and aquaculture effluent treatments are being developed, most of them are still on the concept phase (Gorito et al. 2022; Liu et al. 2022; Deng et al. 2023).

Rearing aquatic animals in the natural environment (e.g. in cages or ponds) is well known to lead to a crossover between wild and reared species (Krkosek et al. 2006). The bilateral transfer of pathogens between wild and reared animals has been previously identified as a concern in the aquaculture sector (Krkosek et al. 2006; Petchimuthu et al. 2022). Recent research suggests that the close interaction between reared and wild species can also lead to the dissemination of AMR and ARGs from reared to wild animals and vice versa (Ruzauskas et al. 2021; Lima et al. 2022), leading to AMR spread and amplification throughout the environment.

Finally, probably the most concerning, and the most effective manner of disseminating aquaculture-related AMR and ARGs is through the international trade of reared aquatic products. Encompassing the growth of the aquaculture industry, international trade of aquaculture commodities has grown over 170% between 1976 and 2022 (FAO 2022). While the international trade of reared aquatic products was limited to few high-value species (e.g. salmon) coming from few HICs a couple of decades ago, current market has seen an increase in the number of species internationally traded, and an increase in products coming from LMICs (Anderson et al. 2018; FAO 2022). For example, in 2020, 225 states and territories were involved in trading aquatic product commodities (FAO 2022). While increasing aquaculture trade in LMICs can contribute to global food safety (e.g. through increased incomes and improved nutrition) (Gephart et al. 2021; Gonzalez Parrao et al. 2021), higher trade volumes and connectivity also pose public health concerns. Inappropriate antibiotic administration during aquatic animal rearing (e.g. misuse or overuse of antibiotics, lack of hygiene protocols leading to cross-contamination, insufficient withdrawal time) can lead to trading and selling products for human consumption that contain antibiotics in their tissues (Yipel et al. 2017; Chen et al. 2018; Okocha et al. 2018). The presence of antibiotic residues in aquatic products, even at trace amounts, poses serious public health threats (Okocha et al. 2018). First, products with residual antibiotics also have a higher probability of containing AMR microbes and ARGs, which can pass directly to humans upon consumption (Vivekanandhan et al. 2002). Some AMR bacteria and ARGs from aquaculture products are pathogens for humans (Zhang et al. 2015; Sivaraman et al. 2021), posing serious health risks. Second, global trade has increased the rate of transmission across natural boundaries of both pathogenic and non-pathogenic resistant microbes, facilitating

AMR dissemination and transfer from aquatic animals to humans (Memish et al. 2003; Okocha et al. 2018).

2.3.1.3 Connected Systems: The Role of Cross-Contamination on AMR Amplification and Dissemination Across Systems

As described above, the aquaculture sector actively contributes to global AMR emergence and dissemination. However, the described processes can be significantly enhanced when aquaculture practices are not appropriately managed or isolated from other potentially polluted systems (e.g. via low biosecurity measures and use of contaminated resources). For example, the use of antibiotic and AMR-contaminated water sources and animal feed has been shown to be a major source of AMR emergence, dissemination, and amplification in aquaculture systems both in HICs and LMICs (Su et al. 2017; Novais et al. 2018; Thornber et al. 2022). Pollution from upstream aquaculture or agriculture farms, urban effluents, or run-off from floods can all introduce antibiotics at sub-inhibitory doses and ARGs to aquaculture systems (Zhong et al. 2018; Thornber et al. 2022). Furthermore, it has been shown that a variety of pollutants such as heavy metals and pesticides induce selection of AMR bacteria by promoting the transmission of ARGs and associated MGEs (Chenia and Jacobs 2017; Xing et al. 2020; Gupta et al. 2022), further exacerbating AMR selection and dissemination in polluted or cross-contaminated environments.

Integrated and multitrophic aquaculture systems such as earthen ponds where different types of organisms (e.g. livestock (fish/shellfish) or fish/shellfish in agriculture) are reared together is highly prevalent in LMICs. These systems are generally considered more efficient and sustainable, as they allow higher nutrient retention through the transformation of waste products into food (Granada et al. 2016; Nederlof et al. 2022). However, despite the advantages of such systems, increasing evidence show that integrated culture systems, especially those integrating different animals, act as important hotspots for AMR emergence and spread. Integrated animal aquaculture provides multiple opportunities for pathogen and AMR exchange between compartments, leading to higher AMR microbes, ARGs, and associated MGEs in the aquaculture systems (Petersen and Dalsgaard 2003; Yuan et al. 2019). In such environments, farmers tend to reuse organic material such as dead animals or animal manure as feed for the reared aquatic animals, introducing residual antibiotics and AMR from livestock rearing into the aquatic system (Shah et al. 2012; Bouelet Ntsama et al. 2018). In aquaculture–agriculture integrated systems, pond water is reused without further treatment for agricultural farming, such as rice fields (Guan et al. 2022; Thornber et al. 2022). Although research on whether these integrated systems also facilitate AMR is scarce, the recent discovery that pesticides induce AMR selection (Xing et al. 2020) rises some concerns over these practices.

Finally, high correlations exist between human and aquaculture antibiotic resistance levels, which are also positively correlated with socioeconomic indices such as the HSBC climate vulnerability index (CVI, combines climatic and socioeconomic information to estimate the countries' vulnerability to climate change, with lower levels indicating higher vulnerability) (Reverter et al. 2020). This not only indicates that the aquatic and human resistome are highly connected (i.e. cannot be compartmentalized) but also highlights that countries with less capacity to respond to climate change will also be most affected by AMR. OneHealth approaches that target reducing AMR global levels in all contributing sectors are therefore paramount to halt the current global AMR rise.

2.3.2 AMR in Different Aquaculture Sub-sectors

Aquaculture is one of the fastest-growing food sectors in the world, which experienced significant production increases in the last decade (2000–2019: finfish 2%, crustaceans 7%, and other aquatic animals production 0.5% increases). Due to the constant production growth in reared aquatic animals production, intensive practices are increasingly used (FAO 2022; Verdegem et al. 2023). High density of animals, which are often accompanied with decreasing water conditions, create stressful conditions, under which animals become more susceptible to contagious diseases (Santos and Ramos 2018). Higher disease rates increase prophylactic and therapeutic use of antimicrobials and chemical disinfectants to control these infections, creating multiple opportunities for AMR emergence (Cabello et al. 2016). Given the high diversity of organisms reared (494 species belonging to over 5 phyla) and the disparity in the aquaculture practices (e.g. offshore vs. inshore, intense vs. extense, monoculture vs. polyculture), the opportunities and mechanisms of AMR emergence may differ (FAO 2022). Below, we review how different aquaculture practices contribute to AMR emergence and dissemination.

2.3.2.1 AMR in Aquaculture of Different Organisms

AMR Emergence and Dissemination in Crustacean Aquaculture

Global production of crustaceans increased 5.2 times between 2000 and 2019, with marine shrimp reared in brackish-water ponds, being the most reared crustacean (FAO 2022). The rapid growth of shrimp farms is being driven by their high market value. In 2019, the price of crustaceans was 7.3 US$/kg compared to 2.6 US$/kg for finfish (FAO 2022; Verdegem et al. 2023). Asia is the leading continent in the production of crustaceans, accounting for 98% of global production (FAO 2022). Due to the growing shrimp production and demand, stocking densities are very high, which increase shrimp stress levels and bacterial densities, making the stock more susceptible to infections (Ali et al. 2022; Thornber et al. 2022). Shrimp bacterial diseases are one of the biggest problems, for example, hepatopancreatic necrosis disease (AHPND) (Restrepo et al. 2018).

Crustaceans such as shrimps do not have an acquired immune system, which means vaccines are not a possibility to prevent disease outbreaks in crustacean aquaculture systems (Smith et al. 2003; Thornber et al. 2020). Antibiotics are the most commonly substance used in shrimp farming to treat diseases, but other antimicrobials, including heavy metals, fungicides and antiparasitics are also frequently applied (Thornber et al. 2020; Li et al. 2021). As such, antibiotic residues and ARGs have been detected in effluents and sediments surrounding shrimp farms. For example, Khan et al. (2022) detected five multidrug-resistant *Enterobacterales* in sediments from *Penaeus monodon* and *Macrobrachium rosenbergii* farms in Bangladesh. In Thailand, AMR resistance against amoxicillin, oxytetracycline, and enrofloxacin and ARGs (*sul*1 and *sul*2) were found in eight farming pond waters (Suyamud et al. 2021). Although most Asian countries have national plans to control antibiotic use, a study performed in 2018 found that 17% of shrimp farmers in Vietnam used antibiotics as prevention, and that one farm used banned antibiotics (Luu et al. 2021).

AMR Emergence and Dissemination in Mollusc Aquaculture

World aquaculture mollusc production grew by about 70 million tonnes between 1992 and 2022 (FAO 2022). The great development of bivalve aquaculture has been widely distributed along the coasts all over the world (Kijewska et al. 2022). Since most bivalves (e.g. mussels and oysters) are filter feeders, they feed on particles suspended in the water column including phytoplankton and bacteria (Grevskott et al. 2017). As most coasts are heavily under anthropogenic influence, reared bivalves can be exposed to pathogenic bacteria (e.g. from wastewater) or other pollutants, including antibiotics and heavy metals, which may cause disease outbreaks and lead to AMR emergence and dissemination (Bernard 1989; Kijewska et al. 2022; Pavón et al. 2022). A previous study detected four antibiotics from the tetracycline group (tetracycline, oxytetracycline, doxycycline, and chlortetracycline) in the tissue of cultured mussels and clams (Chiesa et al. 2018). Oysters and other bivalves are frequently eaten raw, and therefore AMR bacteria can be passed onto the consumer's gut, where through HGT may pass ARGs to the resident consumer gut flora (Rees et al. 2015; Grevskott et al. 2017). As such, mollusc aquaculture can be a risk to global human health and the global AMR pandemic (Pavón et al. 2022). The potential risks of mollusc aquaculture to human health, in regard to bacterial infections and toxin accumulation, are generally well controlled and mitigated (Turner et al. 2014; O'Mahony 2018). However, mollusc monitoring national plans need to incorporate AMR monitoring, as mollusc aquaculture presents a different and potentially more severe threat to AMR dissemination to humans than other aquatic cultured species.

AMR Emergence and Dissemination in Finfish Aquaculture

Fishes (freshwater and marine) are the most consumed aquatic animals globally, with global finfish production increasing from 20.8 million Mt in 2000 to 56.3 million Mt in 2019, an increase of 170% (FAO 2022; Verdegem et al. 2023). However, the rapid expansion and intensification of the fish industry is translating into an increased use of antibiotics to treat disease outbreaks (Asche et al. 2013). Most marine fish farms (e.g. salmon, sea bass, sea bream) use either flow-through systems that pump seawater to the inshore fish tanks, discharging it back to the ocean, or offshore cages, where the fish are contained through nets in shallow marine environments (FAO 2022). Residual antibiotics leeched out from fish feeds are therefore often directly discarded into the ocean (Pepi and Focardi 2021). Some studies show that fish farm effluents contained higher amounts of antibiotic resistant bacteria and genes (ARGs), compared to non-fish farm areas (Jang et al. 2018; Jo et al. 2021). Prevalent ARGs against beta-lactam, tetracycline, and aminoglycosides were found (Jo et al. 2021).

2.3.2.2 AMR in Different Aquaculture Systems

AMR Emergence and Dissemination in Open Aquaculture Systems

Open aquaculture systems are those where the organisms are reared in facilities that are interconnected with the natural waterways (i.e. the farm structures are not isolated from the natural environment). Most aquaculture facilities, especially those in LMICs, use open systems (Garza et al. 2022). Open systems can be both extensive and intensive and can be adapted to culture of a wide variety of species in different environments, including freshwater and brackish water open ponds, and coastal aquaculture and mariculture facilities using ropes, sticks, racks, pens, and cages (FAO 2022). Antimicrobial use in open aquaculture systems increases the risk of impacting the nearby ecosystems by dispersing and accumulating antibiotics and facilitating dissemination of AMR bacteria and ARGs (Rico et al. 2012; Song et al. 2016; Garza et al. 2022). Undigested food and faeces could retain the antimicrobial residues and remain in the water and sediments around fish and shellfish farms for extended periods of time or even be transported by the water flow or currents to other ecosystems (Herbeck and Unger 2013; Song et al. 2016). By selection pressure and genetic fluctuation, these environmental antimicrobial residues can alter the environmental microbial communities and act as reservoirs of ARGs (Cabello et al. 2013; Nakayama et al. 2017).

AMR Emergence and Dissemination in Recirculating Aquaculture Systems (RAS)

Recirculating aquaculture uses enclosed systems, where the water is treated and reused within the facilities. In this type of aquaculture, animal rearing does not interact directly with the surrounding environment and thus has been considered a less environmentally impactful practice than open aquaculture systems (Martins et al. 2010; Rurangwa and Verdegem 2015; Bergman et al. 2020). These practices not only lead to a reduction in water consumption but also avoid the direct release of faeces, feed surplus, and antimicrobial residues into the environment (Martins et al. 2010). Establishing barriers between compartments is thought to limit cross-contamination and AMR dissemination (e.g. by reducing contact between wild and reared animals and limiting element exchanges), meaning that recirculating aquaculture systems (RAS) may be more suitable to reduce aquaculture contribution to AMR (Ahmed and Turchini 2021). RAS tend to have more technological advances than open systems leading to better disease surveillance and biosecurity, which could lead to reductions in the use of antibiotics and integration of in situ antibiotic water removal systems (Ahmed and Turchini 2021; Sha et al. 2022). However, recent research suggests that contaminated feed and reared organisms can introduce antibiotic residues and AMR bacteria into free-antibiotic RAS leading to further AMR emergence and dissemination (Kampouris et al. 2022; Liu et al. 2022). Also, wastewater treatment and solid removal (e.g. particulated organic matter coming from faeces and uneaten feed) from RAS can produce a sludge containing antimicrobial residues and AMR bacteria. If this solid waste is not appropriately treated (e.g. effective removal of antibiotics and AMR bacteria) and is instead directly discarded onto soil or used as fertilizer in agriculture, this can amplify AMR emergence and dissemination (Cripps and Bergheim 2000; Watts et al. 2017).

2.4 Summary of Global Challenges in Reducing AMR Emergence and Dissemination in Aquaculture

AMR emergence and dissemination in the aquaculture sector is highly complex. First, the high diversity of species reared accompanied by the wide range of aquaculture systems used translates into different possible mechanisms for AMR emergence and dissemination. Second, the high connectivity between aquaculture systems and surrounding environments, especially in LMICs, increases the difficulty of isolating and reducing AMR emergence in aquaculture without intervening at a global level. Actions targeting the aquaculture sector are necessary, for example, increasing awareness of antibiotics and AMR among rural aquatic farmers, reinforcing national regulations to avoid misuse of antibiotics (e.g. use of banned antibiotics and use of antibiotics as prophylactic treatments) and reinforcing hygiene protocols (e.g. quarantine procedures, wastewater treatment). However, evidence suggest that these actions alone will not be sufficient because the aquatic resistome is highly connected with the human resistome, and therefore OneHealth approaches must be prioritized. Research has shown that countries with less response capacity will be most badly affected by global threats such as AMR and climate change, posing real global human security risks. Therefore, international efforts are urgently needed to ensure the establishment of appropriate OneHealth actions aiming to (1) regulate and, when possible, decrease antimicrobial use and (2) implement appropriate control barriers to avoid cross-contamination between environmental compartments.

References

Aarestrup FM, Carstensen B (1998) Effect of tylosin used as a growth promoter on the occurrence of macrolide-resistant enterococci and staphylococci in pigs. Microb Drug Resist 4:307–312

Abraham EP, Chain E (1940) An Enzyme from Bacteria able to Destroy Penicillin. Nature 146:837–837

Abraham EP, Chain E, Fletcher CM, Gardner AD, Heatley NG, Jennings MA, Florey HW (1941) Further observations on penicillin. The Lancet 238:177–189

Ahmed N, Turchini GM (2021) Recirculating aquaculture systems (RAS): environmental solution and climate change adaptation. J Clean Prod 297:126604

Ali H, Rahman MM, Jaman A, Basak SK, Eltholth M, Murray F (2022) Economic performance characterization of intensive shrimp (Penaeus monodon) farming systems in Bangladesh. Aquac Fish Fish 2:57–70

Anderson JL, Asche F, Garlock T (2018) Globalization and commoditization: the transformation of the seafood market. J Commod Mark 12:2–8

Andersson DI, Hughes D (2014) Microbiological effects of sublethal levels of antibiotics. Nat Rev Microbiol 12:465–478

Andrieu M, Rico A, Phu TM, Huong DTT, Phuong NT, Van den Brink PJ (2015) Ecological risk assessment of the antibiotic enrofloxacin applied to Pangasius catfish farms in the Mekong Delta, Vietnam. Chemosphere 119:407–414

Asche F, Roll KH, Sandvold HN, Sørvig A, Zhang D (2013) Salmon aquaculture: larger companies and increased production. Aquac Econ Manag 17:322–339

Babić S, Periša M, Škorić I (2013) Photolytic degradation of norfloxacin, enrofloxacin and ciprofloxacin in various aqueous media. Chemosphere 91:1635–1642

Bayliss SC, Verner-Jeffreys DW, Bartie KL, Aanensen DM, Sheppard SK, Adams A, Feil EJ (2017) The promise of whole genome pathogen sequencing for the molecular epidemiology of emerging aquaculture pathogens. Front Microbiol 8:121

Bergman K, Henriksson PJG, Hornborg S, Troell M, Borthwick L, Jonell M, Philis G, Ziegler F (2020) Recirculating aquaculture is possible without major energy tradeoff: life cycle assessment of warmwater fish farming in Sweden. Environ Sci Technol 54:16062–16070

Bernard FR (1989) Uptake and elimination of coliform bacteria by four marine bivalve mollusks. Can J Fish Aquat Sci 46:1592–1599

Björklund H, Bondestam J, Bylund G (1990) Residues of oxytetracycline in wild fish and sediments from fish farms. Aquaculture 86:359–367

Bly JE, Quiniou SM, Clem LW (1997) Environmental effects on fish immune mechanisms. Dev Biol Stand 90:33–43

Bouelet Ntsama IS, Tambe BA, Tsafack Takadong JJ, Medoua Nama G, Kansci G (2018) Characteristics of fish farming practices and agrochemicals usage therein in four regions of Cameroon. Egypt J Aquat Res 44:145–153

Boutin S, Bernatchez L, Audet C, Derôme N (2013) Network analysis highlights complex interactions between pathogen, host and commensal microbiota. PLoS One 8:e84772

Brotzu G (1948) *Ricerche su di un nuovo antibiotico*. Lavori dell'Istituto d'Igiene di Cagliari

Brudeseth BE, Wiulsrød R, Fredriksen BN, Lindmo K, Løkling KE, Bordevik M, Steine N, Klevan A, Gravningen K (2013) Status and future perspectives of vaccines for industrialised fin-fish farming. Fish Shellfish Immunol 35:1759–1768

Cabello FC (2006) Heavy use of prophylactic antibiotics in aquaculture: a growing problem for human and animal health and for the environment. Environ Microbiol 8:1137–1144

Cabello FC, Godfrey HP, Tomova A, Ivanova L, Dölz H, Millanao A, Buschmann AH (2013) Antimicrobial use in aquaculture re-examined: its relevance to antimicrobial resistance and to animal and human health. Environ Microbiol 15:1917–1942

Cabello FC, Godfrey HP, Buschmann AH, Dölz HJ (2016) Aquaculture as yet another environmental gateway to the development and globalisation of antimicrobial resistance. Lancet Infect Dis 16:e127–e133

Chambers HF (2001) The changing epidemiology of Staphylococcus aureus? Emerg Infect Dis 7:178–182
Charney J, Fisher WP, Curran C, Machlowitz RA, Tytell AA (1953) Streptogramin, a new antibiotic. Antibiot Chemother (Northfield) 3:1283–1286
Chen H, Liu S, Xu XR, Diao ZH, Sun KF, Hao QW, Liu SS, Ying GG (2018) Tissue distribution, bioaccumulation characteristics and health risk of antibiotics in cultured fish from a typical aquaculture area. J Hazard Mater 343:140–148
Chen J, Yang Y, Jiang X, Ke Y, He T, Xie S (2022) Metagenomic insights into the profile of antibiotic resistomes in sediments of aquaculture wastewater treatment system. J Environ Sci (China) 113:345–355
Chenia HY, Jacobs A (2017) Antimicrobial resistance, heavy metal resistance and integron content in bacteria isolated from a South African tilapia aquaculture system. Dis Aquat Org 126:199–209
Chiesa LM, Nobile M, Malandra R, Panseri S, Arioli F (2018) Occurrence of antibiotics in mussels and clams from various FAO areas. Food Chem 240:16–23
Collignon PJ, Conly JM, Andremont A, McEwen SA, Aidara-Kane A, Agerso Y, Andremont A, Collignon P, Conly J, Dang Ninh T, Donado-Godoy P, Fedorka-Cray P, Fernandez H, Galas M, Irwin R, Karp B, Matar G, McDermott P, McEwen S, Mitema E, Reid-Smith R, Scott HM, Singh R, DeWaal CS, Stelling J, Toleman M, Watanabe H, Woo GJ (2016) World Health Organization ranking of antimicrobials according to their importance in human medicine: a critical step for developing risk management strategies to control antimicrobial resistance from food animal production. Clin Infect Dis 63:1087–1093
Collignon P, Beggs JJ, Walsh TR, Gandra S, Laxminarayan R (2018) Anthropological and socioeconomic factors contributing to global antimicrobial resistance: a univariate and multivariable analysis. Lancet Planet Health 2:e398–e405
Cripps SJ, Bergheim A (2000) Solids management and removal for intensive land-based aquaculture production systems. Aquac Eng 22:33–56
Davies J (2006) Where have All the Antibiotics Gone? Can J Infect Dis Med Microbiol 17:287–290
Deng C, Liu X, Li L, Shi J, Guo W, Xue J (2020) Temporal dynamics of antibiotic resistant genes and their association with the bacterial community in a water-sediment mesocosm under selection by 14 antibiotics. Environ Int 137:105554
Deng Y-Y, Zou M-Y, Liu W, Lian Y-L, Guo Q-M, Zhang X-M, Dan A (2023) Antibiotic removal and microbial response mechanism in constructed wetlands treating aquaculture wastewater containing veterinary drugs. J Clean Prod 394:136271
Dubos RJ (1939) Studies on a bactericidal agent extracted from a soil Bacillus: i. Preparation of the agent. Its activity in vitro. J Exp Med 70:1–10
Duggar BM (1948) Aureomycin; a product of the continuing search for new antibiotics. Ann N Y Acad Sci 51:177–181
EFSA Panel on Biological Hazards (BIOHAZ), Koutsoumanis K, Allende A, Álvarez-Ordóñez A, Bolton D, Bover-Cid S, Chemaly M, Davies R, De Cesare A, Herman L, Hilbert F, Lindqvist R, Nauta M, Ru G, Simmons M, Skandamis P, Suffredini E, Argüello H, Berendonk T, Cavaco LM, Gaze W, Schmitt H, Topp E, Guerra B, Liébana E, Stella P, Peixe L (2021) Role played by the environment in the emergence and spread of antimicrobial resistance (AMR) through the food chain. EFSA J 19:e06651
Ehrlich P, Bertheim A (1912) Über das salzsaure 3.3′-Diamino-4.4′-dioxy-arsenobenzol und seine nächsten Verwandten. Ber Dtsch Chem Ges 45:756–766
Ehrlich J, Bartz QR, Smith RM, Joslyn DA, Burkholder PR (1947) Chloromycetin, a New Antibiotic From a Soil Actinomycete. Science 106:417
FAO (2022) The state of world fisheries and aquaculture 2022. Towards blue transformation. FAO, Rome
Feng Y, Lu Y, Chen Y, Xu J, Jiang J (2023) Microbial community structure and antibiotic resistance profiles in sediments with long-term aquaculture history. J Environ Manag 341:118052
Fleming A (1929) On the Antibacterial Action of Cultures of a Penicillium, with Special Reference to their Use in the Isolation of B. influenzæ. Br J Exp Pathol 10:226–236

Fleming, A. (1945) Sir Alexander Fleming - Nobel Lecture. https://www.nobelprize.org/prizes/medicine/1945/fleming/lecture/
Flynn EH, Sigal MV Jr, Wiley PF, Gerzon K (1954) Erythromycin. I. Properties and Degradation Studies1. J Am Chem Soc 76:3121–3131
Ford CW, Zurenko GE, Barbachyn MR (2001) The discovery of linezolid, the first oxazolidinone antibacterial agent. Curr Drug Targets Infect Disord 1:181–199
Garza M, Mohan CV, Brunton L, Wieland B, Häsler B (2022) Typology of interventions for antimicrobial use and antimicrobial resistance in aquaculture systems in low- and middle-income countries. Int J Antimicrob Agents 59:106495
Gephart JA, Golden CD, Asche F, Belton B, Brugere C, Froehlich HE, Fry JP, Halpern BS, Hicks CC, Jones RC, Klinger DH, Little DC, McCauley DJ, Thilsted SH, Troell M, Allison EH (2021) Scenarios for global aquaculture and its role in human nutrition. Rev Fish Sci Aquac 29:122–138
Gonzalez Parrao C, Moratti M, Shisler S, Snilstveit B, Eyers J (2021) PROTOCOL: aquaculture for improving productivity, income, nutrition and women's empowerment in low- and middle-income countries: a systematic review and meta-analysis. Campbell Syst Rev 17:e1188
Gorito AM, Ribeiro ARL, Rodrigues P, Pereira MFR, Guimarães L, Almeida CMR, Silva AMT (2022) Antibiotics removal from aquaculture effluents by ozonation: chemical and toxicity descriptors. Water Res 218:118497
Gosio B (1896) *Ricerche batteriologiche e chimiche sulle alterazioni del mais: contributo all'etiologia della pellagra*. Tip. delle mantellate
Goutard FL, Bordier M, Calba C, Erlacher-Vindel E, Góchez D, de Balogh K, Benigno C, Kalpravidh W, Roger F, Vong S (2017) Antimicrobial policy interventions in food animal production in South East Asia. BMJ 358:j3544
Graham DW, Bergeron G, Bourassa MW, Dickson J, Gomes F, Howe A, Kahn LH, Morley PS, Scott HM, Simjee S, Singer RS, Smith TC, Storrs C, Wittum TE (2019) Complexities in understanding antimicrobial resistance across domesticated animal, human, and environmental systems. Ann N Y Acad Sci 1441:17–30
Granada L, Sousa N, Lopes S, Lemos MFL (2016) Is integrated multitrophic aquaculture the solution to the sectors' major challenges? a review. Rev Aquac 8:283–300
Grevskott DH, Svanevik CS, Sunde M, Wester AL, Lunestad BT (2017) Marine bivalve mollusks as possible indicators of multidrug-resistant Escherichia coli and other species of the Enterobacteriaceae family. Front Microbiol 8:24
Griffith RS (1981) Introduction to vancomycin. Rev Infect Dis 3(suppl):S200–S204
Guan W, Li K, Li K (2022) Bacterial communities in co-cultured fish intestines and rice field soil irrigated with aquaculture wastewater. AMB Express 12:132
Guardiola FA, Cerezuela R, Meseguer J, Esteban MA (2012) Modulation of the immune parameters and expression of genes of gilthead seabream (Sparus aurata L.) by dietary administration of oxytetracycline. Aquaculture 334–337:51–57
Gullberg E, Cao S, Berg OG, Ilbäck C, Sandegren L, Hughes D, Andersson DI (2011) Selection of resistant bacteria at very low antibiotic concentrations. PLoS Pathog 7:e1002158
Gupta S, Graham DW, Sreekrishnan TR, Ahammad SZ (2022) Effects of heavy metals pollution on the co-selection of metal and antibiotic resistance in urban rivers in UK and India. Environ Pollut 306:119326
Harrison F, Roberts Aled EL, Gabrilska R, Rumbaugh Kendra P, Lee C, Diggle Stephen P (2015) A 1,000-Year-Old Antimicrobial Remedy with Antistaphylococcal Activity. mBio 6. https://doi.org/10.1128/mbio.01129-01115
Hatosy SM, Martiny AC (2015) The ocean as a global reservoir of antibiotic resistance genes. Appl Environ Microbiol 81:7593–7599
Hektoen H, Berge JA, Hormazabal V, Yndestad M (1995) Persistence of antibacterial agents in marine sediments. Aquaculture 133:175–184
Henriksson PJG, Rico A, Troell M, Klinger DH, Buschmann AH, Saksida S, Chadag MV, Zhang W (2018) Unpacking factors influencing antimicrobial use in global aquaculture and their implication for management: a review from a systems perspective. Sustain Sci 13:1105–1120

Herbeck LS, Unger D (2013) Pond aquaculture effluents traced along back-reef waters by standard water quality parameters, δ15N in suspended matter and phytoplankton bioassays. Mar Ecol Prog Ser 478:71–86

Hiller CX, Hübner U, Fajnorova S, Schwartz T, Drewes JE (2019) Antibiotic microbial resistance (AMR) removal efficiencies by conventional and advanced wastewater treatment processes: a review. Sci Total Environ 685:596–608

Hornick KM, Buschmann AH (2018) Insights into the diversity and metabolic function of bacterial communities in sediments from Chilean salmon aquaculture sites. Ann Microbiol 68:63–77

Jang HM, Kim YB, Choi S, Lee Y, Shin SG, Unno T, Kim YM (2018) Prevalence of antibiotic resistance genes from effluent of coastal aquaculture, South Korea. Environ Pollut 233:1049–1057

Jessen O, Rosendal K, Bülow P, Faber V, Eriksen KR (1969) Changing staphylococci and staphylococcal infections. A ten-year study of bacteria and cases of bacteremia. N Engl J Med 281:627–635

Jo H, Raza S, Farooq A, Kim J, Unno T (2021) Fish farm effluents as a source of antibiotic resistance gene dissemination on Jeju Island, South Korea. Environ Pollut 276:116764

Jones D, Metzger HJ, Schatz A, Waksman SA (1944) Control of Gram-negative bacteria in experimental animals by streptomycin. Science 100:103–105

Kampouris ID, Klümper U, Kramer L, Sorum H, Wedekind H, Berendonk TU (2022) Dissemination of antibiotic resistance in antibiotic-free recirculating aquaculture systems. J Hazard Mater Adv 8:100201

Khan M, Paul SI, Rahman MM, Lively JA (2022) Antimicrobial resistant bacteria in shrimp and shrimp farms of Bangladesh. Water 14:3172

Kijewska A, Koroza A, Grudlewska-Buda K, Kijewski T, Wiktorczyk-Kapischke N, Zorena K, Skowron K (2022) Molluscs-A ticking microbial bomb. Front Microbiol 13:1061223

Kirchhelle C (2018) Pharming animals: a global history of antibiotics in food production (1935–2017). Palgrave Commun 4:96

Krkosek M, Lewis MA, Morton A, Frazer LN, Volpe JP (2006) Epizootics of wild fish induced by farm fish. Proc Natl Acad Sci USA 103:15506–15510

Lalumera GM, Calamari D, Galli P, Castiglioni S, Crosa G, Fanelli R (2004) Preliminary investigation on the environmental occurrence and effects of antibiotics used in aquaculture in Italy. Chemosphere 54:661–668

Lau SH, Hsu JT, Chen YJ, Li ZF, Chao WL, Yeh SL, Ying C (2022) Sub-lethal concentration of sulfamethoxazole affects the growth performance of milkfish (Chanos chanos), the microbial composition of antibiotic-resistant bacteria and the prevalence of sulfonamide-resistance genes in mariculture. Mar Pollut Bull 182:113989

Laxminarayan R, Matsoso P, Pant S, Brower C, Røttingen JA, Klugman K, Davies S (2016) Access to effective antimicrobials: a worldwide challenge. Lancet 387:168–175

Lesher GY, Froelich EJ, Gruett MD, Bailey JH, Brundage RP (1962) 1,8-naphthyridine derivatives. A new class of chemotherapeutic agents. J Med Pharm Chem 91:1063–1065

Li F, Huang J, Wang M, Chen L, Xiao Y (2021) Sources, distribution and dynamics of antibiotics in Litopenaeus vannamei farming environment. Aquaculture 545:737200

Lima LS, Proietti-Junior AA, Rodrigues YC, da Silva Vieira MC, Lima L, de Oliveira Souza C, Dias Gonçalves V, de Oliveira Lima M, Dos Prazeres Rodrigues D, Lima KVB (2022) High genetic diversity and antimicrobial resistance in Escherichia coli highlight Arapaima gigas (Pisces: Arapaimidae) as a reservoir of quinolone-resistant strains in Brazilian Amazon Rivers. Microorganisms 10:808

Limbu SM, Zhou L, Sun SX, Zhang ML, Du ZY (2018) Chronic exposure to low environmental concentrations and legal aquaculture doses of antibiotics cause systemic adverse effects in Nile tilapia and provoke differential human health risk. Environ Int 115:205–219

Ling LL, Schneider T, Peoples AJ, Spoering AL, Engels I, Conlon BP, Mueller A, Schäberle TF, Hughes DE, Epstein S, Jones M, Lazarides L, Steadman VA, Cohen DR, Felix CR, Fetterman KA, Millett WP, Nitti AG, Zullo AM, Chen C, Lewis K (2015) A new antibiotic kills pathogens without detectable resistance. Nature 517:455–459

Liu H, Yao Y, Ye W, Qian R, Chen H, Liang J, Ye J (2022) Enhanced removal of antibiotics and decreased antibiotic resistance genes in the photo-sequencing batch reactor during the aquaculture wastewater treatment. Environ Technol 43:3608–3619

Lulijwa R, Rupia EJ, Alfaro AC (2020) Antibiotic use in aquaculture, policies and regulation, health and environmental risks: a review of the top 15 major producers. Rev Aquac 12:640–663

Luu QH, Nguyen TBT, Nguyen TLA, Do TTT, Dao THT, Padungtod P (2021) Antibiotics use in fish and shrimp farms in Vietnam. Aquac Rep 20:100711

Martins CIM, Eding EH, Verdegem MCJ, Heinsbroek LTN, Schneider O, Blancheton JP, d'Orbcastel ER, Verreth JAJ (2010) New developments in recirculating aquaculture systems in Europe: a perspective on environmental sustainability. Aquac Eng 43:83–93

McPhearson RM, DePaola A, Zywno SR, Motes ML, Guarino AM (1991) Antibiotic resistance in Gram-negative bacteria from cultured catfish and aquaculture ponds. Aquaculture 99:203–211

Memish ZA, Venkatesh S, Shibl AM (2003) Impact of travel on international spread of antimicrobial resistance. Int J Antimicrob Agents 21:135–142

Metwaly AM, Ghoneim MM, Eissa IH, Elsehemy IA, Mostafa AE, Hegazy MM, Afifi WM, Dou D (2021) Traditional ancient Egyptian medicine: A review. Saudi J Biol Sci 28:5823–5832

Millanao BA, Barrientos HM, Gómez CC, Tomova A, Buschmann A, Dölz H, Cabello FC (2011) Injudicious and excessive use of antibiotics: public health and salmon aquaculture in Chile. Rev Med Chile 139:107–118

Murray CJL, Ikuta KS, Sharara F, Swetschinski L, Robles Aguilar G, Gray A, Han C, Bisignano C, Rao P, Wool E, Johnson SC, Browne AJ, Chipeta MG, Fell F, Hackett S, Haines-Woodhouse G, Kashef Hamadani BH, Kumaran EAP, McManigal B, Achalapong S, Agarwal R, Akech S, Albertson S, Amuasi J, Andrews J, Aravkin A, Ashley E, Babin F-X, Bailey F, Baker S, Basnyat B, Bekker A, Bender R, Berkley JA, Bethou A, Bielicki J, Boonkasidecha S, Bukosia J, Carvalheiro C, Castañeda-Orjuela C, Chansamouth V, Chaurasia S, Chiurchiù S, Chowdhury F, Clotaire Donatien R, Cook AJ, Cooper B, Cressey TR, Criollo-Mora E, Cunningham M, Darboe S, Day NPJ, De Luca M, Dokova K, Dramowski A, Dunachie SJ, Duong Bich T, Eckmanns T, Eibach D, Emami A, Feasey N, Fisher-Pearson N, Forrest K, Garcia C, Garrett D, Gastmeier P, Giref AZ, Greer RC, Gupta V, Haller S, Haselbeck A, Hay SI, Holm M, Hopkins S, Hsia Y, Iregbu KC, Jacobs J, Jarovsky D, Javanmardi F, Jenney AWJ, Khorana M, Khusuwan S, Kissoon N, Kobeissi E, Kostyanev T, Krapp F, Krumkamp R, Kumar A, Kyu HH, Lim C, Lim K, Limmathurotsakul D, Loftus MJ, Lunn M, Ma J, Manoharan A, Marks F, May J, Mayxay M, Mturi N, Munera-Huertas T, Musicha P, Musila LA, Mussi-Pinhata MM, Naidu RN, Nakamura T, Nanavati R, Nangia S, Newton P, Ngoun C, Novotney A, Nwakanma D, Obiero CW, Ochoa TJ, Olivas-Martinez A, Olliaro P, Ooko E, Ortiz-Brizuela E, Ounchanum P, Pak GD, Paredes JL, Peleg AY, Perrone C, Phe T, Phommasone K, Plakkal N, Ponce-de-Leon A, Raad M, Ramdin T, Rattanavong S, Riddell A, Roberts T, Robotham JV, Roca A, Rosenthal VD, Rudd KE, Russell N, Sader HS, Saengchan W, Schnall J, Scott JAG, Seekaew S, Sharland M, Shivamallappa M, Sifuentes-Osornio J, Simpson AJ, Steenkeste N, Stewardson AJ, Stoeva T, Tasak N, Thaiprakong A, Thwaites G, Tigoi C, Turner C, Turner P, van Doorn HR, Velaphi S, Vongpradith A, Vongsouvath M, Vu H, Walsh T, Walson JL, Waner S, Wangrangsimakul T, Wannapinij P, Wozniak T, Young Sharma TEMW, Yu KC, Zheng P, Sartorius B, Lopez AD, Stergachis A, Moore C, Dolecek C, Naghavi M (2022) Global burden of bacterial antimicrobial resistance in 2019: a systematic analysis. The Lancet 399:629–655

Muziasari WI, Pärnänen K, Johnson TA, Lyra C, Karkman A, Stedtfeld RD, Tamminen M, Tiedje JM, Virta M (2016) Aquaculture changes the profile of antibiotic resistance and mobile genetic element associated genes in Baltic Sea sediments. FEMS Microbiol Ecol 92:fiw052

Nadimpalli M, Delarocque-Astagneau E, Love DC, Price LB, Huynh BT, Collard JM, Lay KS, Borand L, Ndir A, Walsh TR, Guillemot D (2018) Combating global antibiotic resistance: emerging one health concerns in lower- and middle-income countries. Clin Infect Dis 66:963–969

Nakayama T, Tuyet Hoa TT, Harada K, Warisaya M, Asayama M, Hinenoya A, Lee JW, Phu TM, Ueda S, Sumimura Y, Hirata K, Phuong NT, Yamamoto Y (2017) Water metagenomic analysis reveals low bacterial diversity and the presence of antimicrobial residues and resistance genes in a river containing wastewater from backyard aquacultures in the Mekong Delta, Vietnam. Environ Pollut 222:294–306

Navarrete P, Mardones P, Opazo R, Espejo R, Romero J (2008) Oxytetracycline treatment reduces bacterial diversity of intestinal microbiota of Atlantic salmon. J Aquat Anim Health 20:177–183

Nederlof MAJ, Verdegem MCJ, Smaal AC, Jansen HM (2022) Nutrient retention efficiencies in integrated multi-trophic aquaculture. Rev Aquac 14:1194–1212

Newton GG, Abraham EP (1955) Cephalosporin C, a new antibiotic containing sulphur and D-alpha-aminoadipic acid. Nature 175:548

Novais C, Campos J, Freitas AR, Barros M, Silveira E, Coque TM, Antunes P, Peixe L (2018) Water supply and feed as sources of antimicrobial-resistant Enterococcus spp. in aquacultures of rainbow trout (Oncorhynchus mykiss), Portugal. Sci Total Environ 625:1102–1112

O'Mahony M (2018) EU regulatory risk management of marine biotoxins in the marine bivalve mollusc food-chain. Toxins 10:118

O'Neill J (2016) *Tackling drug-resistant infections globally: final report and recommendations*. Government of the United Kingdom

Okocha RC, Olatoye IO, Adedeji OB (2018) Food safety impacts of antimicrobial use and their residues in aquaculture. Public Health Rev 39:21

Opiyo MA, Marijani E, Muendo P, Odede R, Leschen W, Charo-Karisa H (2018) A review of aquaculture production and health management practices of farmed fish in Kenya. Int J Vet Sci Med 6:141–148

Pallares-Vega R, Blaak H, van der Plaats R, de Roda Husman AM, Hernandez Leal L, van Loosdrecht MCM, Weissbrodt DG, Schmitt H (2019) Determinants of presence and removal of antibiotic resistance genes during WWTP treatment: a cross-sectional study. Water Res 161:319–328

Pavón A, Riquelme D, Jaña V, Iribarren C, Manzano C, Lopez-Joven C, Reyes-Cerpa S, Navarrete P, Pavez L, García K (2022) The high risk of bivalve farming in coastal areas with heavy metal pollution and antibiotic-resistant bacteria: a Chilean perspective. Front Cell Infect Microbiol 12:867446

Pepi M, Focardi S (2021) Antibiotic-resistant bacteria in aquaculture and climate change: a challenge for health in the Mediterranean area. Int J Environ Res Public Health 18:5723

Petchimuthu M, Rosalind George M, Riji John K, Santhana Kumar V (2022) Pathogenicity potential of antimicrobial-resistant Photobacterium damselae subsp. damselae isolated from maricultured and wild-caught fish of the southeast coast of India. Aquac Res 53:4505–4520

Petersen A, Dalsgaard A (2003) Antimicrobial resistance of intestinal Aeromonas spp. and Enterococcus spp. in fish cultured in integrated broiler-fish farms in Thailand. Aquaculture 219:71–82

Pham TTH, Rossi P, Dinh HDK, Pham NTA, Tran PA, Ho T, Dinh QT, De Alencastro LF (2018) Analysis of antibiotic multi-resistant bacteria and resistance genes in the effluent of an intensive shrimp farm (Long An, Vietnam). J Environ Manag 214:149–156

Phu TM, Phuong NT, Scippo ML, Dalsgaard A (2015) Quality of antimicrobial products used in striped catfish (Pangasianodon hypophthalmus) aquaculture in Vietnam. PLoS One 10:e0124267

Rammelkamp CH, Maxon T (1942) Resistance of Staphylococcus aureus to the Action of Penicillin. Proc Soc Exp Biol Med 51:386–389

Ray MJ, Tallman GB, Bearden DT, Elman MR, McGregor JC (2019) Antibiotic prescribing without documented indication in ambulatory care clinics: national cross sectional study. BMJ 367:l6461

Rees EE, Davidson J, Fairbrother JM, St Hilaire S, Saab M, McClure JT (2015) Occurrence and antimicrobial resistance of Escherichia coli in oysters and mussels from Atlantic Canada. Foodborne Pathog Dis 12:164–169

Restrepo L, Bayot B, Arciniegas S, Bajaña L, Betancourt I, Panchana F, Reyes Muñoz A (2018) PirVP genes causing AHPND identified in a new Vibrio species (Vibrio punensis) within the commensal Orientalis clade. Sci Rep 8:13080

Reverter M, Sarter S, Caruso D, Avarre J-C, Combe M, Pepey E, Pouyaud L, Vega-Heredía S, de Verdal H, Gozlan RE (2020) Aquaculture at the crossroads of global warming and antimicrobial resistance. Nat Commun 11:1870
Reverter M, Jackson M, Rohde S, Moeller M, Bara R, Lasut MT, Segre Reinach M, Schupp PJ (2021) High taxonomic resolution surveys and trait-based analyses reveal multiple benthic regimes in North Sulawesi (Indonesia). Sci Rep 11:16554
Reygaert WC (2018) An overview of the antimicrobial resistance mechanisms of bacteria. AIMS Microbiol 4:482–501
Rico A, Satapornvanit K, Haque MM, Min J, Nguyen PT, Telfer TC, van den Brink PJ (2012) Use of chemicals and biological products in Asian aquaculture and their potential environmental risks: a critical review. Rev Aquac 4:75–93
Rico A, Jacobs R, Van den Brink PJ, Tello A (2017) A probabilistic approach to assess antibiotic resistance development risks in environmental compartments and its application to an intensive aquaculture production scenario. Environ Pollut 231:918–928
Rosado D, Xavier R, Severino R, Tavares F, Cable J, Pérez-Losada M (2019) Effects of disease, antibiotic treatment and recovery trajectory on the microbiome of farmed seabass (Dicentrarchus labrax). Sci Rep 9:18946
Roth B, Falco EA, Hitchings GH, Bushby SRM (1962) 5-Benzyl-2,4-diaminopyrimidines as Antibacterial Agents. I. Synthesis and Antibacterial Activity in vitro. J Med Pharm Chem 5:1103–1123
Rurangwa E, Verdegem MCJ (2015) Microorganisms in recirculating aquaculture systems and their management. Rev Aquac 7:117–130
Ruzauskas M, Armalytė J, Lastauskienė E, Šiugždinienė R, Klimienė I, Mockeliūnas R, Bartkienė E (2021) Microbial and antimicrobial resistance profiles of microbiota in common carps (Cyprinus carpio) from aquacultured and wild fish populations. Animals (Basel) 11:929
Samuelsen OB, Torsvik V, Ervik A (1992) Long-range changes in oxytetracycline concentration and bacterial resistance toward oxytetracycline in a fish farm sediment after medication. Sci Total Environ 114:25–36
Sanguanrut P, Munkongwongsiri N, Kongkumnerd J, Thawonsuwan J, Thitamadee S, Boonyawiwat V, Tanasomwang V, Flegel TW, Sritunyalucksana K (2018) A cohort study of 196 Thai shrimp ponds reveals a complex etiology for early mortality syndrome (EMS). Aquaculture 493:26–36
Santos L, Ramos F (2018) Antimicrobial resistance in aquaculture: current knowledge and alternatives to tackle the problem. Int J Antimicrob Agents 52:135–143
Schar D, Klein EY, Laxminarayan R, Gilbert M, Van Boeckel TP (2020) Global trends in antimicrobial use in aquaculture. Sci Rep 10:21878
Schar D, Zhao C, Wang Y, Larsson DGJ, Gilbert M, Van Boeckel TP (2021) Twenty-year trends in antimicrobial resistance from aquaculture and fisheries in Asia. Nat Commun 12:5384
Serrano-Grijalva L, Sánchez-Carrillo S, Angeler DG, Sánchez-Andrés R, Álvarez-Cobelas M (2011) Effects of shrimp-farm effluents on the food web structure in subtropical coastal lagoons. J Exp Mar Biol Ecol 402:65–74
Sha S, Dong Z, Gao Y, Hashim H, Lee CT, Li C (2022) In-situ removal of residual antibiotics (enrofloxacin) in recirculating aquaculture system: effect of ultraviolet photolysis plus biodegradation using immobilized microbial granules. J Clean Prod 333:130190
Shah SQ, Colquhoun DJ, Nikuli HL, Sørum H (2012) Prevalence of antibiotic resistance genes in the bacterial flora of integrated fish farming environments of Pakistan and Tanzania. Environ Sci Technol 46:8672–8679
Sivaraman GK, Rajan V, Vijayan A, Elangovan R, Prendiville A, Bachmann TT (2021) Antibiotic resistance profiles and molecular characteristics of extended-spectrum beta-lactamase (ESBL)-producing Escherichia coli and Klebsiella pneumoniae isolated from shrimp aquaculture farms in Kerala, India. Front Microbiol 12:622891
Smith VJ, Brown JH, Hauton C (2003) Immunostimulation in crustaceans: does it really protect against infection? Fish Shellfish Immunol 15:71–90
Song C, Zhang C, Fan L, Qiu L, Wu W, Meng S, Hu G, Kamira B, Chen J (2016) Occurrence of antibiotics and their impacts to primary productivity in fishponds around Tai Lake, China. Chemosphere 161:127–135

Ström GH, Björklund H, Barnes AC, Da CT, Nhi NHY, Lan TT, Magnusson U, Norman Haldén A, Boqvist S (2019) Antibiotic use by small-scale farmers for freshwater aquaculture in the upper Mekong Delta, Vietnam. J Aquat Anim Health 31:290–298

Su H, Liu S, Hu X, Xu X, Xu W, Xu Y, Li Z, Wen G, Liu Y, Cao Y (2017) Occurrence and temporal variation of antibiotic resistance genes (ARGs) in shrimp aquaculture: ARGs dissemination from farming source to reared organisms. Sci Total Environ 607–608:357–366

Suyamud B, Lohwacharin J, Yang Y, Sharma VK (2021) Antibiotic resistant bacteria and genes in shrimp aquaculture water: identification and removal by ferrate(VI). J Hazard Mater 420:126572

Tarkhani R, Imani A, Hoseinifar SH, Ashayerizadeh O, Sarvi Moghanlou K, Manaffar R, Van Doan H, Reverter M (2020) Comparative study of host-associated and commercial probiotic effects on serum and mucosal immune parameters, intestinal microbiota, digestive enzymes activity and growth performance of roach (Rutilus rutilus caspicus) fingerlings. Fish Shellfish Immunol 98:661–669

Thornber K, Verner-Jeffreys D, Hinchliffe S, Rahman MM, Bass D, Tyler CR (2020) Evaluating antimicrobial resistance in the global shrimp industry. Rev Aquac 12:966–986

Thornber K, Bashar A, Ahmed MS, Bell A, Trew J, Hasan M, Hasan NA, Alam MM, Chaput DL, Haque MM, Tyler CR (2022) Antimicrobial resistance in aquaculture environments: unravelling the complexity and connectivity of the underlying societal drivers. Environ Sci Technol 56:14891–14903

Tiseo K, Huber L, Gilbert M, Robinson TP, Van Boeckel TP (2020) Global Trends in Antimicrobial Use in Food Animals from 2017 to 2030. Antibiotics (Basel) 9

Turner AD, Stubbs B, Coates L, Dhanji-Rapkova M, Hatfield RG, Lewis AM, Rowland-Pilgrim S, O'Neil A, Stubbs P, Ross S, Baker C, Algoet M (2014) Variability of paralytic shellfish toxin occurrence and profiles in bivalve molluscs from Great Britain from official control monitoring as determined by pre-column oxidation liquid chromatography and implications for applying immunochemical tests. Harmful Algae 31:87–99

Verdegem M, Buschmann AH, Latt UW, Dalsgaard AJT, Lovatelli A (2023) The contribution of aquaculture systems to global aquaculture production. J World Aquac Soc 54:206–250

Vivekanandhan G, Savithamani K, Hatha AA, Lakshmanaperumalsamy P (2002) Antibiotic resistance of Aeromonas hydrophila isolated from marketed fish and prawn of South India. Int J Food Microbiol 76:165–168

Wainwright M, Rally L, Ali TA (1992) The scientific basis of mould therapy. Mycologist 6:108–110

Watts JEM, Schreier HJ, Lanska L, Hale MS (2017) The rising tide of antimicrobial resistance in aquaculture: sources, sinks and solutions. Mar Drugs 15:158

Xing Y, Wu S, Men Y (2020) Exposure to environmental levels of pesticides stimulates and diversifies evolution in Escherichia coli toward higher antibiotic resistance. Environ Sci Technol 54:8770–8778

Xiong W, Sun Y, Zhang T, Ding X, Li Y, Wang M, Zeng Z (2015) Antibiotics, antibiotic resistance genes, and bacterial community composition in fresh water aquaculture environment in China. Microb Ecol 70:425–432

Xu M, Huang X-H, Shen X-X, Chen H-Q, Li C, Jin G-Q, Cao J-S, Xue Z-X (2022) Metagenomic insights into the spatiotemporal responses of antibiotic resistance genes and microbial communities in aquaculture sediments. Chemosphere 307:135596

Yipel M, Kürekci C, Tekeli İO, Metli M, Sakin F (2017) Determination of selected antibiotics in farmed fish species using LC-MS/MS. Aquac Res 48:3829–3836

Yuan J, Ni M, Liu M, Zheng Y, Gu Z (2019) Occurrence of antibiotics and antibiotic resistance genes in a typical estuary aquaculture region of Hangzhou Bay, China. Mar Pollut Bull 138:376–384

Zhang J, Yang X, Kuang D, Shi X, Xiao W, Zhang J, Gu Z, Xu X, Meng J (2015) Prevalence of antimicrobial resistance of non-typhoidal Salmonella serovars in retail aquaculture products. Int J Food Microbiol 210:47–52

Zhong Y, Chen ZF, Dai X, Liu SS, Zheng G, Zhu X, Liu S, Yin Y, Liu G, Cai Z (2018) Investigation of the interaction between the fate of antibiotics in aquafarms and their level in the environment. J Environ Manag 207:219–229

Chapter 3
AMR in Freshwater and Marine Environments

Ramchandran Ishwarya, Subramanian Sivakumar, Baskaralingam Vaseeharan, and Preetham Elumalai

Abstract Since more than 10 years ago, antimicrobial resistance (AMR) has been a major and complicated problem as well as a significant societal concern, with significant implications for the water surroundings that gives to its appearance and extend. The assessment of antibiotic resistance genes (ARGs) present in human populations, the identification of key capital for the adaptation and spread of resistance, the notify of epidemiological and public healthiness threat evaluation forms, and the quantification of elimination efficiencies by domestic wastewater infrastructure are all examples of how environmental observing can offer crucial information for containing the increase of AMR. Aquatic ecosystems are crucial for preserving the biosphere's elevated stages of diversity of life, subsistence, and creativity. Antibiotic-resistant pathogens, including superbugs, are a growing danger to the surroundings and human healthiness, so the occurrence of antibiotic-resistant bacteria (ARB) and ARGs in the water surroundings is causing great anxiety. It is acknowledged that aquatic ecosystems serve as important storages and conduits for the propagation of antibiotic/antimicrobial resistance. Additionally, freshwater ecosystems are vulnerable to pollution from lingering antibiotics that can be released through a variety of channels, including agricultural runoff, sewage discharges, and farmland leaching. Thus, freshwater environments have the potential to act as reser-

R. Ishwarya
Mandapam Regional Centre, ICAR—Central Marine Fisheries Research Institute, Mandapam, Tamil Nadu, India

S. Sivakumar
Department of Biochemistry, Sri Sankara Arts and Science College, Enathur, Kanchipuram, Tamil Nadu, India

B. Vaseeharan
Biomaterials and Biotechnology in Animal Health Lab, Department of Animal Health and Management, Alagappa University, Karaikudi, Tamil Nadu, India

P. Elumalai (✉)
Department of Marine Biology, Microbiology and Biochemistry, School of Marine Sciences, Cochin University of Science and Technology, Kochi, Kerala, India
e-mail: preetham@cusat.ac.in

P. Elumalai, S. Lakshmi (eds.), *Antimicrobial Resistance in Aquaculture and Aquatic Environments*, https://doi.org/10.1007/978-981-97-7320-6_3

voirs for microbes that are affected by antibiotics, which is a significant public health problem. Processes of deprivation and intensity are crucial for determining the true danger of antibiotic resistance spreading from freshwater storages. Although human and animal samples have been the main focus of research on antimicrobial resistance (AMR), the AMR situation in water habitats can vary and have complex patterns depending on their environment.

Keywords Fresh water · Marine water · Antibiotics · Antimicrobials · Resistance

Abbreviations

AMR	Antimicrobial resistance
API	Active pharmaceutical ingredient
ARB	Antibiotic-resistant bacteria
ARGs	Antibiotic resistance genes
CDC	Centers for Disease Control and Prevention
EUR	Europe
FAO	Food and Agriculture Organization
FPP	Finished pharmaceutical products
HGT	Horizontal gene transfer
HMRGs	Heavy metal resistance genes
kg	Kilogram
MDR	Multi-drug resistance
mL	Milliliter
ng	Nanogram
NORMs	Naturally occurring radioactive material
OHA	One Health Approach
OIE	Office International des Epizooties
PhACs	Pharmaceutically active compounds
tetR	Tetracycline resistance
USA	United States of America
USD	United States dollar
WHO	World Health Organization
WWTPs	Wastewater treatment plants

3.1 Introduction

Environmental antimicrobial resistance (AMR) is among the most largest dangers to human healthiness worldwide. Clinical interventions and productivity losses have an economic cost of 9 billion euros and 20 billion dollars in Europe and the United States, correspondingly, and it is predicted that by 2050, there will be an increase to ten million patients globally, costing a total of USD 100 trillion (Amarasiri et al. 2020). Antibiotic-resistant bacteria alone ideal roughly 2.8 million diseases and 35,000 fatalities each year in the United States. The World Health Organization (WHO) claims that AMR has an impact on people that goes beyond health. AMR might result in a 28 million rise in the amount of poor individuals in low- and medium-income nations by 2030, along with the WHO, because of health inequities and the high rate of treating resistant diseases. It has appeared as the most important role in people strength challenges of the twenty-first century, overshadowing the effectiveness of obtainable medicines against a large range of diseases that are appropriate further resistant to conventional antibiotics (Prestinaci et al. 2015). The AMR issue is growing quickly and is becoming more urgent by the day. Most new antibiotics are frequently defeated by microorganisms that cause a variety of prevalent ailments within 5–10 years of their introduction to the market. As bacteria, fungi, viruses, and parasites adapt to reproduce in the presence of antibiotics, AMR is seen in these organisms. According to the WHO, we are rapidly running out of therapeutic options as a result of rising AMR (WHO 2021). The WHO has put up a six-point plan known as the "One Health Approach" (OHA) in response to the mounting threat posed by AMR (Mazimba et al. 2021). For the most part, One Health (people, animals, and environment) has been incorporated into global and national action plans to combat AMR; nonetheless, it is becoming increasingly clear that environmental factors require further consideration. The One Health Alliance (OHA) is envisioned as "involvement of human health, animal health, and environmental health and focus on those infectious disease-related issues that undermine overall health and well-being" (Mazimba et al. 2021). The US National Action Plan for Combating Antibiotic-Resistant Bacteria places a strong focus on the necessity of better understanding how the environment contributes to the antibiotic-resistant bacteria (ARB) and antibiotic resistance genes (ARGs) across and among individuals, groups of organisms, and geographic regions. Expanded surveillance efforts are one of the plan's main objectives for the 2020 update. Given this context, it is clear how any element that raises the occurrence of resistance factors in the surroundings or in animal diseases may have an impact on public healthiness.

Consequently, in the rising practice of antibiotics in healthcare facilities, nursing homes, and farming, the effects of AMR on human health will only get worse. Antibiotic abuse, overuse, and unrestricted release into waste are all factors that contribute to antimicrobial resistance in the surroundings. AMR is as well a result of the extensive usage of antibiotics in developing countries to care for infectious illness brought on by insufficient hygienic conditions, water quality, and sanitation. According to Chokshi et al. (2019), AMR may be the outcome of the general utilization of antibiotics in healthcare centers in wealthy countries to cure infectious illness or prevent infections from happening through procedures, cleanliness, and hygiene. Antibiotics and other antimicrobials generate selection pressure that hastens the emergence of ARBs that are resistant to them and related substances. Every antibiotic that is regularly used ends up in the environment, hastening the growth of AMR. The ARGs are now frequently discovered with ARB in both the environment and in people who have never had antibiotic therapy (Alduina 2020). There are some newly discovered ARGs such as NDM-1 and McR-1, which are garnering significant attention due to their property of exhibiting resistance to polymyxins, which is one of the latest classes of antibiotics (Liu et al. 2016). NDM-1 has also displayed resistivity towards β-lactic acid drugs as well as other antibiotics (Krempaska et al. 2018). NDM-1 and McR-1 (ARGs) are prevalent in both the input and biological treatment stages of wastewater treatment plants (WWTPs), and they can also be detected in significant amounts in the WWTP outflow (Ruonan et al. 2019). ARGs in environmental bacteria and soil put human health at risk even when horizontal gene transfer (HGT) pathways encourage the spread of resistance factors in infective bacteria. Medical centers, agriculture, aquacultures, and WWTPs are regarded as "hotspot environments" where bacteria can be chosen prior to being discharged into the normal surroundings, are subjected to large and repetitive amount of antibiotics, and have elevated growth rates since the nutrients are abundant. As a result, waste emitted from these "hotspots" can contain pathogenic-resistant bacteria and antimicrobial compounds, and the final effluent can then enter aquatic environments. Heavy metal resistance genes (HMRGs) and the class 1 integron, a potentially mobile genetic element involved in conjugatively mediated gene transfer, are frequently linked to ARGs (Di Cesare et al. 2016). AMR and/or resistance genes came to be discovered in a mixture of aquatic surroundings, as well as wastewater, recreational coastal waters, and surface waters as well as environments and animals most probably impervious through anthropogenic issues, like glaciers and marine animals (Sucato et al. 2021). Consequently, seawater is a significant contributor to the rise in resistance genes among bacterial species, whether or not they are related, and aquatic environments are considered to be sources of antibiotic resistance. ARGs and multi-resistant bacteria may be important indicators of anthropogenic environmental contamination, particularly in coastal areas where there is high human activity, the presence of effluents, and the existence of wastewater treatment facilities.

Significant anthropogenic antibiotic residues have a negative impact on the ecosystem. Thiamphenicol, for instance, has been shown to accelerate the emission of the greenhouse gas N_2O and to hinder nitrate reduction mechanisms (Yin et al. 2016). Due to disrupted microbial colonization in the digestive tract, newborn antibiotic exposure is closely associated with the child's poor growth over the first 6 years of life (Uzan-Yulzari et al. 2021). ARGs are enriched based on the widespread utilization of antibiotics, which places particular force on bacteria. According to Wilson et al. (2020), ARGs primarily deactivate antibiotics during cellular defense, enzymatic degradation, and antibiotic deactivation. They are able to reduce the effectiveness of antibiotics against bacteria, making it difficult for therapeutic treatments to effectively stop infection. In the past, antibiotics and ARGs have frequently been found in groundwater (Tong et al. 2020), lakes (Li et al. 2012), rivers (Rodriguez-Mozaz et al. 2015), and estuaries and coastal environments (Zhu et al. 2017). In particular, estuarine and coastal ecosystems are places where freshwater and saltwater coexist with one another (Ward et al. 2020). Estuarine and coastal ecosystems offer crucial ecosystem services because of the region's particular geographic location. The elevated stages of diversity of life, subsistence, and production of the biosphere depend heavily on aquatic ecosystems (Hassan et al. 2020). The occurrence of antimicrobials, ARB, and ARGs in the water surroundings is an increasing source of alarm because antibiotic-resistant pathogens, including superbugs, are posing an increasing danger to the surroundings and human healthiness. Aquatic habitats are most important primary storages and diffusion ways to increase AMR, according to Amarasiri et al. (2020). Antibiotics and ARGs, however, are posing enormous problems for human health, ecological stability, and biogeochemical cycling in estuarine and coastal ecosystems. As a result, this chapter examines the sources, prevalences, and control strategies for AMR in freshwater and marine ecosystems (Table 3.1).

Table 3.1 Antimicrobial resistance genes (ARGs) obtained from different microbes and freshwater and marine water habitats

Antibiotic resistance genes (ARGs)	Biological source	Environmental source	References
tet(A)	*Alcaligenes* spp.	Soil	Agersø and Sandvang (2005)
	E. coli strain	Pigsty soil	
	Enterobacter sp.	Pigsty soil	
tet(B)	*E. coli* strain	Pigsty soil	
tet(C)	*Pseudomonas* spp.	Soil	
Methicillin resistance			
mecA gene	*Staphylococcus aureus*	Freshwater beach	Levin-Edens et al. (2012)
mecA gene	*Staphylococcus aureus*	River	Hatcher et al. (2016)
mecA gene	*Staphylococcus aureus*	Freshwater recreational beach	Thapaliya et al. (2017)
Vancomycin resistance			
vanA	*Enterococci faecalis*	River	Novais et al. (2005)
vanC	*E. casseliflavus*	River	Nishiyama et al. (2017)
VanC	*Enterococcus* spp.	River	Nam et al. (2013)
vanA	*Enterococcus faecium*	River	Morris et al. (2012)
vanC1	*Enterococcus casseliflavus/ gallinarum*	River	Roberts et al. (2009)
vanC2/3	*Enterococcus casseliflavus/ gallinarum*	River	Roberts et al. (2009)

(continued)

Table 3.1 (continued)

Antibiotic resistance genes (ARGs)	Biological source	Environmental source	References
Extended-spectrum β-lactamase–producing *Enterobacteriaceae*			
$bla_{CTX-M-15}$	*E. coli*, *K. pneumoniae*, *E. cloacae*	Stream	Miyagi and Hirai (2019)
$bla_{CTX-M-14-like}$	*E. coli*, *K. pneumoniae*, *E. cloacae*	Stream	Miyagi and Hirai (2019)
bla_{SHV}	*E. coli*, *C. freundii*	River water	Ye et al. (2017)
bla_{SHV-1} in association with bla_{CTX-M}	*E. coli*, *K. pneumoniae*	River water	Ye et al. (2017)
bla_{TEM-1}	*K. pneumoniae*	River water	Ye et al. (2017)
bla_{CMY-2}	*E. coli*, *C. freundii*	River water	Ye et al. (2017)
$bla_{CTX-M-14}$	*E. coli*	River	Dhanji et al. (2011)
bla_{VEB}	*Aeromonas* spp.	River	Girlich et al. (2011)
bla_{SHV-12}			
bla_{PER-1}			
bla_{TLA-2}			
bla_{GES-7}			
bla_{CTX-M}, bla_{TEM}, $bla_{OXA\ 21}$	*A. hydrophila*	River	Tacão et al. (2012)
Integrase genes intI1	*Escherichia coli*, *Pseudomonas* sp., *Acinetobacter* sp.	River	Tacão et al. (2012)
Carbapenemase–producing *Enterobacteriaceae*			
bla_{KPC-2}	*E. coli*, *K. pneumoniae*, *Enterobacter cloacae*	River sediments	Piedra-Carrasco et al. (2017)
bla_{IMI-2}	*Enterobacter cloacae*	River sediments	Piedra-Carrasco et al. (2017)
bla_{KPC-2} and bla_{VIM-1}	*Klebsiella oxytoca*	River sediments	Piedra-Carrasco et al. (2017)
tetB, tetX	*Rhodopirellula*, *Crocinitomix*, *Ilumatobacter*, *PAUC26f*	Coastal RAS (recirculating aquaculture system) farm, China	Lu et al. (2019)
sul1, sul2, tetG, ermF, qnrS	*Mycobacterium*, *Pir4_lineage*, *Marinobacter*, *SM1A02*, *Altererythrobacter*	Coastal RAS (recirculating aquaculture system) farm, China	Lu et al. (2019)

3.2 Antimicrobial Resistance Source, Emergence and Spread in Water Environment

The necessity for life-saving antibiotics has grown over time, which has led to a widespread use of different pharmaceutically active compounds (PhACs) to make medicines all over the world. As a result, there are now more pharmaceutical manufacturing waste matter being produced, many of which come from factories that produce together active pharmaceutical ingredients (API) and finished pharmaceutical products (FPP), both of which have high levels of antibiotic residues (Kotwani et al. 2021). Owing to the occurrence of significant amounts of ARGs from pharmaceutical effluents, recent studies have identified places with a pharmaceutical sector as the main cause of antibiotics, which are attractive for the development of AMR in the surroundings, among humans, and among animals. Pharmaceutical companies deliver antibiotics and the chemicals they contain in a variety of environments where AMR can occur. Antibiotic residues and ARB are transported through environmental sections using food, animals, agricultural settings, wastewater, healthcare facilities, and these substances. During drain water and the straight dumping of feces into outside waters, ARB-infected humans and animals have the potential to spread resistance across the ecosystem. The environment affects the destiny and movement of antibiotic residues and resistant microorganisms (Bengtsson-Palme and Larsson 2016). For example, eagerly water-soluble antibiotics can break up in runoff water, possibly lowering their capability to reason AR before reaching surface water. Extreme environmental circumstances can make some ARB more vulnerable, which lowers their chances of surviving in the wild. Surface water AMR development and dissemination may be facilitated by antibiotic residues and ARB that are not easily influenced by unfavorable situation that continue through transit. Large quantity of remaining antibiotics and ARGs from veterinary and public channels reach river systems because of the widespread use of antibiotics in terrestrial regions before being released into estuarine and coastal environments. The great quantity of antibiotics and ARGs typically declines after rivers enter the ocean due to seawater dilution.

Large amounts of wastewater discharged from residential, educational, commercial, industrial, hospital, and agricultural contexts finally make their mode into waterways. Humans that use antibiotics release them inside dirt systems and sewage disposal tanks, which are home to a mixture of bacteria species. According to Anderson et al. (2006), a considerable part of the active components in antibiotics are freed into the surroundings during animal and human waste. The emission of stimulants, antibiotics, and drug metabolites in household urine and feces results in the formation of antibiotic compounds in municipal wastewater systems (Velpandian et al. 2018). Before being expelled from the body, human-ingested antibiotics go through a number of changes to create antibiotic molecules or metabolites. To facilitate the identify and analyze the quantities of particular antibiotics in wastewater, the metabolites are utilized as biomarkers. Up to 50% of over-the-counter drugs worldwide comprise antibiotics, which has led to an increase in the discharge of antibiotic metabolites inside drain water. AMR has been connected to the improper use and distribution of antibiotics in homes (Esimone et al. 2007). Many healthcare facilities across the world release untreated wastewater into sanitary sewage systems, which may contain antibiotic residues and microorganisms resistant to antibiotics. Similar to this, ARGs are widely present in pharmaceutical industry wastewater. Wastewater and water chemistry have an impact on the presence and level of ARGs. The multiplicity of pathogens and antibiotic resistance in exterior waterways is facilitated by the wastewater produced by pharmaceutical industries and slaughterhouses. The wastewater generated by slaughterhouses and pharmaceutical enterprises contributes to the proliferation of infections and antibiotic resistance in external waterways. Additionally, discovered that no pollution source, such as rainfall, affects antibiotic resistance in aquatic systems.

Furthermore, it has been shown that aquaculture effluent is a major source of antibiotics and ARGs. The aquaculture industry has grown in coastal areas because of the healthy aquatic environment and the dietary choices of the local population (Froehlich et al. 2018). In the next few years, aquaculture is expected to provide more than half of the world's seafood addiction because to the depletion of natural fisheries. However, there is growing evidence that suggests that such expansion may not be sustainable because aquaculture has unfavorable effects on the ecosystem, including habitat degradation, eutrophication, and pollution with pesticides and antimicrobials. Antibiotics are frequently included in fishery feed as a means of preventing disease or as a therapeutic treatment. High levels of antibiotic residues were consequently extensively found in the areas close to aquaculture grounds. The acquirement and spread of AMR/AR are found to be best in aquatic environments (Leonard et al. 2015). Tetracycline, macrolides, and aminoglycosides are just a few of the numerous antibiotics engaged in aquaculture that are essential for human medicine (World Health Organization 2021). Through culture-independent research, several researches exposed the circuitous transmission of ARGs from antibiotic-resistant microorganisms of fish source to public diseases such *E. coli*, *Salmonella* spp., and *Aeromonas* spp. (Amarasiri et al. 2020). ARGs of special classes have been discovered in oceanic sediments by metagenomics or culture-independent investigations, indicating a crucial function for them in lateral gene transmission.

3.2.1 Antimicrobial Resistance in Freshwater

Among the natural regions that could be contaminated by antibiotics are freshwater habitats. Antibiotics are disseminated by a number of sources, including agricultural runoff, dirt liberation, and runoff from near farmhouse. The availability of antibiotics in the surroundings and the superior concentration of active bacteria found in freshwater both give the appearance of antibiotic resistance genes. As a result, freshwater environments have become a capital for ARG HGT and are therefore areas where resistance is starting to develop. The potential for developing infections from organisms that are resistant to therapy can increase because ARBs and ARGs can survive in freshwater habitats. ARGs can endure in the environment and then resurface as diseases in both people and animals. A reservoir for drinking water and even chances for recreation are provided by the same body of freshwater that also receives effluent. ARBs and ARGs may also be developed in an elevated ARG pool with ecological bacteria, which may promote the spread of resistance to both familiar and recently emerging illnesses.

3.2.1.1 Sources

Antimicrobials/antibiotics, ARBs, and ARGs are continuously introduced into freshwater bodies like tributary, watercourse, channels, and ponds through a variety of sources including waste matter from WWTPs, chemical production facilities, livestock farming, aquaculture, etc. (Fig. 3.1). Diverse antibiotics and ARGs have been found in exterior and groundwater resources, according to numerous researches (Matongo et al. 2015; Madikizela et al. 2017; Danner et al. 2019). Some essential services to human well-being are provided by aquatic ecosystems, especially river waters. However, because these ecosystems serve as AMR reservoirs, they are also thinking to be a significant pathway for the increase of resistance elements. Worldwide, antibiotic residues in normal river water have been identified (Charuaud

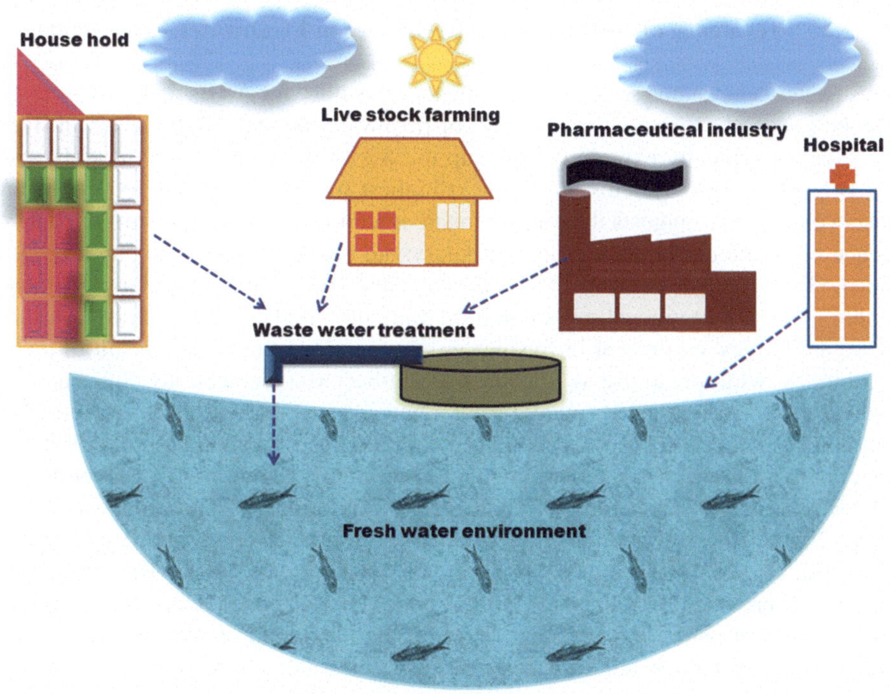

Fig. 3.1 Release of antimicrobial residues into the freshwater environment

et al. 2019; Singh et al. 2019), and their levels are rising rapidly, primarily due to the widespread use of antibiotics in people and veterinary remedy. Antibiotics in feces cause water contamination from medicines and ARBs. The type of antibiotic being used and the body's metabolism both have an impact on the amount of antibiotics and ARG free into the surroundings. According to a recent analysis, >80% of the antibiotics given out are eradicated in urine and feces in an energetic state (FAO and WHO 2019). Due to this, the greater part of analysis on antibiotic pollution of exterior water (Szymańska et al. 2019) concentrate on WWTP runoff. The occurrence of a diverse microbial community, with members derived from peoples and animals, is supported by the organic carbon content of WWTP effluents, along with other nutrients and elements that bacteria can attract to and the relatively constant heat and pH (Pazda et al. 2019). The fact that ARGs and antibiotics are frequently used for irrigation raises still another issue with the usage of these substances in WWTP effluents. According to scientific data, plant-based foods can expose consumers to germs that are resistant to antibiotics (FAO and WHO 2019). Concerning handlings that are victorious at eradicating antibiotics and associated resistance genes yet also affordable and simple to realize in the modern WWTPs, there are major acquaintance break about the input of antibiotics into surface water. In addition, the transformation products of antibiotics must also be taken into account. Future studies might focus on combining techniques to remove ARGs effectively.

3.2.1.2 Prevalence of Antimicrobial Resistance in Freshwater Ecosystems

Freshwater ecosystems serve as a significant repository for ARBs and ARGs. They can be seen to be common in a variety of freshwater habitats. ARB levels in river systems are typically extremely high (up to 98% of the total bacteria tested), and ARB levels in lake systems are the next highest (up to 77% of the total bacteria discovered). This suggests that opportunistic infections are susceptible to the emergence of antibiotic resistance since tributary and ponds act as substantial storages for it. Given the prevalence of ARBs in rivers and lakes that are resistant to cephalothin, penicillin, tetracycline, ampicillin, and chloramphenicol, it is obvious that these antibiotics are present in freshwater systems. This can be as a result of these antibiotics' widespread use as fundamental antibacterial medications. Since antibiotics are used so frequently, many bacteria have developed resistance to them. The amount of ARBs in freshwater habitats can be inclined by a numeral of variables, together with the region's geography and prescription regulations. Low occurrence of ARB has been recorded in areas with tight prescription and disposal laws (Tacão et al. 2015). However, the quantity of ARBs in the lake, pond, and dam (1%) is very low. However, dam and pond systems can be overlooked as ARB reservoirs. Antibiotic prescription and consumption data are used to track antibiotic usage. The amount of antibiotics in rivers relies on the likelihood of their breakdown or biodegradation in the water. Recent studies have shown that lipophilic chemicals are largely sorbed onto solid phases, while antibiotics with low molecular weights and low octanol/water panel volume dissolve fast in water bodies. Aminoglycosides, amphenicols, lactamases, lincosamides, quinolones, sulfonamides, and tetracyclines are a few examples of antibiotics that became discovered to be constant and intractable (Krzeminski et al. 2019). Since penicillins quickly degrade in environmental media (Timm et al. 2019), despite being the majority main antibiotic families utilized in people and veterinary remedy, penicillin volumes in tributary are lowered down. River waters have been shown to contain mean values of 0.40–2.7 ng/L (Carvalho and Santos 2016). Though beta-lactam ring hydrolysis decreases the efficiency of antibiotics, the presence of penicillin conversion products is rarely addressed (Li et al. 2008; Timm et al. 2019). Tetracyclines are a class of antibiotics with an expansive array of applications for treating both Gram-positive and Gram-negative bacteria. Despite their extensive use in veterinary and public drugs, in addition to their use as preservatives in animal foods, they are frequently detectable in water medium at extremely small concentrations (10 ng/L) due to the fact that they appear as a steady compound with floating material (Carvalho and Santos 2016; Wernersson et al. 2015). Both public and veterinary drugs use quinolones and fluoroquinolones (Janecko et al. 2016). Examples of aminoglycosides that have not been extensively seen in aqueous ecological medium, particularly in freshwaters, are streptomycin, gentamicin, and amikacin (Felis et al. 2020). This is probably the case because of their limited use, possible risks, unfavorable consequences, and residues in food animals. Only gentamicin (21 ng/L) was discovered in ground water. In animal husbandry, sulfonamides, one of the oldest antibiotics still in use today, are commonly utilized as feed supplements. Sulfonamides are included in the pharmaceutical chemical classes that are most commonly investigated in aquatic medium in Europe, with a usual quantity in river water of less than 100 ng/L (Carvalho and Santos 2016; Kergoat et al. 2021). They are now combined (5:1) with trimethoprim (a diaminopyrimidine), which broadens their spectrum and boosts their antibacterial action (Fig. 3.2).

For instance, the Italian Medicines Agency's Annual Report (Agenzia Italiana del Farmaco 2020), which is released by the National Observatory on the Use of Medicines, provides statistics on drug consumption and spending in Italy. According to Kotwani and Holloway (2011), penicillins ranked third among the most often approved antibiotics in India in 2008 (28%), after quinolones (34%) and cephalosporins (32%), and were pursued by macrolides (14%) and tetracyclines (6%). According to Van Boeckel et al. (2014), China and Thailand also showed a similar tendency. Due to their widespread subtherapeutic antibiotic usage and abuse in agriculture, unethical antibiotic recommending practice, strong incentives for overrecommending, and other factors, the problem of the rapid introduction of ARBs and ARGs is principally urgent in China (Qiao et al. 2018). Even though it was expected that 54,000 tons of antibiotics out of 92,700 tons of the total antibiotic use entered the receiving environment (Zhang et al. 2015), this type of pollutant is not taken into account in the supervising study that are helpful for classifying the quality of river water. The presence of 4 classes of 17 frequently used public and veterinary antibiotics was examined by Zhou et al. (2011) in the sediments of the rivers in China (i.e., Yellow, Hai, and Liao). The most often found antibiotics in all rivers examined were norfloxacin, ofloxacin, ciprofloxacin, and oxytetracycline. The Yangtze River

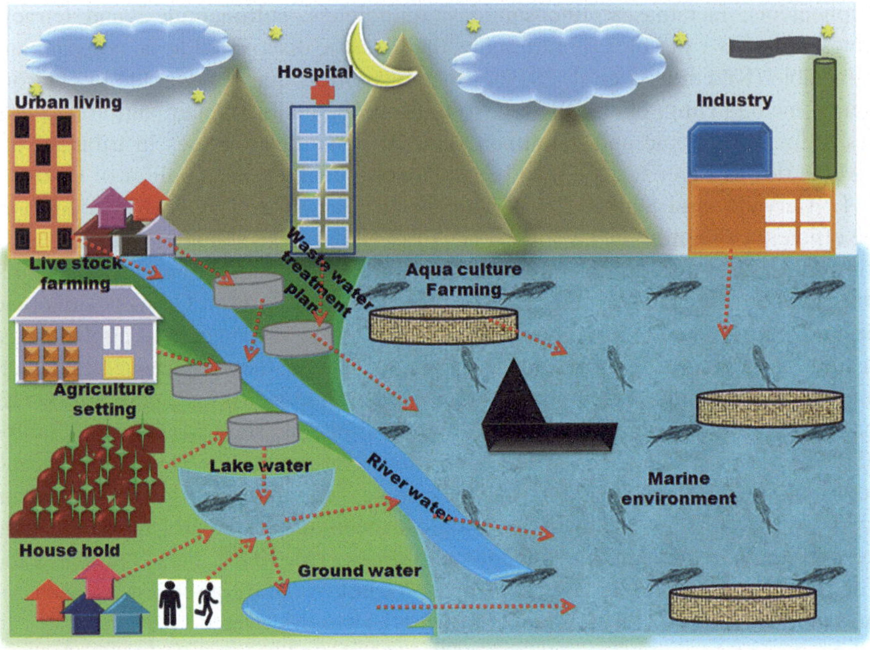

Fig. 3.2 Different sources of antimicrobial resistance genes and antibiotics in the water environment. The pathways are shown as red arrows. For each of the human activity sectors represented in the figure, a list is provided of the main AMR-driving substances (i.e. antimicrobial residues, antimicrobial resistance genes, heavy metals, biocides, inorganic fertilizers) emitted by that sector

Delta's drinking water source reservoir contains 12 antimicrobial substances that belonged to 4 types, which are tetracyclines, fluoroquinolones, macrolides, and sulfonamides, according to Cui et al. (2018). In North China, Cheng et al. (2019) found quinolones, tetracyclines, amphenicols, and sulfonamides at amounts ranging from 0.2 to 421 ng/L in river water samples. Antimicrobials were examined in water and sediment samples from seven significant rivers and four oceans in China by Li et al. (2018). In the water and sediments, they found 12 chemicals frequently present, with median amounts <100 ng/L for the sulfonamides, tetracyclines, fluoroquinolones, and macrolides. According to Liu et al. (2017) antibacterial agents, 12 that were illegal, were utilized between 1996 and 2013 and belonged to 9 different groups: aminoglycosides, quinolones, sulfonamides, sulfonamides, β-lactams, sulfonamides, macrolides, nitrofurans, and chloramphenicols. According to the Centers for Disease Control and Prevention (CDC), penicillins are the most commonly recommended antibiotics for human consumption in the United States, pursued by macrolides, cephalosporins, β-lactams, and tetracycline. Sulfonamides, one of the oldest antibiotic families still in use today, are frequently utilized as feed additives in animal husbandry. With a standard quantity in tributary water of less than 100 ng/L, sulfonamides are also one of the pharmaceutical chemical families that are most frequently studied in aquatic media in Europe (Carvalho and Santos 2016; Kergoat et al. 2021). Teglia et al. (2019) found 0.97–22.1 μg/L of many fluoroquinolone antibiotics in farmhouse drain water from different areas in Argentina, including ofloxacin, enrofloxacin, enoxacin, difloxacin, and ciprofloxacin. The presence of frequently used veterinary antibacterial drugs was examined in 159 freshwater, 443 fish, and 150 sediment samples from Polish rivers and lakes by Gbylik-Sikorska et al. (2014). Sulfamethoxazole, sulfamethazine, trimethoprim, erythromycin, azithromycin, clarithromycin, and roxithromycin were discovered by Managaki et al. (2007) at concentrations of 4–448 ng/L in the urban Tamagawa river in Japan and 7–360 ng/L in both city and rustic regions of the Mekong Delta in Vietnam. The scientists also demonstrated the occurrence of sulfamethazine in wastewater from a pig farm at elevated amounts (19.2 and 18.5 μg/L). In the effluents of the pharmaceutical companies, antimicrobial chemicals may be present, according to Bielen et al. (2017). They found amounts of up to 30 g/L in the recipient river and up to 10.5 mg/L of the macrolide antibiotic azithromycin and macrolide by-products in the treated wastewater from an antibiotic-producing company in Croatia. Fluoroquinolones, trimethoprim, sulfonamides, and tetracyclines were found in effluents of another pharmaceutical business up to 200 μg/L, and quantities of these substances in the receiver stream ranged from below the limit of detection to a range of μg/L. The effluents also contained high concentrations of bacteria that were resistant to sulfamethazine (90%), azithromycin (83%), and oxytetracycline (50%), at high frequency.

3.2.2 Antimicrobial Resistance in Marine Environments

Approximately 40% of the world's population lives within 100 km of the coast. It is well recognized that the world's highly developed coastlines have drained contaminants from the land inside the offshore surroundings. Metals, natural pollutants, naturally occurring radioactive material (NORMs), medicines, microparticles, and other materials are released, poured out, and percolated into the aquatic environment from ports, ships, oil plants, desalination plants, drain water treatment plants, and other industries (FAO 2016). Additionally, dust, a persistent issue in arid climates, transports toxins over a long distance. One of the most productive ecosystems is the coastal environment, which also supports a diverse microbial community. Antibiotic resistance in bacteria has sparked worries due to the occurrence of drug chemicals, besides antibiotics, in sea waters and drain water overflows. Recent studies have confirmed that sediments serve as a repository for ARGs that can spread throughout marine habitats. It is well recognized that the aquatic sediments provide a momentous environmental matrix for the diffusion of ARGs. ARGs can spread among bacterial species, including pathogens and fecal indicators like enterococci, after they are introduced into the environment (Agersø et al. 2006). In addition, maximum removal efficiencies of only 50.8–88.5% have been observed (Rodriguez-Mozaz et al. 2015), indicating that they cannot be entirely removed from WWTPs.

Sewer overflows frequently release pathogenic bacteria from WWTPs harboring ARGs into the marine environment. As a result, harmful bacteria carrying ARGs may spread to people through marine settings. It is crucial to examine and comprehend the diversity and delivery of ARGs in marine settings because of the serious threat they pose to human health. Microbes have been found to be resistant to antibiotics in estuaries, coastal sediments, and deep marine sediments. In contrast to freshwater and wastewater ecosystems, marine environments may have various ARB and ARG occurrence processes.

3.2.2.1 Sources

According to a survey, 28% of ARGs are found in marine habitats (Hatosy and Martiny 2015). The offshore discharge of ARBs from the earthly surroundings is the main cause of the increased incidence of AMR in the marine environment (Hatosy and Martiny 2015). The idea of a two-way exchange of ARGs between these aquatic habitats and individuals cannot be entirely lined out. ARBs are widely present in aquatic ecosystems, especially in open oceans, which are contaminated worldwide, according to research by means of both culture-dependent and culture-independent methods (Fekadu et al. 2019). Due to induced mutagenesis caused by low antibiotic concentrations, microorganisms in the normal marine surroundings are able to increase AMR (Kohanski et al. 2010). Even though the proportion of resistant mutants is quite little, the prolonged presence of antimicrobials might expedite the choice of ARBs over inventions. Aquatic habitats should therefore be taken into account as a solution to the establishment and increase of ARBs and ARGs. Antibiotics, ARBs, and ARGs may behave differently in marine environments than they do in freshwater and wastewater. One pathway for the appearance of antibiotic resistance in marine ecosystems has been discovered as coastal discharge of ARB from terrestrial sources (Hatosy and Martiny 2015). Another way is that operations associated with marine aquaculture might entail a choice that is connected to discharge of anthropogenic antibiotics. For example, 363.4 tons of antibiotics used via Chilean marine salmonid fish farm in 2016 unaccompanied might have acted as a collection force for the emergence of antibiotic resistance in the coastal environment (Miranda et al. 2018). Metagenomic analysis of the intestinal inside of fish from farms in the Baltic Sea revealed ARGs found in fish intestines were identical to those supplemented in farm sediments, demonstrating that fish feces still give to ARG enhancement in the absence of concurrent antibiotic treatments. ARG produced in a water environment, as suggested, might be transmitted to the fish intestine. The production of antibiotics by marine bacteria is another process that leads to the development of antibiotic resistance (Fig. 3.3).

Aquaculture is the area of animal food production that is rapidly expanding. From 5% in 1962 to 44.1% in 2014, it contributed to the whole quantity of fish destined for public utilization. An estimated 110 million metric tons of fish are produced globally each year; of this amount, aquaculture production accounts for around 50%. In reaction to the overfishing and climate change-related depletion of

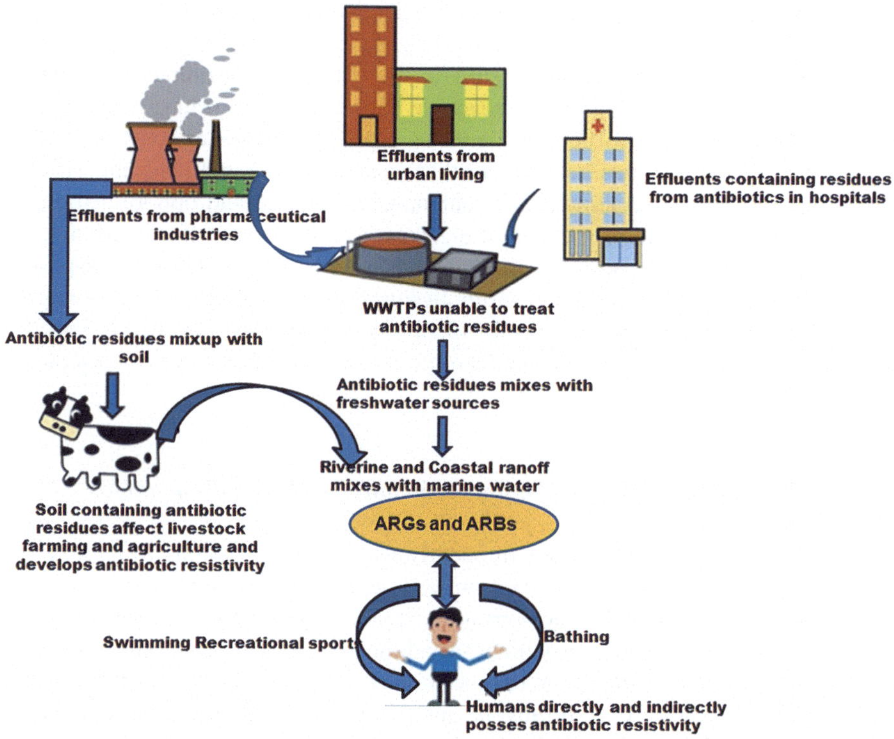

Fig. 3.3 Schematic diagram depicting the overall concept of antibiotic resistivity in freshwater and marine water environments.This figures shows the source of antibiotic and antibiotic resistance genes, Factors affecting the antibiotics and ARGs, impacts or pressure of antibiotic and ARGs in the both fresh water and marine water environment

fish stocks, a quick shift from a fishing to a production model took place. However, extensive utilization of antibiotics in intense aquaculture primarily for the making of fish, prawns, and shellfish leads to exceptionally high levels of antibiotics entering the aquatic surroundings. Drugs used as fish food might remain for an extremely long period in the aquatic ecology. Due to the potential for selection pressure in many ecosystems, the practice of employing antibiotics in fish farming has major negative effects on the surroundings (Burridge et al. 2010). Accordingly, sediments and watercourses contain large concentrations of bacteria resistant to antibiotics and can produce ARGs for infections that affect fish. Out of 58 marine microorganisms examined by Rosenfeld and Zobell (1947), four *Bacillus* species, three *Micrococcus* species, one *Actinomyces* species, and four *Serratia* species established antibiotic action on non-marine forms. The most aggressive hostile marine genera among them were *Micrococcus* and *Bacillus* (Rosenfeld and Zobell 1947). The emergence of ARB in aquatic habitats, especially open oceans, has been suggested by recent reports on worldwide antibiotic pollution in water ecosystems (Fekadu et al. 2019). At extremely little antibiotic amounts in normal aquatic habitats, bacteria can have mutations that make them resistant to antibiotics (Kohanski et al. 2010). The selection of ARB can happen over generations even while the part of resistant mutants is relatively small. Sub-MIC choice results in the extended determination of ARB, while low dosages of antibiotics are adequate to sustain the present resistant bacteria. Water habitats need to be considered as a possible cause of ARB and ARG danger. Furthermore, it has been found that aquaculture-related resistant bacteria can increase to terrestrial animals, the human environment, and opportunistic diseases that impact people or animals. It has been discovered that the healthiness of people and other earthly animals is affected when ARBs and ARGs are transported from aquatic ecology. None of the antibiotics used in aquaculture, which primarily provide preventive and metaphylactic functions, are designed with this specific usage in mind. They are actually medications created for different veterinary medical specialties.

3.2.2.2 Prevalence of Antimicrobial Resistance in Marine Ecosystems

In aquaculture, antimicrobials are frequently used as metaphylaxis to treat entire populations of people, including well, unwell, and carrier persons, and at dosages that are typically superior than persons utilized in the production of earthly animals. The use of florfenicol, oxytetracycline, sulfadimethoxine, and ormetoprim in aquaculture is permitted in the USA (Romero et al. 2012), whereas most European nations permit the use of erythromycin, sarafloxacin, trimethoprim, sulfonamides, and ormetoprim (Kümmerer 2009). Up until 2010, China used the most antibiotics and produced the most of them. At the same time, China was the biggest exporter of fish and fisheries goods, with 67% of the world's aquaculture. However, according to Maron et al. (2013), China does not demand veterinary recommendations for the administration of antibiotics to animals. China produces 8% of the world's aquaculture, followed by India (where the use and sale of antibiotics are similarly unregulated). ARGs are accumulated in the microbiome of peoples and animals. According to Anderson et al. (2006), bacteria that are in the intestinal system and conjugate with other bacteria can develop resistance to antibiotic and end in both public and animal feces. Antibiotic resistance in intestinal bacteria may emerge as a result of selective pressure brought on by antibiotic consumption. Dairy calves administered milk containing penicillin G (0–50 mL/kg) experienced a dose-dependent development of penicillin resistance that remained throughout the research (Langford et al. 2003). Chicken dung compost that was taken from farms that utilize gentamicin, kanamycin, cephalexin, and amoxicillin contained between 21% and 66% of the whole bacterial isolates. According to Yang et al. (2013), 8–20% of the ARB isolates demonstrated multidrug resistance. Antibiotics, ARBs, and ARGs are established in drain water, and further organic resources stimulate wastewater and other biological materials created in healthcare centers, livestock farming, houses, and other locations often enclose antibiotics, ARBs, and ARGs. In accordance with a recent research, sheep fed tylosin excreted the antibiotic for up to 4 days in their feces and urine released in hospitals, farms, residences, and further spaces.

According to Ishikawa et al. (2018), 11% on average of the tylosin that was supplied was excreted in urine and feces. Along with a recent metagenomics examination of 79 WWTPs that extend over 60 nations, North America, Europe, and Oceania differ from Asia, Africa, and South America in terms of ARG diversity and large quantity. While the Asia–Africa–South America cluster had ARGs related to sulfonamides and phenicols, the North America–Europe and Oceania cluster featured a small number of ARGs encoding macrolide resistance in high abundance. The research also found that Vietnam, India, and Brazil had the most different ARG distribution, and it suggested that these three nations would be potential capital for the evolution of new antibiotic resistance mechanisms. Different sources contribute to the microbiological and chemical loads that WWTPs receive. HGT can take place inside WWTPs under the selection pressure of antibiotics, metals, and other chemical compounds (Karkman et al. 2018). Numerous investigations have shown that WWTPs are not capable to entirely take away antibiotics, ARBs, and ARGs (Ben et al. 2017). Since environmental microorganisms outside water, in groundwater, in coastal surroundings, and in earth can interrelate with ARB and ARGs, discharging WWTP effluent into these habitats runs the risk of causing the microbial life in those ecosystems to become resistant to antibiotics.

First found in aquatic bacteria, some antibiotic resistance genes were later established in public and animal infections. A new fosfomycin resistance gene was isolated from aquatic surroundings, along with some quinolone resistance genes found in *Vibrio*, *Shewanella*, and *Aeromonas*, as well as β-lactamase genes found in *Photobacterium damselae* and *Oceanobacillus iheyensis* and chloramphenicol resistance genes found in aquatic *Photobacterium*, *Vibrio*, and *Shewanella*. The prevalence of microorganisms resistant to oxytetracycline, quinolones, sulfa/trimethoprim, florfenicol, and amoxicillin significantly increased in aquaculture farms where antibiotics were utilized (Cabello et al. 2013). The genomes of *Renibacterium salmoninarum* and *Stenotrophomonas maltophilia* both contained resistance gene variations for beta-lactams, aminoglycosides, tetracyclines, and macrolides (Crossman et al. 2008; Wiens et al. 2008). One of the antibiotics most frequently used in aquaculture is oxytetracycline, a universal antibiotic from the tetracycline class. According to Leal et al. (2019), this antibiotic is predominantly utilized in Canada, Norway, and the United States to treat fish bacterial illnesses brought on by *Aeromonas salmonicida*, *Aeromonas hydrophila*, *Aeromonas sobria*, *Pseudomonas*, *Lactococcus garvieae*, and *Vibrio anguillarum*. Aquaculture facilities that use oxytetracycline have a superior occurrence of tetracycline resistance (tetR) genes than those that do not, according to Seyfried et al. (2010). Tamminen et al. (2010) also established the existence of tetracycline resistance genes including tetA, tetC, tetH, and tetM in fish farm surroundings. Florfenicol use is allowed in Canada and Norway, and quinolone use in aquaculture is allowed in Norway. The two antibiotics that were most commonly used in Chilean salmon farms in 2016 were oxytetracycline and florfenicol. These drugs were mostly used to treat *Piscirickettsia salmonis*, which is now thought to be the primary bacterial pathogen of salmon (Miranda et al. 2018). Transferable resistance plasmids are present in bacterial cells, which causes the multidrug resistance of the pathogenic strain of *A. salmonicida* that affects fish. Akinbowale et al. (2006) demonstrated that β-lactams, first-generation cephalosporins, and macrolides were widely resistant among Gram-negative and Gram-positive bacteria isolated from farmed fish, crustaceans, and water from various parts of Australia. The research performed by Akinbowale et al. (2006) supports the idea that consuming aquaculture products has a risk of transferring resistant germs to people. According to the results of the study by Xiong et al. (2015), fish ponds in Guangdong, China, were ARG storage, with tetracyclines, sulfonamides, and fluoroquinolone amounts reaching as high as 446 μg/kg and 98.6 ng/L in sediment and water samples, respectively. Tetracycline resistance genes, sulfonamide resistance genes, and plasmid-mediated quinolone resistance genes were all shown to be present among the resistance genes. The persistence of ARGs at fish farms in the lack of antibiotic choice was discovered by Tamminen et al. (2010), and this is due to gene transport via mobile genetic elements that last a lengthy period in aquatic environments. When antibiotics are introduced into aquatic environments, environmental bacteria and fish pathogens are more likely to develop antibiotic resistance. Antibiotic-resistant bacteria can spread to the public from aquatic environments. Additionally, antibiotics might hasten the horizontal transfer of AGRs, which raises the likelihood that ambient bacteria will transfer their resistance genes to public pathogenic bacteria.

3.3 Prevention of the Development and Spread of Antimicrobial Resistance in Waterbodies

The removal of organic matter, nutrients, and solids was the original purpose of WWTPs; however, today's WWTPs are also required to eliminate antimicrobials/antibiotics, ARBs, and ARGs. The efficiency of the handling in getting rid of ARBs and ARGs is still mostly unknown. In order to assist the progress of the plan to WWTPs and the methodology employed, research is required to close a significant awareness space in this field. In order to reach a large reduction in the number of particular objective microorganisms that exhibit intra- and inter-process fluctuations, water supply systems and WWTPs use one or a mixture of multiple treatment processes (Amarasiri et al. 2017). Antibiotics, ARBs, and ARGs should all be reduced to a minimum in water and wastewater, according to specific, complete criteria. In comparison to traditional treatment techniques such as activated sludge, oxidation drain, sand filtration, rotary organic contactor, or sedimentation, membrane bioreactor treatment plants exhibited much larger reductions of ARB and ARGs (Su et al. 2018). Before searching for local treatments to prevent the increase of these contaminants inside the surroundings, it is essential to be aware of the exact remains and the AMR factors there in wastewater. Public health research must move faster in order to keep up with and even advance beyond water sustainability technology. Additionally, more study is required to apply efficient handling and sanitize methods for the total elimination of ARBs in WWTPs due to the substantial dangers to the immediate environment and public health. We may easily improve our ability to identify the origin of AMR and ARGs by using contemporary molecular tools for the characterization of bacterial species. A thorough grasp of the people ecology of organisms and the hereditary multiplicity of organisms that enter water are only a couple of the insights that these could offer (Olivas and Faulkner 2008). In contrast to cultivation-dependent approaches, these methodologies could enable quick and correct AMR cause trailing as well as other genetic mobility stages implicated in AMR dissemination. This would give a considerably more precise picture of the true multiplicity and difficulty of AMR in waterborne bacteria.

In order to determine the best course of action to decrease the danger of resistance adaptation linked with people waste streams, it is essential to recognize wherever choice is most likely to occur. The relevant protection goal in relation to the growth of antibiotic resistance is bacteria, not (aquatic) wildlife or people, in contrast to controlling hazards with most other contaminants. Environments like open drain and drain water treatment facilities are house to complex, thick bacterial communities that regularly contain diseases. Additionally, because they frequently contain higher quantities of antibiotics than receiving waters, they may serve as a better breeding ground for resistance evolution than recipient rivers. So, in addition to controlling resistance selection, interventions should also include selection risks that exist before wastewater release. Sewage epidemiology is silent in its early years, as a minimum in terms of how effectively the local scientific resistance condition can be envisaged. Whether ecological resistance data are based on isolates or metagenomes, it is critical to compare it to trustworthy clinical data. But to incorporate environmental assessments of resistance into therapeutic policy, more than only strong correlations in the data may be required.

3.4 Conclusion

Aquatic environments have been found to contain a wide range of antimicrobial chemicals, as well as antibiotics and chemotherapeutics. Antimicrobials are employed in aquaculture; however, they primarily derive from earthly sources, particularly the effluents of swine and poultry farms and, to a lesser extent, human medicine. The AMR problem is becoming into a global health disaster as a result of ongoing antimicrobial abuse, overuse, and environmental pollution caused by unchecked antimicrobial contamination. It is known from the literature that antibiotics, ARBs, and ARGs are present in a variety of aquatic environment locations, but it is unclear how much of a concern these substances provide to human health, necessitating more study. The paper also discusses potential causes for antibiotic resistance in different aquatic settings, except further investigation is needed to fully pinpoint and address the origins of this kind of pollution. Antibiotic resistance is currently leaking inside the surroundings through these channels despite efforts to identify ways to stop the transmission of resistance through WWTP. This is due to the fact that all of these agents cannot be completely neutralized by traditional

WWTPs. Additionally, the abundance of nutrients found in WWTPs increases the likelihood of antibiotic-resistant bacteria surviving or even proliferating. To stop the spread of antimicrobial drugs, ABR, and ARGs in the aquatic surroundings, it is imperative to take steps to control this type of contamination at the source. The government should gradually set stricter standards for the quality of water and wastewater treatment in order to achieve this. To address frontiers in the AMR situation, new rules must be established that limit the release of antimicrobial residues into the environment and encourage proper scrutinizing to lessen their building and hasten their clearance. Such initiatives ought to receive support both internationally and locally, as it is common for the existence of a certain concoction of toxins to be a regional issue. It should be mentioned, nevertheless, that there are many other aspects to the phenomena of environmental antimicrobial incidence and increasing drug resistance, and the environment is merely one of them. To improve control tactics, further study is required to comprehend the scope and molecular foundations of AMR development and ARG transfer to other dangerous bacteria. International collaboration is also required to support and help developing countries increase their capabilities and put preventive measures in place. International organizations like the FAO, OIE, and WHO must take the initiative in this effort.

References

Agenzia Italiana del Farmaco (2020) National report on antibiotics use in Italy-2019

Agersø Y, Sandvang D (2005) Class 1 integrons and tetracycline resistance genes in Alcaligenes, Arthrobacter, and Pseudomonas spp. isolated from pigsties and manured soil. Appl Environ Microbiol 71(12):7941–7947

Agersø Y, Pedersen AG, Aarestrup FM (2006) Identification of Tn5397-like and Tn916-like transposons and diversity of the tetracycline resistance gene tet(M) in enterococci from humans, pigs and poultry. J Antimicrob Chemother 57:832–839

Akinbowale OL, Peng H, Barton MD (2006) Antimicrobial resistance in bacteria isolated from aquaculture sources in Australia. J Appl Microbiol 100:1103–1113

Alduina R (2020) Antibiotics and environment. Antibiotics 9:202

Amarasiri M, Kitajima M, Nguyen TH, Okabe S, Sano D (2017) Bacteriophage removal efficiency as a validation and operational monitoring tool for virus reduction in wastewater reclamation: review. Water Res 121:258–269

Amarasiri M, Sano D, Suzuki S (2020) Understanding human health risks caused by antibiotic resistant bacteria (ARB) and antibiotic resistance genes (ARG) in water environments: current knowledge and questions to be answered. Crit Rev Environ Sci Technol 50:2016–2059

Anderson MA, Whitlock JE, Harwood VJ (2006) Diversity and distribution of Escherichia coli genotypes and antibiotic resistance phenotypes in feces of humans, cattle, and horses. Appl Environ Microbiol 72:6914–6922

Ben W, Wang J, Cao R, Yang M, Zhang Y, Qiang Z (2017) Distribution of antibiotic resistance in the effluents of ten municipal wastewater treatment plants in China and the effect of treatment processes. Chemosphere 172:392–398

Bengtsson-Palme J, Larsson DGJ (2016) Concentrations of antibiotics predicted to select for resistant bacteria: proposed limits for environmental regulation. Environ Int 86:140–149

Bielen A, Simatovic A, Kosic-Vuksic J, Senta I, Ahel M, Babic S, Jurina T, Gonzalez Plaza JJ, Milakovic M, Udikovic-Kolic N (2017) Negative environmental impacts of antibiotic-contaminated effluents from pharmaceutical industries. Water Res 126:79–87

Burridge L, Weis JS, Cabello F, Pizarro J, Bostick K (2010) Chemical use in salmon aquaculture: a review of current practices and possible environmental effects. Aquaculture 306:7–23

Cabello FC, Godfrey HP, Tomova A, Ivanova L, Dolz H, Millanao A, Buschmann AH (2013) Antimicrobial use in aquaculture re-examined: its relevance to antimicrobial resistance and to animal and human health. Environ Microbiol 15:1917–1942

Carvalho IT, Santos L (2016) Antibiotics in the aquatic environments: a review of the European scenario. Environ Int 94:736–757

Charuaud L, Jardé E, Jaffrézic A, Liotaud M, Goyat Q, Mercier F, Le Bot B (2019) Veterinary pharmaceutical residues in water resources and tap water in an intensive husbandry area in France. Sci Total Environ 664:605–615

Cheng J, Jiang L, Sun T, Du Z, Lee L, Zhao Q (2019) Occurrence, seasonal variation and risk assessment of antibiotics in the surface water of North China. Arch Environ Contam Toxicol 77:88–97

Chokshi A, Sifri Z, Cennimo D, Horng H (2019) Global contributors to antibiotic resistance. J Glob Infect Dis 11:36–42

Crossman LC, Gould VC, Dow JM, Vernikos GS, Okazaki A, Sebaihia M, Saunders D, Arrowsmith C, Carver T, Peters N (2008) The complete genome, comparative and functional analysis of Stenotrophomonas maltophilia reveals an organism heavily shielded by drug resistance determinants. Genome Biol 9:R74

Cui C, Han Q, Jiang L, Ma L, Jin L, Zhang D, Lin K, Zhang T (2018) Occurrence, distribution, and seasonal variation of antibiotics in an artificial water source reservoir in the Yangtze River delta, East China. Environ Sci Pollut Res 25:19393–19402

Danner MC, Robertson A, Behrends V, Reiss J (2019) Antibiotic pollution in surface fresh waters: occurrence and effects. Sci Total Environ 664:793–804

Dhanji H, Murphy NM, Akhigbe C, Doumith M, Hope R, Livermore DM, Woodford N (2011) Isolation of fluoroquinolone-resistant O25b: H4-ST131 Escherichia coli with CTX-M-14 extended-spectrum β-lactamase from UK river water. J Antimicrob Chemother 66(3):512–516

Di Cesare A, Eckert EM, D'Urso S, Bertoni R, Gillan DC, Wattiez R, Corno G (2016) Co-occurrence of integrase 1, antibiotic and heavy metal resistance genes in municipal wastewater treatment plants. Water Res 94:208–214

Esimone CO, Nworu CS, Udeogaranya OP (2007) Utilization of antimicrobial agents with and without prescription by out-patients in selected pharmacies in South-Eastern Nigeria. Pharm World Sci 29:655–660

FAO (2016) The state of world fisheries and aquaculture. In: Contributing to food security and nutrition for all. FAO, Rome

Fekadu S, Alemayehu E, Dewil R, Van der Bruggen B (2019) Pharmaceuticals in freshwater aquatic environments: a comparison of the African and European challenge. Sci Total Environ 654:324–337

Felis E, Kalka J, Sochacki A, Kowalska K, Bajkacz S, Harnisz M, Korzeniewska E (2020) Antimicrobial pharmaceuticals in the aquatic environment—occurrence and environmental implications. Eur J Pharmacol 866:172813

Food and Agriculture Organization of the United Nations and World Health Organization (2019) Joint FAO/WHO expert meeting in collaboration with OIE on foodborne antimicrobial resistance: role of the environment, crops and biocides—Meeting report

Froehlich H, Gentry R, Halpern B (2018) Global change in marine aquaculture production potential under climate change. Nat Ecol Evol 2:1745–1750

Gbylik-Sikorska M, Posyniak A, Mitrowska K, Gajda A, Bładek T, Sniegocki T, Zmudzki J (2014) Occurrence of veterinary antibiotics and chemotherapeutics in fresh water, sediment, and fish of the rivers and lakes in Poland. Bull Vet Inst Pulawy 58:399–404

Girlich D, Poirel L, Nordmann P (2011) Diversity of clavulanic acid-inhibited extended-spectrum β-lactamases in Aeromonas spp. from the Seine River, Paris, France. Antimicrob Agents Chemother 55(3):1256–1261

Hassan B, Qadri H, Ali MN, Khan NA, Yatoo AM (2020) Impact of climate change on freshwater ecosystem and its sustainable management. In: Fresh water pollution dynamics and remediation. Springer, pp 105–121

Hatcher SM, Myers KW, Heaney CD, Larsen J, Hall D, Miller MB, Stewart JR (2016) Occurrence of methicillin-resistant Staphylococcus aureus in surface waters near industrial hog operation spray fields. Sci Total Environ 565:1028–1036

Hatosy SM, Martiny AC (2015) The ocean as a global reservoir of antibiotic resistance genes. Appl Environ Microbiol 81:7593–7599

Ishikawa NK, Touno E, Higashiyama Y, Sasamoto M, Soma M, Yoshida N (2018) Determination of tylosin excretion from sheep to assess tylosin spread to agricultural fields by manure application. Sci Total Environ 633:399–404

Janecko N, Pokludova L, Blahova J, Svobodova Z, Literak I (2016) Implications of fluoroquinolone contamination for the aquatic environment—a review. Environ Toxicol Chem 35:2647–2656

Karkman A, Do TT, Walsh F, Virta MPJ (2018) Antibiotic-resistance genes in waste water. Trends Microbiol 26:220–228

Kergoat L, Besse-Hoggan P, Leremboure M, Beguet J, Devers M, Martin-Laurent F, Masson M, Morin S, Roinat A, Pesce S, Bonnineau C (2021) Environmental concentrations of sulfonamides can alter bacterial structure and induce diatom deformities in freshwater biofilm communities. Front Microbiol 12:643719

Kohanski MA, DePristo MA, Collins JJ (2010) Sublethal antibiotic treatment leads to multidrug resistance via radical-induced mutagenesis. Mol Cell 37:311–320

Kotwani A, Holloway K (2011) Trends in antibiotic use among outpatients in New Delhi, India. BMC Infect Dis 11:99

Kotwani A, Joshi J, Kaloni D (2021) Pharmaceutical effluent: a critical link in the interconnected ecosystem promoting antimicrobial resistance. Environ Sci Pollut Res 28:32111–32124

Krempaska N, Horňák K, Pernthaler J (2018) Spatiotemporal distribution and microbial assimilation of polyamines in a mesotrophic lake. Limnol Oceanogr 63(2):816–832

Krzeminski P, Tomei MC, Karaolia P, Langenhoff A, Almeida CMR, Felis E, Gritten F, Andersen HR, Fernandes T, Manaia CM, Rizzo L, Fatta-Kassinos D (2019) Performance of secondary wastewater treatment methods for the removal of contaminants of emerging concern implicated in crop uptake and antibiotic resistance spread: a review. Sci Total Environ 648:1052–1081

Kümmerer K (2009) Antibiotics in the aquatic environment: a review: part I. Chemosphere 75:417–434

Langford FM, Weary DM, Fisher L (2003) Antibiotic resistance in gut bacteria from dairy calves: a dose response to the level of antibiotics fed in milk. J Dairy Sci 86:3963–3966

Leal JF, Santos EBH, Esteves VI (2019) Oxytetracycline in intensive aquaculture: water quality during and after its administration, environmental fate, toxicity and bacterial resistance. Rev Aquac 11:1176–1194

Leonard A, Zhang L, Balfour A, Garside R, Gaze W (2015) Human recreational exposure to antibiotic resistant bacteria in coastal bathing waters. Environ Int 82:92–100

Levin-Edens E, Soge OO, No D, Stiffarm A, Meschke JS, Roberts MC (2012) Methicillin-resistant S taphylococcus aureus from Northwest marine and freshwater recreational beaches. FEMS Microbiol Ecol 79(2):412–420

Li D, Yang M, Hu J, Zhang Y, Chang H, Jin F (2008) Determination of penicillin G and its degradation products in a penicillin production wastewater treatment plant and the receiving river. Water Res 42:307–317

Li W, Shi Y, Gao L, Liu J, Cai Y (2012) Occurrence of antibiotics in water, sediments, aquatic plants, and animals from Baiyangdian Lake in North China. Chemosphere 89:1307–1315

Li S, Shi W, LiuW LH, Zhang W, Hu J, Ke Y, Sun W, Ni JA (2018) A duodecennial national synthesis of antibiotics in China's major rivers and seas (2005–2016). Sci Total Environ 615:906–917

Liu YY, Wang Y, Walsh TR, Yi LX, Zhang R, Spencer J, Doi Y, Tian G, Dong B, Huang X, Yu LF (2016) Emergence of plasmid-mediated colistin resistance mechanism MCR-1 in animals and human beings in China: a microbiological and molecular biological study. Lancet Infect Dis 16(2):161–168

Liu X, Steele JC, Meng XZ (2017) Usage, residue, and human health risk of antibiotics in Chinese aquaculture: a review. Environ Pollut 223:161–169

Lu J, Zhang Y, Wu J, Luo Y (2019) Effects of microplastics on distribution of antibiotic resistance genes in recirculating aquaculture system. Ecotoxicol Environ Saf 184:109631

Madikizela LM, Tavengwa NT, Chimuka L (2017) Status of pharmaceuticals in African water bodies: occurrence, removal and analytical methods. J Environ Manag 193:211–220

Managaki S, Murata A, Takada H, Tuyen BC, Chiem NH (2007) Distribution of macrolides, sulfonamides, and trimethoprim in tropical waters: ubiquitous occurrence of veterinary antibiotics in the Mekong Delta. Environ Sci Technol 41:8004–8010

Maron DF, Smith TJS, Nachman KE (2013) Restrictions on antimicrobial use in food animal production: an international regulatory and economic survey. Glob Health 9:11

Matongo S, Birungi G, Moodley B, Ndungu P (2015) Pharmaceutical residues in water and sediment of Msunduzi River, KwaZulu-Natal, South Africa. Chemosphere 134:133–140

Mazimba A, Rupasinghe N, Gomez ACC, Zhao F, Pate M (2021) Landscape analysis of tools to address antimicrobial resistance. World Bank, Washington, DC

Miranda CD, Godoy FA, Lee MR (2018) Current status of the use of antibiotics and the antimicrobial resistance in the Chilean salmon farms. Front Microbiol 9:1–14

Miyagi K, Hirai I (2019) A survey of extended-spectrum β-lactamase-producing Enterobacteriaceae in environmental water in Okinawa prefecture of Japan and relationship with indicator organisms. Environ Sci Pollut Res 26:7697–7710

Morris D, Galvin S, Boyle F, Hickey P, Mulligan M, Cormican M (2012) Enterococcus faecium of the vanA genotype in rural drinking water, effluent, and the aqueous environment. Appl Environ Microbiol 78(2):596–598

Nam S, Kim MJ, Park C, Park JG, Maeng PJ, Lee GC (2013) Detection and genotyping of vancomycin-resistant Enterococcus spp. by multiplex polymerase chain reaction in Korean aquatic environmental samples. Int J Hyg Environ Health 216(4):421–427

Nishiyama M, Ogura Y, Hayashi T, Suzuki Y (2017) Antibiotic resistance profiling and genotyping of vancomycin-resistant enterococci collected from an urban river basin in the provincial city of Miyazaki, Japan. Water 9(2):79

Novais C, Coque TM, Ferreira H, Sousa JC, Peixe L (2005) Environmental contamination with vancomycin-resistant enterococci from hospital sewage in Portugal. Appl Environ Microbiol 71(6):3364–3368

Olivas Y, Faulkner BR (2008) Fecal source tracking by antibiotic resistance analysis on a watershed exhibiting low resistance. Environ Monit Assess 139:15–25

Pazda M, Kumirska J, Stepnowski P, Mulkiewicz E (2019) Antibiotic resistance genes identified in wastewater treatment plant systems—a review. Sci Total Environ 697:159–198

Piedra-Carrasco N, Fàbrega A, Calero-Caceres W, Cornejo-Sánchez T, Brown-Jaque M, Mir-Cros A, Muniesa M, González-López JJ (2017) Carbapenemase-producing enterobacteriaceae recovered from a Spanish river ecosystem. PLoS One 12(4):e0175246

Prestinaci F, Pezzotti P, Pantosti A (2015) Antimicrobial resistance: a global multifaceted phenomenon. Pathog Glob Health 109:309–318

Qiao M, Ying GG, Singer AC, Zhu YG (2018) Review of antibiotic resistance in China and its environment. Environ Int 110:160–172

Roberts MC, Soge OO, Giardino MA, Mazengia E, Ma G, Meschke JS (2009) Vancomycin-resistant Enterococcus spp. in marine environments from the West Coast of the USA. J Appl Microbiol 107(1):300–307

Rodriguez-Mozaz S, Chamorro S, Marti E, Huerta B, Gros M, Sànchez-Melsió A, Borrego CM, Barcelóac D, Balcázar JL (2015) Occurrence of antibiotics and antibiotic resistance genes in hospital and urban wastewaters and their impact on the receiving river. Water Res 69:234–242

Romero J, Gloria Feijoo C, Navarrete P (2012) Antibiotics in aquaculture—use, abuse and alternatives. In: Carvalho ED, David GD, Silva RJ (eds) Health and environment in aquaculture. InTech, Rijeka

Rosenfeld WD, Zobell CE (1947) Antibiotic production by marine microorganisms. J Bacteriol 54:393–398

Seyfried EE, Newton RJ, Rubert KF, Pedersen JA, McMahon KD (2010) Occurrence of tetracycline resistance genes in aquaculture facilities with varying use of oxytetracycline. Microb Ecol 59:799–807

Singh R, Singh AP, Kumar S, Giri BS, Kim KH (2019) Antibiotic resistance in major rivers in the world: a systematic review on occurrence, emergence, and management strategies. J Clean Prod 234:1484–1505

Su HC, Liu YS, Pan CG, Chen J, He LY, Ying GG (2018) Persistence of antibiotic resistance genes and bacterial community changes in drinking water treatment system: from drinking water source to tap water. Sci Total Environ 616:453–461

Sucato A, Vecchioni L, Savoca D, Presentato A, Arculeo M, Alduina R (2021) A comparative analysis of aquatic and polyethylene-associated antibiotic-resistant microbiota in the Mediterranean sea. Biology 10:200

Szymańska U, Wiergowski M, Sołtyszewski I, Kuzemko J, Wiergowska G, Woźniak MK (2019) Presence of antibiotics in the aquatic environment in Europe and their analytical monitoring: recent trends and perspectives. Microchem J 147:729–740

Tacão M, Correia A, Henriques I (2012) Resistance to broad-spectrum antibiotics in aquatic systems: anthropogenic activities modulate the dissemination of bla CTX-M-like genes. Appl Environ Microbiol 78(12):4134–4140

Tacão M, Correia A, Henriques IS (2015) Low prevalence of carbapenem-resistant bacteria in river water: resistance is mostly related to intrinsic mechanisms. Microb Drug Resist 21:497–506

Tamminen M, Karkman A, Lohmus A, Muziasari WI, Takasu H, Wada S, Suzuki S, Virta M (2010) Tetracycline resistance genes persist at aquaculture farms in the absence of selection pressure. Environ Sci Technol 45:386–391

Teglia CM, Perez FA, Michlig N, Repetti MR, Goicoechea HC, Culzoni MJ (2019) Occurrence, distribution and ecological risk of fluoroquinolones in rivers and wastewaters. Environ Toxicol Chem 38:2305–2313

Thapaliya D, Hellwig EJ, Kadariya J, Grenier D, Jefferson AJ, Dalman M, Kennedy K, DiPerna M, Orihill A, Taha M, Smith TC (2017) Prevalence and characterization of Staphylococcus aureus and methicillin-resistant Staphylococcus aureus on public recreational beaches in northeast Ohio. GeoHealth 1(10):320–332

Timm A, Borowska E, Majewsky M, Merel S, Zwiener C, Bräse S, Horn H (2019) Photolysis of four β-lactam antibiotics under simulated environmental conditions: degradation, transformation products and antibacterial activity. Sci Total Environ 651:1605–1612

Tong L, Qin L, Guan C, Wilson ME, Li X, Cheng D, Ma J, Liu H, Gong F (2020) Antibiotic resistance gene profiling in response to antibiotic usage and environmental factors in the surface water and groundwater of Honghu Lake, China. Environ Sci Pollut Res Int 27:31995–32005

Uzan-Yulzari A, Turta O, Belogolovski A, Ziv O, Kunz C, Perschbacher S, Neuman H, Pasolli E, Oz A, Ben-Amram H, Ollila H, Kaljonen A, Isolauri E, Salminen S, Lagstrom H, Segata N, Sharon I, Louzoun Y, Ensenauer R, Rautava S, Koren O (2021) Neonatal antibiotic exposure impairs child growth during the first six years of life by perturbing intestinal microbial colonization. Nat Commun 12:1–12

Van Boeckel TP, Gandra S, Ashok A, Caudron Q, Grenfell BT, Levin SA, Laxminarayan R (2014) Global antibiotic consumption 2000 to 2010: an analysis of national pharmaceutical sales data. Lancet Infect Dis 14:742–750

Velpandian T, Halder N, Nath M, Das U, Moksha L, Gowtham L, Batta SP (2018) Un-segregated waste disposal: an alarming threat of antimicrobials in surface and groundwater sources in Delhi. Environ Sci Pollut Res 25:29518–29528

Ward N, Megonigal J, Bond-Lamberty B, Bailey V, Butman D, Canuel E, Diefenderfer H, Ganju N, Goñi M, Graham E, Hopkinson C, Khangaonkar T, Langley A, McDowell N, Myers-Pigg A, Neumann R, Osburn C, Price R, Rowland J, Sengupta A, Simard M, Thornton P, Tzortziou M, Vargas R, Weisenhorn P, Windham-Myers L (2020) Representing the function and sensitivity of coastal interfaces in Earth system models. Nat Commun 11:1–14

Wernersson AS, Carere M, Maggi C, Tusil P, Soldan P, James A, Sanchez W, Dulio V, Broeg K, Reifferscheid G, Buchinger S, Maas H, Van Der Grinten E, O'Toole S, Ausili A, Manfra L, Marziali L, Polesello S, Lacchetti I, Kase R (2015) The European technical report on aquatic effect-based monitoring tools under the water framework directive. Environ Sci Europe 27:7

Wiens GD, Rockey DD, Wu Z, Chang J, Levy R, Crane S, Chen DS, Capri GR, Burnett JR, Sudheesh PS (2008) Genome sequence of the fish pathogen Renibacterium salmoninarum suggests reductive evolution away from an environmental Arthrobacter ancestor. J Bacteriol 190:6970–6982

Wilson D, Hauryliuk V, Atkinson G, O'Neill A (2020) Target protection as a key antibiotic resistance mechanism. Nat Rev Microbiol 18:1–12

World Health Organization (2021) Antimicrobial resistance. World Health Organization

Xiong W, Sun Y, Zhang T, Ding X, Li Y, Wang M, Zeng Z (2015) Antibiotics, antibiotic resistance genes, and bacterial community composition in fresh water aquaculture environment in China. Microb Ecol 70:425–432

Yang J, Wang C, Shu C, Liu L, Geng J, Hu S (2013) Marine sediment bacteria harbor antibiotic resistance genes highly similar to those found in human pathogens. Microb Ecol 65:975–981

Ye Q, Wu Q, Zhang S, Zhang J, Yang G, Wang H, Huang J, Chen M, Xue L, Wang J (2017) Antibiotic-resistant extended spectrum ss-lactamase-and plasmid-mediated AmpC-producing enterobacteriaceae isolated from retail food products and the pearl river in Guangzhou, China. Front Microbiol 8:96

Yin G, Hou L, Liu M, Zheng Y, Li X, Lin X, Gao J, Jiang X (2016) Effects of thiamphenicol on nitrate reduction and N_2O release in estuarine and coastal sediments. Environ Pollut 214:265–272

Zhang QQ, Ying GG, Pan CG, Liu YS, Zhao JL (2015) Comprehensive evaluation of antibiotics emission and fate in the river basins of China: source analysis, multimedia modeling, and linkage to bacterial resistance. Environ Sci Technol 49:6772–6782

Zhou LJ, Ying GG, Zhao JL, Yang JF, Wang L, Yang B, Liu S (2011) Trends in the occurrence of human and veterinary antibiotics in the sediments of the Yellow River, Hai River and Liao River in northern China. Environ Pollut 159:1877–1885

Zhu YG, Zhao Y, Li B, Huang CL, Zhang SY, Yu S, Chen YS, Zhang T, Gillings MR, Su JQ (2017) Continental-scale pollution of estuaries with antibiotic resistance genes. Nat Microbiol 2:16270

Chapter 4
Antimicrobial Resistance (AMR) in Farmed Aquatic Organisms

Eakapol Wangkahart, Po-Tsang Lee, and Chou-Min Chong

Abstract The aquaculture sector has recognized antimicrobial resistance (AMR) as a significant challenge in recent years. The increase in antibiotic resistance among farmed fish is a notable concern for the aquaculture industry. Due to the frequent bacterial infections in fish populations, antibiotics must be regularly administered, resulting in their ongoing presence in the aquatic environment. Hence, this condition promotes the quick spread of antibiotic-resistant bacteria. It is apparent that a significant number of disease-causing microorganisms can develop resistance to antimicrobial substances. AMR within aquaculture systems can be transferred to medically important strains in the natural environment through horizontal gene transfer, which has the potential to affect the entire ecosystem. Among cultured fish, including ornamental varieties, a broad spectrum of pathogens displays resistance to multiple antibiotics. A thorough understanding of gene transfer processes such as plasmids, transposons, and gene cassettes is crucial for unraveling the complexities of antimicrobial resistance in the context of aquaculture. The control of AMR dissemination in aquaculture involves the initiation of continuous monitoring initiatives, the rapid detection of resistant bacterial strains, and the enforcement of regulations. This chapter offers an outline of antibiotic use and the presence of AMR genes in emerging multidrug resistant pathogens, as well as the prevalence of infectious diseases. It also provides into antimicrobial use within aquaculture systems and explores the management of pathogen infections in the field of aquaculture.

E. Wangkahart (✉)
Laboratory of Fish Immunology and Nutrigenomics, Applied Animal and Aquatic Sciences Research Unit, Division of Fisheries, Faculty of Technology, Mahasarakham University, Kantharawichai, Maha Sarakham, Thailand
e-mail: eakapol.w@msu.ac.th

P.-T. Lee
Department of Aquaculture, National Taiwan Ocean University, Keelung, Taiwan

C.-M. Chong
Laboratory of Immunogenomics, Department of Aquaculture, Faculty of Agriculture, Universiti Putra Malaysia, Serdang, Selangor, Malaysia

P. Elumalai, S. Lakshmi (eds.), *Antimicrobial Resistance in Aquaculture and Aquatic Environments*, https://doi.org/10.1007/978-981-97-7320-6_4

Keywords Antimicrobial resistance · Sustainable aquaculture · Antibiotic drugs · Antimicrobial utilization · Resistance genes

4.1 Introduction

Aquaculture has emerged as one of the world's fastest-growing food industries in recent years, helping food security and boosting global economic status. In aquaculture, the successful production of fish is frequently hampered by various bacterial diseases. The presence of this bacterium in cultured fish raises significant public health concerns, affecting humans and leading to the frequent use of antibiotics in aquaculture. This use of antibiotics for both treatment and prevention in aquaculture has given rise to AMR among bacterial pathogens associated with fish (Caputo et al. 2023). A wide range of aquatic species, such as carp, salmon, tilapia, catfish, and globally distributed crustaceans like shrimp, mussel, and abalone, have been identified as potential carriers of antimicrobial-resistant pathogens. Moreover, farmed fish may also carry antimicrobial-resistant bacteria, some of which can pose a threat to human health (Preena et al. 2022). The continuous application of antibiotics in aquaculture can result in their persistence in the environment. There is some concern that even at very low levels they may exert selective pressure on microbial communities. This selective pressure can lead to the transformation of antimicrobial-resistant bacteria into reservoirs of antibiotic-resistant genes within aquatic environments.

With its rapid growth, aquaculture currently plays a significant role in providing almost half of the fish consumed by people worldwide. As the global demand for aquatic goods rises, there has been a notable escalation in intensive aquaculture methods. Large quantities of fish are cultured through intensive and semi-intensive methods, but they are regularly affected by disease outbreaks, leading to the common practice of using antimicrobial agents to manage and contain these outbreaks. However, the presence of these drugs at subtherapeutic levels in water and sediments over extended periods creates favorable conditions for the development and selection of antibiotic-resistant bacterial strains, as well as for encouraging horizontal gene transfer. These practices frequently necessitate the application of antibiotics and other antimicrobial agents for the prevention and treatment of diseases. It is now widely acknowledged that the transfer of AMR genes and resistant bacteria between aquatic and terrestrial animal farming environments and the human habitat can have adverse impacts on the health of both humans and animals, as well as on aquatic ecosystems. The regular utilization of antibiotics in aquaculture, whether for treatment or prevention, has resulted in the development of AMR. Hence, AMR poses a significant and growing concern in the aquaculture industry, impacting the health of farmed aquatic organisms, the sustainability of the sector, and potential public health risks. AMR is a situation where an antimicrobial agent loses its effectiveness in fully or partially inhibiting bacterial growth. AMR in aquaculture raises public health concerns as resistant bacteria can potentially transfer from farmed organisms to humans through consumption or environmental exposure (Preena et al. 2020a). The dissemination of AMR genes from aquaculture to healthcare settings raises concerns about the treatment of human infections. The key factor behind AMR in aquaculture is the transmission and diffusion of AMR genes within bacterial communities. These genes have the capacity to be disseminated through horizontal gene transfer, resulting in the conversion of once-susceptible bacteria into resistant forms. The presence of effective monitoring and surveillance systems is pivotal for appraising the prevalence and changes in AMR among cultured aquatic species (Santos and Ramos 2018). An international initiative is necessary to halt the excessive use of antimicrobials in aquaculture and to promote the adoption of alternative disease prevention strategies. Charting a different course is essential in order to mitigate the growing menace of AMR. As part of these programs, it is crucial to regularly carry out assessments for antibiotic residues and genetic analyses. The World Health Organization (WHO) and the Food and Agriculture Organization (FAO), both worldwide organizations, are playing an active part in efforts to combat AMR in aquaculture. The Terrestrial and Aquatic Animal Health Codes established by the World Organization for Animal Health (WOAH, formerly known as OIE) set global standards aimed at enhancing animal health, welfare, and veterinary public health. These codes include guidelines for ensuring the safe international trade of terrestrial and aquatic animals and their products. Additionally, the manuals offer a standardized method for diagnosing the diseases outlined in the Terrestrial and Aquatic Codes (Morales et al. 2023). Regulations, protocols, and optimal approaches are under development to encourage the responsible application of antibiotics within the sector. In aquaculture, there is ongoing research into substitute methods for disease management, such as vaccines, probiotics, optimized farming techniques, and selective breeding focused on disease resistance (Wang et al. 2023; Wangkahart et al. 2023a). A significant focus is on minimizing the dependence on antibiotics. The effective management of AMR in cultivated aquatic organisms in the future relies on a mixture of substitute approaches, regulatory steps, vigilant monitoring, educational efforts, and international cooperation. Implementing sustainable approaches and using antibiotics in a responsible manner are essential for reducing the dangers related to AMR and safeguarding the lasting well-being of aquatic ecosystems and human communities. In general, the problem of antimicrobial resistance in cultivated aquatic organisms is multifaceted, with implications for food security, environmental health, and the overall well-being of society. Meeting this challenge entails embracing a holistic approach that highlights sustainability, conscientious antibiotic utilization, and global cooperation.

4.2 Prevalence of Infectious Diseases in Farmed Aquatic Organisms

Aquaculture has become a significant source of seafood production globally. However, this industry faces challenges related to disease outbreaks among fish and shellfish species, leading to the use of antimicrobial agents, including antibiotics, to manage these infections (Leon et al. 2020). The high prevalence of these infectious diseases can lead to considerable economic losses environmental issues. The aquaculture industry's rapid expansion has led to the frequent emergence of diseases caused by bacteria, viruses, fungi, and protozoa in fish and shrimp farming. Bacterial diseases can quickly and extensively propagate owing to the short life cycle and rapid multiplication of bacteria.

In recent years, the aquaculture sector has encountered significant economic setbacks resulting from clinical symptoms triggered by bacteria, with variations in the specific pathogenic strains. These diseases are primarily caused by Gram-negative organisms like *Aeromonas hydrophila*, *A. salmonicida*, *A. veronii*, *Vibrio anguillarum*, *V. harveyi*, *V. parahaemolyticus*, *Flavobacterium psychrophilum*, *F. columnare*, *Francisella orientalis*, *F. noatunensis*, *Edwardsiella tarda*, *E. ictaluri*, *Citrobacter freundii*, *Pseudomonas fluorescens*, and *Yersinia ruckeri*. Diseases of this kind are typically attributed to Gram-positive bacteria, including *Streptococcus agalactiae*, *S. iniae*, and various *Staphylococcus* strains. In fish farming, this disease has the capacity to strike at both the early and later stages, resulting in a mortality rate of over 50%. Among these, *S. agalactiae*, *S. iniae*, and *E. tarda* are notable pathogens that have the potential to infect both human beings and farmed fish (Raabe and Shane 2019). It has the potential to result in human conditions like sepsis, urinary tract infections, skin and soft tissue infections, as well as pneumonia. Clinical signs can differ based on the affected organ, potentially including fever, chills, confusion, respiratory difficulties, and localized signs of infection. *Vibrio harveyi* has the capacity to infect not only fish but also numerous invertebrates, such as shrimps and shellfish (Boopathi et al. 2023).

4.3 Antibiotics Used in Aquaculture and Their Mode of Action

Antibiotics have become an essential component of animal farming and aquaculture, primarily due to the increasing need for animal-derived food items. Overuse of antibiotics in aquaculture can accelerate the emergence of antibiotic resistance within the aquaculture environment and among cultivated aquatic species. Additionally, consuming fish that have been treated with antibiotics may have harmful consequences for human health. In general, the different classes of antibiotics used in aquaculture include beta-lactams, chloramphenicol, glycopeptides, rifamycin, aminoglycosides, quinolones, macrolides, nitrofurans, tetracyclines, and sulfonamides. They are predominantly utilized because of their affordability, excellent solubility in water, and their ability to float effectively, facilitating their transportation and distribution in aquatic settings. The mechanism by which antimicrobial agents work can be classified according to the specific functions they impact. These typically involve the following: hindering cell wall or nucleic acid synthesis, impeding ribosome function, affecting cell membrane operation, and disrupting folate metabolism. According to the information regarding antibiotic used in aquaculture, the antibiotic drugs that are grouped based on mode of action are given in Table 4.1, and their modes of action are provided in Fig. 4.1.

Table 4.1 Antibiotics class, examples and their mode of action

Antibiotic class	Example of antibiotics	Mechanism of action
Beta lactams	Amoxicillin, ampicillin Penicillin, cephalexin Cefotaxime, benzylpenicillin	Inhibit cell wall synthesis
Chloramphenicol	Thiamphenicol, florfenicol chloramphenicol	Inhibit protein synthesis by binding to 50S subunit of the 70S ribosomal subunit
Glycopeptides	Vancomycin	Inhibit cell wall synthesis
Rifamycin	Rifampicin	Inhibit bacterial RNA transcription
Aminoglycosides	Kanamycin, gentamicin Neomycin	Inhibit protein synthesis by binding to 30S ribosomal subunit and 50S ribosomal subunit
Quinolones	Ciprofloxacin, enrofloxacin, flumequine, norfloxacin levofloxacin, ofloxacin, moxifloxacin	Inhibit nucleic acid synthesis
Macrolides	Azithromycin clarithromycin erythromycin Roxithromycin	Inhibit protein synthesis by binding to 50S ribosomal subunit
Nitrofurans	Nitrofurantoin	Inhibit protein and cell wall synthesis
Nitroimidazoles	Metronidazole	Inhibit protein and cell wall synthesis
Lincosamides	Clarithromycin Clarithromycin erythromycin, clindamycin Lincomycin	Inhibit protein synthesis by binding to 50S ribosomal subunit
Tetracyclines	Oxytetracycline Doxycycline, tetracycline	Inhibit protein synthesis by binding to 30S ribosomal subunit
Sulfonamides	Sulfamethoxazole Trimethoprim	Inhibit metabolic pathways

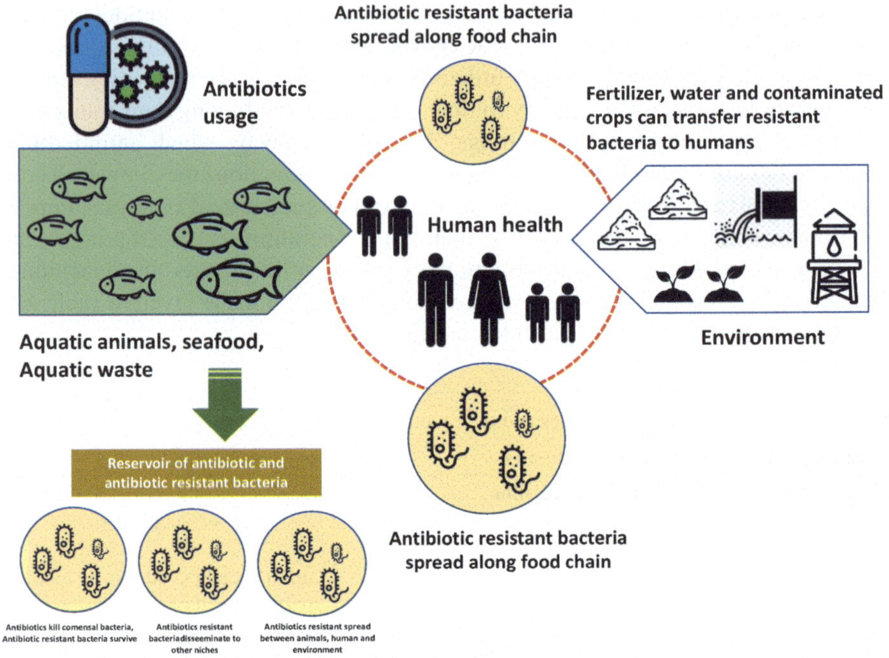

Fig. 4.1 Mechanism of action and resistance of commonly used antimicrobial agents used in aquaculture

4.3.1 Beta-Lactam and Mechanism of Action in Farmed Aquatic Organisms

Beta (β)-lactam antibiotics are among the earliest types of antibacterial medications. Beta-lactam antibiotics hold significant clinical relevance in both human and veterinary medicine since they serve for both preventive and therapeutic functions against various gram-positive, gram-negative, and anaerobic bacteria. The β-lactam antibiotics, including penicillins, cephalosporins, monobactams, and carbapenems, belong to a class of antibiotics characterized by the presence of a beta-lactam ring in their structure (Kong et al. 2010). Their mechanism involves the inactivation of penicillin-binding proteins (PBPs), which are crucial for bacterial cell wall formation, leading to the suppression of bacterial proliferation. They function as a permanent inhibitor of the enzyme transpeptidase, an enzyme employed by bacteria in the construction of their cell walls (Cho et al. 2014). The last phase of transpeptidation in the peptidoglycan synthesis is facilitated by an enzyme referred to as PBPs. PBPs attach to the D-Ala-D-Ala found at the extremity of muropeptides, which are the precursors of peptidoglycan, to create crosslinks in the peptidoglycan structure.

4.3.2 Chloramphenicol and Mechanism of Action in Farmed Aquatic Organisms

Chloramphenicol is widely used in the management of various bacterial infections due to its cost-effectiveness and efficacy. Chloramphenicol is considered a potential carcinogen, and because of its association with the development of aplastic anemia in humans, it is prohibited for use in animals intended for food production in various countries, including the European Union (EU), the United States, Canada, Australia, Japan, and China. It has been found in applications in aquaculture to treat and prevent bacterial infections in farmed aquatic species, contributing to environmental contamination. In situations where aquatic organisms are afflicted by pathogenic bacteria, chloramphenicol may be utilized to address the infection. Occasionally, chloramphenicol has been introduced into the water as a method for treating bacterial infections on a broader scale (Nguyen et al. 2022). This method can assist in managing infections that impact numerous aquatic organisms concurrently. Chloramphenicol is an antibiotic that functions by obstructing the synthesis of bacterial proteins. Its mechanism of action includes attaching to the 50S ribosomal subunit, precisely at the peptidyl transferase center (Syroegin et al. 2022). This attachment disrupts the usual operation of the ribosome. Chloramphenicol also hampers the translocation process, which is crucial for the ribosome's movement along the mRNA strand. The cumulative consequences of binding to the ribosome, preventing peptide bond formation, and impeding translocation lead to the halt of bacterial protein production (Nguena-Dongue et al. 2023). Therefore, the microorganism becomes incapable of generating vital proteins required for its own growth and reproduction.

4.3.3 Glycopeptides and Mechanism of Action in Farmed Aquatic Organisms

Glycopeptides, which encompass antibiotics like vancomycin and teicoplanin, are utilized in aquaculture to address and prevent bacterial infections in cultivated aquatic animals. Glycopeptides function by disrupting the production of bacterial cell walls, which results in the debilitation or elimination of the bacteria causing the infection. This group of antibiotics demonstrates their antimicrobial effects by hindering the formation of bacterial cell walls. Bacterial organisms, particularly Gram-positive ones, possess a defensive outer layer known as the cell wall, which serves as a source of structural support and protection (Stokes et al. 2016). The cell wall comprises peptidoglycan, which is an intricate network consisting of sugar and amino acid chains. Glycopeptides have a particular affinity for the terminal D-Alanine-D-Alanine peptide chains located within the developing peptidoglycan framework of bacterial cell walls (Singh et al. 2017). When the cell wall is compromised, bacterial cells are unable to maintain their structural integrity. Consequently, they become vulnerable to lysis, which is the bursting of the cell caused by the osmotic pressure disparity between the cell's interior and the outside surroundings. Ultimately, this lysis results in the demise of bacterial cells.

4.3.4 Rifamycin and Mechanism of Action in Farmed Aquatic Organisms

In aquaculture, rifamycins have been used to address bacterial infections brought about by susceptible bacterial strains. Rifampicin, being among the most renowned members of this group, possess a distinct mode of action that renders them efficacious against a range of bacterial species (Levin and Hatfull 1993). Rifamycins have a particular affinity for the bacterial RNA polymerase enzyme. RNA synthesis plays a crucial role in transcribing genetic information and producing vital proteins within the bacterial cell. Rifamycins interact with the RNA polymerase enzyme at a specific site called the beta subunit. This interaction inhibits the enzyme's ability to initiate the transcription of bacterial genes (Mosaei et al. 2018). Rifamycins are classified as bactericidal antibiotics, indicating their capacity to eliminate bacteria rather than just inhibiting their proliferation. Through the inhibition of RNA synthesis, they interfere with the generation of essential bacterial proteins, eventually leading to cell death.

4.3.5 Aminoglycosides and Mechanism of Action in Farmed Aquatic Organisms

In aquaculture, aminoglycosides are employed periodically to combat bacterial infections in fish and other aquatic organisms. Their mode of action sets them apart, rendering them efficient against a broad spectrum of bacterial species (Jaimee and Halami 2016). Aminoglycosides predominantly focus on bacterial ribosomes, the cellular components responsible for protein production. Protein synthesis is vital for the growth and reproduction of bacteria. Aminoglycosides specifically attach to bacterial ribosomes, especially the 30S subunit. This interaction disrupts the ribosomes' capacity to precisely interpret the genetic code and create proteins. Apart from their impact on protein synthesis, aminoglycosides can also compromise the structural integrity of the bacterial cell membrane (Poirel et al. 2012). This disturbance facilitates the antibiotics' entry into the bacterial cell, thereby increasing their ability to kill bacteria.

4.3.6 Quinolones and Mechanism of Action in Farmed Aquatic Organisms

Quinolones primarily focus on bacterial enzymes called DNA gyrase and topoisomerase IV, crucial for bacterial DNA replication and maintenance. Negative supercoiling is a crucial process for DNA replication and transcription. Quinolones disrupt this process by inhibiting DNA gyrase, which hinders the correct unwinding of the DNA double helix during replication. Quinolones have an impact on topoisomerase IV, an enzyme crucial for the separation of DNA strands during cell division. Inhibition of DNA gyrase and topoisomerase IV by quinolones results in the accumulation of double-strand breaks in bacterial DNA, impeding replication (Poirel et al. 2012). Double-strand DNA breaks disrupt various vital processes in bacterial cells, including replication, transcription, and repair (Bhatt and Chatterjee 2022). It has the potential to cause lethal mutations, disrupt DNA repair mechanisms, and ultimately lead to the death of the cell. Typically, quinolones demonstrate a bactericidal action, which implies their capacity to eradicate bacteria, as opposed to solely restraining their growth.

4.3.7 Macrolides and Mechanism of Action in Farmed Aquatic Organisms

In aquaculture, macrolides like erythromycin, azithromycin, and clarithromycin are occasionally used to combat bacterial infections in aquatic organisms (Lenz et al. 2021). They primarily work by binding to the bacterial ribosome, a key component in protein synthesis. Specifically, macrolides attach to the 50S subunit of the bacterial ribosome in susceptible bacteria, disrupting the elongation phase of protein synthesis. This disruption involves preventing the ribosome's movement along the mRNA strand, blocking the addition of amino acids to the growing protein structure. It is important to note that while macrolides are highly effective against Gram-positive bacteria, they also exhibit some activity against specific Gram-negative strains.

4.3.8 Nitrofurans and Mechanism of Action in Farmed Aquatic Organisms

Nitrofurans, such as nitrofurazone and furazolidone, are categorized as prodrugs, meaning they lack activity in their primary state. Their activation requires metabolic processes to take place within the bacterial or protozoal cell. After entering the targeted microorganism, nitrofurans undergo enzymatic reduction of the nitro groups ($-NO_2$) that form part of their chemical structure. The nitro groups are transformed, resulting in the creation of extremely reactive intermediates, including derivatives like nitroso and hydroxylamine compounds. These compounds are harmful to the microorganisms and are central to the antimicrobial effectiveness of nitrofurans (Defoirdt et al. 2011). It is essential to note that the highly reactive compounds produced from nitrofurans have the capacity to impair various cellular elements within the microorganism, including proteins, DNA, and cell membranes. Depending on factors such as antibiotic concentration and microbial susceptibility, nitrofurans can display either bactericidal (bacteria-killing) or bacteriostatic (bacteria-inhibiting) properties. When present in higher concentrations, they tend to exhibit bactericidal properties.

4.3.9 Tetracyclines and Mechanism of Action in Farmed Aquatic Organisms

Tetracyclines were first identified in the 1940s and displayed effectiveness against a broad spectrum of microorganisms, encompassing both Gram-positive and Gram-negative bacteria (Amangelsin et al. 2023). Tetracyclines have been widely utilized in both the prevention and therapy of infections in humans and animals, and they have also been incorporated into animal feed in subtherapeutic doses to stimulate growth. Within aquaculture, these antibiotics have been utilized to combat bacterial infections through a mode of action that disrupts bacterial protein synthesis, effectively impeding bacterial proliferation. It is widely known that tetracyclines hinder bacterial protein production by blocking the interaction between aminoacyl-tRNA and the bacterial ribosome. Their binding target is the 30S ribosomal subunit found in the bacterial ribosome. This results in the disturbance of bacterial protein synthesis due to tetracyclines. Hence, for these compounds to engage with their targets, they must traverse one or more membrane systems, depending on whether the susceptible organism is classified as Gram-positive or Gram-negative. When this mechanism is disrupted, it inhibits bacterial growth. Tetracyclines exhibit a wide-ranging effectiveness and can act against a diverse array of bacteria, including both Gram-positive and Gram-negative varieties.

4.3.10 Sulfonamides and Mechanism of Action in Farmed Aquatic Organisms

Sulfonamides are extensively used to promote the growth of industry, livestock farming, and aquaculture and commonly suggest sulfamethoxazole or the combination of sulfamethoxazole and trimethoprim (Lou et al. 2022). Sulfonamides are broad-spectrum antimicrobial agents that can inhibit the growth of bacteria, toxoplasma, and other protozoal agents. However, their ability to combat bacterial infections has been severely compromised by the extensive resistance that has evolved over a period spanning seven decades. Sulfonamides competitively inhibit dihydropteroate synthase, an enzyme essential for the synthesis of bacterial tetrahydrofolic acid (THF). THF is a critical component in the manufacture of nucleic acids and amino acids. Sulfonamides have a similar structure to para-aminobenzoic acid (PABA), which is needed for THF synthesis. Sulfonamides, resembling PABA, enter into a competitive binding interaction with dihydropteroate synthase, preventing the natural substrate from binding. Sulfonamides inhibit the active site of dihydropteroate synthase upon binding, causing bacteria to lose their capacity to convert PABA into dihydropteroic acid, which is essential for THF synthesis. In bacterial cells, THF is a critical component necessary for the generation of nucleic acids (DNA and RNA) and particular amino acids. When the production of THF is interfered with, bacteria cannot replicate their DNA or produce essential proteins.

4.4 AMR Genes in Farmed Aquatic Organisms

Genetic elements known as AMR genes are present in bacteria within aquaculture environments, providing them with resistance against antimicrobial agents, including antibiotics (Preena et al. 2020a). The existence of AMR genes in aquaculture is troubling as it has the potential to foster the emergence of antibiotic-resistant bacteria, potentially endangering both aquatic organisms and public health. Tetracycline has frequently been employed in fish and shrimp farming for the treatment of bacterial infections (Xu et al. 2021). In aquaculture system, bacteria have been identified with Tet genes like tetA and tetE, which provide resistance to tetracycline and similar antibiotics (Sheykhsaran et al. 2019). Sulfonamides are also employed as antibiotics in fish and shrimp farming. Within aquaculture ecosystems, bacteria-carrying Sul genes like sul1 and sul2 have been identified, which confer resistance to sulfonamide antibiotics (Hoa et al. 2008). Enrofloxacin and oxolinic acid are common fluoroquinolone antibiotics administered to treat bacterial infections in fish. The qnr genes, such as qnrS and qnrB, can provide resistance to fluoroquinolone antibiotics (De Silva et al. 2018). In some countries, aminoglycosides, which are among the oldest classes of antimicrobials, like gentamicin and penicillin, are used as antibiotics to treat severe bacterial infections in aquaculture (Nowacka-Kozak et al. 2023). Genes like aac(6′)-Ib can confer resistance to aminoglycosides and are found in bacteria associated with fish farming (Celik et al. 2023). Beta-lactam antibiotics, including penicillins, cephalosporins, and carbapenems, are one of the widely used veterinary drugs (Park et al. 2023). Bacteria-carrying bla genes, such as blaTEM and blaCTX-M, can produce enzymes that break down these antibiotics (Celik et al. 2023). The presence of prohibited carcinogenic antibiotic drug residues, such as nitrofuran metabolites and chloramphenicol, in aquaculture products has emerged as a significant food safety issue for countries engaged in export (Mottier et al. 2003). The cat genes, such as catA and catB, can provide resistance to chloramphenicol. It is essential to monitor and study AMR genes in aquaculture to understand their prevalence, distribution, and potential impact on public health through the food chain (Algammal et al. 2022). Reducing the use of antibiotics in aquaculture, implementing responsible farming practices, and promoting antibiotic stewardship are essential steps in mitigating the emergence and spread of AMR in aquatic environments.

4.5 Virulence Factors That Cause Disease in Farmed Aquatic Organisms

Virulence factors are molecular components or attributes possessed by bacteria and other pathogens that enable them to cause disease. These factors play a critical role in the pathogenicity of bacteria, influencing their ability to establish infections, evade the host's immune defenses, and cause tissue damage. The bacteria possess various virulence factors and their pathogenicity is known to be multifactorial (Zhang and Li 2021). The reported clinical signs were due to the pathogen's ability to invade host epithelial cells, overcome serum immunity, and produce bioactive substances such as hemolysin, catalase, and extracellular toxins. Bacterial adhesion to host cells or tissues, a critical initial step in infections, is made possible by adhesion factors. These adhesins possess the ability to bind to specific receptors on host cells, promoting successful colonization. Surface elements like fimbriae and pili serve as an illustration of how bacteria achieve adherence to mucosal surfaces, contributing to their attachment (Sarkar et al. 2021). Invasion factors facilitate the entry and penetration of bacteria into host cells or tissues, frequently employing secretion systems and effector proteins to manipulate host cell functions. The type III secretion system found in specific Gram-negative bacteria serves as a case in point for an invasion factor (Hauser 2009). Virulence factors that are strong include bacterial toxins (Deris et al. 2020). The consequences of their actions encompass host cell damage, immune response interference, and the emergence of various disease symptoms. Toxins can be divided into exotoxins, which are generated by bacteria and then secreted, and endotoxins, which are integral components of bacterial cell walls (Rungrassamee et al. 2013).

The capacity of pathogens to induce diseases depends on a diverse array of virulence elements and genetic factors that grant them the capability to infect different host species, ultimately departing from the host once infection occurs. Analyzing these essential virulence elements and genetic components can provide valuable understandings into the processes of disease progression and the potential for interspecies transmission. Consequently, this contributes to the enhancement of more efficient and sophisticated preventive strategies (Meng et al. 2023). The rising prevalence of diseases in aquaculture has resulted in an increased dependence on antibiotic drugs, consequently leading to the emergence of antimicrobial resistance AMR. Evaluating AMR in aquaculture bacteria is critical aspect that requires regular updates to assess both public health risks and sustainable farming practices

(Algammal et al. 2022). Antibiotic selection for the assessment should be guided by the specific bacterial genera and species being investigated, underlining its crucial role. Antibiotics to which the target species/genera are naturally resistant should be excluded from the analysis to prevent misinterpretations. Moreover, understanding these virulence factors is crucial for developing strategies to combat bacterial infections. Researchers often target these factors to develop vaccines or antimicrobial agents that disrupt pathogenic processes and reduce the severity of bacterial diseases.

4.6 Antimicrobial Agents and AMR in Farmed Aquatic Organisms

Antimicrobial use in aquaculture is a common practice aimed at preventing and treating bacterial infections in farmed fish and shrimp. These aquatic species in aquaculture are subjected to widespread, stress-inducing practices that compromise their immune systems, making them more susceptible to bacterial colonization and infection (Avendaño-Herrera 2021). Therefore, the use of preventive antibiotics is a common practice, especially in regions with limited access to alternative preventive strategies. In aquaculture, antimicrobials are administered to entire populations, including both healthy and sick individuals, using a method known as metaphylaxis. In addition, the impacts of this practice are extensive and concerning because antibiotics in fish feed can persist in aquatic environments for prolonged periods, spreading rapidly through excretion into water systems and exerting selective influences on ecosystems. Inefficient antibiotic metabolism in fish results in a substantial release of active components into the environment via their fecal matter. According to several investigations, an estimated 70–80% of antibiotics applied in aquaculture end up entering water systems (Chuah et al. 2016).

Antimicrobial agents have been employed in aquaculture for various purposes, including disease prevention, control, and growth promotion. The administration of antibiotics, such as tetracyclines, sulfonamides, and quinolones, is common in aquaculture operations to combat bacterial infections (Leng et al. 2020). Additionally, antimicrobial compounds have been used in feed formulations to enhance the growth and overall health of farmed aquatic species (Olmos et al. 2020). Several factors contribute to the widespread use of antimicrobials in aquaculture. The repetitive use of antibiotics in aquaculture settings can lead to the selection and proliferation of AMR bacteria. Resistant strains may then enter the environment through aquaculture effluents, potentially impacting human health and terrestrial ecosystems. Moreover, antibiotics and resistant bacteria can persist in aquatic environments, leading to waterborne contamination (Hossain et al. 2022). The transfer of AMR genes from aquatic bacteria to human pathogens is a potential public health concern. The consumption of seafood harboring AMR bacteria may pose risks to human health (Leonard et al. 2022). The regulation of antibiotic use in aquaculture

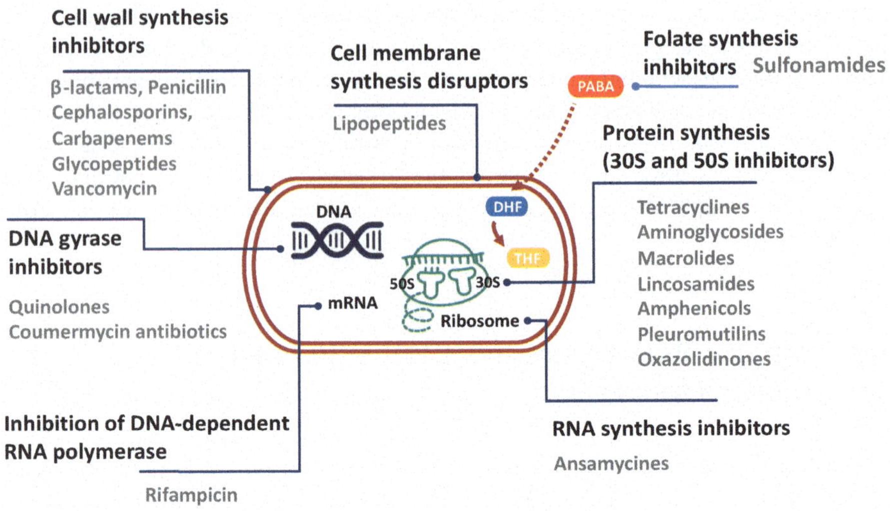

Fig. 4.2 Schematic diagram of probable sources of antibiotic contamination in aquaculture

varies globally, making it challenging to establish consistent guidelines for responsible antibiotic use, as described in Fig. 4.2.

4.7 AMR in Aquaculture Systems

Closed containment aquaculture systems involve techniques that separate the farming process from the natural environment and regulate system parameters like oxygen levels, temperature, and photoperiods (Crouse et al. 2021). These systems can range from single-pass flow-through water systems to comprehensive near-zero discharge recirculating setups. Due to their focus on process control and water recycling, closed systems are often seen as an eco-friendly approach for intensive aquaculture, significantly reducing the release of waste, antibiotics, and chemicals into the environment. However, closed flow-through systems generate wastewater containing suspended solids, nitrogen, phosphorous, and a high microbial load (Zhao et al. 2023). This wastewater can go through various treatment stages, such as entering municipal wastewater systems, passing through constructed wetlands, or undergoing treatment to produce sludge for use as land fertilizer (Iber and Kasan 2021). The utilization of aquaculture sludge in this manner raises concerns about the concentration and spread of AMR genes into food crops and the soil system (Zhao et al. 2021).

The recirculating aquaculture systems (RASs) are designed to efficiently rear high-density fish populations while minimizing their environmental impact. These systems, applicable in both freshwater and marine environments, aim to carefully manage, collect, and treat waste. Ideally, they require minimal water replenishment, primarily to compensate for evaporation losses. RASs employ a combination of mechanical and biological filtration methods to create an efficient, productive, and ecologically responsible habitat. Moreover, in these systems, microbial biofilms play a pivotal role in coordinating biological processes, forming on integrated filtration media, tank surfaces, and even the fish themselves, nourished by nutrient inputs. Because RASs largely recycle water and interact minimally with the external surroundings under optimal conditions, there exists the potential for antibiotics introduced through feeding to amass within the system. This accumulation could potentially promote the proliferation of AMR bacteria, which may exist in the host, sediment, and the RAS biofilter community. Qi et al. (2022) revealed that biofilms within mixed bed biofilters in RAS systems can serve as reservoirs for antibiotic resistance genes. However, biofilms typically exhibit resistance to antibiotic penetration, making pathogen treatment challenging (Blancheton et al. 2013). Gaikowski et al. (2015) elevated levels of oxytetracycline residue in sediment, biofilter, and fish muscle after medicated feed treatment. However, these oxytetracycline levels significantly declined over time following the withdrawal of the drug, eventually becoming nearly undetectable. Information regarding the presence of AMR bacteria in RASs remains limited, with studies suggesting a potential link to the use of infected fish stocks.

4.8 Antibiotic-Resistant Bacteria in Farmed Aquatic Organisms

At present, more and more AMR problems have occurred due to the abuse of antimicrobials in the treatment and control of pathogens infection in aquaculture (Preena et al. 2020b). Antibiotics are largely used for therapeutic and prophylactic purposes on aquaculture. The widely use of antibiotics has helped selection for resistant bacterial strains, and increased antibiotic resistance among the environmental. Antibiotic-resistant bacteria may represent a potential threat to human health due to direct transmission through the food chain or by transferring the AMR genes to human pathogens by mobile genetic elements (Hossain et al. 2022).

Traditionally, antibiotics have been employed for the treatment and control of disease-causing agents in aquaculture, with the potential consequence led to antibiotic-resistant bacterial strains in aquatic habitats. Oral antibiotic delivery is the preferred method for controlling disease outbreaks since there are currently no viable commercial vaccines or effective personalized vaccines on the market (Roiha et al. 2010). Despite the availability of a standardized procedure endorsed by the CLSI for testing the susceptibility of bacteria that afflict fish, fundamental data regarding the cultivation and growth requirements of this pathogen are limited (Thornber et al. 2020; Sutthi et al. 2023; Wangkahart et al. 2023a). This lack of information can be a hindrance in studying these bacteria further or in developing effective treatments against them. In the absence of such specific knowledge, the approach to treatment becomes more generalized. Hence, antibiotics often target essential bacterial functions, such as cell wall synthesis or protein production (Zhang et al. 2023). While this broad approach is often successful, it may not always account for the distinct resistance mechanisms that individual bacterial strains can develop. As a result, over time, certain pathogens can evolve to withstand commonly used antibiotics, leading to the emergence of multidrug-resistant strains. This phenomenon is evident in various aquaculture settings. In shrimp culture, *Vibrio parahaemolyticus*, recognized as one of the principal pathogens, has displayed antibiotic resistance against ampicillin, ciprofloxacin, cephazolin, streptomycin, and cefuroxime sodium (Elmahdi et al. 2016). This growing issue of antibiotic-resistant bacteria is particularly severe for shrimps. Although some functional vaccines have been reported, attributable to innate memory (Chen-Fei et al. 2020), they are generally not considered effective vaccinations due to shrimps lacking adaptive immunity. The rise in antibiotic resistance poses a grave threat to the shrimp industry, making it a pressing concern that demands immediate attention and action.

Moreover, the challenge of antibiotic resistance is not limited to shrimps alone. Other aquatic species are also facing similar threats from resistant bacterial strains. For instance, *E. tarda* isolated from cage-cultured tilapia exhibited elevated levels of resistance to amoxicillin, ampicillin, ceftazidime, oxytetracycline, and sulfamethoxazole/trimethoprim (Niu et al. 2019). In nations including the United States and Norway, there have been restrictions placed on quinolone antibiotics, largely due to their significance in human medical treatments. The overlap between antibiotics used in both human medicine and aquaculture raises concerns about the potential for increased antibiotic resistance in human and terrestrial veterinary medicine pathogens. This concern becomes even more significant when considering the administration of these drugs to aquatic livestock. Internationally, a trend is emerging to discontinue the use of preventive or therapeutic agents in aquaculture that are also used in human medicine. This reflects a global commitment to sustainable and safe practices in aquaculture. However, in countries with expanding aquaculture sectors, unrestricted use of quinolones and other antibiotics continues (Stahlmann and Lode 2013).

The continued detection of bacterial strains resistant to chloramphenicol, an antibiotic no longer in use, underscores the long-term consequences of antibiotic practices. Ali et al. (2022) demonstrated that fish farmers, despite understanding the advantages and risks of antibiotic use, often do not modify their practices, highlighting a disparity between awareness and implementation. This possibly indicates that bans might not be effectively enforced, leading to the illegal use of prohibited yet affordable antibiotics. Additionally, the persistence of antibiotic residues in the ecosystem can contribute to ongoing resistance. Residual antibiotics in the environment and the emergence of resistant bacterial strains necessitate coordinated international efforts and strict regulations (Hossain et al. 2022). Diverse fish and shellfish, associated pathogens, and their resistant antibiotics are summarized in Table 4.2.

Table 4.2 Diverse fish and shellfish, associated pathogens and their resistant antibiotics

Source of fish	Fish pathogens	Resistant antibiotics	References
Catfish	*Aeromonas hydrophila*	Tetracycline, oxytetracycline, chloramphenicol, kanamycin, ampicillin, and nitrofurantoin	McPhearson et al. (1991)
Rainbow trout	*Aeromonas* sp.	Sulfadiazine/trimethoprim	Schmidt et al. (2001)
Rainbow trout	*Aeromonas* sp.	β-Lactam antibiotics	Vega-Sánchez et al. (2014)
Rainbow trout	*Flavobacterium* sp.	Multiple antibiotics	Kum et al. (2008)
Atlantic salmon	*F. psychrophilum*	Florfenicol, oxytetracycline, trimethoprim–sulfamethoxazole and ormetoprim–sulfadimethoxine	Hesami et al. (2010)
Atlantic salmon	*F. psychrophilum*	Quinolones	Shah et al. (2012)
Rainbow trout	*Yersinia ruckeri*	Multiple antibiotics	Schmidt et al. (2000)
Rainbow trout	*Acinetobacter* sp.	Tetracycline	Agersø and Guardabassi (2005)
Rainbow trout	*Edwardsiella tarda*	Chloramphenicol	Xiao et al. (2013)
Nile tilapia	*A. hydrophila*	Tetracycline and amoxycillin	Jacobs and Chenia (2007)
Nile tilapia	*E. tarda*	Penicillins, aminoglycosides, fluoroquinolones, tetracyclines, sulfonamides, and lincosamides	Algammal et al. (2022)
Koi carp	*Aeromonas* sp.	Quinolones, aminoglycosides	Dobiasova et al. (2014)
Yellow swordtails	*Aeromonas* sp.	Quinolones	Chenia (2016)
Yellowtail	*Vibrio* sp.	Oxytetracycline	Nonaka and Suzuki (2002)
Abalone and turbot	*Vibrio* sp.	Oxytetracycline, streptomycin, chloramphenicol, ampicillin	Dang et al. (2007)
Common carp	*Flavobacterium* sp.	Oxytetracycline	Singh et al. (2009)
Abalone and turbot	*Pseudoalteromonas* sp.	Oxytetracycline, streptomycin, chloramphenicol, ampicillin, nalidixic acid	Dang et al. (2007)
Gilthead sea bream	*Y. ruckeri*	Tetracycline, ampicillin, and erythromycin	Preena et al. (2020a)
Cobia, mullet and loach	*Pseudoalteromonas* sp.	Macrolide and tetracycline	Nguyen et al. (2017)

4.9 Alternative Strategies to Combat AMR in Aquaculture

4.9.1 Vaccination

Vaccination offers a compelling strategy to preclude the onset of infections in both humans and farmed animals (Wang et al. 2023). The contribution of vaccines to the AMR battle encompasses both direct and indirect aspects. Vaccination directly impacts resistant pathogens by lowering the rate of infection and, indirectly, by decreasing the transmission of AMR-resistant strains to non-resistant species (Buchy et al. 2020). A decrease in the number of infections is linked to a reduced dependence on antibiotics and a decreased occurrence of secondary infections and superinfections, thus avoiding the need for massive antibiotic treatment (Wangkahart et al. 2023b). The reduced propagation of antibiotic-resistant pathogens emphasizes the value of implementing vaccination approaches and accomplishing herd immunity as pivotal elements in combating AMR (Wang et al. 2021). In recent times, immunization initiatives within the livestock and poultry sectors have led to a significant decline in the use of antibiotics, leading to a simultaneous decrease in AMR within the animal population (Garvey 2022). These efforts have also had valuable implications for human health by limiting the dissemination of AMR characteristics among zoonotic pathogens. Hence, vaccination results in a reduced burden of AMR and consequently reduces the need for antibiotics (Micoli et al. 2021). In contrast, excessive antibiotic use is a significant catalyst for the rise and dissemination of AMR in veterinary, agricultural, and aquaculture. A novel strategy involving the simultaneous use of antibiotics and vaccines to combat AMR is the utilization of a theoretical vaccine with a limited range to specifically address the most resistant strains, thereby promoting their replacement with susceptible ones (Hoelzer et al. 2018). The application of antibiotics may be successful in treating these infections. It is important to note that vaccines are not a substitute for antibiotics, but they do contribute to the mitigation of AMR by reducing the need for antimicrobial treatments, preventing illnesses caused by AMR bacteria, and limiting the transmission of AMR bacteria.

4.9.2 Probiotics

The use of probiotics in aquaculture for managing infectious diseases is common, primarily because of their diverse antibacterial mechanisms, including the production of antibiotics, siderophores, enzymes, and bacteriocins. While bacteriocins offer promise as substitutes for antibiotics, bacteria producing these compounds may also possess AMR genes (Imperial and Ibana 2016). The presence of AMR genes, particularly on mobile genetic components, presents a concern as they have the potential to transfer to pathogens. This underscores the importance of evaluating AMR susceptibility in probiotics (Muñoz-Atienza et al. 2013). Despite the potential for probiotics to serve as substitutes for antimicrobials, they have the capacity to obtain antibiotic resistance genes through horizontal gene transfer, especially in common environments such as the gastrointestinal tract. Probiotic bacteria derived from aquatic animals have been identified as carriers of antibiotic resistance genes. It is necessary to conduct additional studies to understand the enduring impacts of introducing such bacterial populations into aquaculture systems that are already contaminated with AMR genes and antimicrobial substances.

In the twenty-first century, nanomaterials have become a focal point in medical research, contributing to advancements in targeted drug delivery, diagnostics, and vaccine technology, with a notable impact during the COVID-19 pandemic (Park et al. 2020). Research findings indicate that nanoparticles can have a bifunctional impact on the immune system, exhibiting both immunostimulatory and immunosuppressive qualities. Notably, in *Penaeus semisulcatus*, zinc oxide (ZnO) and silver nanoparticles (AgNPs) have been found to induce innate immune responses (Ishwarya et al. 2018) and in the case of *Litopenaeus vannamei* (Sivaramasamy et al. 2016). Shrimp fed diets with silver nanoparticles (AgNPs) have experienced enhanced protection against infections caused by the white spot syndrome virus (WSSV), as well as bacteria like *Vibrio cholerae*, *V. harveyi*, and *V. parahaemolyticus* (Ochoa-Meza et al. 2019). In addition, gold nanoparticles (AuNP) support the immune system in its response against *V. parahaemolyticus* infections (Tello-Olea et al. 2019). Nevertheless, despite their potential, there are concerns regarding the cytotoxicity of metal ions and their environmental impact (Makvandi et al. 2020), which indicate that these nanoparticles are not currently ready to take the place of traditional antibiotics in animal feed (Yen et al. 2021).

In addition to the mentioned materials, carbon-based nanomaterials stand out because of their biocompatibility and distinctive characteristics (Patel et al. 2019). Specifically, carbon quantum dots (CQDs) have shown promise in the fields of disease detection and therapeutic applications (Sharma et al. 2019). Recently, there have been findings indicating that nanomaterials, including carbonized nanoparticles and nanogels derived from lysine (Lys-CNGs), polyamine, and spermidine, possess broad-spectrum antibacterial abilities. This raises the possibility of using them as potential alternatives to traditional antibiotics (Mao et al. 2022). For instance, carbonized nanogels (DAO/DEX-CNGs) synthesized from the polysaccharide dextran 70 and 1,8-diaminooctane maintain structural characteristics and functional groups inherited from their parent compounds. They showcase extensive antibacterial activity against multiple *Vibrio* strains, including drug-resistant variants. Their versatile behavior disturbs bacterial cell membranes, elevates oxidative stress within bacteria, and counteracts PirAB toxins, thereby offering effective protection to shrimp against *Vibrio* infections (Yen et al. 2021).

4.10 Conclusions and Future Perspectives

Effective methods in aquaculture, responsible management protocols, appropriate stocking plans, and good hygiene practices can efficiently handle the introduction of bacterial pathogens, decrease the frequency of bacterial infections, and thereby reduce the need for antimicrobial intervention. To address the challenge of AMR, it is essential to institute continuous monitoring programs following relevant guidelines and enforce impactful policies. Preventing the emergence of AMR in aquaculture can restrict the spread of AMR genes into the natural environment and mitigate its progression as a public health issue. Effective management of AMR in aquaculture's future will depend on a blend of alternative approaches, regulatory actions, monitoring efforts, education, and global cooperation. Vaccination serve as an effective example of alternative strategies aimed at decreasing the utilization of antibiotics. By working together and embracing sustainable practices, stakeholders can mitigate the risks associated with AMR and ensure the long-term health of both aquatic ecosystems and human populations.

Acknowledgement This research project was financially supported by Mahasarakham University.

Disclosure Statement The authors declare that they have no conflicts of interest.

References

Agersø Y, Guardabassi L (2005) Identification of Tet 39, a novel class of tetracycline resistance determinant in Acinetobacter spp. of environmental and clinical origin. Journal of Antimicrobial Chemotherapy 55(4):566–569

Algammal AM, Mabrok M, Ezzat M, Alfifi KJ, Esawy AM, Elmasry N, El-Tarabili RM (2022) Prevalence, antimicrobial resistance (AMR) pattern, virulence determinant and AMR genes of emerging multi-drug resistant *Edwardsiella tarda* in Nile tilapia and African catfish. Aquaculture 548:737643

Ali F, Zainudin ZH, Mahmud MMA, Min CC, Lian ASA, Mansor NH, bin Thabet O (2022) Influence of awareness and knowledge of antimicrobial resistance (AMR) on intention to use vaccines among fish farmers in Malaysia. J Pharm Negat Results 13:5499–5505

Amangelsin Y, Semenova Y, Dadar M, Aljofan M, Bjørklund G (2023) The impact of tetracycline pollution on the aquatic environment and removal strategies. Antibiotics 12(3):440

Avendaño-Herrera R (2021) Salmon aquaculture, *Piscirickettsia salmonis* virulence, and one health: dealing with harmful synergies between heavy antimicrobial use and piscine and human health comment on. Aquaculture 532:736062

Bhatt S, Chatterjee S (2022) Fluoroquinolone antibiotics: occurrence, mode of action, resistance, environmental detection, and remediation—a comprehensive review. Environ Pollut 315:120440

Blancheton JP, Attramadal KJK, Michaud L, d'Orbcastel ER, Vadstein O (2013) Insight into bacterial population in aquaculture systems and its implication. Aquac Eng 53:30–39

Boopathi S, Meenatchi R, Brindangnanam P, Sudhakaran G, Coumar MS, Arockiaraj J (2023) Microbiome analysis of *Litopenaeus vannamei* reveals Vibrio as main risk factor of white faeces syndrome. Aquaculture 576:739829

Buchy P, Ascioglu S, Buisson Y, Datta S, Nissen M, Tambyah PA, Vong S (2020) Impact of vaccines on antimicrobial resistance. Int J Infect Dis 90:188–196

Caputo A, Bondad-Reantaso MG, Karunasagar I, Hao B, Gaunt P, Verner-Jeffreys D, Dorado-Garcia A (2023) Antimicrobial resistance in aquaculture: a global analysis of literature and national action plans. Rev Aquac 15(2):568–578

Celik B, Ergul B, Kekec AI, Halac B, Maslak B, Sigirci BD, Ak S (2023) Beta-lactam, aminoglycoside, and quinolone resistance in *Escherichia coli* strains isolated from shrimps and mussels in the Marmara Sea. Vet Med 68(5):208

Chen-Fei L, Chou-Min C, Jiun-Yan L (2020) Feasibility of vaccination against *Macrobrachium rosenbergii* nodavirus infection in giant freshwater prawn. Fish Shellfish Immunol 104:431–438

Chenia HY (2016) Prevalence and characterization of plasmid-mediated quinolone resistance genes in Aeromonas spp. isolated from South African freshwater fish. International Journal of Food Microbiology 231:26–32

Cho H, Uehara T, Bernhardt TG (2014) Beta-lactam antibiotics induce a lethal malfunctioning of the bacterial cell wall synthesis machinery. Cell 159(6):1300–1311

Chuah LO, Effarizah ME, Goni AM, Rusul G (2016) Antibiotic application and emergence of multiple antibiotic resistance (MAR) in global catfish aquaculture. Curr Environ Health Rep 3:118–127

Crouse C, Davidson J, May T, Summerfelt S, Good C (2021) Production of market-size European strain Atlantic salmon (*Salmo salar*) in land-based freshwater closed containment aquaculture systems. Aquac Eng 92:102138

Dang H, Zhang X, Song L, Chang Y, Yang G (2007) Molecular determination of oxytetracycline-resistant bacteria and their resistance genes from mariculture environments of China. J Appl Microbiol 103(6):2580–2592

De Silva BC, Hossain S, Dahanayake PS, Heo GJ (2018) Frozen white-leg shrimp (*Litopenaeus vannamei*) in Korean markets as a source of *Aeromonas* spp. harboring antibiotic and heavy metal resistance genes. Microb Drug Resist 24(10):1587–1598

Defoirdt T, Sorgeloos P, Bossier P (2011) Alternatives to antibiotics for the control of bacterial disease in aquaculture. Curr Opin Microbiol 14(3):251–258

Deris ZM, Iehata S, Ikhwanuddin M, Sahimi MBMK, Do TD, Sorgeloos P, Wong LL (2020) Immune and bacterial toxin genes expression in different giant tiger prawn, *Penaeus monodon* post-larvae stages following AHPND-causing strain of *Vibrio parahaemolyticus* challenge. Aquac Rep 16:100248

Dobiasova H, Kutilova I, Piackova V, Vesely T, Cizek A, Dolejska M (2014) Ornamental fish as a source of plasmid-mediated quinolone resistance genes and antibiotic resistance plasmids. Vet Microbiol 171(3–4):413–421

Elmahdi S, DaSilva LV, Parveen S (2016) Antibiotic resistance of *Vibrio parahaemolyticus* and *Vibrio vulnificus* in various countries: a review. Food Microbiol 57:128–134

Gaikowski MP, Whitsel MK, Charles S, Schleis SM, Crouch LS, Endris RG (2015) Depletion of florfenicol amine in tilapia (*Oreochromis* sp.) maintained in a recirculating aquaculture system following Aquaflor®-medicated feed therapy. Aquac Res 46(8):1842–1857

Garvey M (2022) Antimicrobial use in animal food production. In: Biodiversity, functional ecosystems and sustainable food production. Springer International Publishing, Cham, pp 183–215

Hauser AR (2009) The type III secretion system of Pseudomonas aeruginosa: infection by injection. Nat Rev Microbiol 7(9):654–665

Hesami S, Parkman J, MacInnes JI, Gray JT, Gyles CL, Lumsden JS (2010) Antimicrobial susceptibility of *Flavobacterium psychrophilum* isolates from Ontario. J Aquat Anim Health 22(1):39–49

Hoa PTP, Nonaka L, Viet PH, Suzuki S (2008) Detection of the sul1, sul2, and sul3 genes in sulfonamide-resistant bacteria from wastewater and shrimp ponds of north Vietnam. Sci Total Environ 405(1–3):377–384

Hoelzer K, Bielke L, Blake DP, Cox E, Cutting SM, Devriendt B, Van Immerseel F (2018) Vaccines as alternatives to antibiotics for food producing animals. Part 2: new approaches and potential solutions. Vet Res 49:1–15

Hossain A, Habibullah-Al-Mamun M, Nagano I, Masunaga S, Kitazawa D, Matsuda H (2022) Antibiotics, antibiotic-resistant bacteria, and resistance genes in aquaculture: risks, current concern, and future thinking. Environ Sci Pollut Res 29:1–22

Iber BT, Kasan NA (2021) Recent advances in Shrimp aquaculture wastewater management. Heliyon 7(11):e08283

Imperial IC, Ibana JA (2016) Addressing the antibiotic resistance problem with probiotics: reducing the risk of its double-edged sword effect. Front Microbiol 7:1983

Ishwarya R, Vaseeharan B, Subbaiah S, Nazar AK, Govindarajan M, Alharbi NS, Al-Anbr MN (2018) Sargassum wightii-synthesized ZnO nanoparticles–from antibacterial and insecticidal activity to immunostimulatory effects on the green tiger shrimp Penaeus semisulcatus. J Photochem Photobiol B Biol 183:318–330

Jacobs L, Chenia HY (2007) Characterization of integrons and tetracycline resistance determinants in Aeromonas spp. isolated from South African aquaculture systems. Int J Food Microbiol 114(3):295–306

Jaimee G, Halami PM (2016) High level aminoglycoside resistance in Enterococcus, Pediococcus and Lactobacillus species from farm animals and commercial meat products. Ann Microbiol 66:101–110

Kong KF, Schneper L, Mathee K (2010) Beta-lactam antibiotics: from antibiosis to resistance and bacteriology. APMIS 118(1):1–36

Kum C, Kirkan S, Sekkin S, Akar F, Boyacioglu M (2008) Comparison of in vitro antimicrobial susceptibility in *Flavobacterium psychrophilum* isolated from rainbow trout fry. J Aquat Anim Health 20(4):245–251

Leng Y, Xiao H, Li Z, Wang J (2020) Tetracyclines, sulfonamides and quinolones and their corresponding resistance genes in coastal areas of Beibu Gulf, China. Sci Total Environ 714:136899

Lenz KD, Klosterman KE, Mukundan H, Kubicek-Sutherland JZ (2021) Macrolides: from toxins to therapeutics. Toxins 13(5):347

Leon R, Ruiz M, Valero Y, Cardenas C, Guzman F, Vila M, Cuesta A (2020) Exploring small cationic peptides of different origin as potential antimicrobial agents in aquaculture. Fish Shellfish Immunol 98:720–727

Leonard AF, Morris D, Schmitt H, Gaze WH (2022) Natural recreational waters and the risk that exposure to antibiotic resistant bacteria poses to human health. Curr Opin Microbiol 65:40–46

Levin ME, Hatfull GF (1993) Mycobacterium smegmatis RNA polymerase: DNA supercoiling, action of rifampicin and mechanism of rifampicin resistance. Mol Microbiol 8(2):277–285

Lou Q, Wu Y, Ding H, Zhang B, Zhang W, Zhang Y, Zhong J (2022) Degradation of sulfonamides in aquaculture wastewater by laccase–syringaldehyde mediator system: response surface optimization, degradation kinetics, and degradation pathway. J Hazard Mater 432:128647

Makvandi P, Wang CY, Zare EN, Borzacchiello A, Niu LN, Tay FR (2020) Metal-based nanomaterials in biomedical applications: antimicrobial activity and cytotoxicity aspects. Advanced Functional Materials 30(22):1910021

Mao JY, Miscevic D, Unnikrishnan B, Chu HW, Chou CP, Chang L, Huang CC (2022) Carbon nanogels exert multipronged attack on resistant bacteria and strongly constrain resistance evolution. J Colloid Interface Sci 608:1813–1826

McPhearson RM, DePaola A, Zywno SR, Motes ML Jr, Guarino AM (1991) Antibiotic resistance in Gram-negative bacteria from cultured catfish and aquaculture ponds. Aquaculture 99(3–4):203–211

Meng X, Chen F, Xiong M, Hao H, Wang KJ (2023) A new pathogenic isolate of *Kocuria kristinae* identified for the first time in the marine fish *Larimichthys crocea*. Front Microbiol 14:1129568

Micoli F, Bagnoli F, Rappuoli R, Serruto D (2021) The role of vaccines in combatting antimicrobial resistance. Nat Rev Microbiol 19(5):287–302

Morales R, Weber-Vintzel L, Awada L, Caceres P, Tizzani P, Meske M (2023) The World Animal Health Information System as a tool to support decision-making and research in animal health. Rev Sci Tech 42:242–251

Mosaei H, Molodtsov V, Kepplinger B, Harbottle J, Moon CW, Jeeves RE, Zenkin N (2018) Mode of action of kanglemycin A, an ansamycin natural product that is active against rifampicin-resistant Mycobacterium tuberculosis. Mol Cell 72(2):263–274

Mottier P, Parisod V, Gremaud E, Guy PA, Stadler RH (2003) Determination of the antibiotic chloramphenicol in meat and seafood products by liquid chromatography–electrospray ionization tandem mass spectrometry. Journal of Chromatography A 994(1-2):75–84

Muñoz-Atienza E, Gómez-Sala B, Araújo C, Campanero C, Del Campo R, Hernández PE, Herranz C, Cintas LM (2013). Antimicrobial activity, antibiotic susceptibility and virulence factors of lactic acid bacteria of aquatic origin intended for use as probiotics in aquaculture. BMC microbiology 13:1–22

Nguena-Dongue BN, Tchamgoue J, Ngandjui Tchangoue YA, Lunga PK, Toghueo KRM, Zeukoo ME, Fekam BF (2023) Potentiation effect of mallotojaponin B on chloramphenicol and mode of action of combinations against methicillin-resistant *Staphylococcus aureus*. PLoS One 18(3):e0282008

Nguyen TTT, Nguyen HT, Tsai MA, Byadgi O, Wang PC, Yoshida T, Chen SC (2017) Genetic diversity, virulence genes, and antimicrobial resistance of *Streptococcus dysgalactiae* isolates from different aquatic animal sources. Aquaculture 479:256–264

Nguyen LM, Nguyen NTT, Nguyen TTT, Nguyen TT, Nguyen DTC, Tran TV (2022) Occurrence, toxicity and adsorptive removal of the chloramphenicol antibiotic in water: a review. Environ Chem Lett 20(3):1929–1963

Niu G, Wongsathein D, Boonyayatra S, Khattiya R (2019) Occurrence of multiple antibiotic resistance and genotypic characterization in *Edwardsiella tarda* isolated from cage-cultured hybrid red tilapia (*Oreochromis* sp.) in the Ping River, Northern Thailand. Aquac Res 50(12):3643–3652

Nonaka L, Suzuki S (2002) New Mg^{2+}-dependent oxytetracycline resistance determinant Tet 34 in *Vibrio* isolates from marine fish intestinal contents. Antimicrob Agents Chemother 46(5):1550–1552

Nowacka-Kozak E, Gajda A, Gbylik-Sikorska M (2023) Analysis of aminoglycoside antibiotics: a challenge in food control. Molecules 28(12):4595

Ochoa-Meza AR, Álvarez-Sánchez AR, Romo-Quiñonez CR, Barraza A, Magallón-Barajas FJ, Chávez-Sánchez A, Mejía-Ruiz CH (2019) Silver nanoparticles enhance survival of white spot syndrome virus infected *Penaeus vannamei* shrimps by activation of its immunological system. Fish Shellfish Immunol 84:1083–1089

Olmos J, Acosta M, Mendoza G, Pitones V (2020) Bacillus subtilis, an ideal probiotic bacterium to shrimp and fish aquaculture that increase feed digestibility, prevent microbial diseases, and avoid water pollution. Arch Microbiol 202:427–435

Park W, Shin H, Choi B, Rhim WK, Na K, Han DK (2020) Advanced hybrid nanomaterials for biomedical applications. Prog Mater Sci 114:100686

Park SY, Kim YR, Lim SJ, Kim JY, Choi JD, Moon GI (2023) Simultaneous detection of residues of 34 beta-lactam antibiotics in livestock and fish samples through liquid chromatography-tandem mass spectrometry. Food Sci Biotechnol 33:1–20

Patel S, Kim J, Herrera M, Mukherjee A, Kabanov AV, Sahay G (2019) Brief update on endocytosis of nanomedicines. Adv Drug Deliv Rev 144:90–111

Poirel L, Cattoir V, Nordmann P (2012) Plasmid-mediated quinolone resistance; interactions between human, animal, and environmental ecologies. Front Microbiol 3:24

Preena PG, Arathi D, Raj NS, Arun Kumar TV, Arun Raja S, Reshma RN, Raja Swaminathan T (2020a) Diversity of antimicrobial-resistant pathogens from a freshwater ornamental fish farm. Lett Appl Microbiol 71(1):108–116

Preena PG, Swaminathan TR, Kumar VJR, Singh ISB (2020b) Antimicrobial resistance in aquaculture: a crisis for concern. Biologia 75:1497–1517

Preena PG, Dharmaratnam A, Swaminathan TR (2022) A peek into mass mortality caused by antimicrobial resistant *Edwardsiella tarda* in goldfish, *Carassius auratus* in Kerala. Biologia 77(4):1161–1171
Qi W, Skov PV, de Jesus Gregersen KJ, Pedersen LF (2022) Estimation of nitrifying and heterotrophic bacterial activity in biofilm formed on RAS biofilter carriers by respirometry. Aquaculture 561:738730
Raabe VN, Shane AL (2019) Group B streptococcus (*Streptococcus agalactiae*). Microbiol Spectrum 7(2):10–1128
Roiha IS, Otterlei E, Litlabø A, Samuelsen OB (2010) Uptake and elimination of florfenicol in Atlantic cod (*Gadus morhua*) larvae delivered orally through bioencapsulation in the brine shrimp *Artemia franciscana*. Aquaculture 310(1–2):27–31
Rungrassamee W, Maibunkaew S, Karoonuthaisiri N, Jiravanichpaisal P (2013) Application of bacterial lipopolysaccharide to improve survival of the black tiger shrimp after *Vibrio harveyi* exposure. Dev Comp Immunol 41(2):257–262
Santos L, Ramos F (2018) Antimicrobial resistance in aquaculture: current knowledge and alternatives to tackle the problem. Int J Antimicrob Agents 52(2):135–143
Sarkar P, Issac PK, Raju SV, Elumalai P, Arshad A, Arockiaraj J (2021) Pathogenic bacterial toxins and virulence influences in cultivable fish. Aquac Res 52(6):2361–2376
Schmidt AS, Bruun MS, Dalsgaard I, Pedersen K, Larsen JL (2000) Occurrence of antimicrobial resistance in fish-pathogenic and environmental bacteria associated with four Danish rainbow trout farms. Appl Environ Microbiol 66(11):4908–4915
Schmidt AS, Bruun MS, Dalsgaard I, Larsen JL (2001) Incidence, distribution, and spread of tetracycline resistance determinants and integron-associated antibiotic resistance genes among motile aeromonads from a fish farming environment. Appl Environ Microbiol 67(12):5675–5682
Shah SQ, Nilsen H, Bottolfsen K, Colquhoun DJ, Sørum H (2012) DNA gyrase and topoisomerase IV mutations in quinolone-resistant *Flavobacterium psychrophilum* isolated from diseased salmonids in Norway. Microb Drug Resist 18(2):207–214
Sharma G, Kumar A, Sharma S, Naushad M, Dwivedi RP, ALOthman ZA, Mola GT (2019) Novel development of nanoparticles to bimetallic nanoparticles and their composites: a review. J King Saud Univ Sci 31(2):257–269
Sheykhsaran E, Baghi HB, Soroush MH, Ghotaslou R (2019) An overview of tetracyclines and related resistance mechanisms. Rev Res Med Microbiol 30(1):69–75
Singh AK, Rathore G, Singh V, Mani I, Singh RK, Mishra SK, Verma OP (2009) Bacterial resistance to oxytetracycline in different life stages of Indian freshwater carp aquaculture system
Singh M, Chang J, Coffman L, Kim SJ (2017) Hidden mode of action of glycopeptide antibiotics: inhibition of wall teichoic acid biosynthesis. J Phys Chem B 121(16):3925–3932
Sivaramasamy E, Zhiwei W, Li F, Xiang JJJNN (2016) Enhancement of vibriosis resistance in Litopenaeus vannamei by supplementation of biomastered silver nanoparticles by Bacillus subtilis. J Nanomed Nanotechnol 7(352):2
Stahlmann R, Lode HM (2013) Risks associated with the therapeutic use of fluoroquinolones. Expert opinion on drug safety 12(4):497–505
Stokes JM, French S, Ovchinnikova OG, Bouwman C, Whitfield C, Brown ED (2016) Cold stress makes Escherichia coli susceptible to glycopeptide antibiotics by altering outer membrane integrity. Cell Chem Biol 23(2):267–277
Sutthi N, Wangkahart E, Panase P, Karirat T, Deeseenthum S, Ma NL, Luang-In V (2023) Dietary administration effects of exopolysaccharide produced by *Bacillus tequilensis* PS21 using riceberry broken rice, and soybean meal on growth performance, immunity, and resistance to *Streptococcus agalactiae* of Nile tilapia (*Oreochromis niloticus*). Animals 13(20):3262
Syroegin EA, Aleksandrova EV, Polikanov YS (2022) Structural basis for the inability of chloramphenicol to inhibit peptide bond formation in the presence of A-site glycine. Nucleic Acids Res 50(13):7669–7679

Tello-Olea M, Rosales-Mendoza S, Campa-Córdova AI, Palestino G, Luna-González A, Reyes-Becerril M, Angulo C (2019) Gold nanoparticles (AuNP) exert immunostimulatory and protective effects in shrimp (*Litopenaeus vannamei*) against *Vibrio parahaemolyticus*. Fish Shellfish Immunol 84:756–767

Thornber K, Verner-Jeffreys D, Hinchliffe S, Rahman MM, Bass D, Tyler CR (2020) Evaluating antimicrobial resistance in the global shrimp industry. Rev Aquac 12(2):966–986

Vega-Sánchez V, Latif-Eugenín F, Soriano-Vargas E, Beaz-Hidalgo R, Figueras MJ, Aguilera-Arreola MG, Castro-Escarpulli G (2014) Re-identification of *Aeromonas* isolates from rainbow trout and incidence of class 1 integron and β-lactamase genes. Vet Microbiol 172(3–4):528–533

Wang LM, Hashiguchi TCO, Cecchini M (2021) Impact of vaccination on carriage of and infection by antibiotic-resistant bacteria: a systematic review and meta-analysis. Clin Exp Vaccine Res 10(2):81

Wang B, Thompson KD, Wangkahart E, Yamkasem J, Bondad-Reantaso MG, Tattiyapong P, Surachetpong W (2023) Strategies to enhance tilapia immunity to improve their health in aquaculture. Rev Aquac 15:41–56

Wangkahart E, Lee PT, Chong CM, Yamamoto F (2023a) Safety and efficacy of vaccines in aquaculture. In: Fish vaccines: health management for sustainable aquaculture. CRC Press, p 155

Wangkahart E, Thongsrisuk A, Vialle R, Pholchamat S, Sunthamala P, Phudkliang J, Secombes CJ (2023b) Comparative study of the effects of Montanide™ ISA 763A VG and ISA 763B VG adjuvants on the immune response against *Streptococcus agalactiae* in Nile tilapia (*Oreochromis niloticus*). Fish Shellfish Immunol 134:108563

Xiao J, Chen T, Liu B, Yang W, Wang Q, Qu J, Zhang Y (2013) *Edwardsiella tarda* mutant disrupted in type III secretion system and chorismic acid synthesis and cured of a plasmid as a live attenuated vaccine in turbot. Fish Shellfish Immunol 35(3):632–641

Xu L, Zhang H, Xiong P, Zhu Q, Liao C, Jiang G (2021) Occurrence, fate, and risk assessment of typical tetracycline antibiotics in the aquatic environment: a review. Sci Total Environ 753:141975

Yen SC, Mao JY, Lin HY, Huang HT, Harroun SG, Nain A, Lin HJ (2021) Multifunctional carbonized nanogels to treat lethal acute hepatopancreatic necrosis disease. J Nanobiotechnol 19:1–15

Zhang W, Li C (2021) Virulence mechanisms of vibrios belonging to the Splendidus clade as aquaculture pathogens, from case studies and genome data. Rev Aquac 13(4):2004–2026

Zhang J, Zhang X, Zhou Y, Han Q, Wang X, Song C, Zhao S (2023) Occurrence, distribution and risk assessment of antibiotics at various aquaculture stages in typical aquaculture areas surrounding the Yellow Sea. J Environ Sci 126:621–632

Zhao Y, Yang QE, Zhou X, Wang FH, Muurinen J, Virta MP, Zhu YG (2021) Antibiotic resistome in the livestock and aquaculture industries: status and solutions. Crit Rev Environ Sci Technol 51(19):2159–2196

Zhao Y, Xue B, Bi C, Ren X, Liu Y (2023) Influence mechanisms of macro-infrastructure on micro-environments in the recirculating aquaculture system and biofloc technology system. Rev Aquac 15(3):991–1009

Chapter 5
Fish/Shellfish Foodborne and Zoonoses and Their Impact on Antimicrobial Resistance in Aquaculture

Sreeja Lakshmi, Nandha Kumar, Ritam Guha, Neha Omgy, Mehdi Soltani, and Preetham Elumalai

Abstract Antimicrobial resistance (AMR) in aquaculture poses a severe threat to human health and carries significant zoonotic implications. The frequent application of antimicrobials, particularly antibiotics, has led to a noticeable prevalence of multidrug resistance. The absence of reliable drugs and precise preventive measures complicates the task of averting AMR bacterial strains. Updated knowledge is crucial for developing effective treatment strategies and managing infectious diseases caused by AMR bacteria. This review delineates the involvement of zoonoses in the context of antimicrobial resistance within the aquaculture sector.

Keywords Antibiotics · Antibiotic resistance · Foodborne zoonosis

5.1 Introduction

Aquaculture contributes significantly to the advancement of society, food security, poverty reduction, employment, and other socioeconomic activities. According to the Food and Agricultural Organization (FAO), billions of aquatic animals and plants are produced globally (Boyd et al. 2022). The rise in the global human

S. Lakshmi
King Nandhivarman College of Arts and Science, Vandavasi, Tamilnadu, India

N. Kumar · R. Guha · P. Elumalai (✉)
Department of Marine Biology, Microbiology and Biochemistry, School of Marine Sciences, Cochin University of Science and Technology, Kochi, Kerala, India
e-mail: preetham@cusat.ac.in

N. Omgy
SCMS Institute of Bioscience and Biotechnology Research and Development, Ernakulam, Kerala, India

M. Soltani
Department of Aquatic Animal Health, University of Tehran, Tehran, Iran

P. Elumalai, S. Lakshmi (eds.), *Antimicrobial Resistance in Aquaculture and Aquatic Environments*, https://doi.org/10.1007/978-981-97-7320-6_5

population corresponds to an increasing demand for fish, making aquaculture crucial for meeting this demand and bridging the gap alongside sustainable fishing practices. Despite advancements in fishing technology, the demand for fish is expected to surpass the available supply from capture fisheries. Fish, being among the most traded food commodities worldwide, plays a vital role in ensuring food security and providing essential nutrition for the growing population. While fish consumption is generally considered part of a healthy diet, the risk of diseases associated with the consumption of undercooked or raw fish products exists. Diseases induced by various pathogens within cultured fish have emerged as significant challenges in aquaculture, limiting fish production. To mitigate these diseases, farmers and hatchery owners have employed various corrective measures, including the use of antimicrobials and medications (Vignesh et al. 2010).

Antibiotics are the most often used medication in agriculture for managing and maintaining animal health status (Levey and Stiles 1992). The widespread practices used in the aquaculture industry, especially in underdeveloped nations, involve excessive doses of antibiotics to prevent infectious diseases (Cabello 2006). The frequent and excessive use of antibiotics in aquaculture environments may pose several risks to the health of aquatic organisms and consumers.

Antimicrobial resistance (AMR) in bacteria that cause infectious diseases is increasing globally, and the threats this poses to human health are well established. The AMR controversy brought about the most serious health threat of the century (Bardhan and Abraham 2021). The control of diseases in aquaculture relies on a combination of management practices, recommended treatments and vaccines, and other preventive measures (Austin and Austin 2016). The discovery of sulfonamides potential to treat furunculosis in trout by Gutsell (1946) marked the beginning of using antimicrobial substances in aquaculture. Following the advent of sulfonamides, the variety of antibacterial substances used in aquaculture rapidly increased (Austin and Austin 2016). Antibiotic-resistant bacteria are currently regarded as one of the most significant risks to human health, affecting the ability to treat various infections globally (Ahmed et al. 2020). Because of the global development of multidrug-resistant (MDR) bacterial infections, antimicrobial resistance is currently viewed by the World Health Organization (WHO) as one of the major threats to global public health.

Numerous antibiotic-resistant strains are now regularly seen in fish/shellfish and aquaculture environments, which seriously endanger medical treatment alternatives and raise the number of unnecessary fatalities (Watts et al. 2017).

5.2 Fish Human Nutrition

Fish contains all the necessary nutrients crucial for the human body. Human nutrition is quite different where the body system needs macronutrients and micronutrients together to serve all biological functions properly. Among the macronutrients, proteins, carbohydrates, and lipids are the major following vitamins and minerals.

5.2.1 Protein

About 60% of people from developing countries depend upon fish for over 30% of their animal protein supplies. Compared to other dietary protein sources such as chicken, mutton, pork, beef, etc., the unit cost of fish production is much cheaper. All the necessary amino acids found in fish are present, including cysteine, methionine, and sulfur-containing amino acids, which are not found in plant proteins.

Studies show that the fish muscle is more digestible than other proteins from animal sources due to the presence of connective tissue in minimal amounts. Digestibility and concentration of amino acids in fish proteins are usually high, i.e., 85–95% (Pal and Maiti 2018). Fish protein is the main source responsible for synthesizing and managing of muscle tissues, improving immune response and blood quality level.

Fish is a perfect source of animal protein and is crucial in avoiding protein calorie malnutrition (PCM). The protein immunoglobulins in fish act as an important defense factor against pathogenic infections and help maintain water and electrolyte balance in human body fluids. The fish protein also serves as a balancing factor of many body regulatory functions. Recent studies reveal that dietary supplementation of fish benefits for treating many inflammatory-related conditions, such as cardiovascular disease, ulcerative colitis, and hyperlipidemia. Fish protein hydrolysates and peptides are better known for their antioxidant properties. Fish hydrolysates are formed as a breakdown product during the conversion of fish proteins into smaller peptides with structural effects. Amino acid composition and peptide sequences are responsible for their various functional consequences. The peptide fractions with molecular weights ranging from 270 to 1800 Da were reported to show antioxidative activity. Fish have been proven to have therapeutic efficacy in supplementing medicine. Recent observations confirmed that fish and fish-derived compounds could be potential candidates for various health-enhancing human effects (Pal and Maiti 2018).

5.2.2 Fatty Acids

Fish lipids consist of essential long-chain n-3 polyunsaturated fatty acids (PUFA), especially eicosapentaenoic acid (EPA) and docosahexaenoic acid (DHA). Consuming PUFAs is crucial for human nutrition, health, and illness prevention. The effects of omega-3 fatty acids on human health are diverse. It reduces the risk of myocardial infarction and helps lower the blood pressure and concentration of triglycerides in the blood. Fish lipids are found to have a positive response in preventing cardiovascular diseases. Omega-3 fatty acids are essential in fatty fish for enhancing the growth of children and preventing coronary heart disease. The DHA plays a crucial role in children's brain and neuron development, while EPA enhances cardiovascular health improvement (Mozaffarian and Rimm 2006).

Fish is also a rich source of polyunsaturated fatty acids with functional effects (Zuraini et al. 2006). Research evaluating the process of fish lipids' beneficial effects on health enhancement reveals that substantial quantities of LC n-3 PUFA, at a rate of 1–1.5 g daily, are necessary for the favorable effects in lowering the risk of cardiovascular diseases, including reduced blood pressure, platelet proliferation, plasma triacylglycerol levels, and inflammatory response. The mother should supplement the fetus and infants with LC n-3 PUFA during pregnancy and lactation, which is crucial for the development and maturation of the central nervous system. Fatty acids and lipids also act as an important component in regulating membrane-mediated processes, including osmoregulation, nutrient assimilation, and transport of various compounds.

5.2.3 *Minerals*

Fish act as an essential storehouse of micronutrients, which any other available food sources cannot supply through the diet of poor people (Table 5.1). In comparison with other minerals, Ca absorption to the body is insufficient. Only a limited quantity, preferably 25–30% of Ca, is assimilated by our body. Apart from dairy products, fish and fish bones are a rich source of Ca, and it was demonstrated that calcium assimilation from fish sources is equivalent to skimmed milk. Ca is essential for bone formation, and calcium ions play a beneficial role in most of the metabolic processes occurring in our body. Fish is also a rich source of minerals like iodine and selenium. Some fish species consumed as a whole, including their head and bones, constitute small indigenous species (SIS) and are considered an excellent source of minerals such as iodine, calcium, selenium, zinc, phosphorus, iron, and potassium. In smaller quantities, selenium serves as an essential micronutrient for humans, which constitutes the form of selenoproteins as a cofactor for decreasing the concentration of antioxidant enzymes, glutathione peroxidase. Selenium is also responsible for the proper functioning of the thyroid gland. A deficiency of selenium in the body can lead to the risk of myocardial infarcts and increased mortality due to cardiovascular disease and is also related to a high risk of cancer and renal disease. Iron constitutes a major component involved in synthesizing hemoglobin present in RBCs, facilitating oxygen transport to various body parts. Iron deficiency leads to various health problems, including anemia, impaired brain function, poor learning abilities, and infant behavior (Simopoulos 2007).

Table 5.1 The amount of micronutrients in different fish samples (Tenyang et al. 2016)

Fish species	Zn	Cu	Mn	Fe
Mullet	14.37 ± 1.00a	0.63 ± 0.02c	3.41 ± 0.20c	5.47 ± 0.30b
Carp	11.34 ± 0.30a	0.89 ± 0.07a	5.24 ± 0.12b	7.98 ± 1.00a
Catfish	12.28 ± 0.28bc	0.71 ± 0.03b	6.67 ± 0.05a	7.72 ± 0.60a
Kanga	12.75 ± 0.80b	0.71 ± 0.01b	1.64 ± 0.02d	5.72 ± 0.40b

5.2.4 *Vitamins*

The whole vitamins vital for human health are present in reasonable amounts in fish, but their amounts vary depending on the fish species and fish nutrition. Fish are a major source of vitamins A and D and several complex vitamins. Many natural dietary products have significantly low amounts of vitamin D, and fortified food products have inadequate amounts for infant's and adult development. Many fish species serve as a rich source of vitamins A and D stored in the liver. In addition to aiding in the prevention of issues related to impaired eyesight and the treatment of eye ailments, vitamin A is necessary for proper growth, bone and tooth development, and cell creation. When contrasted with vitamin D (ergocalciferol), vitamin D3 (cholecalciferol), which is found in fish, has three times the efficacy. This form of vitamin D is similar to the one that the skin produces from 7-dehydrocholesterol when exposed to UV light (Norman 2008).

Vitamin D deficiency causes rickets, osteomalacia, and low bone mineral density (BMD), leading to increased bone porosity and bone fragility. Vitamin D is also associated with diabetes, and its deficiency leads to impairments such as osteopenia, osteoporosis, and fractures in adults. Fish and fish oils are the foods that are the rich source of vitamin D naturally (Holick et al. 2007).

5.3 Fish/Shellfish Foodborne Pathogens

Fish and their derived products are often linked with many human diseases by consuming raw or undercooked fish and fish products. The accumulation of various species of bacteria, including human pathogenic bacteria in fish, can be associated with direct contact with a contaminated water environment and ingestion of pathogenic bacteria from contaminated feed or sediments. The health and safety of aquatic environments can be detected indirectly by disease-causing bacteria in fish bodies (Table 5.2).

Some bacterial species are zoonotic pathogens that can cause severe diseases in fish and humans. A typical example is *Yersinia ruckeri*, associated with enteric red mouth disease in rainbow trout, resulting in significant commercial losses. Interestingly, this fish pathogen can affect humans, especially immunocompromised people.

Concerning the concern regarding public health, the various bacteria transmitted through fish that have the potential to cause human diseases are important. Various factors, including cultural practices, environmental conditions, processing, distribution, and handling of products, influence the colonization of human pathogenic microbes in fish and derived products. The most common species of bacterial fish pathogens can be generally categorized into two groups of Gram-positive and Gram-negative agents. Among Gram-positive bacteria, species of *Streptococcus iniae*, *S. agalactiae*, *S. dysgalactiae*, *S. uberis*, *S. parauberis*, and *Lactococcus garvieae* are more common zoonotic pathogens. Compared to Gram-positives, more species of Gram-negatives can act as fish zoonotic agents. For instance, several species of *Vibrio*, including *V. parahaemolyticus*, *V. vulnificus*, *V. fluvialis*, *V. harveyi*, *V. damsela*/*Photobacterium damselae*, *V. cholera*, and *V. alginolyticus* are well-known fish zoonotic bacterial pathogens. In addition, several species of

Table 5.2 Some examples of infectious agents that can affect human health via aquaculture activity

Pathogenic agents	Disease caused	References
Bacterial agents		
• *Mycobacterium*—common pathogens of fish zoonoses are *M. gordonae*, *M. chelonae*, *M. avium*, *M. marinum*, *M. ulcerans*, *M. septicum*, and *M. fortuitum*	Fish zoonoses include acute and chronic illnesses. In humans, they cause skin lesions, necrotic lesions, deep tissue infection, osteomyelitis, arthritis, and bronchitis	Delghandi et al. (2020)
• Streptococcaceae—*S. agalactiae*, *S. difficile*, *S. difficilis*, *S. dysgalactiae*, *S. iniae*, and *S. shiloi*	Cellulitis, endocarditis, meningitis, severe systemic infections, suppurating ulcers, septicemia, arthritis, and lymphadenitis	Haenen et al. (2013)
• Erysipelothrix—*E. rhusiopathiae*	Necrotizing dermatitis, myositis, and cellulitis They cause acute sepsis and skin infections	Shane (2011)
• Vibrionaceae—*V. cholerae*, *V. alginolyticus*, *V. vulnificus*, *V. damselae*, *V. hollisae*, *V. metschnikovii*, and *V. parahaemolyticus*	Lesions, septicemia, erythema, and tissue necrosis	Hernández-Cabanyero and Amaro (2020)
Fungal agents		
• Basidiobolomycosis—*Basidiobolus ranarum*	Subcutaneous and gastrointestinal infection	Ageel et al. (2017)
• Sporotrichosis caused by *Sporothrix schenckii*	Ulceration, edema, pain, and purulent discharge in the affected area	Barros et al. (2011)
Viral agents		
Noroviruses	Gastroenteritis including nausea, vomiting, watery diarrhea, and abdominal pain	Li et al. (2014)
Fish-derived parasites		
• Trematodes Liver flukes, i.e., *Clonorchis sinensis*, *Opisthorchis viverrini*, and *Opisthorchis felineus* Lung flukes, e.g., *Paragonimus westermani* and *P. heterotremus*	Cholangitis, choledocholithiasis, pancreatitis, and cholangiocarcinoma	Boerlage et al. (2013)
• Cestodes—*D. dendriticum*, *D. nihonkaiense*, *D. latum*, *Adenocephalus pacificus*, and *Diplogonoporus balaenopterae*	Diarrhea, abdominal pain, anemia, weight loss, and vitamin B12 deficiency	McConnaughey et al. (2020)

motile *Aeromonas*, such as *A. hydrophila*, *Pseudomonas*, *Proteus*, and *Edwardsiella*, can induce infections in humans and fish.

Furthermore, species of *Staphylococcus*, *Listeria*, *Campylobacter*, *Salmonella*, and *E. coli* are serious fish/shellfish foodborne pathogens that can transmitted to humans via consumption of raw or uncooked fish/shellfish.

Various sources of contamination can be broadly divided into (1) pollution caused by organic and fecal matter of the aquatic environment, (2) contamination by the natural aquatic environment, and (3) domestic waste from industry, retail, restaurant, or home processing and preparation. Apart from fish consumed raw, the reported incidences of seafood-related diseases are scarce.

The disease linked with seafood other than consumption of raw molluscan dishes seems to be much lower, and pathological risk may arise due to the cross-contamination of cooked items by raw products due to the absence of ambient time and temperature of cooking to destroy disease-causing organisms. Such incidents occur at the post-processing stage, which applies to all other foods, including seafood products.

Parasitic infections associated with seafood are less common than bacterial and viral infections, with *Anisakis simplex* and cestodes contributing to the most significant public health concern in the United States. As a whole, parasitic infections are confined to specific ethnic groups that favor raw or partially cooked seafood consumption.

All bacteria and parasitic pathogens can be eliminated through cooking but cannot degrade some toxic microbial metabolites like *Staphylococcus* toxins and certain microbial spores that can withstand higher temperatures. Consumption of raw seafood has certain health risks, and such raw seafood has to be avoided. More significant risks are associated with people with immunosuppressed conditions, those who have diabetes, people with defective livers, and the elderly. Such patients have to be warned against consuming raw seafood. Like other food particles, fish and shellfish accommodate a variety of bacteria, viruses, and parasites that are pathogenic and cause disease in consumers. Pathogenic agents occur naturally in seafood and the marine environment, accompanied by sewage contamination of harvesting areas, or can be acquired during seafood harvest, during processing, or while transporting. Usually, pathogens associated with seafood cause only slight risks to humans, whereas some may even cause serious health issues to people with other health-related problems like liver disease, diabetes, and persons with improper immune systems (AL-Mamun et al. 2018).

Diseases associated with pathogenic organisms can be broadly divided into two groups: gastroenteritis, caused by the consumption of pathogenic agents, and infections of wounds related to contamination of wounds by water. Such diseases are reported during late summer and early fall when the concentration of bacteria is the highest in the water. A diverse group of helminths are associated with fish and shellfish, but most helminths are harmless to humans. Pathogenic infestation due to these agents is most commonly accomplished by consuming fish. However, helminths cause much less harm than either bacterial or viral agents—only digenetic trematodes cause human diseases. The mode of transmission of *Clonorchis sinensis*, commonly known as Chinese liver fluke, is through consuming raw or undercooked freshwater fish with cyst stages. Chinese fluke is endemic to East Asia, which extends from Japan to Vietnam. The mode of transmission of intestinal fluke, called *Heterophyes*, is by consuming raw, salted, or dry forms of fresh or brackish-water fish (mostly mullet). This organism is endemic to the lower Nile valley near the Mediterranean coastline and has been reported as present in the Orient and western India. Freshwater trout, which are the inhabitants of the Orient, northern Siberia, and the Balkans, are found to be infected with *Metagonimus yokogawai*, which resembles *H. heterophyes*. Freshwater crabs and crawfish consumption in raw or pickled form is affected by the oriental lung fluke, *Paragonimus westermani*. Oriental liver flukes inhabit East Asian countries like Japan, Korea, and Taiwan. Adequate cooking and freezing of seafood helps prevent the transmission of food-borne infection caused by trematodes. The only cestodes contributing to parasitic illness from seafood are *Diphyllobothrium latum* and *D. pacificum*. Consumption of improperly cooked food and raw fish results in the infectious attack by *Anisakis simplex* in humans. The gutting process carried out soon after the capture of fish results in decreased risk of nematode infection, and salting will also help prevent the disease. Cooking methods like smoking or light salting of fish will not destroy larvae. But methods like freezing fish for about 72 h at 20 °C or heat treatment for 10 s at 55 °C can potentially eliminate adult parasites. Burdens of body cavities and samples cannot be immediately frozen or heated, causing disease. Several parasites have to be checked in such cases. If the number of parasites exceeds 15, the fish are unsuitable for consumption. Consumption of undercooked or raw fish in freshwater may lead to the disease caused by *Gnathostoma spinigerum*. Cases are found in India, Israel, and eastern and southeast Asia (Eiras et al. 2018).

5.4 Antibiotic Resistance Genes in Bacteria

Antibiotic resistance genes (ARGs) are bacterial resistance's cause and molecular basis. Pathogenic bacteria pick up ARGs via plasmid exchange at the gene level and strengthen their antibiotic resistance. Bacterial integrons (In), transposons (Tn), and plasmids carrying ARGs are capable of horizontal gene transfer (HGT) across strains of the same species as well as between species (Jian et al. 2021). Deoxynucleotide enzymes enable exposed DNA harboring ARGs to survive in the environment for a considerable amount of time, even after resistant strains have died. Bacteria can become inherently resistant to some antibiotics. The inherent structural and/or functional characteristics of bacteria that make it resistant to the action of a certain antibiotic are the cause of that bacterium's intrinsic resistance to that antibiotic. While certain antibiotics cannot pass through the outer membrane and cannot reach their target locations, others can be successfully removed by efflux after entering the cells through porin. The lack of sensitive targets for a certain antibiotic is the cause of a strain's inherent resistance. For instance, the fabI-insensitive allele has a different enoyl-ACP reductase enzyme, which is the target of the antibiotic in the sensitive species; hence, the biocide triclosan cannot prevent the growth of Gram-negative *Pseudomonas* spp. Bacteria can develop resistance to antibiotics in various ways in addition to their inherent resistance (Fig. 5.1).

First, due to inadequate bacterial osmosis or antibiotic efflux, the intracellular concentration of antibiotics is reduced (Fernández and Hancock 2012). Furthermore, by altering the target protein at the genetic level or post-translational modification, bacteria can change the targets of antibiotics (Nikaido 2009). Porin gene mutations and the development of genes that control the production of porins are both facilitated by the selection pressure exerted by the antibiotics (Baroud et al. 2013). Many antibiotics are actively transported to the extracellular environment by the bacterial efflux pump, rendering them useless. This is the leading cause of Gram-negative bacteria's natural resistance to several medications that can be used to treat Gram-positive bacterial infections.

Overexpression of the efflux pump has been linked to the development of antibiotic resistance. Numerous efflux systems exist in bacteria, according to a huge amount of data from bacterial genomics, and combinations of several different efflux pumps can lead to MDR and significantly boost bacterial resistance to numerous antibiotics. The MexXY pump system in *Pseudomonas aeruginosa* has been identified as one of the key factors in aminoglycoside resistance (Morita et al. 2012). The transcription of multidrug efflux genes is often inhibited or activated by

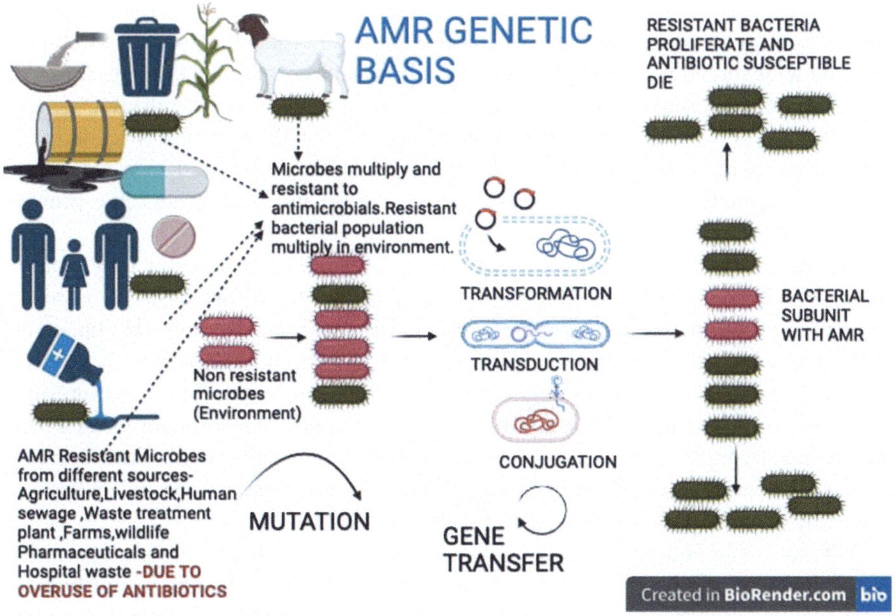

Fig. 5.1 Development of antibiotic-resistant bacteria by the overuse of antibiotics in livestock management, agriculture, pharmaceutical usage, etc. (Created with BioRender)

transcriptional regulators, which control the expression of bacterial multidrug efflux systems (Igbinosa et al. 2012; Jian et al. 2021). High amounts of efflux genes are frequently expressed in MDR bacteria due to regulatory network alterations that govern efflux pump expression (Usui et al. 2013; Deng et al. 2013). Mutations in the regulatory network that regulates efflux pump expression frequently cause excessive numbers of efflux genes to be expressed in MDR bacteria. These mutations may affect intergenic sites, global transcription factors, regional repressors, or local repressors, altering the expression of pump genes or their regulators (Warner et al. 2008; Webber et al. 2005; Olliver et al. 2004).

However, there are typically many different types of pathogens present during infection, and the selection pressure of the antibiotics forces microbes to code for mutations in the genes of antibiotic targets (Lin et al. 2015) leading to the development of antibiotic resistance; strains with such mutations multiply more quickly than those without the ARGs. The dissemination of resistance determinants among bacteria is greatly aided by mobile genetic elements (MGE), such as plasmids, transposons, and integrons carrying gene cassettes (Gao et al. 2010). The main causes of antimicrobial resistance in bacterial populations are horizontal gene transfer through these MGEs and clonal selection brought on by antimicrobial selective pressure (Schmidt et al. 2001).

5.5 AMR Bacteria and Fish Diseases

One of the biggest difficulties in aquaculture is the emergence of AMR in cultured fish. The widespread use of antibiotics and subsequent persistence of those drugs in the aquatic environment as a result of the high frequency of bacterial infections in fish leads to the spread of bacteria that are resistant to those drugs (Preena et al. 2020). Worldwide, antibiotic-resistant infections have been found in a variety of cultured fish, including carp, salmon, tilapia, catfish, and crustaceans like prawns (Watts et al. 2017).

The intestine-associated *Aeromonas* sp. of farm-raised freshwater fish such as *Ctenopharyngodon idella*, *Labeo rohita*, and *Catla catla* showed antibiotic resistance to ampicillin, oxytetracycline, amoxicillin, and novobiocin (Hatha et al. 2005). From a freshwater carp aquaculture system, Singh et al. (2009) recovered more pathogenic *Enterobacteriaceae* and *Flavobacterium* with elevated minimum inhibitory concentration (MIC). Antibiotic-resistant motile aeromonads caused septicemia in catfish. Broad-spectrum co-trimoxazole, chloramphenicol, ciprofloxacin, gatifloxacin, gentamicin, sulfafurazole, and oxytetracycline were not effective against *Aeromonas* spp. found in sick fish such as catfish.

Fish from cold and temperate regions are susceptible to the furunculosis disease caused by *Aeromonas salmonicida*, which is also easily resistant to antibiotics. Recent evidence indicates that multiple resistant *A. salmonicida* strains from various regions of the nation have transferable resistance plasmids (Romero et al. 2012). Similar incidents involving distinct multiresistant bacterial infections have been recorded from other nations. According to Karunasagar et al. (1994), *Vibrio harveyi* multiresistant strains that were resistant to chloramphenicol, cotrimoxazole, erythromycin, and streptomycin caused mass mortality in *Penaeus monodon* larvae. Similar to this, freshwater fish from ponds with few treatments were shown to have multiple antibiotic resistance (MAR). Similar to this, MAR pathogens like *Salmonella, Shigella, Pseudomonas*, and *E. coli* have been found in freshwater fish like catfish and tilapia that have not received any antibiotic treatment. The existing resistance genes survived in the cultured environment rather than because antibiotics were applied directly.

In China's abalone and turbot mariculture farms, *Vibrio splendidus* and *Pseudoalteromonas* serve as the primary reservoir of MDR determinants (Dang et al. 2007). Due to the presence of the resistant determinant cat III, *Edwardsiella tarda*, which caused the largest epidemic of septicemia in turbot (*Scophthalmus maximus*) from a mariculture farm, displayed significant resistance to chloramphenicol (Xiao et al. 2009). MDR aeromonads have been found in ornamental fish since a very early time period, including domestic goldfish and koi carp (Dixon and Issvoran 1993).

5.6 Prevention of AMR Transmission

The issue of antibiotic resistance needs to be addressed using efficient solutions. The use of antibiotics in aquaculture has expanded along with the rapid growth in worldwide aquaculture production of aquatic species (fish, shellfish, prawns, and mollusks) over the past 10 years (Watts et al. 2017).

To evaluate and detect resistant pathogens and associated genes, as advised by the FAO and the WHO, it is crucial to develop and implement reliable measurement methods at the national and international levels (WHO 2006). Following the FAO action plan on AMR 2016–2020, effective policies must be put into place, including appropriate regulations and enforcement; information gathering on disease diagnosis, risk assessment, disease control, and management; capacity development at every step of the aquaculture production chain; sensible application of antibiotics to prevent disease outbreaks; and efficient biosecurity practices. As a result, it is critical to maintain an aquatic system that is safe, healthy, and free of disease to stop the spread of germs, reduce the need for antibiotics, and advance antibiotic resistance. Every time a disease epidemic occurs, appropriate multistage antibiotic resistance monitoring is necessary. Understanding the specifics of the resistance mechanism will allow for effective adjustments as well as the right selection and application of antibiotics in the relevant fish farms. The right adsorption techniques, filtration, biological processes, sedimentation, and flocculation can be used to remove the antibiotic residues that have persisted in the aquatic environment as a result of their excessive dose and excess feed (Homem and Santos 2011). In aquaculture, preventive techniques like vaccination are employed to slow the spread of disease. There is a need for novel techniques to control infectious illnesses in aquaculture to stop the emergence and spread of antibiotic resistance in aquatic food animals. For instance, making the most use of already available vaccinations may be a good choice (Assefa and Abunna 2018). Through the creation of humoral antibodies, oral fish vaccines are effective against a variety of aquatic infections (Newaj-Fyzul and Austin 2015). Sea bass was reported to be protected by the oral vaccination created using the porin gene of *V. anguillarum* (Kumar et al. 2008). Due to the lack of *Piscirickettsia salmonis* vaccines, large quantities of quinolones were used to treat such illnesses in Chile (Cabello 2004).

It may be possible to promote fish growth and reduce infectious diseases by improving hygiene (Austin and Austin 2016), and using enzymes, probiotics, prebiotics, and acids to improve health (Sayes et al. 2018; Hoseinifar et al. 2018), and using bacteriocins, antimicrobial peptides, and bacteriophages as antibiotic substitutes (Rao and Lalitha 2015). Probiotics are also being utilized more frequently in the management of aquatic diseases in addition to immunizations due to their positive effects on health. By binding to intestinal mucus, probiotics have the clear ability to combat pathogens like *Vibrio harveyi*, a serious hazard to aquaculture (Chabrillon et al. 2005).

It is important to develop widely recognizable standards for the use of antibiotics in aquaculture and for monitoring programs to track the spread of AMR bacteria worldwide. It is crucial to cut down on aquaculture's use of antibiotics (Bardhan and Abraham 2021). A crucial tactic is to constantly work to educate people about antibiotic resistance. It is also important to improve sanitation and hygiene practices in aquaculture (Romero et al. 2012; Stratev and Odeyemi 2017). Understanding environmental hotspots for AMR gene exchange, like as aquaculture systems, and figuring out how they might move to clinically similar strains, are crucial.

5.7 Conclusion

AMR is highly debated recently as a growing threat to human and animal health. The rise of the AMR rate in aquaculture mainly seems to affect the low- and middle-income countries in both livelihood and economic safety. Food security as well as nutrition should be considered significantly following the resistance of pathogens in aquaculture industry. Disease transmission from animal to human via zoonoses is now a major global concern. The indiscriminate use of antibiotics must be controlled in agriculture and aquaculture industries. Therefore, strengthening surveillance to delineate the connection between AMR and healthy and sustainable aquaculture and fisheries industry as well as other rural food producing environments is utmost prominent.

References

Ageel HI, Arishi HM, Kamli AA, Hussein AM, Bhavanarushi S (2017) Unusual presentation of gastrointestinal basidiobolomycosis in a 7-year-old Child – Case Report. American J Med Case Rep 5(5):131–134

Ahmed ASS, Hossain MB, Semme SA, Babu SMOF, Hossain K, Moniruzzaman M (2020) Accumulation of trace elements in selected fish and shellfish species from the largest natural carp fish breeding basin in Asia: a probabilistic human health risk implication. Environ Sci Pollut Res 27(30):37852–37865. https://doi.org/10.1007/s11356-020-09766-1

AL-Mamun M, Chowdhury T, Biswas B, Absar N (2018) Chapter 11—Food poisoning and intoxication: a global leading concern for human health. In: Food safety and preservation. Academic, pp 307–352. https://doi.org/10.1016/B978-0-12-814956-0.00011-1

Assefa A, Abunna F (2018) Maintenance of fish health in aquaculture: review of epidemiological approaches for prevention and control of infectious disease of fish. Vet Med Int 2018:5432497

Austin B, Austin DA (2016) Aeromonadaceae representatives (motile aeromonads). In: Bacterial fish pathogens. Springer, Cham, Switzerland, pp 161–214

Bardhan A, Abraham TJ (2021) Antibiotic-resistance in Motile Aeromonas spp. of Indian major carps sold in retail markets of Peri-urban Kolkata India. J Aqua Food Product Technol 30(7):786–793. https://doi.org/10.1080/10498850.2021.1945178

Barros MBDL, de Almeida Paes R, Schubach AO (2011) Sporothrix schenckii and Sporotrichosis SUMMARY. Clin Microbiol Rev 24(4):633–654. https://doi.org/10.1128/CMR.00007-11

Baroud M, Dandache I, Araj GF, Wakim R, Kanj S, Kanafani Z et al (2013) Underlying mechanisms of carbapenem resistance in extended-spectrum β-lactamase-producing Klebsiella pneumoniae and Escherichia coli isolates at a tertiary care centre in Lebanon: role of OXA-48 and NDM-1 carbapenemases. Int J Antimicrob Agents 41:75–79

Boerlage AS, Graat EA, Verreth JA, de Jong MC (2013) Effect of control strategies on the persistence of fish-borne zoonotic trematodes: a modelling approach. Aquaculture 408:106–112. https://doi.org/10.1016/j.aquaculture.2013.05.026

Boyd CE, McNevin AA, Davis RP (2022) The contribution of fisheries and aquaculture to the global protein supply. Food Secur 14:805–827. https://doi.org/10.1007/s12571-021-01246-9

Cabello FC (2004) Antibiotics and aquaculture in Chile: implications for human and animal health. Rev Med Chil 132:1001–1006. https://doi.org/10.4067/s0034-98872004000800014

Cabello FC (2006) Heavy use of prophylactic antibiotics in aquaculture: a growing problem for human and animal health and for the environment. Environ Microbiol 8(7):1137–44

Chabrillon M, Rico RM, Arijo S, Diaz-Rosales P, Balebona MC, Morinigo MA (2005) Interactions of microorganisms isolated from gilthead sea bream, Sparus aurata L., on Vibrio harveyi, a pathogen of farmed Senegalese sole, Soleasenegalensis (Kaup). J Fish Dis 28:531–537

Dang HY, Zhang XX, Song LS, Chang YQ, Yang GP (2007) Molecular determination of oxytetracycline-resistant bacteria and their resistance genes from mariculture environments of China. J Appl Microbiol 103:2580–2592

Delghandi MR, El-Matbouli M, Menanteau-Ledouble S (2020) Mycobacteriosis and infections with non-tuberculous mycobacteria in aquatic organisms: a review. Microorganisms 8(9):1368. https://doi.org/10.3390/microorganisms8091368

Deng Z, Shan Y, Pan Q, Gao X, Yan A (2013) Anaerobic expression of the gadE-mdtEF multidrug efflux operon is primarily regulated by the two-component system ArcBA through antagonizing the H-NS mediated repression. Front Microbiol 4:194

Dixon BA, Issvoran G (1993) Antibacterial drug resistance in Aeromonas spp. isolated from domestic goldfish and koi from California. J World Aquacult Soc 24:102–104

Eiras JC, Pavanelli GC, Takemoto RM, Nawa Y (2018) An overview of fish-borne nematodiases among returned travelers for recent 25 years—unexpected diseases sometimes far away from the origin. Korean J Parasitol 56(3):215–227. https://doi.org/10.3347/kjp.2018.56.3.215. Epub 2018 Jun 30.

Fernández L, Hancock R (2012) Adaptive and mutational resistance: role of porins and efflux pumps in drug resistance. Clin Microbiol Rev 25:661–681

Gao W, Chua K, Davies JK, Newton HJ, Seemann T, Harrison PF, Holmes NE, Rhee HW, Hong JI, Hartland EL, Stinear TP, Howden BP (2010) Two novel point mutations in clinical Staphylococcus aureus reduce linezolid susceptibility and switch on the stringent response to promote persistent infection. PLoS Pathog 6:e1000944

Gutsell JS (1946) Sulfa drugs and the treatment of furunculosis in trout. Science 104(2691):85–86. https://doi.org/10.1126/science.104.2691.85

Haenen D, Zhang J, da Silva CS, Bosch G, van der Meer IM, van Arkel J, van den Borne JJ, Gutiérrez OP, Smidt H, Kemp B, Müller M (2013) A Diet high in resistant starch modulates microbiota composition SCFA Concentrations and gene expression in pig intestine. J Nutr 143(3):274–283. https://doi.org/10.3945/jn.112.169672

Hatha M, Vivekanandhan AA, Joice GJ, Christol (2005) Antibiotic resistance pattern of motile aeromonads from farm raised freshwater fish. Int J Food Microbiol 98:131–134

Hernández-Cabanyero C, Amaro C (2020) Phylogeny and life cycle of the zoonotic pathogen Vibrio vulnificus Summary. Environ Microbiol 22(10):4133–4148. https://doi.org/10.1111/1462-2920.15137

Homem V, Santos L (2011) Degradation and removal methods of antibiotics from aqueous matrices—a review. J Environ Manag 92:2304–2347

Holick MF, Chen TC, Lu Z, Sauter E (2007) Vitamin D and skin physiology: a D-Lightful story abstract. J Bone Miner Res 22(S2):V28–V33. https://doi.org/10.1359/jbmr.07s211

Hoseinifar SH, Sun Y, Wang A, Zhou Z (2018) Probiotics as means of diseases control in aquaculture, a review of current knowledge and future perspectives. Front Microbiol 9:2429

Igbinosa IH, Igumbor EU, Aghdasi F, Tom M, Okoh AI (2012) Emerging Aeromonas species infections and their significance in public health. Sci World J 12:625023

Jian Z, Zeng L, Xu T, Sun S, Yan S, Yang L et al (2021) Antibiotic resistance genes in bacteria: occurrence, spread, and control. J Basic Microbiol 61:1049–1070

Karunasagar I, Pai R, Malathi GR, Karunasagar I (1994) Mass mortality of Penaeus monodon larvae due to antibioticresistant Vibrio harveyi infection. Aquaculture 128(3-4):203–209. https://doi.org/10.1016/0044-8486(94)90309-3

Kumar SR, Ahmed VPI, Parameswaran V, Sudhakaran R, Babu VS, Hameed AS (2008) Potential use of chitosan nanoparticles for oral delivery of DNA vaccine in Asian sea bass (Lates calcarifer) to protect from Vibrio (Listonella) anguillarum. Fish Shellfish Immunol 25:47–56

Levey DJ, Stiles FG (1992) Evolutionary precursors of long-distance migration: resource availability and movement patterns in neotropical landbirds. Am Nat 140(3):447–476. https://doi.org/10.1086/285421

Li J, Ni J, Li J, Wang C, Li X, Wu S, Zhang T, Yu Y, Yan Q (2014) Comparative study on gastrointestinal microbiota of eight fish species with different feeding habits. J Appl Microbiol 117(6):1750–1760. https://doi.org/10.1111/jam.12663

Lin J, Nishino K, Roberts MC, Tolmasky M, Aminov RI, Zhang L (2015) Mechanisms of antibiotic resistance. Front Microbiol 6:34

McConnaughey RA, Hiddink JG, Jennings S, Pitcher CR, Kaiser MJ, Suuronen P, Sciberras M, Rijnsdorp AD, Collie JS, Mazor T, Amoroso RO (2020) Choosing best practices for managing impacts of trawl fishing on seabed habitats and biota Abstract. Fish Fish 21(2):319–337. https://doi.org/10.1111/faf.v21.2. https://doi.org/10.1111/faf.12431

Morita Y, Tomida J, Kawamura Y (2012) MexXY multidrug efflux system of Pseudomonas aeruginosa. Front Microbiol 3. https://doi.org/10.3389/fmicb.2012.00408

Mozaffarian D, Rimm EB (2006) Fish intake, contaminants, and human health: evaluating the risks and the benefits. JAMA 296:1885–1899

Newaj-Fyzul A, Austin B (2015) Probiotics, immunostimulants, plant products and oral vaccines, and their role as feed supplements in the control of bacterial fish diseases. J Fish Dis 14:937–955. https://doi.org/10.1111/jfd.12313

Nikaido H (2009) Multidrug resistance in bacteria. Annu Rev Biochem 78:119–146

Norman AW (2008) From vitamin D to hormone D: fundamentals of the vitamin D endocrine system essential for good health. Am J Clin Nutr 88:491S–499S

Olliver A, Vallé M, Chaslus-Dancla E, Cloekaert A (2004) Role of an acrR mutation in multidrug resistance of in vitro-selected fluoroquinolone-resistant mutants of Salmonella enterica serovar Typhimurium. FEMS Microbiol Lett 238:267–272

Pal D, Maiti SK (2018) Seasonal variation of heavy metals in water sediment and highly consumed cultured fish (Labeo rohita and Labeo bata) and potential health risk assessment in aquaculture pond of the coal city Dhanbad (India). Environ Sci Pollut Res 25(13):12464–12480. https://doi.org/10.1007/s11356-018-1424-5

Preena PG, Swaminathan TR, Kumar VJR, Singh ISB (2020) Antimicrobial resistance in aquaculture: a crisis for concern. Biologia 75:1497–1517

Rao BM, Lalitha KV (2015) Bacteriophages for aquaculture: are they beneficial or inimical. Aquaculture 437:146–154

Romero J, Feijoó CG, Navarrete P (2012) Antibiotics in aquaculture—use, abuse and alternatives. In: Health and environment in aquaculture. IntechOpen. https://www.intechopen.com/books/health-andenvironment-in-aquaculture/antibiotics-inaquaculture-use-abuse-and-alternatives

Sayes C, Leyton Y, Riquelme C (2018) Probiotic bacteria as a healthy alternative for fish aquaculture. In: Antibiotic use in animals. IntechOpen, London, pp 115–132. https://www.intechopen.com/books/antibiotic-usein-animals/probiotic-bacteria-as-an-healthyalternative-for-fish-aquaculture

Schmidt AS, Bruun MS, Dalsgaard I, Larsen JL (2001) Characterization of class 1 integrons associated with R-plasmids in clinical Aeromonas salmonicida isolates from various geographical areas. J Antimicrob Chemother 47:735–743. https://doi.org/10.1093/jac/47.6.735

Shane B (2011) Zoonoses associated with fish. Vet Clin N Am: Exot Anim Pract 14(3):427–438. https://doi.org/10.1016/j.cvex.2011.05.003

Simopoulos A (2007) Nutritional aspects of fish, in seafood from producer to consumer, integrated approach to quality, and V. Mamaladze. Effectiveness and safety of vitamin D in relation to bone health. Evid Rep Technol Assess 158:1–235

Singh AF, Rathore G, Singh V, Mani I, Singh RK, Mishra SK, Mishra BN, Verma OP (2009) Bacterial resistance to oxytetracycline in different life stages of Indian freshwater carp aquaculture system. Int J Microbiol Res 1:25–34

Stratev D, Odeyemi OA (2017) An overview of motile Aeromonas septicaemia management. Aquacult Int 25(3):1095–1105

Tenyang N, Ponka R, Tiencheu B, Womeni HM, Tonfack F (2016) Proximate composition, fatty acid and mineral contents of four freshwater fish from Maga Lake (far north region of Cameroon). Am J Food Sci Technol 4(3):64–69

Usui M, Nagai H, Hiki M, Tamura Y, Asai T (2013) Effect of antimicrobial exposure on AcrAB expression in Salmonella enterica subspecies enterica serovar choleraesuis. Front Microbiol 4:53

Vignesh S, Raja A, James RA (2010) Marine drugs: implication and future studies. Int J Pharmacol 7(1):22–30. https://doi.org/10.3923/ijp.2011.22.30

Warner DM, Shafer WM, Jerse AE (2008) Clinically relevant mutations that cause depression of the Neisseria gonorrhoeae MtrC–MtrD–MtrE efflux pump system confer different levels of antimicrobial resistance and in vivo fitness. Mol Microbiol 70:462–478

Watts JEM, Schreier HJ, Lanska L, Hale MS (2017) The rising tide of antimicrobial resistance in aquaculture: sources, sinks and solutions. Mar Drugs 15:e158

Webber MA, Talukder A, Piddock LJV (2005) Contribution of mutation at amino acid 45 of AcrR to acrB expression and ciprofloxacin resistance in clinical and veterinary Escherichia coli isolates. Antimicrob Agents Chemother 49:4390–4392

World Health Organization (WHO) (2006) Report of a Joint FAO/OIE/WHO Expert Consultation on antimicrobial use in aquaculture and antimicrobial resistance. Republic of Korea, Seoul. http://www.who.int/topics/foodbornediseases/aquaculturerep1316june2006.pdf

Xiao JF, Wang QY, Liu Q, Wang X, Liu H, Zhang YX (2009) Isolation and identification of fish pathogen Edwardsiella tarda from mariculture in China. Aquac Res 40:13–17. https://doi.org/10.1111/j.1365-2109.2008.02101

Zuraini A, Somchit MN, Solihah MH, Goh YM, Arifah AK, Zakaria MS, Somchit N, Rajion MA, Zakaria ZA, Mat Jais AM (2006) Fatty acid and amino acid composition of three local Malaysian Channa spp. fish. Food Chem 97(4):674–678. https://doi.org/10.1016/j.foodchem.2005.04.031

Chapter 6
Screening and Monitoring of AMR in Wastewater Facilities, Recreation Facilities and Other Aquatic Environments

Nayomi John, Saumya Kumaresan, Gopika Gopal, P. Nisha, M. Thangavel, Harsha Mohan, Febiya Anna Fedin, and M. S. Jisha

Abstract One of the biggest risks to human health is the growth of antimicrobial resistance (AMR). Aquatic environments are crucial to maintain high levels of biodiversity and biosphere production. Antibiotic resistance develops in the aquatic environment as a result of the use of antibiotics and other drugs in aquaculture to promote fish growth, treatment of disease, and anti-fouling. Up to 80% of antibiotics given to fish are actively eliminated in their urine and feces. When water from aquaculture facilities escapes, spills, or mixes with surface waters, antibiotic-resistant bacteria can spread from aquaculture systems to natural waterways. Wastewater continues to be a significant source of AMR in the environment because it enables AMR bacteria to survive and spread antibiotic-resistant genes (ARGs). It might be possible to protect water bodies against AMR contamination by wastewater treatment plants (WWTPs). Monitoring the environment is vital for obtaining essential data to control the rise of AMR bacteria. Coordinated environmental surveillance efforts are necessary to halt the spread of AMR. Efficient screening and monitoring of AMR bacteria in wastewater facilities, recreational spaces, and various aquatic environments are crucial for identifying potential sources of AMR and ultimately reducing its impact on human health.

N. John (✉)
Department of Microbiology, Mar Athanasius College (Autonomous), Kothamangalam, Kerala, India

S. Kumaresan
Cardiff Metropolitan University, Wales, UK

G. Gopal
Avisa Biotech Pvt Ltd., KRIBS BIONEST, Kalamassery, Kerala, India

P. Nisha
Department of Biosciences, MES College, Marampally, Kerala, India

M. Thangavel
School of Life Sciences, Nehru College of Arts & Sciences, Coimbatore, Tamil Nadu, India

H. Mohan · F. A. Fedin · M. S. Jisha
School of Biosciences, Mahatma Gandhi University, Kottayam, Kerala, India

P. Elumalai, S. Lakshmi (eds.), *Antimicrobial Resistance in Aquaculture and Aquatic Environments*, https://doi.org/10.1007/978-981-97-7320-6_6

Keywords Antibiotics · Antimicrobial resistance · Aquaculture · Monitoring · Screening

Abbreviations

AMR	Antimicrobial resistance
ARB	Antibiotic-resistant bacteria
ARGs	Antibiotic-resistant genes
AST	Antimicrobial susceptibility testing
CLSI	Clinical and Laboratory Standards Institute
CSOs	Combined sewer overflows
DD	Disk diffusion
ddPCR	Digital droplet PCR
ESBL	Extended-spectrum beta-lactamases
EUCAST	European Committee on Antimicrobial Susceptibility Testing
HGT	Horizontal gene transfer
HPLC	High-performance liquid chromatography
HT-qPCR	High throughput real-time PCR
IFC	Integrated fluidic circuit
LAMP	Loop-mediated isothermal amplification
LC-MS	Liquid chromatography–mass spectrometry
MALDI-TOF MS	Matrix-assisted lasers desorption ionization–time of flight mass spectrometry
MGEs	Mobile genomic elements
MIC	Minimum inhibitory concentration
RAST	Rapid antibiotic susceptibility test
SERS	Surface-enhanced Raman scattering
TAC	TaqMan Array Card
WGS	Whole genome sequencing
WWTPs	Wastewater treatment plants

6.1 Introduction

Antibiotic contamination is of increasing concern globally. Due to the growing and indiscriminate use of antibiotics as growth promoters in animal husbandry, antibiotics have become a new class of environmental contaminants. Antibiotic contamination of aquatic environments can have detrimental effects on the microbial flora of these water bodies and contribute to the emergence and dissemination of antibiotic-resistant genes (ARGs). Antimicrobial resistance (AMR) poses a significant threat to global health by reducing the effectiveness of antimicrobial drugs, such as antibiotics, and making infections more difficult to treat. The emergence and spread of AMR is a complex issue influenced by various factors, including the release of antimicrobial agents into the environment (Hernando-Amado et al. 2019). The movement of ARGs between bacteria through mobile genetic elements (MGEs) is a considerable concern. It highlights the need to address their transfer alongside antibiotic-resistant bacteria (ARB) and ARGs in the environment. Pollution leads to the direct release of ARB, ARGs, and MGEs into the environment. Antimicrobials are also found in pollution, which can lead to selection pressure, maintenance, and amplification of ARB and ARGs and promote the emergence of new types of antibiotic resistance (Larsson and Flach 2022).

Since the emergence of AMR from wastewater has become a major source for the spread of various pathogenic species, including bacteria, fungi, etc., their monitoring and assessment are important. Screening and monitoring for AMR in these aquatic environments is crucial in understanding the extent of AMR dissemination, identifying sources of resistance, and implementing appropriate interventions to mitigate its impact. Wastewater is an ideal environment for horizontal gene transfer (HGT) because of the high microbial biomass and the availability of nutrients (Guardabassi et al. 2002). The release of antimicrobial products and resistant genes and bacteria into aquatic environments occurs primarily through agricultural runoff, illegal discharges, and wastewater treatment plants (WWTPs) (Baquero et al. 2008). Treated or untreated sewage discharges contain anthropogenic pollutants that have been shown to co-select for AMR (Baker-Austin et al. 2006).

WWTPs remove sewage, including a confluence of antibiotics, ARB, ARGs, MGEs, and human bacterial pathogens from residential, industrial, and clinical sources (Baquero et al. 2008; Guo et al. 2017). WWTPs have a significant potential to reduce the transmission of AMR to water bodies. However, this is not their intended use.

It has been proposed that specific measures may be required for the pretreatment of hospital sewage prior to its release or for the additional treatment of wastewater containing a higher concentration of clinically significant AMR strains (Novo and Manaia 2010). Recycled water, which has undergone further treatment to make it suitable for potable or non-potable uses (such as irrigation of fields or crops), is of particular concern since there may be less chance for AMR to lessen before usage.

The traditional method of treating wastewater combines primary, secondary, and tertiary processes. Before wastewater is released into the environment, it must first be disinfected using disinfectants such as chlorine, UV radiation, and ozonization (Kumar and Pal 2018). Reducing the spread of waterborne infections is one of the main goals of wastewater treatment. Antibiotic resistance genes are released into the environment, and wastewater treatment is essential in preventing this. WWTP contains various chemical components, including disinfectants, metals (such as copper and zinc), different medications (such as antibiotics), and other organic compounds. Different selection pressures, including chemical ones, have an impact on how ARGs are transmitted, expressed, and mobilized and how ARB originate, persist, and proliferate (Nguyen et al. 2021). The fact that ARGs can be transmitted between bacteria via MGEs is a significant difficulty, necessitating consideration of their movement with ARB and ARGs in the environment (Wang et al. 2020). According to current thinking, the majority of ARGs found in pathogens were initially sourced from the enormous and varied genetic pool held by ambient bacteria. The direct release of ARB, ARGs, and MGEs into the environment is caused by pollution. Antimicrobials are also directly found in pollution, which can lead to selection pressure, maintain and amplify ARB and ARGs, and promote the emergence of new types of resistance (Freeland et al. 2021; Liguori et al. 2022).

Concerns about the regeneration of ARB in distribution systems, particularly when using recycled nonpotable water, have also been raised by research (Garner et al. 2018; Zhu et al. 2021). As a result, major potential monitoring locations for AMR in the environment have been identified as wastewater, recycled water, and affected receiving waters. The field must advance towards a joint knowledge of environmental occurrence, human exposures, efficient mitigation techniques, and eventually a dose-response assessment for human health risk assessments to confront the growing danger of drug-resistant illnesses and diseases (Amarasiri et al. 2020). The creation of standardized techniques that may be applied locally and internationally to create baseline environmental occurrence data unique to all pertinent water habitats is the first step in reaching such aims.

In wastewater influent, effluent, and biosolids or sludge, a large number of ARGs, including varieties and subvarieties of nearly all types of commonly used antibiotics, have been found. HGT, which is anticipated to be found during wastewater treatment, is a significant pathway for ARG proliferation (Karkman et al. 2018). Since bacterial multiplication and degradation are dynamic processes, variations in these rates could impact the behavior of ARGs and ARBs in a particular environment. Before effluent discharge, the primary objective of wastewater treatment is to deactivate pathogens effectively. However, given that commensal, non-pathogenic bacteria might serve as significant reservoirs for plasmids and other MGEs harboring ARGs, this remit needs to be reviewed in the context of AMR (Xie et al. 2018; Liguori et al. 2022).

Recreational water activities such as swimming, boating, or fishing can contribute to AMR dissemination. People carrying antibiotic-resistant bacteria on their skin or digestive tract can release them into the water during these activities (Leonard et al. 2015). Improper disposal of antibiotics or personal products containing antimicrobial ingredients can introduce AMR-related compounds into the water (Nicholas et al. 2013). Other aquatic environments include oceans, seas, rivers, lakes, ponds, wetlands, estuaries, drinking water, and even underground aquifers. In aquatic environments, it is common to find antibiotics such as macrolides (e.g., clarithromycin, erythromycin, roxithromycin), aminoglycosides (amikacin, gentamicin, kanamycin, neomycin, netilmicin, streptomycin, tobramycin), tetracyclines (tetracycline, chlortetracycline, oxytetracycline), demeclocycline (doxycycline), sulfonamides (sulfadimethoxine, sulfanilamides) and quinolones (nalidixic acid, ciprofloxacin, lomefloxacin, norfloxacin, ofloxacin, levofloxacin, sparfloxacin, tosufloxacin, clinafloxacin, gemifloxacin, moxifloxacin, sitafloxacin) (Hartmann et al. 1998; Lindberg et al. 2004; Brown et al. 2006; Thomas et al. 2007; Duong et al. 2008; Martins et al. 2008). In this context, AMR can occur in microorganisms found in water, sediments, and aquatic animals and can have significant implications for human and animal health, as well as the health of aquatic ecosystems.

To effectively address the challenge of AMR in the aquatic environment, it is crucial to establish standardized monitoring methods that can provide important information on the circulation of ARGs among human populations, as well as identify areas of high resistance evolution and dissemination, assess associated risks to human health, and evaluate the effectiveness of wastewater treatment infrastructure in removing antibiotics and reducing the spread of AMR. An important first step in creating better removal tactics is the quick and affordable identification of antibiotics in wastewater and during wastewater treatment operations. The development of reliable analytical methods for antibiotic identification and quantification has received a lot of attention (Östman et al. 2017). Studies have shown that among different environmental sectors, water environments stand out as the most significant pathway through which AMR can evolve and disseminate (Liguori et al. 2022). Therefore, monitoring the levels of AMR in aquatic environments is crucial for understanding the extent of the problem and developing effective strategies to address it (Huang et al. 2023).

6.2 Sources of AMR

Bacteria found in the water can be foreign, transient, and sporadically present due to shedding from animal, vegetal, or soil surfaces. Alternatively, they may be indigenous to aquatic habitats (John and Thangavel 2017). More than 90% of bacteria that has marine origins are resistant to two or more antibiotics, and 20% are resistant to five or more. Antibiotic resistance research in native aquatic species is crucial because it may reveal how much human activity has altered aquatic ecosystems. Several studies showed AMR occurrence as a natural phenomenon (Baquero et al. 2008). Besides naturally occurring ARGs, AMR was examined by many studies in untreated wastewater from hospitals, residential areas, and municipal WWTPs. AMR can occur in household and hospital wastewater, which in turn occurs in municipal WWTPs. Moreover, heightened levels of AMR were observed in effluents from abattoirs and antibiotic production facilities in China and India. In addition to WWTPs, combined sewer overflows (CSOs) can cause an increased release of AMR into recipient streams during periods of heavy rainfall. Agricultural runoff that has come into contact with manure and land-applied biosolids, which are currently used worldwide, is another route by which anthropogenically produced resistances enter surface waterways (Hiller et al. 2019). The major sources of AMR in aquatic environments are listed below:

6.2.1 Contaminated Water Sources

Various sources of water contamination can contribute to antimicrobial resistance.

- **Industrial wastewater**: Discharge of wastewater from industries, such as pharmaceutical manufacturing or chemical production, which may contain antimicrobial agents and contribute to resistance.

- **Municipal wastewater**: Wastewater from households, hospitals, and other urban sources that carry antibiotics and resistant bacteria into aquatic environments.
- **Agricultural runoff**: Runoff from agricultural fields carrying fertilizers, pesticides, and antimicrobial agents used in farming practices, which can contaminate water bodies. Agricultural activities in the vicinity of recreation facilities and aquatic environments can introduce antimicrobial resistance through the runoff of agricultural practices.
- **Livestock waste**: Antibiotics and other antimicrobial agents administered to livestock can enter the environment through runoff or leaching, reaching water sources associated with recreation facilities. This contamination may contribute to the selection and spread of antimicrobial resistance in aquatic ecosystems (Hao et al. 2014).

6.2.2 Aquaculture

Farming aquatic organisms like fish or shellfish, where antimicrobial use can lead to the development and spread of resistance. Excessive use of antibiotics in aquaculture to prevent or treat diseases in farmed organisms promotes the development of resistance. The release of antibiotic-treated feed in aquaculture systems potentially leads to the contamination of surrounding waters.

6.2.3 Recreational Activities

Activities involving human contact with aquatic environments that may contribute to antimicrobial resistance. Human activity in recreation facilities, such as gyms, swimming pools, and spas, contributes to disseminating antimicrobial resistance. People visiting these facilities may carry resistant microorganisms on their skin, gastrointestinal tract, or other parts of their body. These microorganisms can contaminate surfaces, equipment, and water sources within the facilities. Inadequate disinfection practices, improper maintenance of water quality, and inadequate filtration systems can create favorable conditions for the survival and spread of resistant microorganisms. Human activities, including body contact, improper hygiene practices, and overcrowding, contribute to the transmission of resistant strains among individuals. Boating and fishing can introduce antimicrobial agents and resistant bacteria through various means, such as boat discharges or the disposal of waste.

6.2.4 Overuse of Disinfectants

Disinfectants are commonly used in recreation facilities and aquatic environments to maintain cleanliness and prevent the spread of infections. However, the excessive and inappropriate use of disinfectants can lead to the development of antimicrobial resistance. Over time, microorganisms present in these environments may acquire resistance mechanisms to withstand the effects of the disinfectants. Furthermore, the use of broad-spectrum disinfectants can indiscriminately eliminate both harmful and beneficial microorganisms, disrupting the natural microbial balance and potentially creating a niche for resistant strains to thrive.

6.2.5 Inadequate Recreational Water Treatment

Recreational facilities and aquatic environments often receive wastewater discharges from various sources, including residential, commercial, and industrial sources. Wastewater and sewage can contain significant amounts of antimicrobial agents and resistant bacteria from human and animal sources. Antimicrobial-resistant bacteria can enter the environment due to improper wastewater treatment and the discharge of untreated or partially treated effluents into water bodies. Water treatment is necessary in aquatic environments like swimming pools, hot tubs, and water parks to stop the growth and spread of germs. Poor management of water quality parameters or inadequate chlorination of the water might result in an environment favorable to the survival and spread of resistant microbes. Then, others who come into touch with the polluted water may contract these resistant strains. When antimicrobial drugs are used exclusively to treat water, this selection pressure can encourage the survival and spread of resistant strains. Poor water treatment system monitoring and upkeep might result in inadequate disinfection, which can lead to suboptimal disinfection, creating conditions for resistant microorganisms to thrive.

6.2.6 Wildlife and Environmental Reservoirs

Wildlife, including birds, rodents, and insects, can serve as carriers of antimicrobial-resistant microorganisms. These animals may come into contact with recreation facilities and aquatic environments, introducing resistant strains into these settings. Additionally, the natural environment, such as rivers, lakes, and coastal areas, can harbor ARBs due to the presence of pollutants, including human and animal waste. Recreational activities that involve direct contact with these environments can increase the likelihood of acquiring and spreading antimicrobial-resistant microorganisms.

Different types of antibiotic-resistant determinants are present among bacteria in aquatic environments (Table 6.1). Efforts to mitigate the sources of antimicrobial resistance in recreation facilities and aquatic environments require a multifaceted approach. This includes implementing and enforcing proper water treatment and disinfection protocols, promoting responsible antimicrobial use in aquaculture and agriculture, improving wastewater treatment systems, and raising awareness among individuals about appropriate hygiene practices in recreational water settings. Collaboration between public health agencies, environmental organizations, and the

Table 6.1 Antimicrobial resistance determinants found in bacteria within the aquatic environment

Antimicrobial resistance determinant	Description	Examples
Extended-spectrum beta-lactamases (ESBL)	Enzymes that confer resistance to beta-lactam antibiotics, including penicillins and cephalosporins, by breaking down the antibiotics	TEM, SHV, CTX-M
Carbapenemases	Enzymes that hydrolyze carbapenem antibiotics, rendering them ineffective	KPC, NDM, OXA-48, VIM, IMP, GES
AmpC beta-lactamases	Beta-lactamase enzymes that confer resistance to penicillins, cephalosporins, and monobactams by inactivating the antibiotics	CMY, ACT, FOX, DHA, ACC, MIR, EBC, MOX
Fluoroquinolone resistance genes	Genes that encode proteins that protect bacterial DNA gyrase and topoisomerase IV from inhibition by fluoroquinolone antibiotics	qnr, aac(6′)-Ib-cr, qepA, oqxAB, qepB
Aminoglycoside-modifying enzymes	Enzymes that modify aminoglycoside antibiotics, preventing them from binding to the bacterial ribosome and inhibiting protein synthesis	AAC, ANT, APH, AAD, Aph(3′), AAD(2′), ANT(2″)
Macrolide resistance genes	Genes that encode proteins that modify the target site or actively pump out macrolide antibiotics	erm, mef, mph, msr, lnu, vga
Tetracycline resistance genes	Genes that produce proteins that actively efflux tetracycline antibiotics or protect the bacterial ribosome from tetracycline binding	tet(A), tet(B), tet(M), tet(Q), tet(O)
Sulfonamide resistance genes	Genes that encode enzymes that modify or inactivate sulfonamide antibiotics	sul1, sul2, sul3
Trimethoprim resistance genes	Genes that encode enzymes that modify or inactivate trimethoprim antibiotics	dfrA, dfrB, dfrC, dfrD, dfrE
Chloramphenicol resistance genes	Genes that encode enzymes that modify or inactivate chloramphenicol antibiotics	cat, cml, flo
Glycopeptide resistance genes	Genes that produce modified cell wall precursors that reduce the affinity of glycopeptide antibiotics	vanA, vanB, vanC, vanD, vane
Colistin resistance genes	Genes that produce proteins that modify the bacterial outer membrane, reducing the binding of colistin antibiotics	mcr-1, mcr-2, mcr-3, mcr-4, mcr-5

recreational industry is crucial to address this growing threat and preserve the effectiveness of antimicrobial agents for future generations (Taneja and Sharma 2019).

6.3 Importance of Screening and Monitoring AMR in Aquatic Environments

1. ***Preventing the spread of antibiotic resistance***: Aquatic environments, including wastewater facilities and recreation facilities, can serve as reservoirs for ARB and genes. By screening and monitoring for AMR, we can identify and track the presence and prevalence of resistant strains, allowing for targeted interventions to prevent the spread of AMR to humans and other organisms.
2. ***Assessing the efficacy of wastewater treatment***: Wastewater facilities play a crucial role in removing pollutants, including antibiotics and resistant bacteria, from wastewater before it is released into the environment. Regular screening and monitoring of AMR can help assess the effectiveness of wastewater treatment processes in removing antibiotics and reducing the levels of resistant bacteria, ensuring that these facilities are functioning optimally.
3. ***Protecting human health***: Recreation facilities such as swimming pools, beaches, and water parks are popular sites for human activities. Regular screening and monitoring of AMR in these aquatic environments can help identify potential sources of antibiotic resistance and protect human health. By understanding the prevalence of AMR, appropriate measures can be implemented to reduce the risk of infection and transmission of resistant bacteria to individuals using these facilities.
4. ***Environmental impact assessment***: Aquatic environments are interconnected ecosystems, and the presence of AMR in these environments can have far-reaching consequences. Screening and monitoring AMR can provide insights into the extent and impact of antibiotic contamination on aquatic organisms and ecosystems. This information can guide the development of sustainable management strategies and promote the conservation of aquatic biodiversity.
5. ***Early detection of emerging resistance***: Continuous screening and monitoring of AMR in wastewater facilities and other aquatic environments enable the early detection of emerging antibiotic resistance patterns. Identifying new resistance genes or novel mechanisms of resistance can help researchers and healthcare professionals stay ahead of the curve and develop timely strategies to combat the spread of resistance, thereby preserving the effectiveness of existing antibiotics.
6. ***Informing policy and decision-making***: Data obtained from screening and monitoring AMR in aquatic environments can serve as valuable evidence for policy development and decision-making processes. It can aid in formulating guidelines and regulations regarding the use of antibiotics, wastewater management, and environmental protection. By utilizing evidence-based approaches,

policymakers can enact measures to mitigate the risks associated with AMR and ensure the sustainable use of antibiotics and the protection of aquatic ecosystems.

6.4 Screening and Monitoring of AMR in Aquatic Environments

The key aspects of AMR in wastewater are characterization, surveillance, monitoring, and assessment that can be broadly classified into culture-dependent and culture-independent methods (Fig. 6.1). Culture-dependent methods are more commonly used to monitor wastewater. However, this type of method is less efficient for the detection of some active sources of antimicrobial genes and extracellular DNA. Non-culturable and non-pathogenic ARB species may contribute to these AMR genes. This method is ineffective on account of the actual microbial load in the water. Instead, several culture-independent methods like real-time PCR, ultrafiltration-based DNA detection, and metagenomic screening are used for identification of ARGs. Different monitoring programs for AMR are summarized in Fig. 6.2 (Liguori et al. 2022).

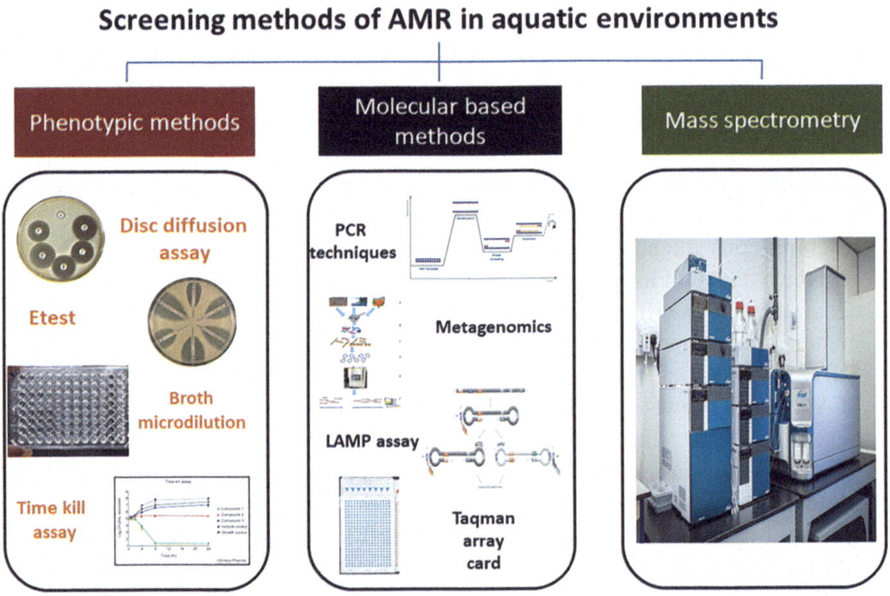

Fig. 6.1 Screening methods of AMR in aquatic environments

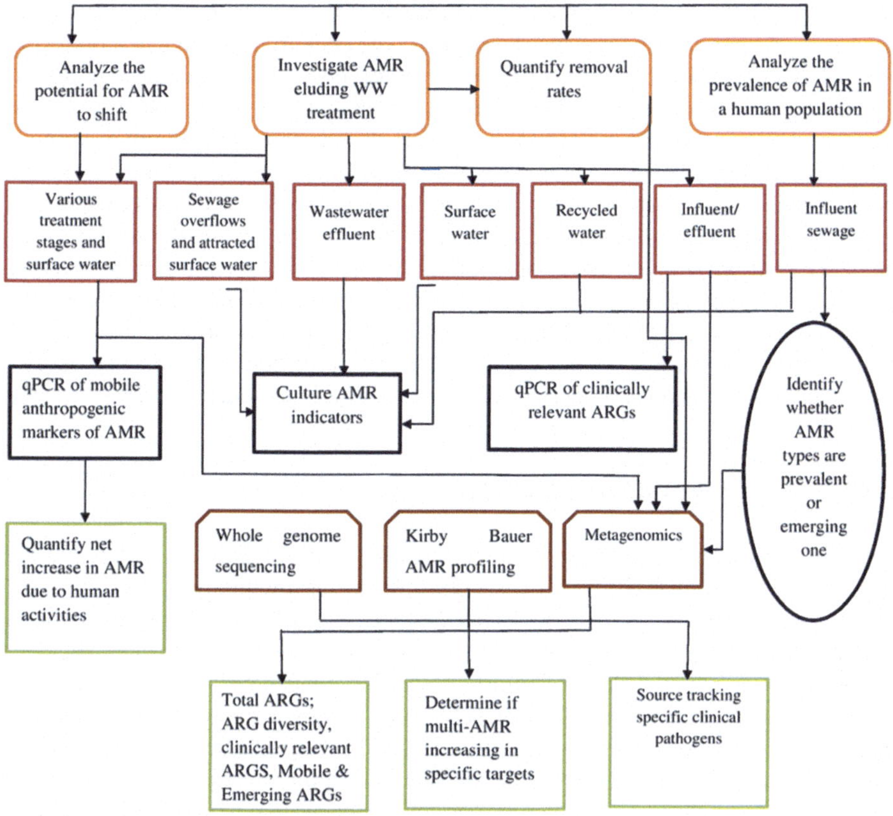

Fig. 6.2 AMR monitoring program

6.5 AMR Testing by Phenotypic Methods

6.5.1 Culture-Based Methods

These are widely accepted for AMR monitoring because they can be used to select bacterial targets with clinical relevance, and the methods are well standardized for defining clinical resistance levels. Additionally, once the isolate has been recovered, it can be subjected to further analysis, such as multidrug-resistance tests, whole genome sequencing, etc., which helps identify ARGs carrying MGEs. These methods are considered most suitable to inform human health risk assessments because they provide direct evidence of the presence and levels of antibiotic resistance in bacterial strains that are most likely to cause human illness (Gopal and John 2020). While culture-based methods are attractive for monitoring AMR, they face significant challenges when applied to wastewater and surface water environments due to the presence of a diverse range of microorganisms. Despite the potential for numerous genera/species in these environments to serve as informative targets, the typically high levels of background microorganisms can make it difficult to isolate the target and increase the risk of interference. The primary challenge in monitoring AMR lies in the vast diversity of species thriving in wastewater environments. At times, these species impede the isolation process intended for specific targets. Another bottleneck with culture-based methods are the large number of non-culturable bacteria that are viable microbial species that threaten the human population. Furthermore, most environmental microbes cannot be easily cultured, which means that the use of culture-based methods may overlook the broader pool of resistance present in a given environmental reservoir (Rizzo et al. 2013). Overall, culture-based methods are valuable for AMR monitoring because they can provide information on the prevalence, patterns, and mechanisms of resistance in clinically relevant bacterial strains, which can be used to inform human health risk assessments and guide the development of effective strategies to address AMR.

6.5.1.1 Agar Disk Diffusion Method

Agar disk diffusion testing was developed by Bauer (1966) and is considered as the official and traditional antibiotic susceptibility testing method practiced in many clinical microbiology laboratories. This method involves inoculating agar plates with a standardized amount of the microorganism under investigation and then placing small filter paper disks, containing a specific concentration of the test compound, onto the surface of the agar. The Petri dishes are then incubated under appropriate conditions, allowing the antimicrobial agent to diffuse into the agar and inhibit the growth of the microorganism. The size of the resulting inhibition zones can then be measured by the naked eye or automated system to assess the effectiveness of the test compound. This thus provides qualitative results by categorizing bacteria as susceptible, intermediate or resistant (Reller et al. 2009). Although the method is widely accepted, there are some limitations that need to be addressed. These include the challenges associated with determining the minimum inhibitory concentration (MIC) and potential delays in obtaining results. Additionally, it is important to take into account the biological properties of bacterial growth during the lag and log phases, as well as their expression in response to antibiotics (Idelevich et al. 2017). Reduction in incubation time for disk diffusion (DD) can be tackled with a method combining DD and rapid antibiotic susceptibility test (RAST), defined by EUCAST, and MALDI-TOF MS for the identification of bacteria (Idelevich et al. 2017). However, DD offers many advantages such as low cost, ease of performance, applicability of numerous bacterial species and antibiotics, and the ease of interpreting results (Balouiri et al. 2016). It can also detect important resistance phenotypes, such as ESBL (extended spectrum beta-lactamase), carbapenemases, and inducible resistance to macrolides. Moreover, it can identify the presence of heteroresistant bacterial populations (Gajic et al. 2022). Given these benefits, it is likely that the disk diffusion method will continue to be widely used for antimicrobial susceptibility testing in the future.

6.5.1.2 Antimicrobial Gradient Test Method (E Test)

An innovative technique for determining the MIC value of an antimicrobial agent is the antimicrobial gradient method, which combines the principles of dilution and diffusion methods. This technique involves creating a concentration gradient of the antimicrobial agent in an agar medium, using a strip impregnated with increasing concentrations of the agent from one end to the other. A commercial version of this technique, known as the E test, is widely used in clinical microbiology laboratories. This involves placing the strip on the surface of the agar previously inoculated with the microorganism of interest (Balouiri et al. 2016). The antimicrobial gradient method is a versatile technique that can be used to determine the MIC of various types of antimicrobial agents, including antibiotics, antifungals, and antimycobacterials (Hausdorfer et al. 1998). The MIC value is determined by observing the point of intersection between the strip and the growth inhibition ellipse on the agar surface, which indicates the concentration of the antimicrobial agent required to inhibit the growth of the microorganism being tested (Reller et al. 2009).

The antimicrobial gradient tests offer several advantages, including ease of use, flexibility in testing various antibiotic combinations, and the absence of any special expertise or technologies required. Additionally, they are particularly suitable for testing a limited number of antibiotics. However, it should be noted that each strip's cost is relatively high compared to antibiotic disks, which is why gradient tests are typically reserved for testing a few antibiotics per strain. One potential drawback of gradient tests is their relatively long incubation period of 16–24 h, which can be a disadvantage compared to the faster and more automated systems available for determining MIC (Gajic et al. 2022).

6.5.1.3 Dilution Methods

Besides the disk diffusion method, the most widely practiced antimicrobial susceptibility test includes broth macro, micro, and agar dilution. Dilution methods are considered the most suitable method for MIC determination since they offer the possibility to estimate the concentration of the tested antimicrobial agent in the agar (agar dilution) or broth medium (macro dilution or micro dilution). The MIC value represents the minimum concentration of the antimicrobial agent required to inhibit the growth of the microorganism being tested and is typically expressed in units of µg/mL or mg/L. To ensure consistent and reliable results, various approved guidelines exist for performing dilution antimicrobial susceptibility testing on both fastidious and non-fastidious bacteria, as well as yeast and filamentous fungi. Among these guidelines, those provided by the Clinical and Laboratory Standards Institute (CLSI) and the European Committee on Antimicrobial Susceptibility Testing (EUCAST) are widely recognized as the gold standard (Reller et al. 2009).

Broth Dilution Method

This procedure involves the preparation of twofold dilutions of the antimicrobial agent in a liquid growth medium dispensed in tubes containing a minimum volume of 2 mL in case of macro dilution or with smaller volumes using 96-well micro titration plate for micro dilution. Each tube is then inoculated with a microbial inoculum prepared in the same medium and incubated under appropriate conditions depending upon the test microorganisms. The MIC is defined as the lowest concentration of the antimicrobial agent that can completely inhibit the growth of the microorganism being tested, as determined by visual observation of tubes or microdilution wells. Moreover, several colorimetric methods based on the use of dye reagents have been developed for MIC endpoint determination (Liang et al. 2012).

Agar Dilution Method

To determine the MIC of antimicrobial substances, the nonselective medium is prepared with various concentrations of the substances prior to solidification. The standardized bacterial inoculum is then applied onto the surface of the agar plates and incubated overnight. The presence or absence of bacterial growth at the inoculated sites is visually evaluated, with the lowest concentration of antimicrobial substances that inhibit bacterial growth considered as the MIC. This method allows for the simultaneous testing of multiple bacterial strains (Mirajkar and Gebhart 2016) and is also applicable for both antibacterial and antifungal susceptibility testing. For efficient and standardized testing of bacterial inoculum, commercially available inoculum replicators can transfer up to 32–60 different bacterial strains onto a single agar plate. The agar dilution method is commonly recommended for testing fastidious organisms.

6.5.1.4 Time-Kill Assay

A time-kill test is a strong tool for determining the bactericidal or fungicidal effect. It reveals a time-dependent or a concentration-dependent antimicrobial effect (Pfaller et al. 2004). Bacterial suspension of 5×10^5 CFU/mL is prepared in broth culture medium using three tubes: the first and second tubes containing the molecule or extract being tested, usually at concentrations of $0.25 \times$ MIC and $1 \times$ MIC, while the third tube serves as the growth control. The tubes are incubated under suitable conditions for varying time intervals (0, 4, 6, 8, 10, 12, and 24 h). The number of living cells (CFU/mL) in each tube is determined using the agar plate count method, and the percentage of dead cells is calculated relative to the growth control. A bactericidal effect is typically achieved with a lethality percentage of 90% within 6 h, equivalent to 99.9% lethality within 24 h (White et al. 1996; Pfaller et al. 2004; Konaté et al. 2012). This method can also be used to determine synergism or antagonism between drugs in combination and has been applied in studying several antifungal substances (Klepser et al. 1998).

6.6 AMR Testing by Molecular-Based Methods

6.6.1 High-Throughput Real-Time PCR (HT-qPCR)

High-throughput real-time PCR (HT-qPCR) is a more recent technology that offers a quick and more efficient way to conduct several experiments on a single sample while simultaneously examining numerous ARGs. In order to identify and characterize patterns of anthropogenic influence, as well as to evaluate the removal of ARGs during wastewater treatment or other water treatments, to inform modeling, such as environmental fate and human health risk assessment, qPCR can provide very sensitive detection and quantification over several logs (Liguori et al. 2022). The amount of DNA from the aquatic sample is much lower when compared to other environmental matrices. HT-qPCR can operate at a nanoliter scale and concurrently examine a variety of AMR gene targets (Sander and Kalff 1993). The high throughput qPCR contains hundreds of ARG targets, hence no longer needed to select the correct ARG gene target (Jacobs and Chenia 2007). Additional markers, including MGE and pathogen-specific genes, can be included. However, the lack of widespread availability of HT-qPCR equipment, the lack of established quality assurance protocols, and the requirement for the addition of a pre-amplification step to achieve sensitive detection limits via specific target amplification are all disadvantages for environmental monitoring (Ishii et al. 2013). Additionally, guidance would still be required for determining which occurrence patterns of ARGs are of concern from high throughput qPCR data. Digital droplet PCR (ddPCR), hailed as a more sensitive and less inhibited variant of qPCR, is gaining popularity in the field (Liguori et al. 2022).

HT-qPCR has certain drawbacks, however, such as the difficulty of simultaneously optimizing the assay for each gene of interest because each test must use the same PCR conditions. The specificity of HT-qPCR can be improved by using probes rather than SYBR green or other fluorescent detection tests. Another disadvantage is the limited sample size, which may result in a higher limit of quantification and detection, which is troublesome for low amounts of environmental contaminants around ARGs. Additionally, it is difficult or even impossible to recover nanoliter scale reaction to sequence the amplified products, although this is commonly possible with traditional PCR or qPCR. The requirement to know the sequences of target genes in advance for proper selection or construction of primers and probes is yet another significant drawback of this approach. However, HT-qPCR has certain advantages over metagenomics sequencing due to its lower detection limit, which has been found to be 10–4 ARGs/16S rRNA (Muziasari et al. 2016, 2017).

6.6.2 Whole-Genome Sequencing (WGS)

Recent studies have investigated whether whole-genome sequencing (WGS) may be used to provide better knowledge of the total resistance existing within a bacterial strain (Hendriksen et al. 2019). When employing solely phenotypic AST, WGS can theoretically detect all the AMR-associated genes inside a single bacterial isolate without any constraints on the class of genes or antibiotics, which is typically a limitation. WGS can predict phenotypic and multidrug resistance within a single isolate with greater precision than phenotypic analysis (McDermott et al. 2016; Ellington et al. 2017). AMR reference gene database can also be updated and made more complete by adding undiscovered genes related to AMR. The common problems and drawbacks of cultivating environmental bacteria to find AMR are minimized using WGS.

WGS gives a better knowledge of the transmission of ARB and ARGs throughout the environment and into human and animal populations; it is hoped that it will improve the surveillance of AMR in surface waters and aid risk assessment operations. AMR genes can be found on chromosomes or MGEs in the bacterial genome and transmits these genes through either vertical or HGT method. The mode of transmission and potential health hazards associated with a gene can be better comprehended by knowing where the gene is located in the cell, but sequencing them is challenging (Madec and Haenni 2018). WGS can be used to subtype bacteria utilizing in silico techniques, which offer lower costs, higher resolution, increased robustness, and quicker results than conventional subtyping methods. It can distinguish between two bacterial isolates of the same species and infer or disprove their epidemiological association (Tassios and Moran-Gilad 2018; Schürch et al. 2018).

Recent studies have started using WGS in addition to phenotypic and other genotypic methods to analyze AMR in surface water (Gomi et al. 2017). Although the application of WGS for analyzing AMR in surface waters is still limited, its widespread adoption could advance resistome surveillance, where AMR could be predicted only from genetic data (McDermott et al. 2016). According to the studies, WGS has a true positive rate of >87% and a high analytical sensitivity for distinguishing AMR determinants in bacterial isolates with an AMR phenotype. In order to reliably identify the absence of AMR determinants within phenotypically established AMR susceptible isolated, the analytical specificity or true negative rate of WGS has been reported to be even greater at >98% (Gordon et al. 2014; McDermott et al. 2016; Shelburne et al. 2017).

6.6.3 Metagenomics

Compared to PCR, metagenomics is a molecular technique that entails sequencing every microbial genome in an environmental sample. It provides detailed genomic data and a higher level of taxonomic resolution. When analyzing AMR in the environment, metagenomic techniques are a rich source of data that enables the identification of all aspects of environmental resistome and associated impacts in surface waters. It enables the identification of new ARGs, it is regarded as an open format strategy that is appropriate for exploratory discovery investigations (Zhou et al. 2015). The amount of DNA extracted from environmental samples must be sufficient to produce a library that gives enough coverage to completely characterize the micro biome in order to do metagenomic research on environmental samples. A larger sample size may be required to carry out a metagenomic study due to the low population of bacteria per sample volume in surface waters.

Direct genomic DNA extraction, fragmentation, and application of NGS are used for metagenomic sequencing to produce millions of reads representing the bacterial community in ambient samples (Kristiansson et al. 2011). Initially, metagenomics was utilized to investigate river sediments of severely influenced surface waters exposed to pharmaceutical wastewater discharges (particularly antibiotics). Researchers discovered large concentrations of integrons, transposons, plasmids, and ARGs (sul2, aph, tet, qnr, and erm), which are risk factors for HGT as a mechanism of AMR propagation in water. Metagenomics is now widely applied for examining shifts in the resistome (i.e., the collection of all ARGs across an environmental sample) through WWTPs and identifying MGEs in order to estimate the extent of HGT events (Yang et al. 2014; Ju et al. 2019; Majeed et al. 2021). It has been demonstrated that variations in sampling techniques (composite versus grab sampling), sample storage conditions, DNA extraction procedures, and sequencing depths and coverages skew the creation and comparison of metagenomic libraries (Knudsen et al. 2016; Gweon et al. 2019).

6.6.3.1 Smart Chip Real-Time PCR

For studying environmental samples, the Smart chip is more frequently used than HT-qPCR because it allows for a customized number of samples versus assay configurations, which may be chosen depending on the design of each experiment (Lamas et al. 2016).

6.6.3.2 Bio Mark Dynamic Array PCR with Integrated Fluidic Circuit

Fluidigm BioMark HD and Integrated Fluidic Circuit (IFC) have been used widely to investigate ARGs in environmental samples. This 96-well platform with high throughput capability for microfluidic qPCR and digital PCR has three PCR sample-assay combinations. By combining samples and PCR assays into individual reaction chambers utilizing a number of channels that are controlled by pressurization of Nanoflex TM valves, the IFC microfluidic device makes HT-qPCR possible. This IFC microfluidic system's precise response volume delivery achieves high reproducibility rates, which removes the variability generally associated with dispensing liquids (Sandberg et al. 2018).

6.6.3.3 Applied Biosystem Open Array

Another HT-qPCR device with a respectable capacity and a 32 high-throughput configuration with 3072 reaction wells is the Open Array from Applied Biosystems. While requiring less sample volume and laboratory reagents than other high throughput systems like Smart chip, it may have better analytical sensitivity than the Fluidigm system due to its 33 nL reaction volume (Shahraki et al. 2019).

6.6.3.4 LAMP Assay

Loop-mediated isothermal amplification (LAMP) is a nucleic acid amplification technique that was first introduced in 2000 by Notomi and colleagues (2000). It has since gained popularity in various nucleic acid research applications and clinical settings as a screening tool due to its rapid, sensitive, and efficient nature. One of the unique features of LAMP is that it can be performed in a single tube and completed within an hour (Nagamine et al. 2002). This is because LAMP uses a strand-displacing DNA polymerase and a set of four to six primers that recognize six to eight distinct regions on the target DNA. This results in a highly efficient amplification process that can produce a large amount of amplification products quickly and accurately. LAMP has been shown to be highly specific and sensitive, with a low risk of false positives or negatives. Numerous studies have demonstrated its robustness as a diagnostic tool, showing outstanding results compared to other preexisting molecular techniques (Dhama et al. 2014). Additionally, LAMP is less prone to interference from inhibitors and can be performed using simple equipment, making it an ideal source for diagnosis in resource-limited settings.

The reaction mixture for LAMP assay includes primers, dNTPs, LAMP buffer solution, Betaine (optional), Fluorescent DNA dye (optional), and template DNA, which are mixed and incubated under an isothermal temperature of 60–65 °C. Using four primers for amplification takes about an hour, while using six primers, including two loop primers, can shorten the amplification time to about 30 min. The assay result can be detected immediately by naked-eye visualization of the turbidity resulting from a positive reaction (Mori et al. 2001). Adding DNA dyes such as SYBR green, HNB, Picogreen, ethidium bromide, or calcein metal ion indicator can enhance the visibility of the turbidity (Thekisoe et al. 2007).

The design of LAMP primers can be a limiting factor due to its complicated nature. Unlike traditional PCR primers that consist of two complementary sequences, LAMP primers require the design of four to six primers that recognize six to eight distinct regions of the target DNA sequence. However, this is now simplified with the introduction of LAMP primer design software, by the availability of lyophilized LAMP kits (Dhama et al. 2014).

6.6.3.5 TaqMan Array Card

A TaqMan Array Card (TAC) is a microfluidic card that detects multiple nucleic acid targets in a single reaction. It is a 384-well single plex real-time PCR format and a powerful tool for high-throughput gene expression analysis and pathogen detection (Liu et al. 2013). The TAC is a small plastic card containing a series of reaction chambers, each containing a set of primers and probes specific to a particular target sequence. The probes are labeled with fluorescent dyes, which emit a signal when they bind to their target sequence. The TAC can be used with a variety of real-time PCR instruments, which detect and quantify the fluorescent signal in each reaction chamber. Simpler assay design, better quantification and sensitivity, and shorter turnaround time are advantages over conventional methods, but repeats are expensive (Liu et al. 2013).

6.7 AMR Testing by Analytical Methods

6.7.1 Mass Spectrometry

Several matrix-assisted laser desorption ionization-time of flight mass spectrometry (MALDI-TOF MS)-based methods have been applied for the rapid detection of antibiotic resistance. Today, one of the most frequently suggested and in-demand applications in the field of antimicrobial resistance detection are the direct detection of resistance markers in mass spectra (Sparbier et al. 2012). The main uses that have been successfully tested using MALDI-TOF MS are (1) the detection and expression of antibiotic resistance mechanisms, (2) specific mass peak profiles within spectra, (3) the detection of stable isotope-labeled biomarkers, and (4) the estimation of the effect of antibiotics on microorganism growth. Among these, the detection of antibiotic resistance is the most explored as the degradation of antibiotics produces intracellular metabolites, generating specific peaks on spectra (Sparbier et al. 2016).

6.7.2 Raman Spectroscopy

It is a rapid and cost-effective method for the detection of antibiotics in wastewater. Although there are numerous instrument-based techniques, such as capillary electrophoresis, high-performance liquid chromatography (HPLC), liquid chromatography–mass spectrometry (LC-MS), and liquid chromatography–tandem mass spectrometry (LC-MS/MS), their acceptance is lower than that of other techniques because of the high cost of instrumentation and the intricate, time-consuming sample preparation procedures. Nowadays various analytical methods such as electrochemical, colorimetric, and fluorescence sensors are getting more attention in antibiotic monitoring of the environment. Some of the advantages of this are the short response time, ease of use, and portability. The level of sensitivity and accuracy is very good. In this context, vibrational spectroscopy technique like surface-enhanced Raman scattering (SERS) is of more importance. Its working principle is purely based on the vibrational spectra of molecular bonds which provide information about the molecular structure. This process is too sensitive and accurate and uses a label-free online sensor for the identification of antibiotics. Wastewater samples collected were used as the background sample matrix to examine the SERS signals intensification and interferences from organic and inorganic molecules in sewage. The wastewater was treated by primary sedimentation, activated sludge process. Self-assembled SERS substrates were fabricated in microfluidic channels with capacitor architecture. Wastewater samples were spiked with quinoline in the same concentration range to determine the sensitivity and interference of chemical species in wastewater matrices for SERS (Huang et al. 2023). Raman measurements and UPLC−MS/MS analysis include preprocessing, analysis, and visualization using Python. Slight deviations can be compared and analyzed through the preprocessing. In situ molecular identification is rapid and precise in Raman spectroscopy. It also aids in the identification of molecules at the single-molecule scale. A major limitation in this method is the signal to ratio which causes the result to fluctuate. There are so many advantages and disadvantages for methods used to detect AMR (Table 6.2).

6.8 Challenges in Screening and Monitoring of AMR

6.8.1 Heterogeneous AMR Sources

Recreation facilities and aquatic environments host diverse sources of AMR, including human visitors, animals, and environmental reservoirs. Each source contributes to the complex pool of ARBs and genetic elements. Identifying and characterizing the specific sources of AMR within these environments poses a significant challenge, as they can vary in composition and dynamics over time.

6.8.2 Complex Microbial Interactions

Recreation facilities and aquatic environments harbor diverse microbial communities that interact in intricate ways. These ecosystems include a wide range of bacteria, viruses, fungi, and parasites, which can exchange genetic material and promote the spread of resistance genes. Identifying and characterizing specific resistance mechanisms and their transferability among microorganisms is a complex task, requiring advanced molecular techniques and comprehensive surveillance strategies.

Table 6.2 Advantages and disadvantages of various screening methods of AMR

Methods	Advantages	Disadvantages
Disk diffusion	Simplicity Versatility Cost effective Standardized method	Slow turnaround time Results can be subjective Unable to determine MIC
Antimicrobial gradient test method	High accuracy Objectivity Flexibility Wide range of antimicrobial agents	Cost Technical expertise Time consuming
Broth dilution	High accuracy in MIC determination Standardized method Versatility Quantitative measure	Time consuming Technical expertise
Agar dilution	Applicable in testing large number of isolates Objective results	Limited concentration of antimicrobial agents Time consuming
Time-kill assay	Comprehensive results Reproducibility Flexibility	Lack of standardization Time consuming
HT qPCR	Limited amount of DNA sample Consistent	Inability to optimize the assay for each gene Higher limit of detection and limit of quantification
Smart chip RT-PCR	More frequently used than HT-qPCR for profiling ARGs	High cost Limited flexibility and complex workflow
BioMark dynamic array PCR with integrated fluidic circuit	High-throughput analysis Precise and accurate quantification	High cost in initial set-up Chip compatibility and availability
Applied biosystem open Array	Genome-wide coverage Data quality and reproducibility	Limited dynamic range Cross-hybridization and background noise
Metagenomics	Used for environmental sample analysis Identify new resistant gene	PCR bias can affect the accuracy High cost, laborious, and poor repeatability
Whole-genome sequencing	Used to trace AMR bacteria with specific allelic profiles	Need accurate reference database
LAMP assay	Highly specific and sensitive Robustness	Primer design Complexity in data interpretation Risk of contamination

(continued)

Table 6.2 (continued)

Methods	Advantages	Disadvantages
TaqMan Array card	High throughput Multiplexing capability Better quantification Automation	High cost Limited flexibility and sensitivity
Mass spectrometry	High sensitivity and specificity Rapid turnaround time	High initial cost Complexity in data interpretation Limitations in detecting unknown resistance mechanisms

6.8.3 Detection of Emerging Resistance Mechanisms

Screening and monitoring programs should be equipped to detect not only known AMR genes but also emerging resistance mechanisms. However, the rapid evolution of resistance genes and the emergence of novel resistance mechanisms pose a significant challenge in keeping screening techniques current (Tiwari et al. 2022). Continuous research and development efforts are necessary to stay ahead of the evolving landscape of AMR.

6.8.4 Detection Methods and Targets

Screening and monitoring AMR in recreation facilities and aquatic environments require reliable and sensitive detection methods. Traditional culture-based methods for identifying resistant bacteria are time-consuming and may underestimate the true prevalence of AMR. Additionally, the range of antimicrobial resistance genes and mechanisms complicates the identification of relevant targets for screening. Developing rapid and accurate detection techniques that cover a wide spectrum of AMR mechanisms is crucial. Simplification and standardization of screening protocols are necessary to enable broader implementation. Developing harmonized guidelines and protocols would enhance the comparability of results, enabling a more comprehensive understanding of AMR dynamics in these environments (Pinto Ferreira et al. 2022).

6.8.5 Sampling Strategies

Designing effective sampling strategies is essential to capture the diversity and dynamics of AMR in recreation facilities and aquatic environments. These settings are often characterized by large volumes of water, making obtaining representative samples challenging. Sampling at different spatial and temporal scales is necessary to account for variations in microbial communities, AMR profiles, and potential hotspots. Standardizing sampling protocols and optimizing sample processing and storage methods are crucial for reliable and comparable results.

6.8.6 Detection of Non-traditional Resistance Mechanisms

Traditional susceptibility testing methods primarily focus on detecting resistance to common antibiotics. However, the emergence of non-traditional resistance mechanisms, such as efflux pumps and biofilm formation, complicates the screening and monitoring process. Incorporating advanced molecular techniques, such as whole-genome sequencing and metagenomics, can help identify and understand these non-traditional resistance mechanisms in recreation facilities and aquatic environments.

6.8.7 Data Integration and Analysis

Screening and monitoring AMR generate vast amounts of complex data that require integration and analysis to derive meaningful insights. Combining information from various sources, such as microbial community composition, antimicrobial usage patterns, and environmental factors, is essential to understanding the drivers of AMR in recreation facilities and aquatic environments. Effective data integration and analysis require interdisciplinary collaboration among experts from various fields, including microbiology, epidemiology, and environmental science. This collaboration can be challenging to establish and maintain due to differences in expertise and data management approaches. Developing robust bioinformatics tools and analytical frameworks is crucial to extract valuable information from multi-omics data and facilitate data interpretation (Iskandar et al. 2021).

6.8.8 Implementation of Interventions

Identifying AMR in recreation facilities and aquatic environments is only the first step. Translating the surveillance data into effective interventions poses another challenge. Implementing strategies to mitigate AMR spread, such as improved hygiene practices, proper waste management, and judicious use of antimicrobials, requires collaboration among multiple stakeholders, including facility managers, public health authorities, and policymakers. Overcoming logistical, economic, and behavioral barriers to intervention implementation is crucial for effective AMR control.

6.8.9 Public Awareness and Education

Creating public awareness and promoting responsible behavior are vital for reducing the spread of AMR. However, disseminating information and educating the public about AMR risks and preventive measures in recreation facilities and aquatic environments can be challenging. Effective communication strategies tailored to different target audiences are essential to bridge the gap between scientific knowledge and public understanding.

6.9 Conclusion

AMR poses a serious risk to both human and environmental health. Therefore, it is critical to put in place effective techniques for detecting and containing it. Before wastewater is discharged back into the environment, wastewater facilities are essential for handling and purifying it. Due to the presence of numerous antimicrobial agents and residues in the wastewater, these facilities might act as hubs for the development and spread of AMR. The incidence and dynamics of resistance bacteria, genes, and the processes underlying them can be useful insights gained through routinely screening and monitoring AMR in wastewater. The spread of AMR can be stopped, and the hazards to human and environmental health can be reduced by determining the origins and modes of transmission. Similar to recreational centers, aquatic landscapes, including lakes, rivers, and coastal regions, may act as AMR reservoirs. These areas are frequently subjected to a variety of pollution sources, including agricultural runoff, industrial discharges, and human and animal waste, all of which can aid in the development of AMR. Understanding the possible dangers connected with recreational activities and the overall influence on ecosystem health depends on monitoring the prevalence and abundance of resistant bacteria and genes in these habitats. Policymakers, researchers, and environmental authorities can acquire valuable information on AMR's prevalence, distribution, and transmission patterns by developing screening and monitoring programs. This knowledge may be used to construct evidence-based AMR surveillance, risk assessment, and management methods. The ability to deploy efficient infection control measures and treatment regimens is also made possible by the early diagnosis of developing resistance strains. Additionally, information gleaned through screening and monitoring activities can aid in creating laws and policies that will limit the discharge of antimicrobial chemicals into the environment. The spread of AMR can be controlled, and its effects can be reduced by implementing targeted therapies after determining its origins and routes. Determining the scope of the issue, spotting potential hazards, and implementing efficient countermeasures to the rising threat of antimicrobial resistance depend on screening and monitoring AMR in wastewater facilities, recreation centers, and other aquatic environments. These initiatives are essential for ensuring future generations may utilize antimicrobial compounds in a sustainable manner while also protecting the environment and public health.

References

Amarasiri M, Sano D, Suzuki S (2020) Understanding human health risks caused by antibiotic resistant bacteria (ARB) and antibiotic resistance genes (ARG) in water environments: current knowledge and questions to be answered. Crit Rev Environ Sci Technol 50(19):2016–2059

Baker-Austin C, Wright MS, Stepanauskas R, McArthur JV (2006) Co-selection of antibiotic and metal resistance. Trends Microbiol 14(4):176–182

Balouiri M, Sadiki M, Ibnsouda SK (2016) Methods for in vitro evaluating antimicrobial activity: a review. J Pharm Anal 6(2):71–79

Baquero F, Martínez JL, Cantón R (2008) Antibiotics and antibiotic resistance in water environments. Curr Opin Biotechnol 19(3):260–265

Bauer AW (1966) Antibiotic susceptibility testing by a standardized single diffusion method. Am J Clin Pathol 45:493–496

Brown KD, Kulis J, Thomson B, Chapman TH, Mawhinney DB (2006) Occurrence of antibiotics in hospital, residential, and dairy effluent, municipal wastewater, and the Rio Grande in New Mexico. Sci Total Environ 366(2–3):772–783

Dhama K, Karthik K, Chakraborty S, Tiwari R, Kapoor S, Kumar A, Thomas P (2014) Loop-mediated isothermal amplification of DNA (LAMP): a new diagnostic tool lights the world of diagnosis of animal and human pathogens: a review. Pak J Biol Sci 17(2):151–166

Duong HA, Pham NH, Nguyen HT, Hoang TT, Pham HV, Pham VC et al (2008) Occurrence, fate and antibiotic resistance of fluoroquinolone antibacterials in hospital wastewaters in Hanoi, Vietnam. Chemosphere 72(6):968–973

Ellington MJ, Ekelund O, Aarestrup FM, Canton R, Doumith M, Giske C et al (2017) The role of whole genome sequencing in antimicrobial susceptibility testing of bacteria: report from the EUCAST Subcommittee. Clin Microbiol Infect 23(1):2–22

Freeland G, Hettiarachchy N, Atungulu GG, Apple J, Mukherjee S (2021) Strategies to combat antimicrobial resistance from farm to table. Food Rev Intl 39:1–14

Gajic I, Kabic J, Kekic D, Jovicevic M, Milenkovic M, Mitic Culafic D et al (2022) Antimicrobial susceptibility testing: a comprehensive review of currently used methods. Antibiotics 11(4):427

Garner E, Chen C, Xia K, Bowers J, Engelthaler DM, McLain J et al (2018) Metagenomic characterization of antibiotic resistance genes in full-scale reclaimed water distribution systems and corresponding potable systems. Environ Sci Technol 52(11):6113–6125

Gomi R, Matsuda T, Matsumura Y, Yamamoto M, Tanaka M, Ichiyama S, Yoneda M (2017) Whole-genome analysis of antimicrobial-resistant and extraintestinal pathogenic Escherichia coli in river water. Appl Environ Microbiol 83(5):e02703–e02716

Gopal G, John N (2020) A comparative study on the antibiotic resistance pattern of soil bacterial communities in flood affected and unaffected areas of Kothamangalam municipality. Mater Today Proc 25:252–256

Gordon NC, Price JR, Cole K, Everitt R, Morgan M, Finney J et al (2014) Prediction of Staphylococcus aureus antimicrobial resistance by whole-genome sequencing. J Clin Microbiol 52(4):1182–1191

Guardabassi L, Wong DMLF, Dalsgaard A (2002) The effects of tertiary wastewater treatment on the prevalence of antimicrobial resistant bacteria. Water Res 36(8):1955–1964

Guo J, Li J, Chen H, Bond PL, Yuan Z (2017) Metagenomic analysis reveals wastewater treatment plants as hotspots of antibiotic resistance genes and mobile genetic elements. Water Res 123:468–478

Gweon HS, Shaw LP, Swann J, De Maio N, AbuOun M, Niehus R et al (2019) The impact of sequencing depth on the inferred taxonomic composition and AMR gene content of metagenomic samples. Environ Microbiome 14(1):1–15

Hao H, Cheng G, Iqbal Z, Ai X, Hussain HI, Huang L et al (2014) Benefits and risks of antimicrobial use in food-producing animals. Front Microbiol 5:288

Hartmann A, Alder AC, Koller T, Widmer RM (1998) Identification of fluoroquinolone antibiotics as the main source of umuC genotoxicity in native hospital wastewater. Environ Toxicol Chem 17(3):377–382

Hausdorfer J, Sompek E, Allerberger F, Dierich MP, Rüsch-Gerdes S (1998) E-test for susceptibility testing of Mycobacterium tuberculosis. Int J Tuberc Lung Dis 2(9):751–755

Hendriksen RS, Bortolaia V, Tate H, Tyson GH, Aarestrup FM, McDermott PF (2019) Using genomics to track global antimicrobial resistance. Front Public Health 7:242

Hernando-Amado S, Coque TM, Baquero F, Martínez JL (2019) Defining and combating antibiotic resistance from One Health and Global Health perspectives. Nat Microbiol 4(9):1432–1442

Hiller CX, Hübner U, Fajnorova S, Schwartz T, Drewes JE (2019) Antibiotic microbial resistance (AMR) removal efficiencies by conventional and advanced wastewater treatment processes: a review. Sci Total Environ 685:596–608. https://doi.org/10.1016/j.scitotenv.2019.05.315

Huang YH, Wei H, Santiago PJ, Thrift WJ, Ragan R, Jiang S (2023) Sensing antibiotics in wastewater using surface-enhanced Raman scattering. Environ Sci Technol 57(12):4880–4891

Idelevich EA, Hoy M, Görlich D, Knaack D, Grünastel B, Peters G et al (2017) Rapid phenotypic detection of microbial resistance in gram-positive bacteria by a real-time laser scattering method. Front Microbiol 8:1064

Ishii S, Segawa T, Okabe S (2013) Simultaneous quantification of multiple food- and waterborne pathogens by use of microfluidic quantitative PCR. Appl Environ Microbiol 79(9):2891–2898

Iskandar K, Molinier L, Hallit S, Sartelli M, Hardcastle TC, Haque M et al (2021) Surveillance of antimicrobial resistance in low- and middle-income countries: a scattered picture. Antimicrob Resist Infect Control 10(1):1–19

Jacobs L, Chenia HY (2007) Characterization of integrons and tetracycline resistance determinants in Aeromonas spp. isolated from South African aquaculture systems. Int J Food Microbiol 114(3):295–306

Nayomi J, Thangavel M (2017) Stenotrophomonas maltophilia: a novel plant growth promoter and biocontrol agent from marine environment. Int J Adv Res 5(4):207–214. https://doi.org/10.21474/IJAR01; https://doi.org/10.21474/IJAR01/3797

Ju F, Beck K, Yin X, Maccagnan A, McArdell CS, Singer HP et al (2019) Wastewater treatment plant resistomes are shaped by bacterial composition, genetic exchange, and upregulated expression in the effluent microbiomes. ISME J 13(2):346–360

Karkman A, Do TT, Walsh F, Virta MP (2018) Antibiotic-resistance genes in waste water. Trends Microbiol 26(3):220–228

Klepser ME, Ernst EJ, Lewis RE, Ernst ME, Pfaller MA (1998) Influence of test conditions on antifungal time-kill curve results: proposal for standardized methods. Antimicrob Agents Chemother 42(5):1207–1212

Knudsen BE, Bergmark L, Munk P, Lukjancenko O, Priemé A, Aarestrup FM, Pamp SJ (2016) Impact of sample type and DNA isolation procedure on genomic inference of microbiome composition. mSystems 1(5):e00095–e00016

Konaté K, Mavoungou JF, Lepengué AN, Aworet-Samseny RR, Hilou A, Souza A et al (2012) Antibacterial activity against β-lactamase producing Methicillin and Ampicillin-resistants Staphylococcus aureus: fractional Inhibitory Concentration Index (FICI) determination. Ann Clin Microbiol Antimicrob 11:1–12

Kristiansson E, Fick J, Janzon A, Grabic R, Rutgersson C, Weijdegård B et al (2011) Pyrosequencing of antibiotic-contaminated river sediments reveals high levels of resistance and gene transfer elements. PLoS One 6(2):e17038

Kumar A, Pal D (2018) Antibiotic resistance and wastewater: correlation, impact and critical human health challenges. J Environ Chem Eng 6(1):52–58

Lamas A, Franco CM, Regal P, Miranda JM, Vázquez B, Cepeda A (2016) High-throughput platforms in real-time PCR and applications. In: Polymerase chain reaction for biomedical applications, vol 15. IntechOpen

Larsson DJ, Flach CF (2022) Antibiotic resistance in the environment. Nat Rev Microbiol 20(5):257–269

Leonard AF, Zhang L, Balfour AJ, Garside R, Gaze WH (2015) Human recreational exposure to antibiotic resistant bacteria in coastal bathing waters. Environ Int 82:92–100

Liang H, Xing Y, Chen J, Zhang D, Guo S, Wang C (2012) Antimicrobial activities of endophytic fungi isolated from Ophiopogon japonicus (Liliaceae). BMC Complement Altern Med 12:1–6

Liguori K, Keenum I, Davis BC, Calarco J, Milligan E, Harwood VJ, Pruden A (2022) Antimicrobial resistance monitoring of water environments: a framework for standardized methods and quality control. Environ Sci Technol 56(13):9149–9160

Lindberg R, Jarnheimer PÅ, Olsen B, Johansson M, Tysklind M (2004) Determination of antibiotic substances in hospital sewage water using solid phase extraction and liquid chromatography/mass spectrometry and group analogue internal standards. Chemosphere 57(10):1479–1488

Liu J, Gratz J, Amour C, Kibiki G, Becker S, Janaki L et al (2013) A laboratory-developed TaqMan Array Card for simultaneous detection of 19 enteropathogens. J Clin Microbiol 51(2):472–480

Madec JY, Haenni M (2018) Antimicrobial resistance plasmid reservoir in food and food-producing animals. Plasmid 99:72–81

Majeed HJ, Riquelme MV, Davis BC, Gupta S, Angeles L, Aga DS et al (2021) Evaluation of metagenomic-enabled antibiotic resistance surveillance at a conventional wastewater treatment plant. Front Microbiol 12:657954

Martins AF, Vasconcelos TG, Henriques DM, Frank CDS, König A, Kümmerer K (2008) Concentration of ciprofloxacin in Brazilian hospital effluent and preliminary risk assessment: a case study. CLEAN Soil Air Water 36(3):264–269

McDermott PF, Tyson GH, Kabera C, Chen Y, Li C, Folster JP et al (2016) Whole-genome sequencing for detecting antimicrobial resistance in nontyphoidal Salmonella. Antimicrob Agents Chemother 60(9):5515–5520

Mirajkar NS, Gebhart CJ (2016) Comparison of agar dilution and antibiotic gradient strip test with broth microdilution for susceptibility testing of swine Brachyspira species. J Vet Diagn Invest 28(2):133–143

Mori Y, Nagamine K, Tomita N, Notomi T (2001) Detection of loop-mediated isothermal amplification reaction by turbidity derived from magnesium pyrophosphate formation. Biochem Biophys Res Commun 289(1):150–154

Muziasari WI, Pärnänen K, Johnson TA, Lyra C, Karkman A, Stedtfeld RD et al (2016) Aquaculture changes the profile of antibiotic resistance and mobile genetic element associated genes in Baltic Sea sediments. FEMS Microbiol Ecol 92(4):fiw052

Muziasari WI, Pitkänen LK, Sørum H, Stedtfeld RD, Tiedje JM, Virta M (2017) The resistome of farmed fish feces contributes to the enrichment of antibiotic resistance genes in sediments below Baltic Sea fish farms. Front Microbiol 7:2137

Nagamine K, Hase T, Notomi TJMCP (2002) Accelerated reaction by loop-mediated isothermal amplification using loop primers. Mol Cell Probes 16(3):223–229

Nguyen AQ, Vu HP, Nguyen LN, Wang Q, Djordjevic SP, Donner E et al (2021) Monitoring antibiotic resistance genes in wastewater treatment: current strategies and future challenges. Sci Total Environ 783:146964

Nicholas J, Alejandro A, Thomas A, Peter B, Kristian KB, Peter B, Anja C, Rita C, William HF, Thomas G, John RH, Joakim LDG, Scott AL, James JM, Jens R, Peter S, Jason RS, Christel S, den Eede Edward V, Topp (2013) Human health risk assessment (HHRA) for environmental development and transfer of antibiotic resistance environmental health perspectives 121(9):993–1001. https://doi.org/10.1289/ehp.1206316

Notomi T, Okayama H, Masubuchi H, Yonekawa T, Watanabe K, Amino N, Hase T (2000) Loop-mediated isothermal amplification of DNA. Nucleic Acids Res 28(12):e63–e63

Novo A, Manaia CM (2010) Factors influencing antibiotic resistance burden in municipal wastewater treatment plants. Appl Microbiol Biotechnol 87:1157–1166

Östman M, Lindberg RH, Fick J, Björn E, Tysklind M (2017) Screening of biocides, metals and antibiotics in Swedish sewage sludge and wastewater. Water Res 115:318–328

Pfaller MA, Sheehan DJ, Rex JH (2004) Determination of fungicidal activities against yeasts and molds: lessons learned from bactericidal testing and the need for standardization. Clin Microbiol Rev 17(2):268–280

Pinto Ferreira J, Battaglia D, Dorado García A, Tempelman K, Bullon C, Motriuc N et al (2022) Achieving antimicrobial stewardship on the global scale: challenges and opportunities. Microorganisms 10(8):1599

Reller LB, Weinstein M, Jorgensen JH, Ferraro MJ (2009) Antimicrobial susceptibility testing: a review of general principles and contemporary practices. Clin Infect Dis 49(11):1749–1755

Rizzo L, Manaia C, Merlin C, Schwartz T, Dagot C, Ploy MC et al (2013) Urban wastewater treatment plants as hotspots for antibiotic resistant bacteria and genes spread into the environment: a review. Sci Total Environ 447:345–360

Sandberg KD, Ishii S, LaPara TM (2018) A microfluidic quantitative polymerase chain reaction method for the simultaneous analysis of dozens of antibiotic resistance and heavy metal resistance genes. Environ Sci Technol Lett 5(1):20–25

Sander BC, Kalff J (1993) Factors controlling bacterial production in marine and freshwater sediments. Microb Ecol 26:79–99

Schürch AC, Arredondo-Alonso S, Willems RJL, Goering RV (2018) Whole genome sequencing options for bacterial strain typing and epidemiologic analysis based on single nucleotide polymorphism versus gene-by-gene–based approaches. Clin Microbiol Infect 24(4):350–354

Shahraki AH, Heath D, Chaganti SR (2019) Recreational water monitoring: nanofluidic qRT-PCR chip for assessing beach water safety. Environ DNA 1(4):305–315

Shelburne SA, Kim J, Munita JM, Sahasrabhojane P, Shields RK, Press EG et al (2017) Whole-genome sequencing accurately identifies resistance to extended-spectrum β-lactams for major gram-negative bacterial pathogens. Clin Infect Dis 65(5):738–745

Sparbier K, Schubert S, Weller U, Boogen C, Kostrzewa M (2012) Matrix-assisted laser desorption ionization–time of flight mass spectrometry-based functional assay for rapid detection of resistance against β-lactam antibiotics. J Clin Microbiol 50(3):927–937

Sparbier K, Schubert S, Kostrzewa M (2016) MBT-ASTRA: a suitable tool for fast antibiotic susceptibility testing? Methods 104:48–54

Taneja N, Sharma M (2019) Antimicrobial resistance in the environment: the Indian scenario. Indian J Med Res 149(2):119–128

Tassios PT, Moran-Gilad J (2018) Bacterial next generation sequencing (NGS) made easy. Clin Microbiol Infect 24(4):332–334

Thekisoe OM, Kuboki N, Nambota A, Fujisaki K, Sugimoto C, Igarashi I et al (2007) Species-specific loop-mediated isothermal amplification (LAMP) for diagnosis of trypanosomosis. Actatropica 102(3):182–189

Thomas KV, Dye C, Schlabach M, Langford KH (2007) Source to sink tracking of selected human pharmaceuticals from two Oslo city hospitals and a wastewater treatment works. J Environ Monit 9(12):1410–1418

Tiwari A, Kurittu P, Al-Mustapha AI, Heljanko V, Johansson V, Thakali O et al (2022) Wastewater surveillance of antibiotic-resistant bacterial pathogens: a systematic review. Front Microbiol 13:977106

Wang S, Ma X, Liu Y, Yi X, Du G, Li J (2020) Fate of antibiotics, antibiotic-resistant bacteria, and cell-free antibiotic-resistant genes in full-scale membrane bioreactor wastewater treatment plants. Bioresour Technol 302:122825

White RL, Burgess DS, Manduru M, Bosso JA (1996) Comparison of three different in vitro methods of detecting synergy: time-kill, checkerboard, and E test. Antimicrob Agents Chemother 40(8):1914–1918

Xie J, Jin L, He T, Chen B, Luo X, Feng B et al (2018) Bacteria and antibiotic resistance genes (ARGs) in PM2.5 from China: implications for human exposure. Environ Sci Technol 53(2):963–972

Yang Y, Li B, Zou S, Fang HH, Zhang T (2014) Fate of antibiotic resistance genes in sewage treatment plant revealed by metagenomic approach. Water Res 62:97–106

Zhou J, He Z, Yang Y, Deng Y, Tringe SG, Alvarez-Cohen L (2015) High-throughput metagenomic technologies for complex microbial community analysis: open and closed formats. MBio 6(1):e02288–e02214

Zhu NJ, Ghosh S, Edwards MA, Pruden A (2021) Interplay of biologically active carbon filtration and chlorine-based disinfection in mitigating the dissemination of antibiotic resistance genes in water reuse distribution systems. Environ Sci Technol 55(12):8329–8340

Chapter 7
Pros and Cons of Diagnostic Methods Used for AMR Surveillance in Aquaculture

Lokesh Pawar, Nayan Chouhan, Arya Singh, Mayuri Nag, Bandi Sumanth Kumar Reddy, Arzoo Khan, Soibam Khogen Singh, and Victor Awugo

Abstract Antimicrobial resistance (AMR) is a global public health risk that could be exacerbated by aquaculture. To effectively address this issue, it is essential to implement AMR surveillance within aquatic systems. The sensitivity, specificity, and traceability of modern molecular diagnostic technologies for AMR make it possible to undertake targeted control measures, such as detecting and quantifying specific infections and associated resistance genes. However, these techniques have limitations, including the need for highly trained technicians to carefully operate expensive and time-consuming equipment that carries the danger of producing false-negative and false-positive results. Effective diagnostic approaches are essential for surveillance to understand the origins and spread of AMR. Despite the promising future of these, we must carefully assess their benefits and drawbacks to find the best approach consistent with the goals of each surveillance program. Insights gained from this will ultimately enhance mitigation methods, helping us to fight AMR in aquaculture. The present chapter aims to summarize culture-based, molecular, and phenotypic methodologies included in AMR surveillance programs for aquaculture, recognizing that the use of specific diagnostic tools for such purpose is not without its own merits and limitations.

L. Pawar · N. Chouhan · S. K. Singh (✉)
College of Fisheries (Central Agricultural University-Imphal), Agartala, Tripura, India

A. Singh · M. Nag
ICAR—Central Institute of Fisheries Education, Mumbai, Maharashtra, India

B. S. K. Reddy
Department of Aquaculture, Kerala University of Fisheries and Ocean Studies, Kochi, Kerala, India

A. Khan
College of Fisheries, Govind Ballabh Pant University of Agriculture and Technology, Pantnagar, Uttarakhand, India

V. Awugo
Federal University of Technology, Akure, Ondo, Nigeria

P. Elumalai, S. Lakshmi (eds.), *Antimicrobial Resistance in Aquaculture and Aquatic Environments*, https://doi.org/10.1007/978-981-97-7320-6_7

Keywords Antimicrobial resistance · AMR in aquaculture · Diagnostic methods · AMR surveillance

7.1 Introduction

Aquaculture plays a vital role in global food production, but the intensification of practices has led to increased antimicrobial use, contributing to the emergence and spread of antimicrobial resistance (AMR) (Reverter et al. 2020). This has significant implications for human and animal health. Surveillance and control measures in aquaculture are crucial due to the intersection of global warming and antimicrobial resistance (Reverter et al. 2020). The misuse of antibiotics in the aquaculture sector is a key factor driving the dissemination of AMR globally (Thornber et al. 2022; Caputo et al. 2023).

The potential risks of AMR to human health and its movement into different ecosystems, as shown in Fig. 7.1, and the environment are substantial, necessitating

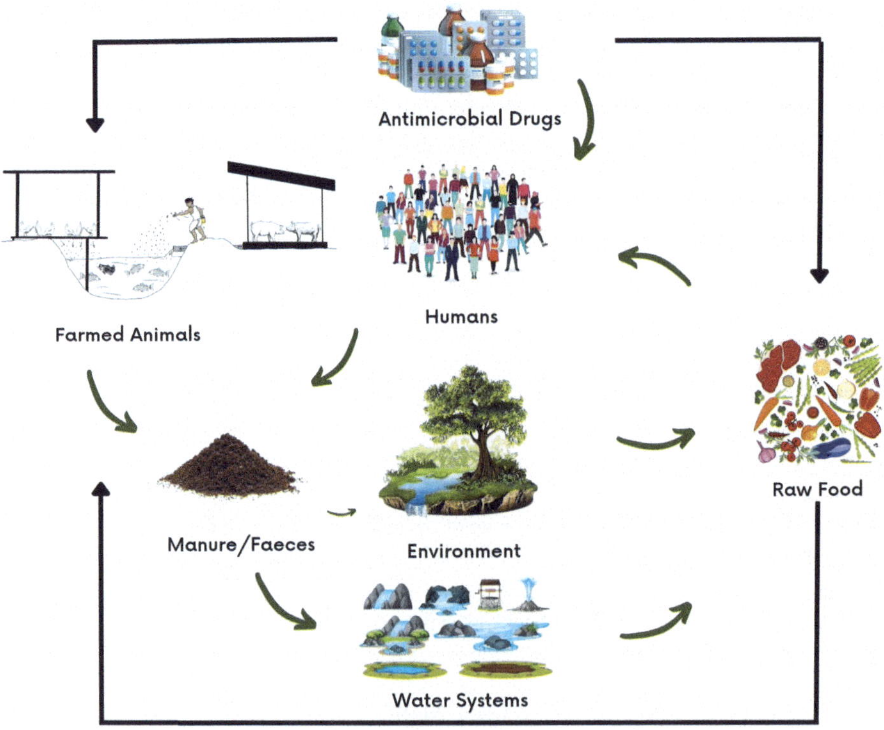

Fig. 7.1 Movement of antibiotic resistance bacterial strains between different ecosystems

effective diagnostic methods to detect and monitor AMR in the ecosystem, especially in aquaculture (Narbonne et al. 2021). However, the aquaculture industry's surveillance and interventions for AMR face challenges, including limited systematic monitoring of aquatic animals, which can impact food and nutritional security (Schar et al. 2021). Enhancing documentation of AMR trends can inform targeted surveillance programs and interventions in the world's most productive aquaculture and fisheries regions (Schar et al. 2021).

AMR surveillance is paramount in aquaculture due to its potential risks to human health and the environment. Effective diagnostic methods are necessary to detect and monitor AMR, while strengthened surveillance efforts are needed to guide interventions by the aquaculture industry.

7.2 Disease Outbreak as a Consequence of Aquaculture Intensification

Aquaculture's expansion has resulted in a wide spectrum of disease outbreaks and has a global impact on industrial productivity, profitability, and sustainability. Its development is anticipated to meet the increasing demand for fish globally. This is due to changes in consumer tastes towards a much broader aquatic food menu and increased spending power and health consciousness. To meet this demand, culture fishery practice is highly reliant as it produces far more fish per unit area than natural waterways. Aquatic animals represent 17% of animal protein consumed globally (FAO 2018). "Blue Transformation" is a key programming area of FAO under its Strategic Framework 2022–2031, developed in response to the 2021 COFI Declaration for Sustainable Fisheries and Aquaculture (COFI 2021). Aquaculture is a promising sector that is important in developing global economy. This sector has been troubled by diseases caused by viruses, bacteria, fungi, parasites, and other undetected and emergent pathogens (Bondad et al. 2005). This is mainly due to the intensification of aquaculture practices, with the conversion of conventional aquaculture farms into intensive ones, where effort is made to maximize fish output from a minimum input in terms of water, land, and area (Hepher 1985).

Attempts to intensify pond aquaculture beyond the common practice of supplemental aeration can be divided into physical/chemical techniques and a broad range of microbial strategies. The former technique includes using in-pond cages and raceways, water blending and shading of the algal community, and direct flocculation and removal of algal and bacteria biomass from ponds. To lower ammonia levels in a traditional pond, a number of microbiological methods can be applied. Nitrification/denitrification, photosynthesis, and heterotrophic bacterial regrowth are examples of these processes. In the later technology, multiple microbial processes react to remove and add ammonia from conventional aquaculture ponds. Examples include nitrification, denitrification, photosynthesis, mineralization, and heterotrophic bacterial regrowth. Ecological intensification of aquaculture systems

(ELIAS) aims to increase the sustainability of aquaculture production systems by rationally combining anthropogenic inputs with aquaculture ecosystem services. Sustainable intensification (SI) is a framework for promoting the production of farmed aquatic animals in Asia and their commerce with Europe.

As aquaculture grows in popularity, there is greater demand to develop more sustainable practices that save resources and decrease waste. The optimization of rearing conditions at the lowest environmental cost in intensive, super-intensive like Recirculatory Aquaculture Systems (Martins et al. 2010; D'Orbcastel et al. 2009) and Aquaponics (Endut et al. 2010; Azad et al. 2016) can lead to cultural water degradation if not appropriately maintained thereby leading to disease in fish stocks. The outbreak of disease is recognized as a barrier to aquaculture production and trade, and the economic growth of the sector in different parts of the world (Tanekhy 2015). The presence of pathogens does not always result in disease, but environmental stresses can upset the balance between potential pathogens and their host (Iqbal et al. 2012). Diseases and problems produced by viruses, bacteria, fungi, parasites, and other undiscovered and emerging pathogens have plagued the aquaculture industry (Shao 2001; Bondad et al. 2005). These outbreaks hamper the vital source of revenue and employment in this sector and pose a major threat to the economic and long-term sustainability of this industry (Pridgeon and Klesius 2012).

Bacteria are one of the most prevalent causative agents of disease and pose a great challenge to aquaculture, particularly in tropical locations (Pridgeon and Klesius 2012). It is estimated that up to half of the farmed fish are lost before selling due to bacterial infection (Algammal et al. 2020). Bacterial disease causes hemorrhagic spots or ulcers along the body's surface and around the eyes and mouth. In addition, they may have an enlarged, fluid-filled abdomen and bulging eyes. Bacterial infections can also be external, causing skin erosion, lesions, and ulceration (Omojowo and Sogbesan 2003). The progression of diseases is complex and influenced by various factors. It is not solely determined by the ability of bacteria to cause health issues; rather, it involves interactions with fish immune responses, environmental conditions, and the virulence attributes of the disease-causing agents (Pekala 2018). Organic discharge can also increase concentrations of bacteria and viruses in water systems (McAllister and Bebak 1997). Aquaculture uses antibiotics in an effort to manage bacterial diseases. Regular use of these chemicals, especially in the absence of visible illness concerns, poses a serious risk. Aquaculture infections acquire resistance as a result of the widespread and regular use of antibiotics in aquaculture. Antibiotics are therefore now ineffective in treating bacterial infections in certain situations. Furthermore, widespread antibiotic use in farm systems can hasten the development of disease-resistant bacteria (Boxall 2004; Hossain et al. 2012).

7.3 Antibiotics in Farmed Animals: Its Role as Therapeutics and Growth Promoter

Antibiotics, whether synthetic or naturally derived, are crucial medications used in modern pharmaceuticals to combat bacterial infections and promote animal health on farms. Their ability to kill bacteria is fundamental in treating infections and supporting the well-being of farmed animals. Antibiotics that are not hazardous to the host are employed as chemotherapeutic agents in the prevention and management of infectious disorders and have a significant role in the battle between microbes and cultured animal organisms. Antibiotics have been primarily used in aquaculture as therapeutics, prophylactics, and growth promoters which significantly outweighs their use in this industry to generate greater economic yield (Bush et al. 2011; Cabello et al. 2013).

Antibiotics are designed to work in one of three ways: by disrupting bacterial cell membranes, by inhibiting DNA or synthesis of proteins, or by interfering with the action of certain enzymes produced by bacteria (Todar 2006). Antibiotics are used in aquaculture to treat bacterial infections (Alderman and Hastings 1998). These are sometimes used to treat diseases, but they are more usually used to prevent infections by treating the water or fish before the development of disease (Christensen et al. 2006).

Antibiotics initially served a purpose in livestock farming to encourage development growth and improve feed efficiency, resulting in higher weight gain and a shorter time to market. It has been revealed that some medicines have growth-promoting properties in addition to their antibacterial activity. Sanchez Martínez et al. (2008) investigated the effect of oxytetracycline supplementation in channel catfish (*Ictalurus punctatus*) diets and observed significantly higher weight gain in treated fish, indicating growth of the fish. A similar experiment was conducted by He et al. (2017) in zebrafish (*Danio rerio*), using olaquindox with increased growth observed in treated fish. Gaskins et al. (2002) established that the effect of oxytetracycline on fish growth was dependent on a decrease in gastrointestinal tract bacteria and concluded that the use of antibiotics can act as growth promoter in addition to its well-described therapeutic properties.

Sub-therapeutic antibiotic dosages may affect the metabolic activity of gut bacteria or disrupt the balance in microbe species present in the gut, a combination which may cause weight gain. Antibiotics may also modify the intestinal flora of animals, resulting in alterations in nutrition absorption and metabolism. Antibiotics work by eradicating the harmful bacteria in a subclinical population. Eliminating this metabolic loss enables more effective nutrient utilization by the fish. Antibiotics modify the nonpathogenic gut flora, resulting in improved digestive processes and more effective nutritional utilization of feeds. They could also impact the immune systems of animals, directing less energy towards fighting illness and more towards growth and development.

7.4 Misuse of Antibiotics by the Aquaculture Sector and Development of Antimicrobial Resistance

Although some research implies that the use of antibiotics as growth promoters can boost weight gain by 1–11% (Lawrence 1998), it is possible that contemporary manufacturing methods hinder the realization of these advantages. Since the initial discovery of antibiotics and their first use in patients, antimicrobial resistance in bacteria has become a serious issue because of their use, misuse, and abuse (Woolhouse et al. 2016). Antibiotic-resistant bacteria are a problem because of antibiotics being used as growth boosters. Antibiotic resistance may develop in the gut flora of animals given antibiotics. Their use has been prohibited in many countries due to concerns about organisms developing resistance to antibiotics, so genetic manipulation of common bacterial communities may not pose a threat to public health. Antibiotic-resistant bacteria can evolve spontaneously when selection pressure to live causes changes in a bacterium's DNA sequence, allowing the bacteria to withstand antibiotic treatment.

The transition of aqua farms to more intensive animal production systems (Steinfeld et al. 2006) is characterized by an emphasis on production volume before considering farm biosecurity, cleanliness, and management standards. In fact, non-therapeutic antimicrobial usage in this setting may seem to contribute to improved growth and serve as a substitute for sound animal husbandry practices. Antimicrobial usage in terrestrial food producing animals currently outnumbers human use (Van Boeckel et al. 2017). The application of antibiotics is increasing in a variety of cultured aquatic species as production intensifies and the frequency of aquatic animal illnesses rises with the development of patterns of resistance.

It is unclear how antimicrobials promote farm animal development or feed efficiency (Barragry 1994). It may be that antimicrobials inhibit certain bacteria sensitive to the antibiotic that compete for nutritional resources with host animal, reducing the influence subclinical infections have on growth (Visek 1978; Stahly et al. 1980; Ribeiro de Lima et al. 1981; Lawrence 1998). Another idea is that growth factors alter hormones, cytokines, and other immunological components in recipient animals, thus improving their immune systems (Schiffrin et al. 1995; Famularo et al. 1997; Cunningham-Rundles et al. 2000; Gorbach 2000).

Furthermore, antibiotics and other pesticides are commonly used in aquaculture, which can contaminate the water and harm aquatic life. Public concern is growing, and legislative requirements are being tightened. Following the implementation of vaccination and improved management and husbandry programs, some countries have seen a marked decrease in antibiotic usage (Henriksson et al. 2018). Beyond exercising extreme caution in applying antibiotics in aquatic settings to avoid their detrimental effects, this practice has given rise to antibiotic-resistant bacteria in fish farms. Moreover, it has contributed to the transmission of resistance-related elements to human pathogens, as well as the presence of antibiotic residues in fish and fish products.

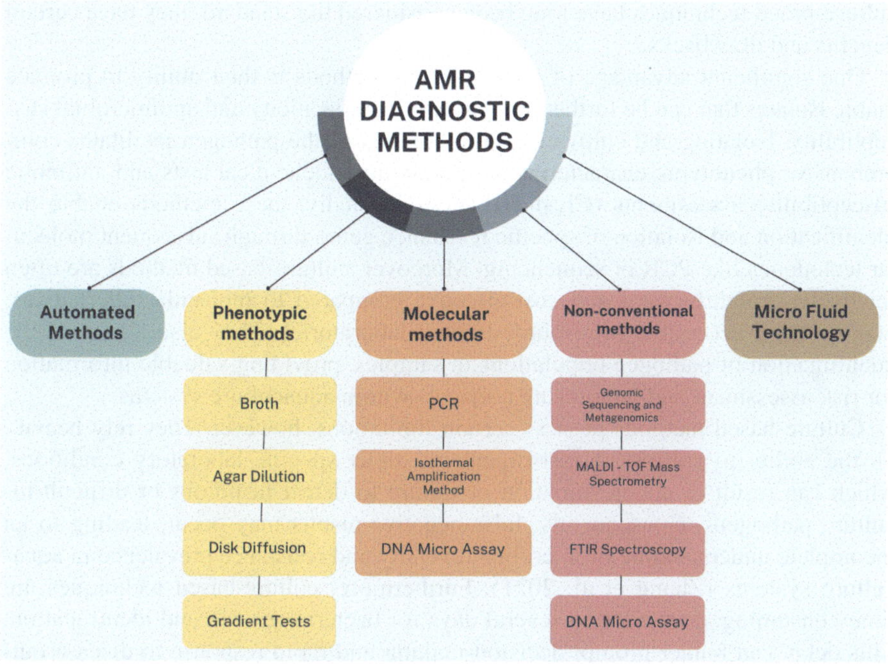

Fig. 7.2 Different diagnostic methods used for surveillance and detection of AMR in aquaculture. *PCR* polymerase chain reaction, *MALDI–TOF* matrix-assisted laser desorption/ionization–time-of-flight, *FTIR* Fourier transform infrared

7.5 Effective Diagnostic Methods Which Are Necessary to Detect and Monitor AMR

AMR detection and monitoring require accurate diagnostic procedures. These approaches span from genetic assays to phenotypic testing, culture-based diagnostic methods, and nonconventional and microfluidic technology, which allow healthcare professionals to identify resistant infections, prescribe appropriate therapies, and follow the progress of AMR, so assisting in the worldwide fight against antibiotic resistance (Fig. 7.2).

7.5.1 Culture-Based Diagnostic Methods

Culture-based diagnostic methods are commonly utilized in aquaculture settings to identify and measure bacterial pathogens and their corresponding resistance genes (Thornber et al. 2022; Caputo et al. 2023). This approach involves cultivating microorganisms on specific media to isolate and identify target pathogens. While culture-based techniques have long been considered the standard, they have certain benefits and drawbacks.

One significant advantage of culture-based methods is their ability to produce viable isolates that can be further analyzed for pathogenicity and antimicrobial susceptibility. Isolating and cultivating pure cultures of the pathogen facilitates comprehensive phenotypic characterization, including biochemical tests and antibiotic susceptibility assessments (CLSI 2018). Additionally, these methods enable the identification and isolation of specific resistance genes through subsequent molecular techniques like PCR or sequencing. Moreover, culture-based methods are often relatively straightforward and cost-effective compared to molecular alternatives, making them accessible to resource-limited laboratories. They also allow for the quantification of pathogen populations in samples, providing valuable information for risk assessment and monitoring purposes within aquaculture systems.

Culture-based methods possess certain limitations, however. They rely heavily on the ability to cultivate target organisms under specific laboratory conditions, which can result in underestimation or failure to detect fastidious or difficult-to-culture pathogens. Consequently, false-negative results may occur, leading to an incomplete understanding of microbial diversity and resistance prevalence in aquaculture systems (Zhang et al. 2021). Furthermore, culture-based techniques are time-consuming, necessitating several days for bacterial growth and identification. This delay can hinder prompt decision-making and rapid response to disease outbreaks or antimicrobial resistance issues in aquaculture settings. Additionally, these methods can be biased towards cultivable organisms, potentially overlooking significant but non-cultivable or nonviable pathogens and their associated resistance genes. Despite these limitations, culture-based diagnostic methods remain valuable for surveillance and monitoring in aquaculture. They provide crucial information regarding the presence, identity, and antimicrobial susceptibility of pathogens. They can be integrated with molecular techniques to achieve a comprehensive understanding of antimicrobial resistance dynamics in aquaculture systems.

7.5.2 Molecular Diagnostic Methods

The process of diagnosing diseases in organisms involves identifying abnormal conditions based on the signs and symptoms exhibited by the host. This information is typically obtained through a combination of the host's history and physical examination (Abdisa and Abdisa 2017). Over time, fish disease diagnosis has undergone significant advancements, moving away from traditional techniques focused on isolating and characterizing pathogens based on their physical attributes to more advanced methods rooted in molecular biology (Austin 2019).

Molecular diagnostics represents the culmination of successful collaboration between laboratory medicine, genomics, and technological knowledge in the field of molecular genetics, particularly with remarkable discoveries in molecular genomic technologies (Dwivedi et al. 2017). Patrinos et al. (2017) also define molecular diagnostics as the detection of genomic variations aimed at facilitating diagnosis, subclassification, and prognosis and monitoring the response to therapy.

In the field of aquaculture, molecular diagnostic methods such as polymerase chain reaction (PCR) and next-generation sequencing have gained increasing prominence for the detection and quantification of specific pathogens and their associated resistance genes (Ador et al. 2022). These methods offer several advantages over traditional techniques. Firstly, they exhibit higher accuracy and speed in identifying septicemic bacterial pathogens compared to conventional approaches. Additionally, molecular methods can detect pathogens from various sample types, including water, sediment, and fish tissues, providing a broader understanding of disease dynamics in aquaculture systems (Abdelsalam et al. 2023). Next-generation sequencing, in particular, offers a comprehensive view of the microbial community in aquaculture settings, allowing for the identification of both cultivable and non-cultivable organisms (Schar et al. 2021).

Despite their advantages, molecular diagnostic methods do have some limitations. They often require specialized equipment, technical expertise, and substantial financial investment. Moreover, the interpretation of molecular data can be complex, necessitating robust bioinformatic pipelines and data analysis tools. Nevertheless, the advantages of molecular methods, such as their sensitivity, specificity, and ability to detect a wide range of pathogens, make them invaluable tools for disease surveillance, risk assessment, and management in aquaculture settings.

Advantages of molecular-based methods for detecting AMR:

- Molecular-based methods have become essential tools in the field of AMR research, offering several advantages for the detection and characterization of AMR genes. These methods enable multiplex targeting, allowing for the precise identification and detection of specific AMR genes (Kaprou et al. 2021). Furthermore, molecular-based assays can be easily adapted to newly introduced resistance factors, ensuring that emerging resistance mechanisms can be quickly identified and monitored (Vasala et al. 2020).
- Among the molecular-based methods, the PCR stands out as a rapid and convenient technique. PCR enables the simultaneous investigation of a larger number of antibiotic resistance genes (ARGs) and is cost-effective compared to other methods (Waseem et al. 2019). Another advantage of PCR is its ability to deliver quick and sensitive detection of AMR genes (Vasala et al. 2020).
- Isothermal amplification methods also offer several advantages. These methods significantly reduce analysis time, as they do not require a thermal cycling process like PCR. Isothermal amplification works faster and exhibits high sensitivity, making it a valuable tool in AMR research (Zanoli and Spoto 2012). Additionally, isothermal amplification methods have minimal power consumption, contributing to their practicality and sustainability (Kaprou et al. 2021).
- DNA microarrays are another molecular-based approach that allows for the simultaneous analysis of thousands of genes. This high-throughput method offers the advantage of high sensitivity, enabling the detection and characterization of AMR genes on a large scale (Wang 2014).

Disadvantages of molecular-based methods:

- Molecular-based methods also present certain disadvantages. Firstly, these methods tend to be more costly compared to traditional culture-based approaches. Additionally, these assays cannot provide information on minimum inhibitory concentration (MIC), a crucial parameter for assessing antimicrobial susceptibility (Kaprou et al. 2021). Moreover, molecular-based methods may not detect all anti-resistant genes, limiting their comprehensiveness in AMR detection (Kaprou et al. 2021).
- The PCR method, despite its advantages, has lower specificity compared to culture and staining methods (Liu et al. 2019). The success rates of multiplexing approaches in isothermal amplification methods can also be relatively lower (Kaprou et al. 2021). Furthermore, DNA microarrays are time-consuming and expensive due to the use of glass slides and fluorescent dyes (Kaprou et al. 2021).

Modern molecular methods, including PCR and next-generation sequencing, have revolutionized the field of fish disease diagnosis in aquaculture. These techniques offer enhanced accuracy, speed, and versatility in detecting and quantifying pathogens and associated resistance genes. By providing valuable insights into disease dynamics and microbial communities, molecular diagnostics play a crucial role in effective disease management and maintaining the health of aquaculture systems.

7.5.3 Phenotypic Methods

Phenotypic methods are widely used in AMR diagnostics. Within the phenotypic approach, there are several submethods, which include manual methods such as broth dilution, agar dilution, disk diffusion, and gradient tests.

Phenotypic methods involve assessing the observable traits associated with antibiotic resistance in bacterial pathogens (Magiorakos et al. 2012). These techniques play a vital role in antimicrobial susceptibility testing (AST) and provide valuable insight into the functional expression of resistance. An advantage of phenotypic methods is their ability to directly measure the resistance of bacterial strains. Common approaches, such as disk diffusion and broth microdilution, allow for the determination of susceptibility or resistance to specific antibiotics (CLSI 2018). MICs are also determined using these methods, providing quantitative data for establishing clinical breakpoints that guide treatment decisions (EUCAST 2021).

Nevertheless, phenotypic methods have limitations. They rely on the growth of bacteria in culture, which can be time-consuming and introduce biases. Furthermore, these techniques do not provide direct information about the genetic mechanisms underlying resistance and the expression of specific resistance genes. Hence, phenotypic methods provide important functional insights into antibiotic resistance in bacterial pathogens. They contribute to AST and treatment decisions. However, their reliance on culture-dependent growth and the absence of direct genetic information can be considered limitations. Integrating phenotypic methods with molecular techniques can enhance our understanding of the genetic basis of antibiotic resistance.

The conventional phenotypic methods used in AMR diagnostics have their own set of advantages and disadvantages. Researchers and clinicians should consider these factors when selecting an appropriate method for their specific needs, considering factors such as cost, time requirements, flexibility, and the ability to detect resistance patterns or provide MIC values.

Advantages of phenotypic methods:

- The broth dilution method has the advantage of being commercially available and easy to perform. It allows for testing many isolates simultaneously, which is beneficial in high-throughput screening (Gajic et al. 2022). The agar dilution method also enables the testing of a significant number of isolates, although it is time-consuming (Gajic et al. 2022). Disk diffusion methods are relatively inexpensive and easy to perform, providing a cost-effective option for AMR testing.
- These methods allow for the detection of resistance patterns and provide high flexibility in assessing antibiotic resistance (Gajic et al. 2022). Gradient tests are convenient to perform and can also detect resistance patterns effectively (Gajic et al. 2022). Lastly, automated methods offer the advantage of delivering fast results, reducing the burden of manual labor, and ensuring superior quality (García 2002).

Disadvantages of phenotypic methods:

- The broth dilution method, while providing reliable results, requires a substantial amount of time for obtaining the final outcomes. Additionally, this method is prone to individual errors, which can affect the accuracy of the results (Gajic et al. 2022). The major drawback with the agar dilution method lies in its limited applicability to a minimum number of antibacterial agents.
- It is also time-consuming, which can be a disadvantage in urgent clinical situations (Gajic et al. 2022). The disk diffusion method, although cost-effective, lacks the ability to test certain antibiotics and does not provide MIC values, which are crucial for determining the extent of antimicrobial resistance (Gajic et al. 2022). Gradient tests, on the other hand, are relatively expensive compared to other phenotypic methods (Gajic et al. 2022). Lastly, automated platforms, despite their advantages in terms of speed and quality, are more expensive and may have technical restrictions when used with certain microorganisms and antimicrobial agents (García 2002).

7.5.4 Nonconventional Method

Nonconventional methods for genetic analysis refers to innovative and alternative approaches used to study and understand genetic information. Unlike traditional techniques that primarily focus on DNA sequencing and genotyping, nonconventional methods encompass a diverse range of cutting-edge tools. These may include advanced bioinformatics algorithms, single-cell sequencing technologies, epigenetic analyses, CRISPR-Cas9 gene editing, and high-throughput techniques. These novel methodologies offer unique insights into gene function, regulatory mechanisms, and complex interactions within the genome, contributing significantly to unraveling the complexities of genetic inheritance, disease susceptibility, and personalized medicine. Embracing nonconventional methods empowers researchers to delve deeper into the genetic landscape, accelerating breakthroughs and enhancing our understanding of life's fundamental building blocks.

The following are the advantages of nonconventional methods:

- Nonconventional methods in genetic analysis offer several advantages over traditional approaches. Genome sequencing and metagenomics techniques provide relatively long and high-quality DNA sequences, enabling detailed genetic analysis (Kaprou et al. 2021). These methods require less laboratory space, making them more feasible for smaller research facilities, and they are also highly portable, allowing for on-site analysis (González et al. 2019).
- Additionally, nonconventional methods such as matrix-assisted laser desorption/ionization–time-of-flight mass spectrometry (MALDI–TOF MS) are cost-effective and user-friendly and offer high accuracy, making them practical for routine genetic analysis (Vrioni et al. 2018). Fourier transform infrared (FTIR) spectroscopy is another non-conventional method that is highly cost-effective, operates at high speed, and is environmentally friendly, making it an attractive choice for genetic analysis (Kaprou et al. 2021).

The following are the disadvantages of nonconventional methods:

- Despite the advantages, nonconventional methods in genetic analysis also have their limitations. Second-generation sequencing techniques, while producing high-quality DNA sequences, are prone to information loss, copying errors, and sequence-dependent bias (Kaprou et al. 2021). This can affect the accuracy and reliability of the obtained genetic information. MALDI–TOF MS, although cost-effective and easy to use, has lower analytical sensitivity when used without prior culture, limiting its applicability in certain contexts (Kaprou et al. 2021).
- Furthermore, FTIR spectroscopy, while offering cost-effectiveness and speed, may incur high maintenance costs, which should be considered when utilizing this method for genetic analysis (Kaprou et al. 2021).

7.5.5 Microfluidic Technologies

Microfluidic technologies are innovative and rapidly evolving systems that manipulate and control minute volumes of fluids, typically on the microscale. By leveraging principles of fluid dynamics and surface chemistry, microfluidics enables the precise handling of liquids, gases, or particles within tiny channels and chambers. These micro-devices have diverse applications across various fields, from healthcare and biotechnology to environmental monitoring and materials science. Offering advantages such as reduced sample consumption, faster analysis times, and enhanced portability, microfluidic technologies hold great promise for revolutionizing diagnostics, drug development, and personalized medicine, paving the way for a more efficient and accessible future in numerous industries.

The following are the advantages of microfluidic technologies:

- Microfluidic technologies offer several advantages in various applications due to their unique characteristics. One of the key advantages is low power consumption. Compared to traditional techniques, microfluidic devices require significantly less power to operate effectively (Yılmaz and Yılmaz 2018). This feature makes microfluidics an attractive option for portable and battery-operated systems, where energy efficiency is crucial.
- Another advantage of microfluidic technologies is their ability to work with smaller sample volumes. Microfluidic devices can manipulate and analyze samples in microliter or even nanoliter volumes, reducing the amount of sample required for analysis (Yılmaz and Yılmaz 2018). This advantage is particularly valuable when dealing with precious or limited samples, as it allows for more efficient sample utilization.
- Furthermore, microfluidic technologies offer cost savings compared to traditional methods. The miniaturization and integration of multiple functions onto a single microfluidic chip can reduce the overall cost of experimental setups (Yılmaz and Yılmaz 2018). Additionally, the smaller sample volumes required in microfluidic devices can lead to cost reductions in terms of reagents and consumables.

The following are the disadvantages of microfluidic technologies:

- While microfluidic technologies have numerous advantages, they also present certain disadvantages that need to be considered. One of the main drawbacks is the restricted availability of strains when using antibiotics in phenotypic microfluidic assays (Kaprou et al. 2021). Phenotypic assays rely on the observation of observable traits or behaviors of microorganisms in response to different conditions, including the presence of antibiotics. However, obtaining a wide range of microbial strains for testing purposes can be challenging, limiting the diversity of strains that can be studied in phenotypic microfluidic assays.
- Additionally, the cost associated with the use of antibiotics in phenotypic microfluidic assays can be a significant disadvantage (Kaprou et al. 2021). Antibiotics can be expensive, especially when considering the large number of replicates and conditions that may need to be tested in microfluidic experiments. This cost can become a limiting factor in the scalability and widespread adoption of phenotypic microfluidic assays.

In conclusion, microfluidic technologies offer advantages such as low power consumption, smaller sample volumes, and cost savings. However, the restricted availability of strains and the high cost of antibiotics in phenotypic microfluidic

Table 7.1 Different diagnostic methods used in aquaculture with their advantages and limitations

Diagnostic methods	Advantages	Disadvantages	References
Phenotypic methods			
(a) Manual			
1. Broth	Commercially available Easy to perform	Time-consuming	Gajic et al. (2022)
2. Agar dilution method	Can test large number of isolates	Can use for a minimum number of antibacterial agents Time-consuming	
3. Disk diffusion methods	Comparatively cheap Easy to perform high flexibility in antibiotic resistance patterns	Cannot be tested by some antibiotics Cannot provide MIC value	
4. Gradient tests	Easy to perform Can also detect resistance patterns	Expensive	
5. Automated methods	Fast result, superior quality	More expensive Some technical restrictions	García (2002)
(b) Molecular based			
1. Molecular-based assay	Multiplex targeting, detection of AMR genes, and also detection of anti-resistant genes quickly	More costly. Cannot define MIC and all anti-resistant genes	Vasala et al. (2020), Kaprou et al. (2021)
2. PCR method	Quick, efficient, cost-effective	Low specificity as compared with culture and staining	Liu et al. (2019), Waseem et al. (2019)
3. Isothermal amplification method	Less analysis time, works faster, very sensitive, and minimal power consumption	Had success rate that is less for multiplexing approaches	Zanoli and Spoto (2012), Kaprou et al. (2021)
4. DNA micro assays	Analyze thousands of genes simultaneously High throughput, high sensitivity	Time-consuming and costly	Wang (2014), Kaprou et al. (2021)
Nonconventional methods			
(a) Genome sequencing and metagenomics	Produce relatively long and high-quality DNA sequences, portability, and affordability	Information loss, copying errors and sequence-dependent bias	Kaprou et al. (2021)
(b) MALDI–TOF mass spectrometry	Cost effective, easy to use, accurate, and environmentally friendly	Low analytical sensitivity without prior culture	Vrioni et al. (2018), Kaprou et al. (2021)

(continued)

Table 7.1 (continued)

Diagnostic methods	Advantages	Disadvantages	References
(c) FTIR spectroscopy	Highly cost effective, environmentally friendly	High maintenance cost	Kaprou et al. (2021)
(d) Microfluidic technologies	Low power consumption, less costly, requires smaller sample volume	Very costly, restricted stain availability	Yılmaz and Yılmaz (2018), Kaprou et al. (2021)

PCR polymerase chain reaction, *MALDI–TOF* matrix-assisted laser desorption/ionization–time-of-flight, *FTIR* Fourier transform infrared

assays represent significant disadvantages that should be considered when considering the implementation of these technologies (Table 7.1).

7.6 Comparison of Diagnostic Methods for AMR Surveillance in Aquaculture

Antimicrobial resistance is a growing concern in aquaculture, and surveillance is essential to guide interventions. Several diagnostic methods are available for AMR surveillance in aquaculture, each with its own advantages and limitations as discussed above. This section will compare and contrast different diagnostic methods for AMR surveillance in aquaculture, including their sensitivity, specificity, cost, technical requirements, and ability to detect emerging resistance mechanisms.

7.6.1 Antimicrobial Susceptibility Testing (AST)

AST is a commonly employed technique for AMR surveillance in aquaculture. It involves evaluating the ability of bacteria to grow in the presence of antibiotics and provides valuable insights into their susceptibility or resistance to specific antibiotics (Caputo et al. 2023). AST can be conducted using various methods, including disk diffusion, broth microdilution, and agar dilution. It offers the advantages of being relatively cost-effective and straightforward to perform. However, it should be noted that AST necessitates the isolation and culture of bacteria, which can be time-consuming and technically demanding. Moreover, AST may have limitations in detecting emerging resistance mechanisms that are not encompassed by routine testing panels (Caputo et al. 2023).

7.6.2 Whole-Genome Sequencing

Whole-genome sequencing (WGS) has been identified as a valuable tool for AMR surveillance in aquaculture. It offers extensive insights into the genetic underpinnings of resistance, encompassing the identification of resistance genes, mutations, and mobile genetic elements. Additionally, WGS has the capacity to detect emerging resistance mechanisms that may not be covered by routine testing panels. However, it is essential to acknowledge that WGS is associated with significant costs, necessitates specialized equipment and expertise, and generates substantial amounts of data that require bioinformatics analysis. This analysis process can be time-consuming and technically demanding (Forde et al. 2022).

7.6.3 Polymerase Chain Reaction

The PCR has been recognized as a rapid and sensitive diagnostic method for the detection of specific resistance genes or mutations (O'Neill et al. 2021). PCR can be executed using various platforms, such as conventional PCR, real-time PCR, and digital PCR. It offers the advantages of being relatively inexpensive and straightforward to conduct. However, it is important to note that PCR requires prior knowledge of the target resistance genes or mutations for accurate detection. Moreover, it may have limitations in identifying emerging resistance mechanisms that are not covered by routine testing panels (O'Neill et al. 2021).

7.6.4 Matrix-Assisted Laser Desorption/Ionization–Time-of-Flight Mass Spectrometry

According to O'Neill et al. (2021), MALDI–TOF MS is a rapid and accurate technique used for bacterial identification and can also provide information about antibiotic resistance. By detecting changes in bacterial protein profiles associated with resistance, MALDI–TOF MS allows for the identification of resistance patterns. This method is cost-effective and straightforward to perform, although it requires specialized equipment and expertise. However, it should be noted that MALDI–TOF MS may have limitations in detecting emerging resistance mechanisms that do not involve observable alterations in protein profiles.

In summary, there are different diagnostic methods available for surveillance of AMR in aquaculture, each with its own strengths and limitations. AST is widely used but may not be effective in detecting emerging resistance mechanisms. WGS is a powerful method but can be costly and requires specialized equipment and expertise. PCR is a rapid and sensitive technique but relies on knowledge of target resistance genes or mutations. MALDI–TOF MS is a rapid and accurate method, but it may have limitations in detecting emerging resistance mechanisms. Combining these methods may offer a more comprehensive approach to AMR surveillance in aquaculture.

7.6.5 Limitations and Challenges of Diagnostic Methods for AMR Surveillance in Aquaculture

The implementation of diagnostic methods for AMR surveillance faces various limitations and challenges. Some of these challenges include the following:

- Potential for unreliable outcomes: Diagnostic techniques may produce false positives or false negatives, which could result in unreliable outcomes. False positives can lead to the overuse of antibiotics, while false negatives can encourage the growth of bacteria that are resistant to treatment (Bayot and Bragg 2023).
- Requirement for specialized personnel and equipment: Certain diagnostic methods necessitate trained personnel and specialized equipment, which may not be readily available in all settings. This limitation can hinder the adoption of these methods, particularly in resource-limited environments (Subasinghe et al. 2023).
- Financial considerations: The cost associated with implementing different diagnostic methods can be a barrier, particularly in resource-limited settings. This includes the expenses related to equipment, reagents, and skilled personnel (Narbonne et al. 2021).
- Limitations of molecular techniques: Molecular techniques like PCR and WGS have limitations in terms of their diagnostic capacity for diseases and their ability to assess pathogen viability (Subasinghe et al. 2023).

7.6.6 Complexity of the Issue

AMR in aquaculture is a complex and multifaceted problem driven by various societal factors that are challenging to address. This complexity makes it difficult to establish effective AMR surveillance programs (Caputo et al. 2023). Addressing these limitations and challenges requires collaborative efforts among stakeholders from different sectors. Developing context-specific AMR surveillance programs that consider the local constraints and prioritize accurate and reliable diagnostic methods is essential for effective AMR management in aquaculture.

7.7 Best Practices for Diagnostic Method Selection

Choosing the right diagnostic methods for AMR surveillance is crucial to ensure the accuracy and reliability of results. Several best practices can guide the selection process:

- Tailoring the chosen method to sample types: Different samples, such as water, sediment, or fish tissue, may require specific diagnostic methods. Consider the sample type when choosing the appropriate method (Preena et al. 2020a, b).
- Implement quality control measures: Maintain quality controls to ensure accurate and reliable diagnostic results. This includes using appropriate positive and negative controls, following standardized protocols, and regularly conducting proficiency testing (Caputo et al. 2023; Preena et al. 2020a, b).
- Consider cost and availability: Evaluate the cost and availability of different diagnostic methods based on the specific setting and available resources (Thornber et al. 2022).
- Properly interpret and report results: Accurate interpretation and reporting of diagnostic results are crucial for effective AMR surveillance. Follow standardized reporting formats and employ appropriate statistical analyses to interpret the data (Schar et al. 2021).

Selecting appropriate diagnostic methods for AMR surveillance in aquaculture involves considering sample types, implementing quality control measures, evaluating cost and availability, interpreting and reporting results accurately, and acknowledging the limitations and challenges of different methods. By adhering to these best practices, reliable and informative results can be obtained to support effective AMR surveillance in aquaculture.

7.8 Ways for AMR Reduction in Aquaculture

Continuing surveillance programs, timely detection of resistant strains, and appropriate regulatory enforcement are required to halt the spread of antimicrobial resistance in aquaculture. Farmers and hatchery operators should engage in scientific farming and follow good aquaculture practices to prevent infection. All farms and hatcheries must be registered and keep records of all inputs used during the breeding period. All stakeholders should take corrective action to reduce bacterial antimicrobial resistance in the fisheries sector (Rao 2021).

1. **Promoting responsible use of antimicrobials**: Enhancing expert advice, prescriptions, and appropriate antibiotic use can help address the problem of antibiotic misuse. Educating stakeholders through better guidance can help reduce inappropriate antibiotic use. The phasing out of the use of antimicrobials to promote animal growth without risk assessment and the judicious use of antimicrobial pesticides in plants will help accelerate the response to AMR.

2. **Preventive measures**: Hatchery should use healthy parent fish that are specifically pathogen-free (SPF) or specifically pathogen resistant (SPR). Only disease-free, fry/shrimp post-larvae are delivered to farmers. Aquaculture farmers should store fish species at appropriate stocking densities, along with proper water and feed management, to avoid the formation of undesirable bacteria. Feed and feed additives used in aquaculture should always be tested for the presence of antibiotics by the manufacturer and clearly marked as antibiotic-free on the packaging. Manufacturers of animal feed and feed supplements must ensure that the ingredients used to manufacture their products are free of antibiotics. Rapid on-site diagnostic tests need to be developed to prevent inappropriate use of antibiotics, such as to treat viral infections.
3. **Alternative strategies**: To regulate antibiotic dependence and the emergence of antibiotic resistance, the aquaculture industry needs to develop effective alternative strategies such as the following:

    (a) **Vaccination**: In aquaculture, preventive measures such as vaccination are used to control disease outbreaks. Oral fish vaccines are effective against many aquatic diseases by producing humoral antibodies (Newaj-Fyzul and Austin 2015). Oral vaccines of *Vibrio anguillarum* was found to protect sea bass. The DNA vaccine of *V. parahaemolyticus* that was incorporated with chitosan nanoparticles was effective for black sea bream (Preena et al. 2020a, b). Very few oral vaccines are commercially available for aquaculture, however.

    (b) **Probiotics**: By attaching to intestinal mucus, probiotics have the potential to target pathogens such as *V. harveyi*, a major threat in aquaculture (Chabrillón et al. 2005). The antibacterial effects of probiotics are due to the production of antibiotics, iron-trapping siderophores, enzymes (such as proteases, amylases, and lysozymes), hydrogen peroxide, and organic acids.

    (c) **Bacteriocins**: Bacteriocins are proteinaceous toxins produced by a variety of bacteria and archaea that possess numerous antibacterial properties that make them ideal candidates for combating pathogens. The high selectivity of bacteriocins makes them attractive alternatives to conventional antibiotics, but studies indicate that bacteriocinogenic bacteria may contain this AMR gene. Marine bacteriocins such as diversin and piciocin produced by *Carnobacterium* in the gut of fish, as well as bacteriocins produced by lactobacillus and indigenous bacteria, have been shown to be effective in treating bacterial infections in aquaculture (Watts et al. 2017).

    (d) **Immunostimulants**: Immunostimulants such as β-1,3-glucans are other agents effective against various aquatic diseases such as vibriosis, erythema, aeromonosis, and pasteurellosis (Ngamkala et al. 2010). Another immune stimulant, lipopolysaccharide, has also been reported to enhance carp bactericidal activity and reduce mortality in Aeromonas-infected *Oncorhynchus mykiss* (rainbow trout) (Nya and Austin 2010; Kadowaki et al. 2013).

4. **Policy framework**: It includes appropriate regulation and enforcement, disease diagnosis, risk assessment, information gathering in the areas of disease control

and management, capacity building in all aspects of the aquaculture production chain, and antimicrobial judicious use to prevent disease outbreaks. Effective implementation of policies, such as employing effective bio-security practices, is required. To control the prevalence of AMR in aquaculture, it is important to develop and implement valid assays for the evaluation and detection of resistant pathogens and associated genes at national and international levels (WHO 2006).

5. **Proper monitoring**: Effective aquaculture operations, safe management practices, good stocking programs, and good hygiene can control the introduction of bacterial pathogens, the development of bacterial infections, and thus the use of antimicrobial agents. To overcome the threat of antimicrobial resistance, continuous surveillance programs should be implemented according to appropriate and effective guidelines. Assessment of fish mortality, evaluation of antibiotic residues, and identification of causative pathogens and determination of their antimicrobial susceptibility should be conducted frequently as part of surveillance programs. Molecular detection of antimicrobial resistance genes that make up gene delivery systems such as integrons and gene cassettes (GCs) is of critical importance to improve therapy and prevent further spread of antibiotic resistance. Preventing the emergence of antimicrobial resistance in aquaculture could limit the influx of AMR genes into the natural environment, creating a public health hazard.
6. **Surveillance program**: Monitoring of AMU and AMR in fisheries needs to be strengthened to allow realistic assessment of the risks posed to human populations by AMU in fisheries. Laboratory capacity for surveillance systems should also be improvised to reduce antimicrobial use and antimicrobial-resistant bacteria.
7. **Enabling good practices**: Good husbandry practices can help reduce the burden of infectious diseases, reduce the need for antibiotics, and reduce the emergence of antimicrobial resistance. Good practice also extends to addressing the spread of antimicrobial resistance in the environment and its transmission through the food chain. There is an opportunity to increase profitability through more efficient farming practices. Measures should be taken to prevent bacterial contamination during harvest, processing, sale, and transportation of fish. To assure the cleanliness and safety of fish food, potable water and excellent quality ice are utilized throughout processing. All aquaculture operations should have better bio-security measures in place to prevent the introduction of bacteria from the outside, such as bird nets, shrimp fences, vehicle disinfection, and on-site hand sanitizing facilities. Fishermen should disinfect their fishing gear with a potassium permanganate solution to reduce bacterial invasion from the outside and prevent bacterial transfer between different fishponds.
8. **Strengthening governance and allocating resources sustainably**: Achieving effective national planning and goals requires cross-sectoral and multisectoral cooperation. Strengthening antimicrobial resistance policies and regulatory frameworks, as well as incentive-based approaches, offers opportunities to accelerate the response to antimicrobial resistance. Research and innovation in

the areas of antimicrobials, alternatives, and diagnostics need support. Economic case for incentives and public and private investment can help mobilize resources to implement national plans. Strict enforcement of drug regulations to control access to antibiotics for aquaculture farmers and hatchery operators is necessary. Awareness of personal hygiene and AMR needs to be increased among fishermen. To prevent the transmission of bacteria from humans to animals and vice versa, measures such as bathing before entering and after leaving pond water and washing hands before handling aquatic animals should be strictly adhered to.

9. **Integrons and gene cassettes**: Integrons play an important role in the dissemination of antimicrobial resistance in aquaculture (Gao et al. 2012). The genetic architecture of integrons possesses special properties to acquire, excise, and express genes embedded in GCs, which are normally stored in mobile genetic elements such as plasmids and transposons, allowing horizontal gene expression. Determinants diffuse through transmission, facilitating translocation within the gene. GCs, non-replicative mobile elements, are typically inserted into integrons by recombination and use the Pc promoter to express genes responsible for antibiotic resistance (Preena et al. 2020a, b).

7.9 Conclusion

In conclusion, this chapter highlights the significant strides made in the field of AMR in aquaculture and its surveillance, shedding light on the diverse diagnostic methods employed for monitoring AMR in aquatic environments. While conventional culture-based techniques continue to provide valuable information, they are limited by their time-consuming nature and potential underestimation of AMR prevalence. Molecular methods, on the other hand, offer rapid and sensitive detection, enabling the identification of specific resistance genes and mobile genetic elements. Nevertheless, the high initial costs and complexity of molecular approaches pose challenges, especially for resource-limited settings. The integration of both conventional and molecular techniques seems promising, capitalizing on their complementary strengths and compensating for their weaknesses. Moreover, emerging technologies such as metagenomics and microfluidics hold great potential in revolutionizing AMR surveillance in aquaculture. Continued research, standardization of methods, and global collaboration are vital for developing effective and sustainable AMR surveillance strategies in aquaculture, safeguarding both human health and the aquatic ecosystem.

References

Abdelsalam M, Elgendy MY, Elfadadny MR, Ali SS, Sherif AH, Abolghait SK (2023) A review of molecular diagnoses of bacterial fish diseases. Aquac Int 31(1):417–434

Abdisa T, Abdisa T (2017) Review on practical guidance of veterinary clinical diagnostic approach. Int J Vet Sci Res 3(1):030–049. https://doi.org/10.17352/ijvsr.000020

Ador MAA, Haque S, Paul SI, Chakma J, Ehsan R, Rahman A (2022) Potential application of PCR based molecular methods in fish pathogen identification: a review. Aquacul Stud 22(1):AQUAST621. https://doi.org/10.4194/2618-6381/AQUAST621

Alderman DJ, Hastings TS (1998) Antibiotic use in aquaculture: development of antibiotic resistance—potential for consumer health risks. Int J Food Sci Technol 33(2):139–155

Algammal AM, Mabrok M, Sivaramasamy E, Youssef FM, Atwa MH, El-Kholy AW, Hozzein WN (2020) Emerging MDR-Pseudomonas aeruginosa in fish commonly harbor oprL and toxA virulence genes and blaTEM, blaCTX-M, and tetA antibiotic-resistance genes. Sci Rep 10(1):1–12. https://doi.org/10.1038/s41598-020-72264-4

Austin B (2019) Methods for the diagnosis of bacterial fish diseases. Mar Life Sci Technol 1(1):41–49. https://doi.org/10.1007/s42995-019-00002-5

Azad KN, Salam MA, Azad KN (2016) Aquaponics in Bangladesh: current status and future prospects. J Biosci Agric Res 7(02):669–677. https://doi.org/10.18801/jbar.070216.79

Barragry T (1994) Veterinary drug therapy. Lea & Febiger, Philadelphia

Bayot ML, Bragg BN (2023) Antimicrobial susceptibility testing. In: StatPearls [Internet]. StatPearls Publishing, Treasure Island, FL. https://www.ncbi.nlm.nih.gov/books/NBK539714/

Bondad MG, Subasinghe RP, Arthur JR, Ogawa K, Chinabut S, Adlard R, Shariff M (2005) Disease and health management in Asian aquaculture. Vet Parasitol 132:249–272. https://doi.org/10.1016/j.vetpar.2005.07.005

Boxall ABA (2004) The environmental side effects of medication. EMBO Rep 5(12):1110–1116. https://doi.org/10.1038/sj.embor.7400307

Bush K, Courvalin P, Dantas G, Davies J, Eisenstein B, Huovinen P, Zgurskaya HI (2011) Tackling antibiotic resistance. Nat Rev Microbiol 9(12):894–896. https://doi.org/10.1038/nrmicro2693

Cabello FC, Godfrey HP, Tomova A, Ivanova L, Dölz H, Millanao A, Buschmann AH (2013) Antimicrobial use in aquaculture re-examined: its relevance to antimicrobial resistance and to animal and human health. Environ Microbiol 15(7):1917–1942. https://doi.org/10.1111/1462-2920.12134

Caputo A, Bondad-Reantaso MG, Karunasagar I et al (2023) Antimicrobial resistance in aquaculture: a global analysis of literature and national action plans. Rev Aquac 15(2):568–578. https://doi.org/10.1111/raq.12741

Chabrillón M, Rico RM, Arijo S, Díaz-Rosales P, Balebona MC, Moriñigo MA (2005) Interactions of microorganisms isolated from gilthead sea bream, Sparus aurata L., on Vibrio harveyi, a pathogen of farmed Senegalese sole, Solea senegalensis (Kaup). J Fish Dis 28(9):531–537

Christensen AM, Ingerslev F, Baun A (2006) Ecotoxicity of mixtures of antibiotics used in aquacultures. Environ Toxicol Chem 25(8):2208–2215. https://doi.org/10.1897/05-415R.1

CLSI (2018) Performance standards for antimicrobial susceptibility testing: CLSI supplement M100. Clinical and Laboratory Standards Institute

COFI (2021) Harvest deferral process will have devastating impacts shutting 14 to 20 sawmills and threatening 18 000 BC jobs. News release 2 Nov 2021. https://www.cofi.org/harvest-deferral-process-will-havedevastating-impacts-shutting-14-to-20-sawmills-and-threatening-18000-bc-jobs/

Cunningham-Rundles S, Ahrne S, Bengmark S (2000) Probiotics and immune response. Am J Gastroenterol 95:S22–S25

D'Orbcastel E, Jean-Paul B, Belaud A (2009) Water quality and rainbow trout performance in a Danish Model Farm recirculating system: comparison with a flow through system. Aquac Eng 40:135–143. https://doi.org/10.1016/j.aquaeng.2009.02.002

Dwivedi S, Purohit P, Misra R, Pareek P, Goel A, Khattri S, Pant KK, Misra S, Sharma P (2017) Diseases and molecular diagnostics: a step closer to precision medicine. Indian J Clin Biochem 32:374–398

Endut A, Jusoh A, Ali N, Nik WW, Hassan A (2010) A study on the optimal hydraulic loading rate and plant ratios in recirculation aquaponic system. Bioresour Technol 101(5):1511–1517. https://doi.org/10.1016/j.biortech.2009.09.040

EUCAST (2021) Breakpoint tables for interpretation of MICs and zone diameters. https://www.eucast.org/clinical_breakpoints/

Famularo G, Moretti S, Marcellini S et al (1997) Stimulation of immunity by probiotics. In: Fuller R (ed) Probiotics 2: applications and practical aspects. Chapman and Hall, London, pp 133–161

FAO (2018) The State of World Fisheries and Aquaculture 2018—meeting the sustainable development goals. FAO, Rome. http://www.fao.org/3/i9540en/I9540EN.pdf. Accessed 10 May 2023

Forde BM, De Oliveira DM, Falconer C, Graves B, Harris PN (2022) Strengths and caveats of identifying resistance genes from whole genome sequencing data. Expert Rev Anti-Infect Ther 20(4):533–547

Gajic I, Kabic J, Kekic D, Jovićević M, Milenkovic M, Mitić-Ćulafić D, Trudic A, Ranin L, Opavski N (2022) Antimicrobial susceptibility testing: a comprehensive review of currently used methods. Antibiotics 11:427. https://doi.org/10.3390/antibiotics11040427

Gao P, Mao D, Luo Y, Wang L, Xu B, Xu L (2012) Occurrence of sulfonamide and tetracycline-resistant bacteria and resistance genes in aquaculture environment. Water Res 46(7):2355–2364

García CP (2002) Ventajas y problemas de los métodos automatizados de estudio de susceptibilidad in vitro. Rev Chilena Infectol 19:96–100. https://doi.org/10.4067/s0716-10182002019200006

Gaskins HR, Collier CT, Anderson DB (2002) Antibiotics as growth promotants: mode of action. Anim Biotechnol 13(1):29–42

González-Escalona NAO, Allard MA, Brown EW, Sharma S, Hoffmann M (2019) Nanopore sequencing for fast determination of plasmids, phages, virulence markers, and antimicrobial resistance genes in Shiga toxin-producing Escherichia coli. PLoS One 14:e0220494. https://doi.org/10.1371/journal.pone.0220494

Gorbach SL (2000) Probiotics and gastrointestinal health. Am J Gastroenterol 95:S2–S4

He S, Wang Q, Li S, Ran C, Guo X, Zhang Z, Zhou Z (2017) Antibiotic growth promoter olaquindox increases pathogen susceptibility in fish by inducing gut microbiota dysbiosis. Sci China Life Sci 60:1260–1270. https://doi.org/10.1007/s11427-016-9072-6

Henriksson PJ, Rico A, Troell M, Klinger DH, Buschmann AH, Saksida S, Zhang W (2018) Unpacking factors influencing antimicrobial use in global aquaculture and their implication for management: a review from a systems perspective. Sustain Sci 13:1105–1120. https://doi.org/10.1007/s11625-017-0511-8

Hepher B (1985) Aquaculture intensification under land and water limitations. GeoJournal 10(3):253–259

Hossain MS, Aktaruzzaman M, Fakhruddin A, Uddin M, Rahman S, Chowdhury M, Alam M (2012) Prevalence of multiple drug resistant pathogenic bacteria in cultured black tiger shrimp (Penaeus monodon Fabricius). Glob J Environ Res 6(3):118–124. https://doi.org/10.5829/idosi.gjer.2012.6.3.822

Iqbal Z, Sheikh U, Mughal R (2012) Fungal infections in some economically important freshwater fishes. Pak Vet J 32(3):422–426

Kadowaki T, Yasui Y, Nishimiya O, Takahashi Y, Kohchi C, Soma GI, Inagawa H (2013) Orally administered LPS enhances head kidney macrophage activation with down-regulation of IL-6 in common carp (Cyprinus carpio). Fish Shellfish Immunol 34(6):1569–1575

Kaprou GD, Bergšpica I, Alexa EA, Alvarez-Ordóñez A, Prieto M (2021) Rapid methods for antimicrobial resistance diagnostics. Antibiotics (Basel) 10(2):209. https://doi.org/10.3390/antibiotics10020209

Lawrence K (1998) Growth promoters in swine. In: Proceedings of the 15th IVPS Congress (Birmingham, England). 5–9 July. International Pig Veterinary Society

Liu HY, Hopping GC, Vaidyanathan U, Ronquillo YC, Hoopes PC, Moshirfar M (2019) Polymerase chain reaction and its application in the diagnosis of infectious keratitis. Med Hypothesis Discov Innov Ophthalmol 8(3):152–155

Magiorakos AP, Srinivasan A, Carey RB, Carmeli Y, Falagas ME, Giske CG, Harbarth S (2012) Multidrug-resistant, extensively drug-resistant and pandrug-resistant bacteria: an international expert proposal for interim standard definitions for acquired resistance. Clin Microbiol Infect 18(3):268–281. https://doi.org/10.1111/j.1469-0691.2011.03570.x

Martins CIM, Eding EH, Verdegem MCJ, Heinsbroek LTN, Schneider O, Blancheton JP, Verreth JAJ (2010) New developments in recirculating aquaculture systems in Europe: a perspective on environmental sustainability. Aquac Eng 43(3):83–93

McAllister PE, Bebak J (1997) Infectious pancreatic necrosis virus in the environment: relationship to effluent from aquaculture facilities. J Fish Dis 20(3):201–207. https://doi.org/10.1046/j.1365-2761.1997.00297.x

Narbonne JA, Radke BR, Price D, Hanington PC, Babujee A, Otto SJG (2021) Antimicrobial use surveillance indicators for finfish aquaculture production: a review. Front Vet Sci 8:595152. https://doi.org/10.3389/fvets.2021.595152

Newaj-Fyzul A, Austin B (2015) Probiotics, immunostimulants, plant products and oral vaccines, and their role as feed supplements in the control of bacterial fish diseases. J Fish Dis 38(11):937–955

Ngamkala S, Futami K, Endo M, Maita M, Katagiri T (2010) Immunological effects of glucan and Lactobacillus rhamnosus GG, a probiotic bacterium, on Nile tilapia Oreochromis niloticus intestine with oral Aeromonas challenges. Fish Sci 76:833–840

Nya EJ, Austin B (2010) Use of bacterial lipopolysaccharide (LPS) as an immunostimulant for the control of Aeromonas hydrophila infections in rainbow trout Oncorhynchus mykiss (Walbaum). J Appl Microbiol 108(2):686–694

O'Neill J, Davies S, Rex JH (2021) The review on antimicrobial resistance: tackling drug-resistant infections globally. Wellcome Trust and UK Government

Omojowo FS, Sogbesan OA (2003) Fish losses due to bacterial flora and infections of fishes in Kainji Lake Area, Nigeria: a review. Nig Vet J 24(2):41–47. https://doi.org/10.4314/nvj.v24i2.3453

Patrinos GP, Katsila T, Viennas E, Tzimas G (2017) Genomic Databases: Emerging Tools for Molecular Diagnostics. In: Molecular Diagnostics. Academic Press, pp 353–369

Pekala SA (2018) Contemporary threats of bacterial infections in freshwater fish. J Vet Res 62:261–267. https://doi.org/10.2478/jvetres-2018-0037

Preena PG, Swaminathan TR, Rejish Kumar VJ, Bright Singh IS (2020a) Unravelling the menace: detection of antimicrobial resistance in aquaculture. Lett Appl Microbiol 71(1):26–38. https://doi.org/10.1111/lam.13292. Epub 2020 Apr 22

Preena PG, Swaminathan TR, Kumar VJR, Singh ISB (2020b) Antimicrobial resistance in aquaculture: a crisis for concern. Biologia 75:1497–1517

Pridgeon JW, Klesius PH (2012) Major bacterial diseases in aquaculture and their vaccine development. CABI Rev 7:1–16

Rao BM (2021) Antimicrobial resistance (AMR) in aquatic products. ICAR-Central Institute of Fisheries Technology

Reverter M, Sarter S, Caruso D, Avarre JC, Combe M, Pepey E, Pouyaud L, Vega-Heredía S, de Verdal H, Gozlan RE (2020) Aquaculture at the crossroads of global warming and antimicrobial resistance. Nat Commun 11(1):1870. https://doi.org/10.1038/s41467-020-15735-6

Ribeiro de Lima F, Stahly TS, Cromwell GL (1981) Effects of copper, with and without ferrous sulfide, and antibiotics on the performance of pigs. J Anim Sci 52:241–247

Sanchez Martínez JG, Pérez-Castañeda R, Rábago Castro JL, Aguirre Guzmán G, Vázquez-Sauceda ML (2008) A preliminary study on the effects on growth, condition, and feeding indexes in channel catfish, Ictalurus punctatus, after the prophylactic use of potassium permanganate and oxytetracycline. J World Aquac Soc 39(5):664–670. https://doi.org/10.1111/j.1749-7345.2008.00195.x

Schar D, Zhao C, Wang Y et al (2021) Twenty-year trends in antimicrobial resistance from aquaculture and fisheries in Asia. Nat Commun 12:5384. https://doi.org/10.1038/s41467-021-25655-8

Schiffrin EJ, Rochat F, Link-Amster H et al (1995) Immunomodulation of human blood cells following the ingestion of lactic acid bacteria. J Dairy Sci 78:491–497. https://doi.org/10.3168/jds.S0022-0302(95)76659-0

Shao ZJ (2001) Aquaculture pharmaceuticals and biologicals: current perspectives and future possibilities. Adv Drug Deliv Rev 50:229–243. https://doi.org/10.1016/S0169-409X(01)00159-4

Stahly TS, Cromwell GL, Monegue HJ (1980) Effects of the dietary inclusion of copper and/or antibiotics on the performance of weanling pigs. J Anim Sci 51:1347–1351. https://doi.org/10.2527/jas1981.5161347x

Steinfeld H, Gerber P, Wassenaar TD, Castel V, de Haan C (2006) Livestock's long shadow: environmental issues and options. FAO, Rome

Subasinghe RP, Bondad-Reantaso MG, McGladdery SE, East I, Mohan CV (eds) (2023) Diseases in Asian aquaculture VI. Fish Health Section. https://doi.org/10.1111/raq.12734

Tanekhy M (2015) Fish health and diseases. Fish immunology and diseases. Faculty of Veterinary Medicine, Alexandria University, Egypt

Thornber K, Bashar A, Ahmed MS, Bell A, Trew J, Hasan M, Hasan NA, Alam MM, Chaput DL, Haque MM, Tyler CR (2022) Antimicrobial resistance in aquaculture environments: unravelling the complexity and connectivity of the underlying societal drivers. Environ Sci Technol 56(21):14891–14903. https://doi.org/10.1021/acs.est.2c00799

Todar K (2006) Todar's online textbook of bacteriology. University of Wisconsin-Madison

Van Boeckel TP, Glennon EE, Chen D, Gilbert M, Robinson TP, Grenfell BT, Levin SA, Bonhoeffer S, Laxminarayan R (2017) Reducing antimicrobial use in food animals. Science 357(6358):1350–1352

Vasala A, Hytönen VP, Laitinen OH (2020) Modern tools for rapid diagnostics of antimicrobial resistance. Front Cell Infect Microbiol 10:308. https://doi.org/10.3389/fcimb.2020.00308

Visek W (1978) The mode of growth promotion by antibiotics. J Anim Sci 46:1447–1469

Vrioni G, Tsiamis C, Oikonomidis G, Theodoridou K, Kapsimali V, Tsakris A (2018) MALDI-TOF mass spectrometry technology for detecting biomarkers of antimicrobial resistance: current achievements and future perspectives. Ann Transl Med 6:240

Wang L (2014) Encyclopedia of food microbiology, 2nd edn. Academic

Waseem H, Jameel S, Ali J, Saleem Ur Rehman H, Tauseef I, Farooq U, Jamal A, Ali MI (2019) Contributions and challenges of high throughput qPCR for determining antimicrobial resistance in the environment: a critical review. Molecules 24:163. https://doi.org/10.3390/molecules24010163

Watts JE, Schreier HJ, Lanska L, Hale MS (2017) The rising tide of antimicrobial resistance in aquaculture: sources, sinks and solutions. Mar Drugs 15(6):158

Woolhouse M, Waugh C, Perry MR, Nair H (2016) Global disease burden due to antibiotic resistance—state of the evidence. J Glob Health 6:010306. https://doi.org/10.7189/jogh.06.010306

World Health Organization (2006) The world health report 2006: working together for health

Yılmaz B, Yılmaz F (2018) Chapter 8—Lab-on-a-chip technology and its applications. In: Barh D, Azevedo V (eds) Omics technologies and bio-engineering. Academic, pp 145–153

Zanoli LM, Spoto G (2012) Isothermal amplification methods for the detection of nucleic acids in microfluidic devices. Biosensors 3:18–43

Zhang S, Li X, Wu J, Coin L, O'Brien J, Hai F, Jiang G (2021) Molecular methods for pathogenic bacteria detection and recent advances in wastewater analysis. Water 13(24):3551. https://doi.org/10.3390/w13243551

Chapter 8
Assessing Antimicrobial Resistance in Aquafeed: A Holistic Screening and Monitoring Strategy

Ajay Valiyaveettil Salimkumar, Shahinur Akter, Ummay Tahura, Ummeh Iffat, Shaharin Salma Romkey, Md Zahid Hasan Nahid, Nabila Nusrat, and Sheikh Aftab Uddin

Abstract Antimicrobial resistance (AMR) in fish feed sources is becoming a major concern due to the potential spread of resistant microbes to humans through contaminated fish products. This jeopardizes the nutritional and economic viability of aquaculture systems, as it disrupts the efficacy of antimicrobials (AMs) and fosters the proliferation of antibiotic-resistant genes. The development and spread of AMR bacteria are influenced by several factors, such as the use of antibiotics in aquaculture, easy access to AMs, use of sub-therapeutic doses of AMs, discharge of effluents from aqua farms into the environment, inappropriate disposal of antibiotics, lack of awareness among farmers, and, to sum up, poor management practices. Therefore, it is important to establish proper screening and monitoring techniques to prevent the spread of AMR. This involves identifying the presence of AMR bacteria in feed samples, monitoring changes in the prevalence of AMR over time, and tracking the spread of AMR between different species. This insight enables us to improve feed production practices, guide the selection of alternative feed sources, and help mitigate the risk of AMR transmission. This chapter reviews the present knowledge on the types of feed sources, their potential for AMR transmission, and interspecies transmission pathways. Moreover, the chapter examines the different

---

A. V. Salimkumar
Faculty of Science and Technology, Research Centre for Experimental Marine Biology and Biotechnology (PIE-UPV/EHU), Plentzia-Biskaia, Basque, Spain

S. Akter
Faculty of Science and Technology, University of the Basque Country, Leioa, Basque, Spain

Department of Oceanography, University of Chittagong, Chattogram, Bangladesh

U. Tahura · U. Iffat · S. S. Romkey · M. Z. H. Nahid
Department of Oceanography, University of Chittagong, Chattogram, Bangladesh

N. Nusrat · S. A. Uddin (✉)
Institute of Marine Sciences, University of Chittagong, Chattogram, Bangladesh
e-mail: aftabims@cu.ac.bd

P. Elumalai, S. Lakshmi (eds.), *Antimicrobial Resistance in Aquaculture and Aquatic Environments*, https://doi.org/10.1007/978-981-97-7320-6_8

methods of screening and monitoring programs with possible challenges associated with their implementation.

Keywords Aquaculture feed sources · Interspecies transmission · Antimicrobial resistance · Screening and monitoring programs · Challenges

Abbreviations

AA	Amino acid
AM	Anti-microbial
AMR	Antimicrobial resistance
AMU	Antimicrobial use
APW	Alkaline peptone water
ARB	Antibiotic-resistant bacteria
ARG	Antibiotic-resistant gene
AST	Antimicrobial susceptibility test
Cd	Cadmium
Cu	Copper
DHA	Docosahexaenoic acid
DNA	Deoxyribonucleic acid
EAA	Essential amino acid
ELISA	Enzyme-linked immunosorbent assay
EPA	Eicosapentaenoic acid
FAO	Food and Agricultural Organization
FCR	Feed conversion ratio
GCHV	Grass carp hemorrhagic virus
GMP	Good management practice
Hg	Mercury
HGT	Horizontal gene transfer
HUFA	Highly unsaturated fatty acid
IFFO	International Fishmeal and Fish-Oil Organization
LC-PUFA	Long-chain omega-3 polyunsaturated fatty acid
MCR-1	Mobilized colistin resistance-1
MCT	Medium-chain triglyceride
MGE	Mobile genetic element
MIC	Minimum inhibitory concentration
MRL	Maximum residue limit
MS	Mass spectrometry
NGO	Non-Governmental Organization
NGS	Next-generation sequencing
Pb	Lead
PBM	Poultry by-product meal
PCR	Polymerase chain reaction
RT-PCR	Reverse transcription polymerase chain reaction
SBM	Soya bean meal

SFA	Saturated fatty acid
TCBS	Thiosulfate citrate bile salts sucrose agar
Tet	Tetracycline resistance gene
VGT	Vertical gene transfer
WHO	World Health Organization
Zn	Zinc

8.1 Introduction

Aquaculture is expected to play a significant role in fostering economic growth, improving nutritional health, ensuring international food safety, and reducing poverty. This can be achieved by cultivating fish with minimal environmental impact while maximizing societal advantages (Subasinghe et al. 2009). Aquaculture has expanded more rapidly than capture fisheries over the past 2 years. According to the Food and Agriculture Organization (FAO), 178 million tons of fish were produced in 2020, with 88 million tons (49%) from aquaculture, compared with 5% in the 1970s, 20% in the 1990s, and 44% in the 2010s (The state of world fisheries and aquaculture 2022 2023). Aquaculture is currently predicted to meet 50% of the world's food fish needs, which is expected to rise to 60–70% by 2030 (Subasinghe et al. 2009). In 2012, global beef production (63 million tons) was surpassed by aquaculture production (66 million tons) (Larsen and Roney 2013). This shows that aquaculture has a higher potential to fulfill the growing demand for animal protein sources. Despite being one of the industries with the fastest growth, aquaculture still has issues with the quality of the fish it produces. As these fishes are typically maintained in stressful, intensive, or semi-intensive culture systems with higher stocking densities, these farming systems are susceptible to infectious diseases such as bacterial infections that have the potential to cause the collapse of fish production through aquaculture (ECDC (European Centre for Disease Prevention and Control) et al. 2017). Thus, the proper prophylactic and therapeutic use of antimicrobials (AMs) and other chemical agents is crucial to prevent further spread of infections. However, there are strict regulations established by national and international authorities for their use. However, many people do not abide by these, and as a result, there is improper use of these chemicals, leading to the development of AM-resistant bacteria within aquaculture systems. In contrast to agricultural farming, aquaculture relies on AMs more frequently. AMs are also easily spread throughout the ecosystem, which supports large populations of ill, healthy, and sub-clinically infected fishes (O'Neill 2015). Some experiments have demonstrated that fish do not efficiently metabolize all the antibiotics, resulting in up to 80% of them being released into the water systems (Burridge et al. 2010). Based on the findings from Schar et al. (2020), global consumption of antimicrobics is expected to increase by 33% between 2017 and 2030 (Schar et al. 2020).

Gram-negative bacteria, such as *Aeromonas hydrophila*, *Aeromonas salmonicida*, *Vibrio anguillarum*, *Vibrio harveyi*, *Flavobacterium psychrophilum*,

Edwardsiella tarda, *Citrobacter freundii*, *Pseudomonas fluorescens*, and *Yersinia ruckeri*, are among the major bacterial pathogens that affect fish production. Gram-positive bacteria, such as Staphylococcus and Streptococcus, can also compromise the health of fish within aquaculture systems (Mekasha and Linke 2021). *A. hydrophila* is the most prevalent bacterial pathogen in aquaculture systems, posing a potential risk to human health (Igbinosa et al. 2012).

The main driving factors of antimicrobial resistance (AMR) in aquaculture are the use of AMs for disease prevention and treatment, poor biosecurity practices, and environmental contamination. AMs are widely used in aquaculture prophylactically or at subtherapeutic doses to prevent disease outbreaks (Ibrahim et al. 2020). AM overuse and misuse can lead to the selection and spread of AMR bacteria in aquaculture environments. This is especially troublesome in intensive aquaculture systems, where high stocking densities with high nutritional input can foster disease transmission (Cabello et al. 2013). Furthermore, the need for more effective control of AM use and monitoring in many countries exacerbates the problem (Cabello et al. 2013). The emergence of AMR in aquaculture can also be attributed to poor biosecurity practices. AMR bacteria can enter a facility through the introduction of new animals, equipment, or from the water supplies. AMR can also spread through the transport of animals between facilities. Furthermore, the use of contaminated feed and the discharge of untreated effluent both have the potential to act as environmental sources for the spread of AMR bacteria (Cabello et al. 2013). The spread of AMR in wild fish populations and other aquatic organisms may be aided by releasing AMR bacteria and antibiotic residues into the environment from aquaculture facilities. AMR can also develop and spread in aquaculture environments due to the use of antibiotics by other sources, including agriculture and human medicines (Bengtsson-Palme et al. 2016). Given the global production of aquaculture commodities and their importance in providing livelihoods, AMR prevention in this industry necessitates immediate attention. This chapter provides the current knowledge on feed sources used in aquaculture, their possible contributions to AMR transmission, and the methods through which AMR can be transmitted among different fish species. Furthermore, the chapter evaluates various AMR screening and monitoring methods while identifying potential implementation issues. As a result, it sets the framework for an integrated strategy to tackle AMR in aquafeeds (Fig. 8.1).

8.2 Current Status of AMR in Feed Sources

Many European nations have laws governing the use of antibiotics in aquaculture, and oxytetracycline, florfenicol, sarafloxacin, erythromycin, and sulfonamides are among those allowed. However, AMR is still widespread in European aquaculture, particularly in intensive systems. Tetracycline and quinolone resistance was reported to be highly prevalent in bacteria obtained from European aquaculture systems in 2019 (Di Cesare et al. 2016). Sulfadimethoxine/ormetoprim, oxytetracycline, and

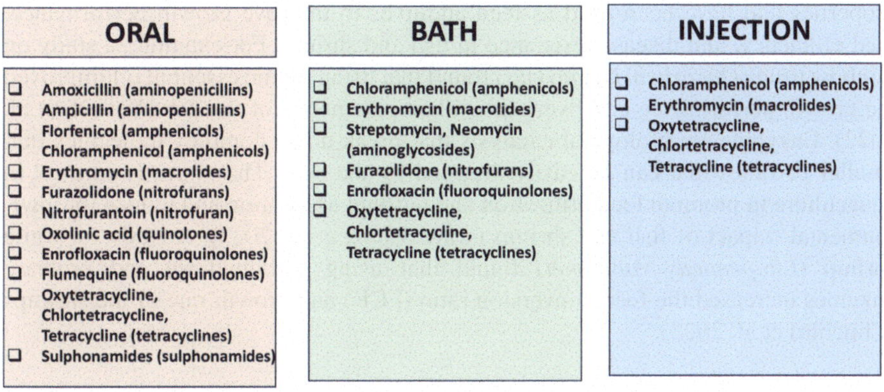

Fig. 8.1 Administration routes of major AMs used in aquaculture

florfenicol are all permitted in the United States (Romero et al. 2012). However, regulations are minimal or nonexistent in major aquaculture-producing nations such as China and India. China accounts for 67% of global aquaculture and is the world's largest producer and exporter of fish and fishery products. There are no veterinarian prescription requirements for antibiotic use in China. The Chinese aquaculture industry is governed by rules; however, the regulatory structure ignores the early production stages where the use of prohibited medicinal drugs might be significant. According to reports, China's enforcement of the law is lax, leading to the use of illegal veterinary medications and antibiotics (Chen et al. 2012). The use of antibiotics in aquaculture may exceed that in human medicine in nations with laxer regulations. After Norway, Chile is the country that produces the most cultured salmon, and it has been noted that a lot of antibiotics are used in the process. With its stringent laws and extremely low levels of antimicrobial use (AMU), Norway is regarded as a role model for regulating AMU in salmon aquaculture (Midtlyng et al. 2011).

Because antibiotic resistance could potentially develop and threaten the well-being of communities, the use of antibiotics in aquaculture has been a source of controversy. In response, there has been an increasing focus on finding alternative approaches to manage and prevent diseases in aquaculture. Among the recent substitute feed additives that have attracted interest are probiotics, prebiotics, essential oils, and enzymes. Live microorganisms known as probiotics have been shown to enhance gut health, disease resistance, and growth performance when given in appropriate amounts. For example, a study on Nile tilapia (*Oreochromis niloticus*) found that using *Bacillus* spp. probiotics improved growth performance and enhanced the immune response of the fish (Fath El-Bab et al. 2022). On the contrary, prebiotics are non-digestible feed ingredients that increase the growth and/or activity of beneficial microbes in the gut (World Health Organization et al. 2006). They have been found to improve growth performance, feed consumption, and disease resistance in a variety of fish and shrimp species (Wee et al. 2023). Essential oils are volatile compounds extracted from plants with AM and anti-inflammatory properties and have been used as feed additives to improve growth performance, feed efficiency, and disease resistance in fish and shrimp. For example, a study on rainbow trout (*Oncorhynchus mykiss*) found that using thyme essential oil improved the growth performance and liver antioxidant parameters of the fish (Yousefi et al. 2022). Enzymes are biological catalysts that break down complex molecules into smaller particles that can be easily digested by the host. They have been used in aquaculture to promote feed utilization and nutrient absorption and reduce the environmental impact of fish and shrimp farms (Liang et al. 2022). A study on white shrimp (*Litopenaeus vannamei*) found that using carbohydrase and protease enzymes increased the feed conversion ratio (FCR) and growth rate of the shrimps (Chuchird et al. 2023).

8.3 Types of Feed Sources

In an aquaculture system, the amount and quality of required nutrients directly impact fish growth, production, and health. The main ingredients in fish feeds are fish meal and fish oil, which are limited resources due to the rising price of these, the sustainability of marine capture fisheries, and unpredictable climatic occurrences. Therefore, the main objective for risk reduction in aquaculture diets is to substitute these ingredients with innovative feed additives to increase productivity (Tacon et al. 2011). For an ingredient to be a good alternative for fish meals in aquafeeds, it needs to be widely available, have a competitive price, and be easy to handle, transport, store, and use during the manufacture of feed. In terms of nutrition, it must have low amounts of fiber, starch, especially non-soluble carbohydrates, and antinutrients. It must also have a relatively high protein content, a good amino acid (AA) profile, high nutrient digestibility, and a good taste (Gatlin et al. 2007). Some research has been conducted on the production of alternative ingredients in aqua diets where they used ingredients from plants, insects, terrestrial animals, microbiological organisms, and genetically engineered organisms, which have great potential for future use by the aquaculture industry (Hua et al. 2019).

8.3.1 Fish Meal and Fish Oil Used in Aquaculture

Fishmeal is produced by heating, pressing, drying, and grinding fresh raw fish with by-products from fish processing. The liquid, a mixture of fish oil, water, and soluble protein, is extracted by pressing. This yellow fish oil is separated by centrifugation and purified several times (Shepherd and Jackson 2013). Normally, long-chain omega-3 polyunsaturated fatty acids (LC-PUFAs) are obtained from fish oil, which has the potential to serve as a natural nutrient reserve. Fish meal is an excellent source of essential amino acids (EAAs) for fish; it also has a high protein content, good micronutrients, and some unique nutrients, such as taurine (Hemre et al.

2009). Complete replacement of fish meals could negatively impact fish health and development and deplete the fish of essential nutrients (Hardy 2010). The manufacturing of fish meal and fish oil is strongly dependent on marine fishing. According to the International Fishmeal and Fish-Oil Organization (IFFO) database, 1 kg of fishmeal takes 4.4–4.6 kg of whole fish, while producing 1 kg of fish oil typically requires around 20 kg (Shepherd and Jackson 2013). Statistics on the manufacturing of fish meals worldwide showed that rates are dropping at a rate of 1.7% each year (Tacon et al. 2011). The global production of fish meal and fish oil faces significant challenges due to intense competition in marine capture fishery, diminishing fish stocks, substantial demand for fishmeal in terrestrial livestock feed, and the use of fish oil in dietary supplements for humans and aquaculture. These factors have prompted the aquaculture industry to actively explore innovative alternatives for feed ingredients (Trushenski et al. 2006).

8.3.2 *Plant-Based Feeds Used in Aquaculture*

Plant ingredients are useful sources for alternative feed ingredients in aquaculture diets due to their greater quantity and lower cost than fish meal. Plenty of plant ingredients are used in the manufacture of aquaculture diets, including oilseeds (such as soybean, sunflower, rapeseed, and cottonseed), grains (such as wheat and corn), and pulses (beans, lupins, and peas, etc.) (Bandara 2018). However, there are some challenges associated with plant ingredients due to the issue of anti-nutritional substances and an imbalance in their nutrient profile (Bandara 2018). The anti-nutrition factors are the metabolic by-products in plants, such as phytochemical compounds for protection against predators (Makkar 1993). Hence, using plant ingredients without proper treatment could negatively impact feed intake and nutritional digestion in fish. Anti-nutritional factors are divided into two groups, heat-labile proteins that are quickly inactivated by heat, including protease, lectins inhibitors, and inhibitors of amylase, and heat-stable substances, including tannins, phytic acids, and phenols that are resistant to heat processing during feed production without being damaged. Antinutritional substances, like lipase and protease inhibitors, can often cause pancreatic enzyme secretion, which can mess up the digestion of protein, carbohydrates, and fats. However, heat-stable anti-nutrients such as phytic acid will remain in the ingredient even after ingredients have been processed (Bandara 2018). Increasing plant-based food intake has been shown to have an impact on fish health, including immunological response, stress, and histological changes. A transcriptomic approach has shown that the complete replacement of fish meals with plant-based diets changes lipid and nitrogen metabolisms and upregulation of hepatic genes (Panserat et al. 2009). The other problem with plant ingredients is an imbalance in their amino acid (AA) and fatty acid profiles. The EAAs required by fish are not obtained from the AA profile of plant ingredients. Therefore, different plant ingredients can be combined to provide a more suitable AA profile for fish, such as a blend of soybean meal (rich in arginine and lysine and low in methionine) and gluten corn (high in methionine and low in arginine and lysine) can complement each other. However, the combination of plant ingredients may also introduce anti-nutritional factors (Bandara 2018).

Plant oils, for example, palm oil, contain larger quantities of saturated fatty acids (SFAs), which reduce energy digestibility in cold water. Plant oils are a great source of medium-chain triglycerides (MCTs) that improve fish performance, although higher inclusion levels typically result in increased fish mortality (Hemre et al. 2009). However, a lack of highly unsaturated fatty acids (HUFAs) may result in a deficiency in essential fatty acids when fish oil is completely replaced in fish diets.

Plant ingredients have many benefits for aquaculture practices despite some drawbacks. Various technologies can overcome the major challenges of plant ingredients. For instance, the protein content of soy protein concentrates (e.g., soybeans and corn gluten meal) can be increased by removing the carbohydrate fraction. Extrusion, heat treatment, and advanced fractionation methods are used to treat both heat-labile and heat-stable anti-nutritional components in plant ingredients (Drew et al. 2007). Plant-based fish feed with enzymatic treatment has been shown to enhance feed intake, growth (Kaushik et al. 1995), increased nutritional quality, and digestibility (Vielma et al. 2000). Soya bean meal (SBM) is one of the most promising alternative protein sources because of its availability, quality, and supply of AAs (Hodar et al. 2020). A combination of soybean and corn gluten meal can provide the EAA for fish and maintain the diet's palatability. For instance, cod growth and feed intake were unaffected by replacing up to 69% of the total dietary protein with this mix (Albrektsen et al. 2006).

Soya bean and wheat gluten can partially or fully replace fishmeal in the diets of halibut, salmon, and trout without affecting their growth or nutrient utilization (Hardy 1996). Methionine levels in canola meal are greater than in SBM. When used at a ratio of 1:2.33, this could replace fishmeal in shrimp diets by up to 60% (Suárez et al. 2009). Algae is a potential replacement for fishmeal because of its remarkable reservoir of EEAs and beneficial omega-3 and omega-6 fatty acids. The use of algae in aquaculture feeds can significantly increase fish growth and disease resistance (AftabUddin et al. 2021). It has been demonstrated to be a great diet for juvenile fish and crustaceans (Bandara 2018).

Vegetable oils are more sustainable than fish oils because they have consistent production, stable costs, high availability, and higher economic value in the aquafeed industry (Fountoulaki et al. 2009). Moreover, fish oil can only be replaced if it contains the essential fatty acids required by fish. Vegetable oils most often used in fish feed include soybean, rapeseed, linseed, sunflower, olive, palm, and canola oils (Bandara 2018). Several studies suggest that 50–60% of fish oil can be replaced by vegetable oil without compromising the growth and feed utilization efficiency of various fish species, such as gilthead sea bream (*Sparus aurata*), Atlantic salmon (*Salmo salar*), rainbow trout (*Oncorhynchus mykiss*), and European sea bass (*Dicentrarchus labrax*) (Bandara 2018; Hodar et al. 2020).

8.3.3 Animal-Based Feeds Used Aquaculture

There are different types of animal-based feeds utilized in Aquaculture described below.

Insect Protein: Cricket food, mealworm meal, black soldier flies, silkworm pupae, termites, Grasshoppers, mealworms, and rat tail maggots are regarded as novel feed elements for most wild fish and crustaceans as natural diets. FAO scientists concluded that insect meal may serve as a viable substitute for soya meal or fishmeal in the animals' diets, ranging between 25% and 100% without causing any adverse effects. Insect meal is richer in EAA (methionine and lysin) and fat than soya meal, which has positive effects on reducing FCRs and overall growth performance (Hodar et al. 2020). The exoskeleton of insects is composed of polysaccharide chitin, which is indigestible to most fish species. Future research should, therefore, concentrate on technological advancements to increase insect output and comprehend the impact of insect meal on fish health (Bandara 2018).

Maggot Meal: Maggots originate from the translucent developmental phase of the housefly *Musca domestica* and are utilized in the processing of magmeal, which contains 39–61.4% crude protein, 5.8–8.2% crude fiber, and 12.5–21% lipid (Hodar et al. 2020). According to previous research reports, magmeal is regarded as a feasible alternative protein source in terms of cost, availability, biological value (free of antinutritional and toxic factors), and FCR (Ajani et al. 2004).

Silkworm Pupae Meal: Silkworm pupae were utilized as a feedstuff in Asian aquaculture, mainly in China and India, in either single or compound diets due to their excellent source of protein (52–72%) and reasonable cost (Nandeesha et al. 1986).

Earthworm Meal: Earthworm has 50–70% protein and 5–10% lipid. Earthworm lipid has potential omega-3 fatty acid, which helps animal immunity (Dynes 2003) and is required in the formulation of several types of fish feed (Hansen and Czochanska 1975). The supplement of the meal in fish diets helps to enhance the growth of common carp, catfish, and all tropical fishes (Hodar et al. 2020).

Terrestrial Animal By-Products: Blood meal, poultry by-product meal (PBM), feather meal, krill meal, and artemia are included in animal by-products. The advantages of animal by-products are high protein and lipid sources as well as anti-nutritional components are not present in terrestrial animal by-products.

Poultry By-Product: Poultry by-products are ground pieces of poultry waste such as legs, necks, undeveloped eggs, and feathers. Poultry by-products are considered digestible, cost-effective, high-protein feed ingredients in aqua diets (Shapawi et al. 2007). In recent years, PBM has been shown to be capable of replacing fishmeal by up to 100% (Gaylord and Rawles 2005). For a variety of fish species, including Chinook salmon (Fowler 1990), rainbow trout (Bureau et al. 2000), Nile tilapia (Fasakin et al. 2005), and so on, these feed elements have been added to the diets as supplements. However, the processing technology used can affect nutrient availability.

Blood Meal: Blood meal is made using only pathogen-free, recently obtained animal blood. Only non-ruminant blood is currently used in the aquaculture business (particularly in Europe) due to rigorous rules and safety concerns (Bandara 2018). Blood meal has attracted significant interest in the aquafeed industry due to its high digestibility, its favorable lysine and histidine content, and the presence of heme iron and other forms of iron that may increase feed component oxidation (El-Haroun and Bureau 2007). Hemoglobin frequently deteriorates during processing, which results in poor flavor and digestion. Spray drying has been demonstrated to offer greater digestibility of blood meal than alternative techniques (El-Haroun and Bureau 2007). It has been shown that fermentation procedures can have positive effects on the nutritional digestibility, quality, and palatability of several feed ingredients in aquaculture diets (Nogueira et al. 2012). The level of fishmeal supplementation varies depending on feeding habits and fish size (Wee 1991). According to previous experiments, blood meal can totally (100%) substitute fishmeal in catfish juveniles, 6–10% in grouper diets (Martins and Guzman 1994) and gilthead sea bream diets (Ribeiro et al. 2011), and 20% in juvenile trout diets (Luzier et al. 1995) without affecting growth, survival, or feed conversion.

Feather Meal: Feather meal is primarily a by-product of the poultry industry, and its incorporation in fish diets depends on production technology. Rainbow trout, *Labeo rohita*, *O. niloticus*, and other teleost species are among the fish species that have been tested using feather meal as a complete or partial replacement for fish meal (Bandara 2018). Feather meal has low EAA, so supplementation of feather meal with EAA is important in fish diets. Feeding techniques may lead to an accumulation of AM compounds in feather meals (Love et al. 2012).

Krill Meal: Normally, krill contains around 60% protein with a balanced nutrition AA, about 25% lipids, and roughly 40% more phospholipids than usual, astaxanthin, and omega-3 fatty acids (eicosapentaenoic acid and docosahexaenoic acid). They can serve as a growth stimulator in shrimp and fish, and their high portion of protein is used as food formulation and food additive (Hodar et al. 2020).

Artemia Biomass: Due to its high nutritional content, artemia may be regarded as a potential element for replacing fishmeal in the diets of crustaceans and fish (Maldonado Montiel and Rodríguez Canché 2005). Additionally, it has a favorable effect on socioeconomic factors in coastal regions.

8.3.4 Animal-Based Oil Used in Aquaculture

Animal fats are mostly used in carnivorous marine fish species such as gilthead seabream and yellowtail kingfish, but it has not been explored extensively in aquafeeds (Pérez et al. 2014). SFAs are found in animal fats in large amounts, ranging from 28.5% in poultry fat to 47.5% in cow tallow, according to Turchini et al. (2009) but low content of PUFA in both poultry and animal fats. The growth performance of rainbow trout did not change when 50% of the fish oil was replaced with rendered fats from terrestrial sources (poultry fat, beef tallow, restaurant grease, or pork lard) (Trushenski et al. 2011). Land animal fats may be more affordable and readily available alternatives to fish oil in the diet that are sustainable (Turchini et al. 2009).

8.4 AMR Transmission in Feed Sources

AMR in microbiological pathogens has resulted in a global push to develop effective alternatives (Mandal et al. 2014). AMR is characterized as ineffective infection-related treatment with a previously effective AM agent (Barbieri et al. 2017). Both intrinsic and extrinsic factors contribute to the rise of AMR. For example, intrinsic factor evolution in bacteria includes the development of structural features (Blair et al. 2015), such as microbial biofilm formation (Zacchino et al. 2017) and the insertion of transposons (Rajagopal et al. 2016). Excessive antibiotic use as a result of injudicious prescribing practices, spurred by growing competition in the pharmaceutical industry's production and marketing of AMs, is an extrinsic contributing factor (Lewis 2007).

8.4.1 Potential for AMR Transmission in Plant-Based Feeds

As mentioned in Sect. 8.3.2, plant-based feeds can transmit AMR and have other challenging issues, such as low protein content, poor AA balance, and anti-nutritional factors (EFSA Panel on Biological Hazards (BIOHAZ) 2014). The production of plant-based feed products can differ due to specific agricultural product requirements, prevailing climate conditions, and cultivation systems used such as open fields, protected, conventional, or alternative cultivation systems. In open cultivation systems, the transmission of AMR occurs through wildlife, polluted irrigation water, and the application of fertilizers obtained from animal waste. In a soilless culture system, there is less potential for transmission of AMR than in open systems as chemical nutrient solutions are used (Bandara 2018). Transmission of AMR also depends on food type; for example, root crops are closer to soil than leafy vegetables, so root crops can receive more contaminants (EFSA Panel on Biological Hazards (BIOHAZ) et al. 2021). In the production of plant-based feed, soil is one of the environmental sources of AMR such as using fecal material of human (sewage sludge, biosolids) or animal (manure) origin as fertilizers. This includes wastewater from buildings and yard run-off from fields contaminated by fecal-based fertilizers (Christou et al. 2019). Dust containing AM residues can represent an additional transmission route (Glaize et al. 2020). Treated or untreated wastewater or polluted water used for irrigation has the potential to contaminate fresh produce (Benami et al. 2016). At harvest, handling, processing, and storage of plant products are widely recognized sources of contamination. Rinsing water, reusable packing cases, humans, machinery, personnel involved in handling the product, and arthropods are the potential sources of AMR in the food-producing environment.

8.4.2 Potential for AMR Transmission in Animal-Based Feeds

Animal-based feed ingredients mainly include fish meal, fish oil, shrimp meal, squid meal, blood meal, wastes of organisms, and so on. The most common is fish meal, which can be obtained from the whole fish or the parts of fish. Most of these feeds are directly added to the systems. However, some of them are used only after preliminary checking. The nutritional value of animal-based feed ingredients is comparatively greater than that of plant-based sources. However, formulated plant-based feed sources with balanced nutritional levels are gaining attention because of their eco-friendly nature and sustainability.

8.4.2.1 Pre-harvest Sector

1. The bacteria can spread in the feed through contamination and favorable growth conditions. For example, *Salmonella* species can multiply in heat-processed feeds when subjected to the cooling process (EFSA Panel on Biological Hazards (BIOHAZ) et al. 2021). This organism will start growing and multiplying whenever conditions become favorable.
2. AMR can be transferred via contaminated parts of machinery and unhygienic storage facilities. When these systems are contaminated, employees can become infected with these microbes and directly or indirectly transfer them to other individuals.
3. Air, water, and dust can also act as the source of AMR, and insects such as flies and beetles can transfer them elsewhere.
4. Animals reared in open outdoor systems are more prone to becoming infected with AMR bacteria than those reared in closed systems.
5. Early onset of feeding can have an impact on the growth of animals. If young animals are fed contaminated or waste feed, it can be the source of AM genes.

8.4.2.2 Post-harvest Sector

1. Microbes can be transferred via contaminated feed, slaughterhouse raw materials, vehicles, or boxes by direct contact.
2. Production systems contaminated with animal feces, blood, and viscera wastes, especially slaughterhouses, are a source of AMR bacteria. Many of these pathogenic microbes can transfer through aerosols. When the animals are killed and processed, they can become infected with bacterial species such as *Salmonella* and *Listeria*.

8.4.3 Potential for AMR Transmission via Aquatic Animals and Other Sources

The aquatic environment, such as lakes, rivers, and coastal water, receives AM residues and resistant bacteria from human, agricultural, and animal sources through various pathways such as sewage, farm wastes, effluent, sludge, stormwater, and groundwater flow. These bacteria can persist and spread in the water or sediments and pose a risk to aquatic organisms if they are pathogenic (Fig. 8.2). Humans can also be exposed to these bacteria by eating aquatic organisms (EFSA Panel on Biological Hazards (BIOHAZ) 2014).

1. Organisms that filter water, such as mollusks and bivalves, can trap and accumulate particles, including organic matter and bacteria.
2. The introduction of new or wild species into aquaculture systems can facilitate the transfer of AMR genes among bacteria.
3. AMR can also be transmitted through unclean or unsanitary post-harvest processing conditions, such as the workplace, machinery, employees, water quality, and so on.

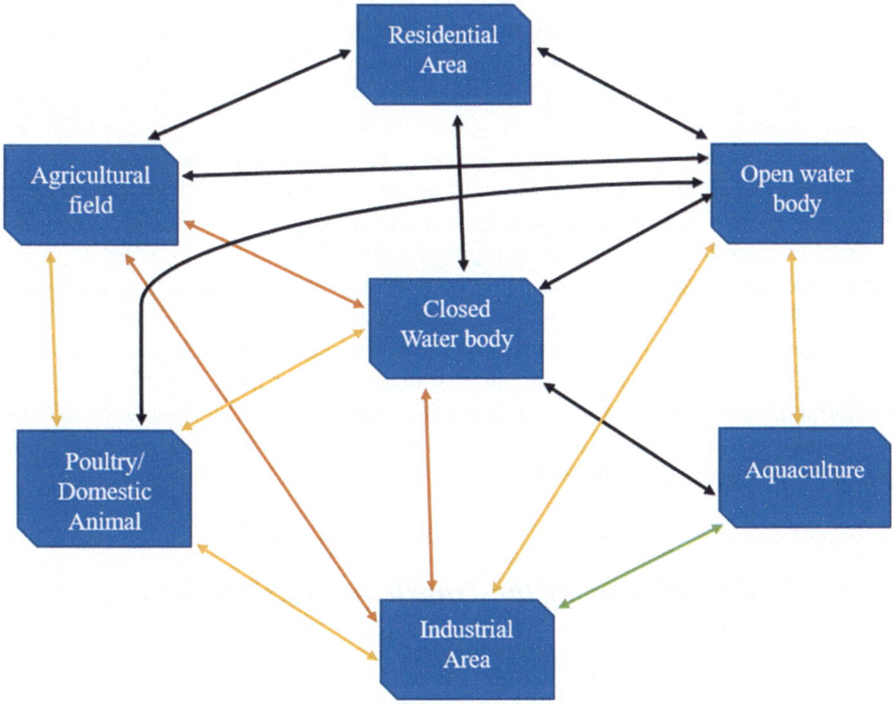

Fig. 8.2 Potential transmission pathways of AMR between different feed sources (Zhao et al. 2020)

8.5 Mechanisms and Factors of AMR Transmission

8.5.1 Mechanism of Transfer of AMR

Microbes use a variety of strategies to combat AM agents, including activating alternative pathways, altering antibiotic targets, cell wall permeability changes, and enzyme-mediated degradation of antibiotics or antibacterial compounds (Verraes et al. 2013). Enzymatic breakdown of AM agents is a typical resistance mechanism in bacteria. As a means of survival, saprophytic microorganisms create AM compounds that prevent the growth of nearby microbes. It is possible to develop resistance by altering the targeted antibiotic so that it no longer binds to the bacterial cell and has no harmful effects. Examples of this kind of mechanism that the quinolone and fluoroquinolone antibiotics target include mutations related to the gyrase and topoisomerase genes (Drlica and Zhao 1997). Bacteria's usage of fluoroquinolone resistance methods includes altering the target (DNA-gyrase), increasing efflux (the cell's removal of the drug outside), inactivating the fluoroquinolone (via aminoglycoside *N*-acetyltransferase), and attacking DNA-binding proteins (Redgrave et al. 2014). Through the persistence and growth of bacteria with inherent or newly acquired resistance (such as mutations), AM exposure increases selective pressure (Aminov 2009). The cell may use two different defense strategies to defend itself from specific antibiotics: changing the permeability of the cell wall and controlling the efflux systems. For instance, genes involved in efflux systems are linked to lower sensitivity to tetracycline (McMurry et al. 1980). Several bacterial genera and species contain intrinsic AMR systems. Infections caused by opportunistic antibiotic-resistant bacteria, such as Trueperella (Arcanobacterium) pyogenes in cattle and *Clostridium difficile* in humans, are of concern because they can be fatal (Viseur et al. 2011). Both horizontal gene transfer (HGT; transmission of genetic material from cell to cell) and vertical gene transfer (VGT; parent to offspring) are methods for acquiring AMR genes. HGT and VGT can both occur at the same time. Through the discovery of novel resistant genes and processes, HGT causes a population of microorganisms to have improved resistance profiles (Acar and Moulin 2006). HGT can take place through the assimilation of extracellular DNA (transformation), bacteriophage (transduction), or pili (conjugation) (Vikesland et al. 2019). Interspecies transmission is made possible by HGT, which also plays a significant role in the propagation of AMR (Holmes et al. 2016).

8.5.2 Factors Influencing the Transfer of AMR Between Feed Sources

The most obvious factor contributing to the spread of AMR in the environment is the use of antibiotics that are sensitive to selection pressure (Fig. 8.4). Over 80% of antibiotics used by people or animals end up in feces in sewage treatment systems.

The prevalence of antibiotics encourages the development of ARB and antibiotic-resistant genes (ARGs) and causes bacteria living in various settings to acquire resistance determinants. The growth of AMR in the soil and aquatic systems is a result of the usage of biofertilizers in agricultural areas. Numerous studies have demonstrated that spreading animal feces directly onto agricultural land can introduce ARGs such as tetO, tetQ, tetW, intl1, and tetY. The resistance to aminoglycosides, chloramphenicol, tetracyclines, trimethoprim, and sulfonamide was also assessed by 13 distinct ARGs from farmed fish waste (Fig. 8.3). First laboratory tests on bacteriophages carrying ARGs revealed that they can transduce genes conferring resistance to ceftazidime, aztreonam, and imipenem in *Pseudomonas aeruginosa*, tetracycline in *Staphylococcus aureus*, and *Actinobacillus*

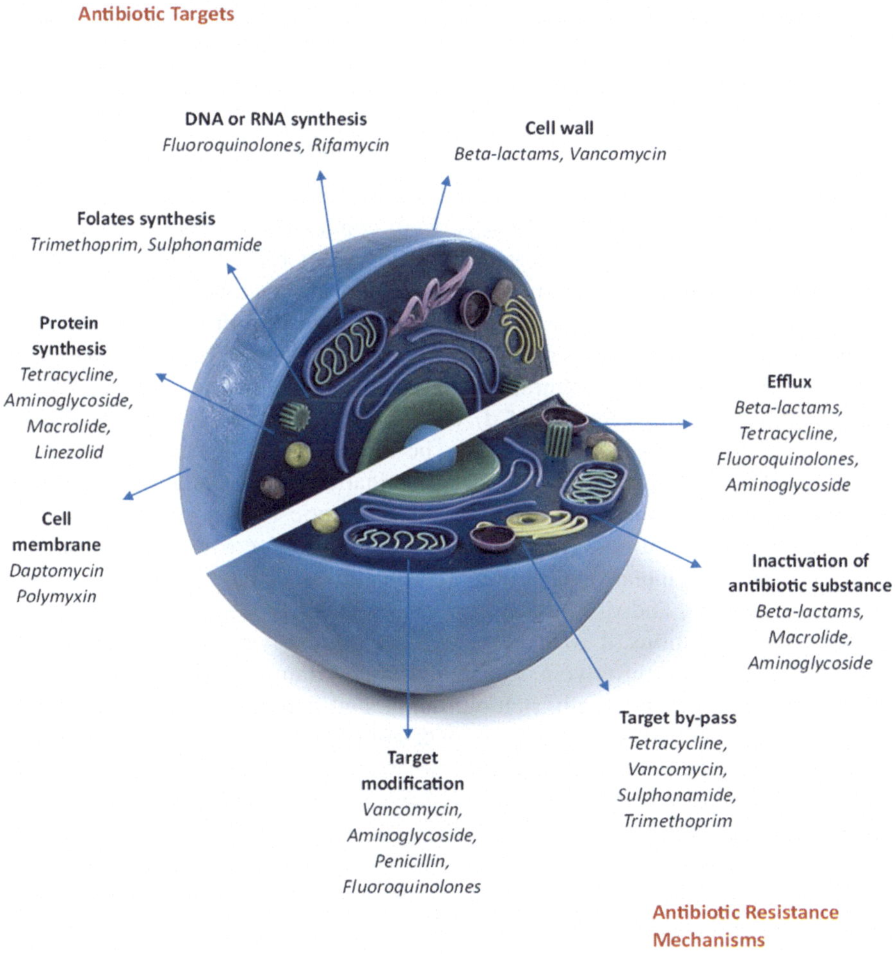

Fig. 8.3 Antibiotic resistance mechanisms with the possible target site (Uluseker et al. 2021)

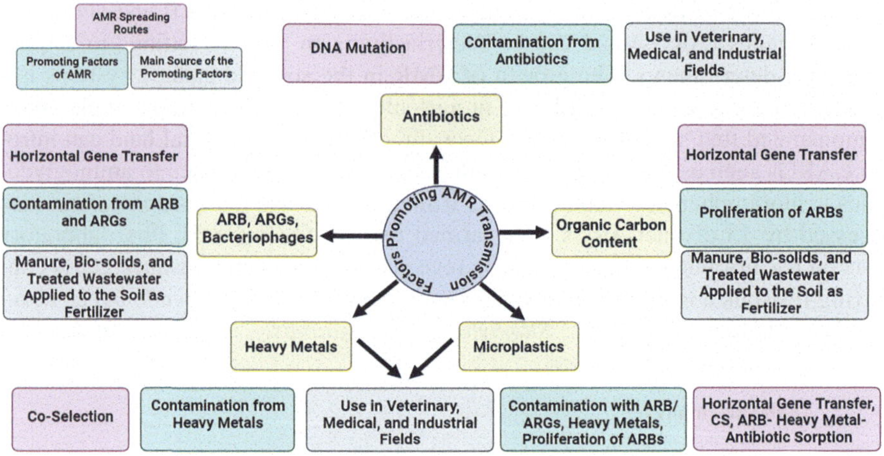

Fig. 8.4 Factors promoting the AMR transmission

actinomycetemomitans and methicillin in *Staphylococcus epidermis*. The ARGs detected in wastewater *mec*A and bla_{TEM} were commonly isolated. ARGs (aadA, strA, strB, and sul1 genes) are stored in soil by soilborne bacteriophages, while DNA (blaTEM, blaCTX-M, and mecA) is isolated from river water.

It is vital to consider bacteriophages as a major element driving the spread of AMR (Fig. 8.4). Heavy metals are a critical factor that should be regarded as promoters of resistance. Agricultural and aquacultural techniques frequently use heavy metals, including cadmium (Cd), copper (Cu), lead (Pb), mercury (Hg), and zinc (Zn), which cause heavy metals to be released into the environment and cause AMR transmission. Numerous investigations have established a link between increased phenotypic or genotypic antibiotic resistance and enhanced heavy metal concentrations. Bacteria have been found to co-select for antibiotic resistance driven by metals in a variety of environments, including the ocean, soil, manure, sediments, and drinking water. Zhang et al. (2018) demonstrated that copper at concentrations of 10 and 100 mg/L substantially improved water bacteria's resistance to erythromycin, kanamycin, and rifampin. ARGs spread more readily in environments with rich nutrients and temperatures above 30 °C (Fig. 8.4). In a nutshell, soil characteristics, soil depth, and temperature play significant roles in the environmental dissemination of AMR. However, soil conditions as a factor favoring AMR spread receive less consideration. Surprisingly, the rhizosphere and certain types of soil have been demonstrated to facilitate the spread of ARGs (Lu et al. 2020). ARGs transfer and abundance were found to be higher in maize rhizosphere soils than in tomato or wheat rhizosphere soils (Goryluk-Salmonowicz and Popowska 2022).

8.6 How Aquaculture Contributes to the Global AMR Crisis

AMR is a growing global health concern, as it renders antibiotics and other AM agents less effective in treating bacterial infections. The extensive use of antibiotics in aquaculture has been identified as a significant contributor to the global AMR crisis. This section delves into how aquaculture practices exacerbate AMR through three main aspects: poor management practices (PMPs), the use of subtherapeutic doses of AMs, and the proximity of aquaculture facilities to each other and to wild aquatic populations.

8.6.1 Poor Management Practices

PMPs in aquaculture contribute to AMR by fostering an environment where bacteria can rapidly develop resistance to antibiotics.

8.6.1.1 Overcrowding and Poor Water Quality

Overcrowding in aquaculture facilities is a common practice to maximize production. In overcrowded conditions, fish are more susceptible to stress and disease. To combat diseases that thrive in these conditions, aquaculturists often use antibiotics preventively. Furthermore, poor water quality resulting from overcrowding can weaken the immune systems of fish, necessitating antibiotic treatment. The continuous use of antibiotics without proper disease diagnosis or regulation promotes the development of AMR in bacteria present in the aquaculture system.

8.6.1.2 Inadequate Biosecurity Measures

Inadequate biosecurity measures, such as the absence of strict quarantine protocols and limited control over the introduction of new fish into the facility, increase the risk of disease outbreaks. To manage these outbreaks, antibiotics are often used indiscriminately. This practice not only promotes AMR but also poses risks to wild aquatic populations if infected fish escape from aquaculture facilities.

8.6.2 Use of Subtherapeutic Doses of AMs

The use of subtherapeutic doses of AMs in aquaculture is another major contributor to AMR. Subtherapeutic use of antibiotics is common in aquaculture for purposes other than therapeutic treatment. Antibiotics are often added to fish feed or water to promote growth and prevent disease, even in the absence of a diagnosed infection. This continuous exposure to low doses of antibiotics creates a selective pressure on bacteria to develop resistance, as only the resistant strains survive and proliferate.

8.6.3 Proximity of Aquaculture Facilities to Each Other and to Wild Aquatic Populations

The spatial arrangement of aquaculture facilities can significantly influence the spread of AMR in the environment.

8.6.3.1 Dense Clustering of Facilities

The clustering of aquaculture facilities in close proximity to one another can facilitate the transmission of AMR bacteria. Bacteria resistant to antibiotics used in one facility may spread to neighboring facilities through waterborne transmission, wind, or the movement of equipment, leading to cross-contamination and the emergence of multidrug-resistant strains.

8.6.3.2 Interaction with Wild Aquatic Populations

The proximity of aquaculture facilities to wild aquatic populations poses another concern. Antibiotic residues and resistant bacteria can escape into natural ecosystems through water discharges and the release of uneaten feed or fish excreta. This can result in the transfer of AMR genes to bacteria in wild aquatic environments, further fueling the global AMR crisis.

8.7 Impacts of AMR Human Health, Animal Health, and the Environment

8.7.1 Increased Treatment Cost

One of the world's most important public health issues is AMR. It increases the cost of health care by extending disease and hospital stays, needing more testing (Maragakis et al. 2008), and necessitating more expensive treatments (Hawkey 2008). Every year, about 700,000 people die as a result of AMR-resistant infections. AMR is estimated to have killed around 10 million people between now and 2050, with an economic cost of 100 trillion USD (O'Neill 2016). A significant proportion of Gram-negative bacteria are recognized for their rapid replication and genetic adaptability (Fig. 8.5) (Prescott 2003). Species such as *Salmonella* can infect the gastrointestinal tract and cause symptoms such as diarrhea, fever, cramps, and so on. They belong to the family of Enterobacteriaceae where most of the species are pathogenic. However, some of the environmental bacteria like *Acinetobacter baumannii* which are usually found in soils, water, and hospital settings are also able to cause human infections such as pneumonia, bloodstream infections, urinary tract infections, and surgical site infections (Fig. 8.6) (Peleg and Hooper 2010).

8.7.2 Reduced Efficacy of AMs

AMR can decrease the effectiveness of other AM drugs on animals. This can increase the illness and mortality rate among animals. When these AM treatments become ineffective, it can increase the need for additional diagnostic tests which eventually increase the overall cost for healthcare. The failure of diagnostic

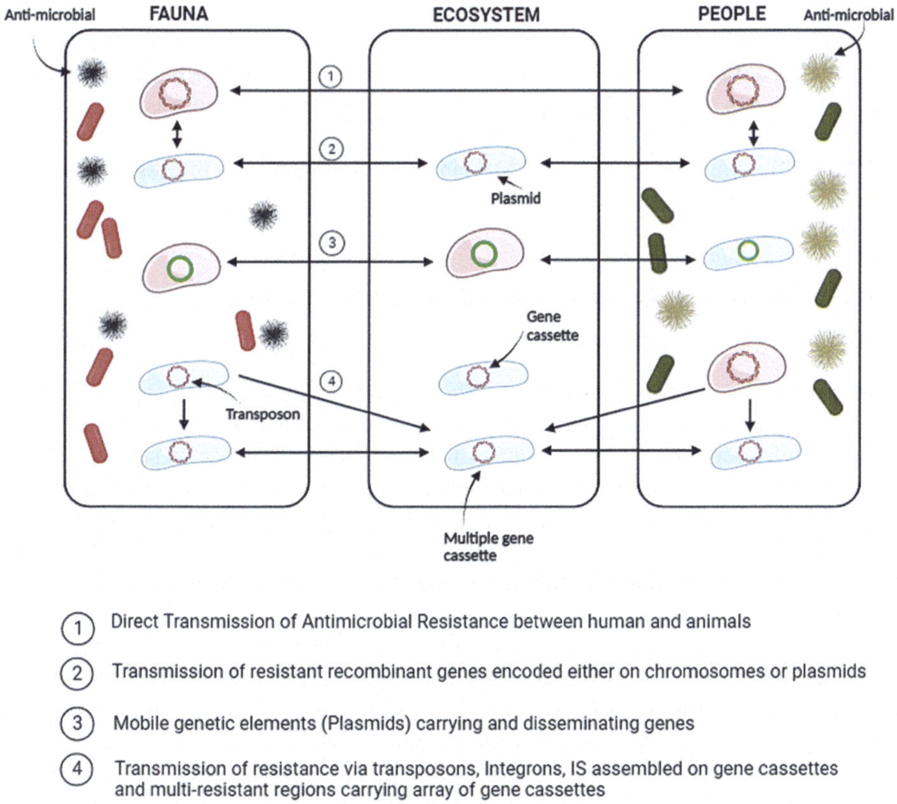

Fig. 8.5 Transmission of AMR between humans and animals

Type	Bacteria
Carbapenem resistance	*Acinetobacter, Escherichia coli, Klebsiella pneumoniae*
Drug resistance	*Neisseria gonorrhoeae, Streptococcus pneumoniae*
Multi-drug resistance	*Mycobacterium tuberculosis, Pseudomonas aeruginosa*
Methicillin resistance	*Staphylococcus aureus*
Vancomycin resistance	*Enterococcus*

Fig. 8.6 Type of emerging resistant bacteria with infectious types

treatments can increase health risks and death rates. In animal farming systems, AMR can negatively impact productivity and economic gain. When these AM residues or resistant genes are discharged into natural environments like waterbodies or soils, they contaminate the ecosystem and disrupt the ecological balance between the organisms in the food chain. Thus, the environment itself can act as the reservoir of these resistant genes, and they can be further horizontally transferred between bacteria, allowing the spread of AMR and the emergence of multidrug-resistant strains.

8.7.2.1 Which Are the Main Infections Becoming Resistant to Antibiotics?

Bacteria are responsible for causing various human illnesses with their ability to develop resistance to one or more antibiotics. In particular, the prevalence of resistance to third-generation cephalosporins has been noticed in species such as *E. coli* and *K. pneumonia*. Most of these bacterial infections need to be treated with alternative antibiotic families, which are often more costly. However, not all bacteria respond to the same antibiotics. Therefore, hospital patients are at higher risk of getting infections from resistant bacteria that live in healthcare settings. These are called nosocomial infections.

The final antibiotic treatment regimen for drug-resistant bacteria should always be personalized to a patient based on the results of an AM susceptibility test (AST) and the experience of a medical practitioner. The following factors influence treatment selection: (1) severity of infection; (2) patterns of local drug susceptibility; (3) age and condition of kidney and liver; (4) sensitivity of the drug; (5) number of times took previous treatments; and (6) cost considerations.

8.7.3 Potential for Resistant Bacteria to Spread to Other Organisms

The dissemination of resistant bacteria can occur through diverse pathways, which are mainly influenced by specific circumstance factors. Out of these, poor hygiene practices, inadequate sanitation measures, and insufficient infection control measures are the main factors responsible for transferring resistant bacteria in healthcare settings and among organisms and surrounding environments. Animals for food production are transferred across borders, groceries are exported from most regions of the world, and bacteria follow. This adds to the intricacy of the antibiotic resistance problem and supports the notion that it is a worldwide problem. It makes no difference where a resistant bacterium arises. If it is successful and grows in popularity, it may travel fast to other regions of the world in our globalized society.

Person to Person: Bacterial infections can rapidly spread from one person to another either by direct contact by touching surfaces contaminated by others or by indirect contact (cough). However, the spread of bacteria to other people depends on the health and hygiene measures they take.

Animals to Humans and Vice Versa: The treatment of infections becomes complex whenever the pathogens develop resistance to medications. Nowadays, pets are sharing our living spaces and becoming an integral part of our daily lives. Humans can encounter close contact with them while petting, playing, or handling them. The same trend was seen in animals raised for food production, where humans, particularly farmers and workers, have consistent contact with animals during feeding, care management, and so on. Moreover, wildlife species can encounter human populations through various activities such as camping, hiking, wildlife observations, and so on. They can act as a reservoir of zoonotic diseases that can be eventually transferred to humans. This kind of transmission can occur via direct or indirect contact with their habitats or contaminated surfaces (Marshall and Levy 2011). The close links between humans, especially farmers and animal populations pertaining to bacterial transmission and related resistance development, were highlighted in Davies and Davies (2010). The occupational risk of spreading bacterial populations to professionals providing medical care to cattle was reported in Garcia-Graells et al. (2012).

Water: Bacteria have the ability to disseminate and potentially infect living beings through different water sources, such as irrigation networks, the cleansing of kitchen utensils, and sanitary facilities. The widespread presence of resistant bacteria in various aquatic environments has been reported previously (Nnadozie and Odume 2019). Infectious diseases can spread when the individual comes in contact with contaminated water. The discharge of untreated water from industries or processing plants to natural waterbodies can act as a source of potentially resistant bacteria.

Spread Within Healthcare Facilities: As these environments are in close proximity to sick people, antibiotic-resistant strains are easily transmitted across healthcare facilities (Clements et al. 2008). Additionally, these bacteria can spread from patients to medical professionals or other visitors when touched by contaminated hands or clothes. This spread may also be accelerated by poor equipment sanitation and hospital hygiene.

8.8 Screening and Monitoring of AMR in Feed

Screening and monitoring of AMR in aquaculture feed is essential to preventing the emergence and spread of AMR. By regularly monitoring the levels of AM residues and resistance genes in feed, producers can detect potential problems and take appropriate actions to mitigate them. Regular screening can also help identify the sources and patterns of AMR in aquaculture, allowing for better-targeted interventions to reduce the prevalence of resistant bacteria in aquatic environments. Furthermore, monitoring the efficacy of antibiotics in aquaculture feed can inform the development of more effective and sustainable alternatives to antibiotics, such as probiotics and prebiotics. In brief, screening and monitoring of AMR in aquaculture feed is crucial for the sustainable development of aquaculture and the protection of public health (Table 8.1). It helps ensure the responsible use of antibiotics and promotes the development of alternative strategies for disease control.

Table 8.1 Advantages and disadvantages of different screening and monitoring methods

Method	Advantages	Disadvantages	References
Culture-based	Inexpensive, easy to perform	Limited ability to detect newer or less commonly used antibiotics	Oliver (2003), Kaprou et al. (2021)
Molecular-based	Highly specific and sensitive, it can detect resistance to a wide range of antibiotics, including those that are newer or less commonly used. They can also reveal the genetics of resistance and track its spread between bacterial strains	Relatively expensive, and requires specialized equipment and expertise. They may also generate a large amount of data that requires careful interpretation	Blair et al. (2015), Liu et al. (2016), Kaprou et al. (2021)
Immunological	Rapid and highly specific, it requires minimal sample preparation	Less sensitive than other methods, limited ability to detect resistance to a wide range of antibiotics	Kaprou et al. (2021)
Biosensors	Rapid and highly specific	Limited ability to detect resistance to a wide range of antibiotics may require specialized equipment and expertise	Kaprou et al. (2021)

8.8.1 Culture-Based Methods for Detecting AMR in Aquaculture Feed

Culture-based methods for detecting AMR involve various laboratory techniques that help identify microorganisms and determine their susceptibility to specific antibiotics. This method relies on culturing the microorganism on either selective or nonselective media and identifying it based on the morphological and biochemical characteristics of the colonies. Selective media promote the growth of target bacteria while inhibiting the growth of competing organisms. Alkaline peptone water (APW) and thiosulfate citrate bile salts sucrose agar (TCBS), for example, have been widely used to enrich and isolate *Vibrio cholerae* and *Vibrio parahaemolyticus* (Oliver 2003). Cellobiose polymyxin colistin agar is the most commonly used selective medium for *Vibrio vulnificus*.

Here are some commonly used traditional AST methods.

Disk diffusion involves placing a paper disk containing a specific AM agent on the surface of a Mueller Hinton agar plate inoculated with a bacterial culture and following standardized guidelines (Bauer et al. 1966). After incubation, the diameter of the zone of inhibition around the disk is measured to determine the susceptibility or resistance of the bacterial strain to the AM agent. It uses a set of specific antibiotics and measures the diameter of the zone of inhibition to determine susceptibility or resistance.

Broth dilution involves exposing a bacterial culture to increasing concentrations of an AM agent in a liquid nutrient medium. The lowest concentration of the AM agent that inhibits bacterial growth is known as the minimum inhibitory concentration (MIC). The MIC is determined as the lowest concentration of AM agent that prevents visible growth of the bacterial strain (Wiegand et al. 2008). It is often performed using broth or agar dilution methods.

While culture-based methods are widely used, they have some limitations, such as their inability to detect non-culturable bacteria and the time required for bacterial isolation and identification. As a result, newer methods such as polymerase chain reaction (PCR)-based assays and metagenomic sequencing are increasingly used to detect AMR in aquaculture.

8.8.2 Molecular-Based Methods for Detecting AMR in Feed

There are various molecular-based methods available for detecting AMR in aquaculture feed. These methods typically involve identifying specific genes or genetic mutations that confer resistance to particular AM agents. Here are some examples of molecular-based methods for detecting AMR in aquaculture feed.

8.8.2.1 Polymerase Chain Reaction

PCR is a technique that allows for the amplification of specific DNA sequences. In the context of AMR detection, PCR can be used to amplify genes that are associated with resistance to particular AM agents. For example, researchers might use PCR to amplify the gene responsible for tetracycline resistance in bacteria found in aquaculture feed. At least 40 tetracycline resistance genes (tet) have been identified in humans, animals, and the aquaculture industry (Roberts 2005). Real-time PCR for AMR gene detection is not well documented. Most studies use the technique to detect bacteria, viruses, and toxins in the food production chain, with interesting results (Galhano et al. 2021).

8.8.2.2 Next-Generation Sequencing

Next-generation sequencing (NGS) is a high-throughput technique that enables the quick sequencing of substantial volumes of DNA. In the context of AMR detection, NGS can be used to identify mutations in genes that are associated with resistance to particular AM agents. For example, researchers might use NGS to sequence the genome of bacteria found in aquaculture feed to identify mutations in genes associated with fluoroquinolone resistance.

8.8.2.3 Microarray Technology

Microarray technology involves the use of a chip that contains a large number of DNA probes that are complementary to specific genetic sequences. In the context of AMR detection, researchers might use a microarray chip to detect the presence of genes that are associated with resistance to particular AM agents. Microarray technology is particularly useful for detecting multiple genes simultaneously.

Overall, molecular-based methods offer a sensitive and specific means of detecting AMR in aquaculture feed. These methods can identify the presence of resistance genes and mutations, which can be used to inform management strategies aimed at reducing the use of AM agents in aquaculture.

8.8.3 Other Methods for Detecting AMR in Feed

AMR in feed can also be detected using enzyme-linked immunosorbent assays (ELISA), biosensors, and mass spectrometry (MS).

ELISA: ELISA is a technique that uses antibodies to detect specific proteins in a sample. The presence of AMR proteins in bacteria can be detected using ELISA.

Biosensors: Biosensors are analytical devices that detect the presence of certain biological molecules in a sample, such as enzymes or antibodies. AMR bacteria or genes can be detected using biosensors in aquaculture feed.

MS: MS is a strong analytical technique that uses mass and charge to discover and identify compounds. Specific AMR compounds in feed samples can be identified using MS.

These alternative approaches have benefits such as sensitivity, specificity, speed, and the capacity to identify a wide range of AMR genes and bacteria. However, these techniques might need specific tools and knowledge and might cost more than conventional methods based on culture.

8.8.4 Challenges Associated with Screening and Monitoring of AMR in Feed

8.8.4.1 Sampling Considerations

When it comes to screening and monitoring AMR in feed, there are several challenges associated with sampling. These challenges include the following:

1. Representative Sampling: It is crucial to ensure that samples taken for analysis are representative of the entire feed batch or population. However, feed is often heterogeneous, and variations can occur within and between batches. Choosing an appropriate sampling strategy that accounts for this variability is essential to obtain accurate and reliable results.
2. Sample Size: It is crucial to determine the optimum sample size to generate relevant results. The sample size should be sufficient to account for feed variability while being manageable and affordable. The issue is to strike the correct balance between these aspects.
3. Sample Collection: Collecting samples in a manner that minimizes contamination and preserves the integrity of the feed is crucial. Contamination during sampling can lead to false-positive or false-negative results, compromising the accuracy of the analysis.
4. Sampling Frequency: Deciding on the frequency of sampling is challenging. AMR patterns can change over time, and it is essential to capture these changes adequately. However, frequent sampling may not always be feasible due to logistical constraints and cost implications.

8.8.4.2 Data Analysis and Interpretation

Once the samples have been collected and analyzed, the data obtained need to be analyzed and interpreted accurately. Some challenges associated with data analysis and interpretation include the following:

1. Complex Data Sets: Screening and monitoring of AMR in feeds can generate large and complex data sets, which can be challenging to analyze. Advanced statistical methods and bioinformatics tools may be required to handle and interpret these data effectively.
2. Data Integration: Data on AMR in the feed may come from various sources, including different farms, regions, and laboratories. Integrating and standardizing these diverse data sets can be challenging, particularly when different methods and protocols are used.
3. Identifying Relevant Indicators: Determining which AMR indicators to prioritize and interpret can be difficult. There are numerous genes, bacteria, and resistance mechanisms associated with AMR, and not all of them may be equally relevant in the context of feed. Identifying the most informative indicators is crucial for effective monitoring and decision-making.

8.8.4.3 Lack of Standardization

Standardization is a key challenge in screening and monitoring AMR in feed. A lack of standardized protocols and methodologies can lead to inconsistencies and variations in data collection, analysis, and interpretation. This lack of standardization hampers comparability between studies and makes it challenging to establish baseline data and monitor trends over time.

1. Sample Processing: Standardizing the methods for sample processing, including DNA extraction, amplification, and sequencing, is essential to ensuring consistency and comparability of results across different laboratories.
2. Resistance Interpretation: There is a need for standardized guidelines and criteria for interpreting AMR data in the context of feed. This includes defining appropriate thresholds for resistance levels, distinguishing between clinically relevant and non-relevant resistance genes, and establishing guidelines for reporting and surveillance.
3. Harmonization of Data Reporting: Standardizing the format and content of data reporting is crucial for effective data sharing, comparability, and integration. Common data elements, metadata standards, and data-sharing platforms can help address this challenge.

Addressing these challenges requires collaboration and coordination among stakeholders, including researchers, regulators, industry, and international organizations. Efforts toward standardization, data sharing, and harmonization are essential for robust screening and monitoring of AMR in feed and for informing effective AM stewardship strategies.

8.9 Mitigation Strategies to Reduce the Transmission of AMR

Based on the data of the World Health Organization (WHO), the post-antibiotic era is about to start already on the verge of that period (Alanis 2005), and the consequence will lead to a situation where treatment for bacterial infection through antibiotics will lose their effectiveness. Based on the studies, it will happen that, in upcoming years' resistance to antibiotics may continue to increase (Oliveira Hashiguchi et al. 2019). To avoid the global consequences of AMR, sustainable mitigation strategies should be adopted.

Feed additives can be an important alternative to antibiotics to accelerate indigenous mechanisms of defense for the expanding aquaculture industry throughout the world (Dawood et al. 2017). It has been hypothesized from a number of studies that inherent immunity could be activated in two ways from feed additives, where the first is to stimulate the inherent immune system directly and the other is to increase the growth of commensal microbiota (Galhano et al. 2021). As an alternative to the widespread use of chemicals and antibiotics, probiotics could be a convenient option (Dawood et al. 2017). Probiotics that defined as a microbial food supplement that can take up and colonize the digestive tract, especially in mucosal epithelium which ultimately reduces the pathogen's ability to produce diseases (Dawood et al. 2017). Non-digestible fibers accelerate the number of beneficial gut bacteria to grow which is defined as prebiotics (Roberfroid 2005). The immune system can be triggered through prebiotics also and which enhance the protection against infection of the host (Dawood et al. 2016). Within the gut intestine tract, the microbial community could be modified through prebiotics which helps to improve non-specific immune responses (Dawood et al. 2017). Having physiological and biochemical functions nucleotides are a key nutrient for fish and shrimp which increase disease resistance for both shrimp and fishes (Roberfroid 2005). Growth performance and digestive enzymes could be increased, notably by using symbiotics, which have a cumulative effect on the efficiency of the digestive system and immune-hematological response in fish (Dawood et al. 2017), where symbiotics include both prebiotics and probiotics (Gatlin et al. 2007; Roberfroid 2005). Dietary AAs are important as energy substrates for fish, which also influence the synthesis of endogenous protein and regulation of metabolic pathways (Dawood et al. 2017).

Infection burdens and AM needs may be reduced by good management practices (GMPs) (FAO 2021). Biosecurity indicates the process through which in culture farms or hatcheries, prevention of pathogen entry has been assured which leads to a reduction of risk for disease outbreaks and boundless use of AMs (Henriksson et al. 2018). Proper equipment management which is used in aquaculture farms or in a hatchery can reduce infectious disease outbreaks which leads to limited use of AM (Henriksson et al. 2018). To prevent transmission of disease, treatment of water could be a convenient option during the exchange of water from local sources, and to use mechanical aerators to introduce more effective biosecurity, which will have a cumulative effect on the use of AM. Screening through PCR and rapid testing of post-larvae can help to identify the presence of pathogens (Henriksson et al. 2018). Coordinated planning can help reduce disease spread when the farm is far enough.

Vaccines have been shown to be effective in controlling finfish disease outbreaks, but not for mollusks and crustaceans (Du Pasquier 2001). Norway has developed successful vaccines to prevent diseases caused by some major bacteria in salmon farming, setting an example for the global aquaculture industry (Henriksson et al. 2018). Despite its extensive use of AM, China has also implemented vaccine treatment for grass carp hemorrhagic virus (GCHV) (Mu et al. 2021).

The aim of the aquaculture extension program is to help farmers through advanced knowledge of farming which is usually arranged by the governments and non-governmental organizations (NGOs) (Garrett et al. 1997). Through the extension program, advanced management of farms can be ensured which includes advanced information pathways and extension agencies can help to provide advanced information on better management processes which leads to limited use of AM (FAO 2005).

Policymakers are in a difficult situation to introduce a guideline for adding antibiotics to animal feed throughout the world (Chattopadhyay 2014). The indiscriminate use of antibiotics globally, especially in major fish-producing countries in Asia is high, which indicates the absence of strict enforcement of policies and regulations, and it has been found that government ministries are influential (Lulijwa et al. 2020). To achieve sustainable aquaculture production and a fair playing field in production and marketing, investments in research, international collaboration, policy coordination, legislation, maximum residue limits (MRLs), and experience exchange are required (Lulijwa et al. 2020). Effective AMR efforts must thus shift upstream from local (farm) surveillance and management of antibiotic consumption and place a larger emphasis on a more proactive, preventative strategy. This has been demonstrated to be a problem for overall AMR policies, with global AMR commitments from 2015 to 2021 remaining largely focused on research, surveillance, and stewardship rather than the more upstream, preventative steps indicated in the 2015 AMR Global Action Plan (Thornber et al. 2022).

8.10 Conclusion

The use of antibiotics as growth promoters in animal feed has been linked to the emergence of AMR in bacterial populations, including those that infect humans. The use of fishmeal and other animal-based protein sources in aquaculture feed has also been linked to the transmission of AMR between fish and other aquatic organisms. Since in the first place, AMs were introduced to the aquaculture system in order to control or prevent disease outbreaks, a good disease management framework as a control strategy is recommended to deal with antibiotic usage. The focus should be on updating management routines, responsible use of AM agents from registration, approval, and controlled application, and strengthening national policy to control the whole antibiotic usage process. Sustainable solutions to solve this crisis need to be exercised. Furthermore, educational and training intervention for the individuals associated with the aqua farms and fish feed business is recommendable to raise awareness and upgrade knowledge and skills on the appropriate use of AMs. It is essential to reduce the use of AMs in fish feed and adopt alternative methods for disease prevention and growth promotion. These methods could consider better management practice, improved nutrition, use of vaccines and probiotics, implementing regulation, and assigning veterinary professionals to guide on the use of AM agents. Developing countries will need support and assistantship from international corporations to ensure capacity building and implementation of sustainable efforts to combat AMR. Therefore, screening and monitoring of AMR in aquaculture feed are integral components of responsible aquaculture management. By implementing effective surveillance programs, promoting responsible AMU, and exploring alternative strategies, we can work toward sustainable aquaculture practices that minimize the risk of AMR development and ensure the health and welfare of both aquatic species and humans.

Acknowledgments The first author acknowledges the Erasmus Mundus Joint Master's Scholarship for support.

Conflict of Interest The authors declare no conflict of interest.

References

Acar JF, Moulin G (2006) Antimicrobial resistance at farm level. Revue Sci Tech 25(2):775–792

AftabUddin S, Siddique MAM, Habib A, Akter S, Hossen S, Tanchangya P, Abdullah Al M (2021) Effects of seaweeds extract on growth, survival, antibacterial activities, and immune responses of *Penaeus monodon* against *Vibrio parahaemolyticus*. Ital J Anim Sci 20(1):243–255

Ajani EK, Nwanna LC, Musa BO (2004) Replacement of fishmeal with maggot meal in the diets of Nile tilapia, *Oreochromis niloticus*. World Aquacult 35(1):52–54

Alanis AJ (2005) Resistance to antibiotics: are we in the post-antibiotic era? Arch Med Res 36(6):697–705. https://doi.org/10.1016/j.arcmed.2005.06.009

Albrektsen S, Mundheim H, Aksnes A (2006) Growth, feed efficiency, digestibility and nutrient distribution in Atlantic cod (*Gadus morhua*) fed two different fish meal qualities at three dietary levels of vegetable protein sources. Aquaculture 261(2):626–640

Aminov RI (2009) The role of antibiotics and antibiotic resistance in nature. Environ Microbiol 11(12):2970–2988. https://doi.org/10.1111/j.1462-2920.2009.01972.x

Bandara T (2018) Alternative feed ingredients in aquaculture: opportunities and challenges. J Entomol Zool Stud 6(2):3087–3094

Barbieri R, Coppo E, Marchese A, Daglia M, Sobarzo-Sánchez E, Nabavi SF, Nabavi SM (2017) Phytochemicals for human disease: an update on plant-derived compounds antibacterial activity. Microbiol Res 196:44–68

Bauer AW, Kirby WM, Sherris JC, Turck M (1966) Antibiotic susceptibility testing by a standardized single disk method. Am J Clin Pathol 45(4):493–496

Benami M, Gillor O, Gross A (2016) Potential microbial hazards from graywater reuse and associated matrices: a review. Water Res 106:183–195

Bengtsson-Palme J, Hammarén R, Pal C, Östman M, Björlenius B, Flach CF, Fick J, Kristiansson E, Tysklind M, Larsson DGJ (2016) Elucidating selection processes for antibiotic resistance in sewage treatment plants using metagenomics. Sci Total Environ 572:697–712. https://doi.org/10.1016/j.scitotenv.2016.06.228

Blair JM, Webber MA, Baylay AJ, Ogbolu DO, Piddock LJ (2015) Molecular mechanisms of antibiotic resistance. Nat Rev Microbiol 13(1):42–51

Bureau DP, Harris AM, Bevan DJ, Simmons LA, Azevedo PA, Cho CY (2000) Feather meals and meat and bone meals from different origins as protein sources in rainbow trout (*Oncorhynchus mykiss*) diets. Aquaculture 181:281–291

Burridge LE, Weis JS, Cabello FC, Pizarro J, Bostick K (2010) Chemical use in salmon aquaculture: a review of current practices and possible environmental effects. Aquaculture 306:7–23

Cabello FC, Godfrey HP, Tomova A, Ivanova L, Dölz H, Millanao A, Buschmann AH (2013) Antimicrobial use in aquaculture re-examined: its relevance to antimicrobial resistance and to animal and human health. Environ Microbiol 15(7):1917–1942. https://doi.org/10.1111/1462-2920.12134

Chattopadhyay MK (2014) Use of antibiotics as feed additives: a burning question. Front Microbiol 5:334. https://doi.org/10.3389/fmicb.2014.00334

Chen Y, Zhang H, Luo Y, Song J (2012) Occurrence and dissipation of veterinary antibiotics in two typical swine wastewater treatment systems in East China. Environ Monit Assess 184(4):2205–2217. https://doi.org/10.1007/s10661-011-2110-y

Christou A, Papadavid G, Dalias P, Fotopoulos V, Michael C, Bayona JM et al (2019) Ranking of crop plants according to their potential to uptake and accumulate contaminants of emerging concern. Environ Res 170:422–432

Chuchird N, Rairat T, Keetanon A, Seguin D, Chotikachinda R, Manomaitis L, Kanjanamayoon C (2023) Effect of feed enzymes and functional immunostimulants supplementation on growth performance and overall health of postlarvae and juvenile Pacific white shrimp, *Penaeus vannamei*, fed soybean-based diets. J World Aquacult Soc 54:814. https://doi.org/10.1111/jwas.12939

Clements A, Halton K, Graves N, Pettitt A, Morton A, Looke D, Whitby M (2008) Overcrowding and understaffing in modern health-care systems: key determinants in meticillin-resistant Staphylococcus aureus transmission. Lancet Infect Dis 8(7):427–434. https://doi.org/10.1016/S1473-3099(08)70151-8

Davies J, Davies D (2010) Origins and evolution of antibiotic resistance. Microbiol Mol Biol Rev 74(3):417–433. https://doi.org/10.1128/MMBR.00016-10

Dawood MA, Koshio S, Ishikawa M, Yokoyama S, El Basuini MF, Hossain MS, Nhu TH, Dossou S, Moss AS (2016) Effects of dietary supplementation of Lactobacillus rhamnosus or/and Lactococcus lactis on the growth, gut microbiota and immune responses of red sea bream, Pagrus major. Fish Shellfish Immunol 49:275–285. https://doi.org/10.1016/j.fsi.2015.12.047

Dawood MA, Koshio S, Esteban MN (2017) Beneficial roles of feed additives as immunostimulants in aquaculture: a review. Rev Aquac 10(4):950–974. https://doi.org/10.1111/raq.12209

Di Cesare A, Eckert EM, D'Urso S, Bertoni R, Gillan DC, Wattiez R, Corno G (2016) Co-occurrence of integrase 1, antibiotic and heavy metal resistance genes in municipal wastewater treatment plants. Water Res 94:208–214. https://doi.org/10.1016/j.watres.2016.02.049

Drew MD, Borgeson TL, Thiessen DL (2007) A review of processing of feed ingredients to enhance diet digestibility in finfish. Anim Feed Sci Technol 138(2):118–136

Drlica K, Zhao X (1997) DNA gyrase, topoisomerase IV, and the 4-quinolones. Microbiol Mol Biol Rev 61(3):377–392. https://doi.org/10.1128/mmbr.61.3.377-392.1997

Du Pasquier L (2001) The immune system of invertebrates and vertebrates. Compar Biochem Physiol B Biochem Mol Biol 129(1):1–15. https://doi.org/10.1016/s1096-4959(01)00306-2

Dynes RA (2003) Earthworms—technology information to enable the development of earthworm production. (Publication No.03/085). RIRDC, Kingston, pp 1–33

ECDC (European Centre for Disease Prevention and Control), EFSA (European Food Safety Authority), and EMA (European Medicines Agency) (2017) ECDC/EFSA/EMA second joint report on the integrated analysis of the consumption of antimicrobial agents and occurrence of antimicrobial resistance in bacteria from humans and food-producing animals—Joint

Interagency Antimicrobial Consumption and Resistance Analysis (JIACRA) report. EFSA J 15(7):4872. https://doi.org/10.2903/j.efsa.2017.4872

EFSA Panel on Biological Hazards (BIOHAZ) (2014) Scientific opinion on the risk posed by pathogens in food of non-animal origin. Part 2 (Salmonella and Norovirus in leafy greens eaten raw as salads). EFSA J 12(3):3600

EFSA Panel on Biological Hazards (BIOHAZ), Koutsoumanis K, Allende A, Álvarez-Ordóñez A, Bolton D, Bover-Cid S et al (2021) Role played by the environment in the emergence and spread of antimicrobial resistance (AMR) through the food chain. EFSA J 19(6):e06651

El-Haroun ER, Bureau DP (2007) Comparison of the bioavailability of lysine in blood meals of various origins to that of L-lysine HCL for rainbow trout (*Oncorhynchus mykiss*). Aquaculture 262:402–409

FAO (2005) Responsible use of antibiotics in aquaculture. FAO Fisheries Technical Paper 469. Food and Agricultural Organization (FAO) of the United Nations, Rome, pp 1–110. ftp://ftp.fao.org/docrep/fao/009/a0282e/a0282e00.pdf

FAO (2021) The FAO action plan on antimicrobial resistance (2021–2025). FAO

Fasakin EA, Serwata RD, Davies SJ (2005) Comparative utilization of rendered animal derived products with or without composite mixture of soybean meal inhybrid tilapia (*Oreochromis niloticus* × *Oreochromis mossambicus*) diets. Aquaculture 249:329–338

Fath El-Bab AF, Majrashi KA, Sheikh HM, Shafi ME, El-Ratel IT, Neamat-Allah ANF, El-Raghi AA, Elazem AYA, Abd-Elghany MF, Abdelnour SA, Abduh MS, Jaremko M, Naiel MAE (2022) Dietary supplementation of Nile tilapia (*Oreochromis niloticus*) with β-glucan and/or *Bacillus coagulans*: synergistic impacts on performance, immune responses, redox status and expression of some related genes. Front Vet Sci 9:1011715. https://doi.org/10.3389/fvets.2022.1011715

Fountoulaki E, Vasilaki A, Hurtado R, Grigorakis K, Karacostas I, Nengas I et al (2009) Fish oil substitution by vegetable oils in commercial diets for gilthead sea bream (*Sparus aurata* L.); effects on growth performance, flesh quality and fillet fatty acid profile: recovery of fatty acid profiles by a fish oil finishing diet under fluctuating water temperatures. Aquaculture 289(3–4):317–326

Fowler LG (1990) Feather meal as a dietary protein source in fall Chinook salmon diets. Aquaculture 89:301–314

Galhano BSP, Ferrari RG, Panzenhagen P, Conte-Junior CA (2021) Antimicrobial resistance gene detection methods for bacteria in animal-based foods: a brief review of highlights and advantages. Microorganisms 9(5):923. https://doi.org/10.3390/microorganisms9050923

Garcia-Graells C, Antoine J, Larsen J, Catry B, Skov R, Denis O (2012) Livestock veterinarians at high risk of acquiring methicillin-resistant Staphylococcus aureus ST398. Epidemiol Infect 140(3):383–389. https://doi.org/10.1017/S0950268811002263

Garrett ES, dos Santos CL, Jahncke ML (1997) Public, animal, and environmental health implications of aquaculture. Emerg Infect Dis 3(4):453–457. https://doi.org/10.3201/eid0304.970406

Gatlin DM, Barrows FT, Brown P, Dabrowski K, Gaylord TG, Hardy RW, Herman E, Hu G, Krogdahl Å, Nelson R, Overturf K, Rust M, Sealey W, Skonberg D, Souza EJ, Stone D, Wilson R, Wurtele E (2007) Expanding the utilization of sustainable plant products in aquafeeds: a review. Aquac Res 38(6):551–579. https://doi.org/10.1111/j.1365-2109.2007.01704.x

Gaylord TG, Rawles SD (2005) The modification of poultry byproduct meal for use in hybrid striped bass Moronechrysops × *M. saxatilis* diets. World Aquac Soc 36:365–376

Glaize A, Gutierrez-Rodriguez E, Hanning I, Díaz-Sánchez S, Gunter C, van Vliet AH et al (2020) Transmission of antimicrobial resistant non-O157 Escherichia coli at the interface of animal-fresh produce in sustainable farming environments. Int J Food Microbiol 319:108472

Goryluk-Salmonowicz A, Popowska M (2022) Factors promoting and limiting antimicrobial resistance in the environment—existing knowledge gaps. Front Microbiol 13:992268. https://doi.org/10.3389/fmicb.2022.992268

Hansen RP, Czochanska Z (1975) The fatty acid composition of the lipids of earthworms. J Sci Food Agric 26:961–971

Hardy RW (1996) Alternate protein sources for salmon and trout diets. Anim Feed Sci Technol 59(1–3):71–80
Hardy RW (2010) Utilization of plant proteins in fish diets: effects of global demand and supplies of fishmeal. Aquac Res 41(5):770–776. https://doi.org/10.1111/j.1365-2109.2009.02349.x
Hawkey PM (2008) The growing burden of antimicrobial resistance. J Antimicrob Chemother 62(Suppl 1):i1–i9. https://doi.org/10.1093/jac/dkn241
Hemre GI, Amlund H, Aursand M, Bakke AM, Olsen RE, Ringø E et al (2009) Criteria for safe use of plant ingredients in diets for aquacultured fish. In: Opinion of the panel of animal feed of the Norwegian Scientific Committee for Food Safety, pp 1–173
Henriksson PJG, Rico A, Troell M, Klinger DH, Buschmann AH, Saksida S, Chadag MV, Zhang W (2018) Unpacking factors influencing antimicrobial use in global aquaculture and their implication for management: a review from a systems perspective. Sustain Sci 13(4):1105–1120. https://doi.org/10.1007/s11625-017-0511-8
Hodar AR, Vasava RJ, Mahavadiya DR, Joshi NH (2020) Fish meal and fish oil replacement for aqua feed formulation by using alternative sources: a review. J Exp Zool India 23(1):13–21
Holmes AH, Moore LS, Sundsfjord A, Steinbakk M, Regmi S, Karkey A, Guerin PJ, Piddock LJ (2016) Understanding the mechanisms and drivers of antimicrobial resistance. Lancet (London, England) 387(10014):176–187. https://doi.org/10.1016/S0140-6736(15)00473-0
Hua K, Cobcroft JM, Cole A, Condon K, Jerry DR, Mangott A, Praeger C, Vucko MJ, Zeng C, Zenger K, Strugnell JM (2019) The future of aquatic protein: implications for protein sources in aquaculture diets. One Earth 1(3):316–329. https://doi.org/10.1016/j.oneear.2019.10.018
Ibrahim M, Ahmad F, Yaqub B, Ramzan A, Imran A, Afzaal M, Mirza SA, Mazhar I, Younus M, Akram Q, Ali Taseer MS, Ahmad A, Ahmed S (2020) Current trends of antimicrobials used in food animals and aquaculture. In: Antibiotics and antimicrobial resistance genes in the environment. Elsevier, pp 39–69. https://doi.org/10.1016/B978-0-12-818882-8.00004-8
Igbinosa IH, Igumbor EU, Aghdasi F, Tom M, Okoh AI (2012) Emerging Aeromonas species infections and their significance in public health. TheScientificWorldJournal 2012:625023. https://doi.org/10.1100/2012/625023
Kaprou GD, Bergšpica I, Alexa EA, Alvarez-Ordóñez A, Prieto M (2021) Rapid methods for antimicrobial resistance diagnostics. Antibiotics (Basel, Switzerland) 10(2):209. https://doi.org/10.3390/antibiotics10020209
Kaushik SJ, Cravedi JP, Lalles JP, Sumpter J, Fauconneau B, Laroche M (1995) Partial or total replacement of fish meal by soybean protein on growth, protein utilization, potential estrogenic or antigenic effects, cholesterolemia and flesh quality in rainbow trout, Oncorhynchus mykiss. Aquaculture 133(3–4):257–274
Larsen J, Roney JM (2013) Farmed fish production overtakes beef. Earth Policy Institute, Washington, DC. https://resiliencesystem.org/farmed-fish-production-overtakes-beef. Accessed 4 Apr 2023
Lewis K (2007) Persister cells, dormancy and infectious disease. Nat Rev Microbiol 5(1):48–56
Liang Q, Yuan M, Xu L, Lio E, Zhang F, Mou H, Secundo F (2022) Application of enzymes as a feed additive in aquaculture. Marine Life Sci Technol 4(2):208–221. https://doi.org/10.1007/s42995-022-00128-z
Liu YY, Wang Y, Walsh TR, Yi LX, Zhang R, Spencer J, Doi Y, Tian G, Dong B, Huang X, Yu LF, Gu D, Ren H, Chen X, Lv L, He D, Zhou H, Liang Z, Liu JH, Shen J (2016) Emergence of plasmid-mediated colistin resistance mechanism MCR-1 in animals and human beings in China: a microbiological and molecular biological study. Lancet Infect Dis 16(2):161–168. https://doi.org/10.1016/S1473-3099(15)00424-7
Love DC, Halden RU, Davis MF, Nachman KE (2012) Feather meal: a previously unrecognized route for reentry into the food supply of multiple pharmaceuticals and personal care products (PPCPs). Environ Sci Technol 46(7):3795–3802
Lu W, Wang M, Wu J, Jiang Q, Jin J, Jin Q, Yang W, Chen J, Wang Y, Xiao M (2020) Spread of chloramphenicol and tetracycline resistance genes by plasmid mobilization in agricul-

tural soil. Environ Pollut (Barking, Essex: 1987) 260:113998. https://doi.org/10.1016/j.envpol.2020.113998

Lulijwa R, Rupia EJ, Alfaro AC (2020) Antibiotic use in aquaculture, policies and regulation, health and environmental risks: a review of the top 15 major producers. Rev Aquac 12(2):640–663. https://doi.org/10.1111/raq.12344

Luzier JM, Summerfelt RC, Ketola HG (1995) Partial replacement of fish meal with spray dried blood powder to reduce phosphorus concentrations in diets for juvenile rainbow trout *Oncorhynchus mykiss*. Aquac Res 26:577–587

Makkar HPS (1993) Antinutritional factors in foods for livestock, vol 16. BSAP Occasional Publication, pp 69–85

Maldonado Montiel TD, Rodríguez Canché LG (2005) Biomass production and nutritional value of Artemia sp. (Anostraca:Artemiidae) in Campeche, Mexico. Review biological tropics. Int J Trop Biol 53:447–454. ISSN 0034 7744

Mandal SM, Roy A, Ghosh AK, Hazra TK, Basak A, Franco OL (2014) Challenges and future prospects of antibiotic therapy: from peptides to phages utilization. Front Pharmacol 5:105

Maragakis LL, Perencevich EN, Cosgrove SE (2008) Clinical and economic burden of antimicrobial resistance. Expert Rev Anti-Infect Ther 6(5):751–763. https://doi.org/10.1586/14787210.6.5.751

Marshall BM, Levy SB (2011) Food animals and antimicrobials: impacts on human health. Clin Microbiol Rev 24(4):718–733. https://doi.org/10.1128/CMR.00002-11

Martins SN, Guzman EC (1994) Effect of drying methods of bovine blood on the performance of growing diets for tambaqui (*Colossomama cropomum*) in experimental culture tanks. Aquaculture 124:335–341

McMurry L, Petrucci RE Jr, Levy SB (1980) Active efflux of tetracycline encoded by four genetically different tetracycline resistance determinants in Escherichia coli. Proc Natl Acad Sci USA 77(7):3974–3977. https://doi.org/10.1073/pnas.77.7.3974

Mekasha S, Linke D (2021) Secretion systems in gram-negative bacterial fish pathogens. Front Microbiol 12:782673. https://doi.org/10.3389/fmicb.2021.782673

Midtlyng PJ, Grave K, Horsberg TE (2011) What has been done to minimize the use of antibacterial and antiparasitic drugs in Norwegian aquaculture? Aquac Res 42:28–34. https://doi.org/10.1111/J.1365-2109.2010.02726.X

Mu C, Zhong Q, Meng Y, Zhou Y, Jiang N, Liu W, Li Y, Xue M, Zeng L, Vakharia VN, Fan Y (2021) Oral vaccination of grass carp (*Ctenopharyngodon idella*) with baculovirus-expressed grass carp reovirus (GCRV) proteins induces protective immunity against GCRV infection. Vaccine 9(1):41. https://doi.org/10.3390/vaccines9010041

Nandeesha MC, Devaraj KV, Sudhakara NS (1986) Growth response of four species of carps to different protein sources in pelleted feeds. Asian Fisheries Forum. Asian Fisheries Society, Manila, Philippines, pp 26–31

Nnadozie CF, Odume ON (2019) Freshwater environments as reservoirs of antibiotic resistant bacteria and their role in the dissemination of antibiotic resistance genes. Environ Pollut 254:113067. https://doi.org/10.1016/j.envpol.2019.113067

Nogueira N, Cordeiro N, Andrade C, Aires T (2012) Inclusion of low levels of blood and feathermeal in practical diets for gilthead seabream (*Sparus aurata*). Turk J Fish Aquat Sci 12(3):641–650

O'Neill J (ed) (2015) Antimicrobials in agriculture and the environment: reducing unnecessary use and waste. The review on antimicrobial resistance. HM Government, London, UK, p 44. https://iiif.wellcomecollection.org/file/b28552556_Antimicrobials%20in%20agriculture%20and%20the%20environment%20%20Reducing%20unnecessary%20use%20and%20waste.pdf. Accessed 9 May 2023

O'Neill J (2016) Tackling drug-resistant infections globally: final report and recommendations. Review on antimicrobial resistance. Wellcome Trust and HM Government. https://amr-review.org/sites/default/files/160525_Final%20paper_with%20cover.pdf

Oliveira Hashiguchi TC, Ouakrim DA, Padget M, Cassini A, Cecchini M (2019) Resistance proportions for eight priority antibiotic-bacterium combinations in OECD, EU/EEA and G20 countries 2000 to 2030: a modelling study. Eurosurveillance 24(20):1800445

Oliver JD (2003) Chapter 17—Culture media for the isolation and enumeration of pathogenic Vibrio species in foods and environmental samples. Prog Ind Microbiol 37:249–269. https://doi.org/10.1016/S0079-6352(03)80020-6

Panserat S, Hortopan GA, Plagnes-Juan E, Kolditz C, Lansard M, Skiba-Cassy S, Corraze G (2009) Differential gene expression after total replacement of dietary fish meal and fish oil by plant products in rainbow trout (*Oncorhynchus mykiss*) liver. Aquaculture 294(1–2):123–131

Peleg AY, Hooper DC (2010) Hospital-acquired infections due to gram-negative bacteria. N Engl J Med 362(19):1804–1813. https://doi.org/10.1056/NEJMra0904124

Pérez JA, Rodríguez C, Bolaños A, Cejas JR, Lorenzo A (2014) Beef tallow as an alternative to fish oil in diets for gilthead sea bream (*Sparus aurata*) juveniles: effects on fish performance, tissue fatty acid composition, health and fleshnutritional value. Eur J Lipid Sci Technol 116(5):571–583

Prescott JF (2003) Veterinary microbiology and microbial disease. Can Vet J 44(12):986

Rajagopal M, Martin MJ, Santiago M, Lee W, Kos VN, Meredith T et al (2016) Multidrug intrinsic resistance factors in Staphylococcus aureus identified by profiling fitness within high-diversity transposon libraries. MBio 7(4):e00950–e00916

Redgrave LS, Sutton SB, Webber MA, Piddock LJ (2014) Fluoroquinolone resistance: mechanisms, impact on bacteria, and role in evolutionary success. Trends Microbiol 22(8):438–445. https://doi.org/10.1016/j.tim.2014.04.007

Ribeiro FB, Lanna EAT, Bomfim MAD, Donzele JL, Quadros M, Cunha PS (2011) True and apparent digestibility of protein and amino acids of feed in Nile tilapia. Rev Bras Zootecnia 40:939–946

Roberfroid MB (2005) Introducing inulin-type fructans. Br J Nutr 93(S1):S13–S25. https://doi.org/10.1079/bjn20041350

Roberts MC (2005) Update on acquired tetracycline resistance genes. FEMS Microbiol Lett 245(2):195–203. https://doi.org/10.1016/j.femsle.2005.02.034

Romero J, Gloria C, Navarrete P (2012) Antibiotics in aquaculture—use, abuse and alternatives. InTech. https://doi.org/10.5772/28157

Schar D, Klein EY, Laxminarayan R, Gilbert M, Van Boeckel TP (2020) Global trends in antimicrobial use in aquaculture. Sci Rep 10(1):1–9. https://doi.org/10.1038/s41598-020-78849-3

Shapawi R, Ng WK, Mustafa S (2007) Replacement of fish meal with poultry by-product meal in diets formulated for the humpback grouper, Cromileptes altivelis. Aquaculture 273(1):118–126

Shepherd CJ, Jackson AJ (2013) Global fishmeal and fish-oil supply: inputs, outputs and markets. J Fish Biol 83(4):1046–1066. https://doi.org/10.1111/jfb.12224

Suárez JA, Gaxiola G, Mendoza R, Cadavid S, Garcia G, Alanis G, Cuzon G (2009) Substitution of fish meal with plant protein sources and energy budget for white shrimp *Litopenaeus vannamei* (Boone, 1931). Aquaculture 289(1–2):118–123

Subasinghe R, Soto D, Jia J (2009) Global aquaculture and its role in sustainable development. Rev Aquac 1:2–9

Tacon AG, Hasan MR, Metian M (2011) Demand and supply of feed ingredients for farmed fish and crustaceans: trends and prospects. FAO Fisheries and Aquaculture technical paper. FAO, p 564

The state of world fisheries and aquaculture 2022. (2023). https://www.fao.org/3/cc0461en/online/sofia/2022/world-fisheries-aquacultureproduction. Accessed 4 Apr 2023

Thornber K, Bashar A, Ahmed MS, Bell A, Trew J, Hasan M, Hasan NA, Alam MM, Chaput DL, Haque MM, Tyler CR (2022) Antimicrobial resistance in aquaculture environments: unravelling the complexity and connectivity of the underlying societal drivers. Environ Sci Technol 56(21):14891–14903. https://doi.org/10.1021/acs.est.2c00799

Trushenski JT, Kasper CS, Kohler CC (2006) Challenges and opportunities in finfish nutrition. N Am J Aquac 68(2):122–140

Trushenski JT, Blaufuss P, Mulligan B, Laporte J (2011) Growth performance and tissue fatty acid composition of rainbow trout reared on feeds containing fish oil or equal blends of fish oil and traditional or novel alternative lipids. N Am J Aquac 73(2):194–203

Turchini GM, Torstensen BE, Ng W-K (2009) Fish oil replacement in finfish nutrition. Rev Aquac 1(1):10–57

Uluseker C, Kaster KM, Thorsen K, Basiry D, Shobana S, Jain M, Kumar G, Kommedal R (2021) A review on occurrence and spread of antibiotic resistance in wastewaters and in wastewater treatment plants: mechanisms and perspectives. Front Microbiol 12:717809. https://doi.org/10.3389/fmicb.2021.717809

Verraes C, Van Boxstael S, Van Meervenne E, Van Coillie E, Butaye P, Catry B, de Schaetzen MA, Van Huffel X, Imberechts H, Dierick K, Daube G, Saegerman C, De Block J, Dewulf J, Herman L (2013) Antimicrobial resistance in the food chain: a review. Int J Environ Res Public Health 10(7):2643–2669. https://doi.org/10.3390/ijerph10072643

Vielma J, Mäkinen T, Ekholm P, Koskela J (2000) Influence of dietary soy and phytase levels on performance and body composition of large rainbow trout (*Oncorhynchus mykiss*) and algal availability of phosphorus load. Aquaculture 183(3–4):349–362

Vikesland P, Garner E, Gupta S, Kang S, Maile-Moskowitz A, Zhu N (2019) Differential drivers of antimicrobial resistance across the world. Acc Chem Res 52(4):916–924. https://doi.org/10.1021/acs.accounts.8b00643

Viseur N, Lambert M, Delmee M, Van Broeck J, Catry B (2011) Nosocomial and non-nosocomial Clostridium difficile infections in hospitalised patients in Belgium: compulsory surveillance data from 2008 to 2010. Eurosurveillance 16(43):20000

Wee KL (1991) Use of non-conventional feedstuff of plant origin as fish feeds—is it practical and economically feasible? In: De Silva SS (ed) Fish nutrition research in Asia. Proceedings of the fourth Asian fish nutrition workshop, Special Publication 5. Asian Fisheries Society, Manila, Philippines, pp 13–32

Wee W, Abdul Hamid NK, Mat K, Khalif RIAR, Rusli ND, Rahman MM, Kabir MA, Wei LS (2023) The effects of mixed prebiotics in aquaculture: a review. Aquacult Fish 9(1):28–34. https://doi.org/10.1016/j.aaf.2022.02.005

Wiegand I, Hilpert K, Hancock RE (2008) Agar and broth dilution methods to determine the minimal inhibitory concentration (MIC) of antimicrobial substances. Nat Protoc 3(2):163–175. https://doi.org/10.1038/nprot.2007.521

World Health Organization, Food and Agriculture Organization of the United Nations & International Office of Epizootics (2006) Report of a joint FAO/OIE/WHO Expert Consultation on antimicrobial use in aquaculture and antimicrobial resistance, Seoul, Republic of Korea, 13–16 June 2006. World Health Organization. https://apps.who.int/iris/handle/10665/133869

Yousefi M, Ghafarifarsani H, Hoseini SM, Hoseinifar SH, Abtahi B, Vatnikov YA, Kulikov EV, Van Doan H (2022) Effects of dietary thyme essential oil and prebiotic administration on rainbow trout (Oncorhynchus mykiss) welfare and performance. Fish Shellfish Immunol 120:737–744. https://doi.org/10.1016/j.fsi.2021.12.023

Zacchino SA, Butassi E, Cordisco E, Svetaz LA (2017) Hybrid combinations containing natural products and antimicrobial drugs that interfere with bacterial and fungal biofilms. Phytomedicine 37:14–26

Zhang M, Chen L, Ye C, Yu X (2018) Co-selection of antibiotic resistance via copper shock loading on bacteria from a drinking water bio-filter. Environ Pollut (Barking, Essex: 1987) 233:132–141. https://doi.org/10.1016/j.envpol.2017.09.084

Zhao Y, Yang QE, Zhou X, Wang F, Muurinen J, Virta MP, Brandt KK, Zhu Y (2020) Antibiotic resistome in the livestock and aquaculture industries: status and solutions. Crit Rev Environ Sci Technol 51:2159–2196

Chapter 9
Public Health Food Safety (Farmed Aquatic Organisms/Live Stocks/Crops)

Sajna Beegum, Mary Carolin Kurisingal Cleetus, Abdullah Bin Abdul Nazar, Eshita Roy Harshitha Erangholi Pattalath, Akshay Thuruthiyil Rajesh, Rohil Forte, and Preetham Elumalai

Abstract Antibiotics are utilized for treating bacterial infections in humans as well as in many agricultural fields, such as aquaculture, plant and crop production, and livestock farming. An important turning point in the development of modern medicine was the invention of antibiotics as a technique for treating and controlling infectious diseases, but humans and animals can become infected with antibiotic-resistant bacteria—the most either through interaction with an affected animal or partly through the food chain by consuming food that has been contaminated. Food safety is a major concern for public health around the world because it is essential to sustaining life and fostering health. The uncontrolled use of antibiotics becomes a potential harm to human health as well as the environment. Antibiotic-resistant bacteria have emerged as a result of the increased usage of antimicrobial medications to treat animal illnesses. Due to the widespread use of agrochemicals such as pesticides and herbicides, weeds are becoming more and more resistant to herbicides. The development of vaccines will lessen the need for antibiotics in systems of an intensive culture for disease management. The World Health Organization has released a worldwide initiative to reduce the risk of antibiotic resistance, which is approved by several Southeast Asian nations.

Keywords Antibiotics · Antibiotic-resistant bacteria · Aquaculture · Antibiotic residue · Pollutant · Public health

S. Beegum · Eshita Roy Harshitha Erangholi Pattalath · A. T. Rajesh · R. Forte · P. Elumalai (✉)
Department of Marine Biology, Microbiology, and Biochemistry, School of Marine Sciences, Cochin University of Science and Technology (CUSAT), Kochi, Kerala, India
e-mail: preetham@cusat.ac.in

M. C. K. Cleetus
Department of Zoology and Animal Cell Biology, Research Center for Experimental Marine Biology and Biotechnology, University of Basque Country (PIE-UPV/EHU), Plentzia, Bizkaia, Spain

A. B. A. Nazar
Kerala University of Fisheries and Ocean Studies, Kochi, Kerala, India

P. Elumalai, S. Lakshmi (eds.), *Antimicrobial Resistance in Aquaculture and Aquatic Environments*, https://doi.org/10.1007/978-981-97-7320-6_9

Abbreviations

ADI	Acceptable daily intake
AGISAR	Advisory Group on Integrated Surveillance of Antimicrobial Resistance
AMR	Antimicrobial resistance
AMRSC	Antimicrobial Resistance Subcommittee
AOPs	Advanced oxidation processes
ARB	Antibiotic-resistant bacteria
ASP	Amnesic shellfish poisoning
CFP	Ciguatera fish poisoning
DDE	Dichlorodiphenyldichloroethylene
DSP	Diarrhetic shellfish poisoning
EFSA	European Food Safety Authority
FAO	Food and Agricultural Organization
FDA	Food and Drug Administration
FIB	Fecal indicator bacteria
GAP	Global action plan
HACCP	Hazard analysis and critical control point
JPIAMR	Joint Programming Initiative on AMR
LTPs	Lipid transfer proteins
NSP	Neurotoxic shellfish poisoning
PAH	Polycyclic aromatic hydrocarbons
PCB	Polychlorinated biphenyls
PCDD	Polychlorinated-*p*-dioxins
PCDF	Polychlorinated dibenzofurans
PBDE	Polybrominated diphenyl ethers
PSP	Paralytic shellfish poisoning
RBC	Red blood cells
SRA	Strategic Research Agenda
TTX	Tetrodotoxin
WBCs	White blood cells
WHO	World Health Organization
WT	Waiting time
WTO	World Trade Organization

9.1 Introduction

The discharge of antibiotics, other selective agents, and antibiotic-resistant bacteria into the environment poses risks to humans and animals. The use of antibiotics has been steadily increasing. As well as medical uses, people now use them for other purposes, increasing their impact on the environment globally. Between 100,000 and 200,000 tons of antibiotics have been used globally, and this wide usage is starting to have a considerable impact on human health. More challenging-to-treat illnesses are caused by resistant bacteria, necessitating the use of more costly and hazardous medications. Antibiotics are frequently used in veterinary and medicinal settings, but they are now being used more frequently in cattle and aquaculture. The use of antibiotics or the residues they leave in organisms, or the environment can have detrimental impacts on public health and, through direct or indirect contact, can have lethal toxicological effects.

Antimicrobials and bioactive metabolites can reach the environment in different ways. They can enter water bodies through agroecosystems, sewage biosolids contaminated with antibiotics, sludge, silt, or manure, as well as fertilization. Antibiotics can enter the environment through three main routes: environmental drug release, human and veterinary drug release, and the use of antibiotics in aquaculture. Additional burdens on wastewater systems have increased due to the significant release of unused or expired drugs into the sewage system. Several approaches for degrading antibiotic residues have been researched as antibiotics are now posing a hazard to the environment and ultimately creating public health implications (refer Fig. 9.1).

9.2 Food Safety Risks Associated with Aquaculture

Aquaculture is one of the ways of cultivating aquatic organisms in a controlled environment and has recently experienced increased production. Over the past decade, it has been under transition and evolving day by day to saturate the demand

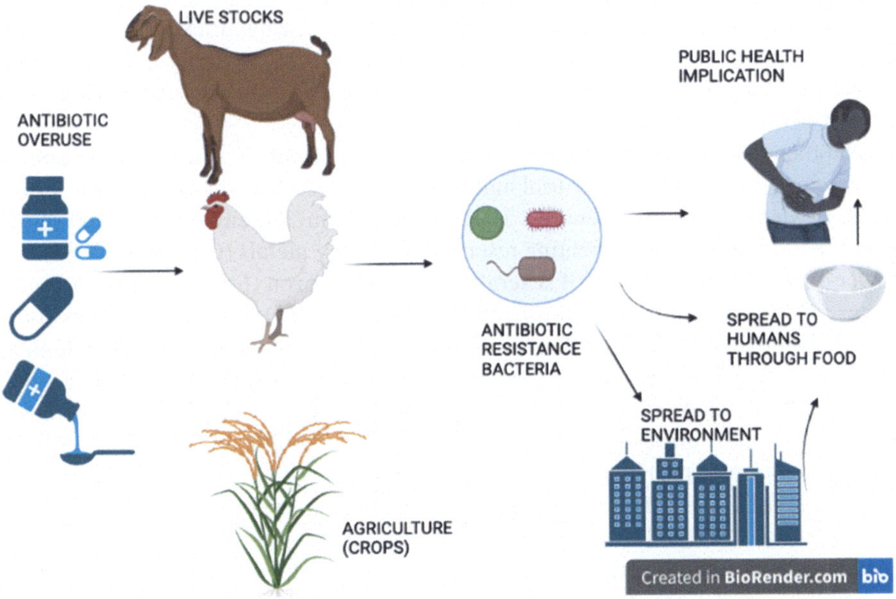

Fig. 9.1 The effect of antibiotic overuse. (Created with BioRender https://biorender.com/)

for nutrition. Global food fish aquaculture production has expanded by 6.9% annually on average since 2008, rising from roughly 13 million tons in 1990 to 67 million tons in 2012 (Food and Agriculture Organization, FAO 2014). As the demand for aquaculture products increases due to a growing population, safe and sustainable production methods are needed. The products from aquaculture face some of the common risks that are involved with the contamination of biological, chemical, and environmental hazards into the products (Reilly et al. 1997). There are many risk elements involved in different stages of production in an aquaculture system. Recent studies have defined institutional risks, market risks, and production and environmental risks, which include adverse conditions such as harmful algal blooms, antibiotic usage, increase in biological oxygen demand, and introduction of pathogens which cause a decrease in production, as well as health issues to the consumer (Ahsan and Roth 2010). Ingredients used in commercial animal feed may likewise pose a risk to food safety in aquaculture. Veterinary drug residues, metals, pesticides, persistent organic pollutants, mercury (Hg), lead (Pb), cadmium (Cd), arsenic (As), and selenium (Se) are the most prevalent types of chemical contaminants. The period of exposure and concentration of these pollutants are factors to be considered (Tacon and Metian 2008). The usage of fertilizers is a common practice, and it has its merits and demerits. It can promote the growth of phytoplankton populations in the water column but can also have negative consequences. The main inorganic fertilizers contain nitrogen, phosphorus, and potassium, along with some trace metals and silicates, and are highly water-soluble salts. They are readily absorbed by plants and other aquatic organisms, eventually when these organisms die and decompose their organic matter, as well as unutilized inorganic fertilizers are deposited along the pond bottom. These inorganic fertilizers get trapped in the pond bottom, and as a result, the pond bottom cannot hold large quantities of nutrients and organic matter forever. Nitrogen fertilizers can increase toxicity by increasing ammonia concentrations in the aquaculture systems if used excessively (Hunt and Boyd 1981).

Another issue of concern is the presence of heavy metals in aquaculture systems. These can originate from industrial manufacturing processes runoffs, mining, fertilizers, E-wastes, and other sources. They can leach into the environment and settle down in bottom sediments creating reservoirs of heavy metals (Chao et al. 2014). At lower quantities, the toxicity of several metals such as iron (Fe), copper (Cu), manganese (Mn), chromium (Cr), nitrogen (N), and zinc (Zn) is restricted. They may even help the fish in these ponds by improving their metabolism. However, at higher concentrations, these essential metals can become toxic and cause stunted growth, mass mortality, reproductive degeneracy, and bodily abnormalities. These have been observed in various species such as *C. carpio*, *L. rohita*, and *O. mykiss* in various aquaculture farms (Van Bussel et al. 2014). However, heavy metals such as Pb, Hg, As, and Cd are often hazardous to the organisms living in that environment. Through direct and indirect exposure, these are taken up through the gills, during feeding and dermal absorption. Once they reach the bloodstream, they are transported through the organism until they reach the liver and kidney where they can accumulate over a period. High concentrations of heavy metals found in tissues, fat bodies, and organs can cause internal damage, pathological changes, changes in osmotic regulation, and enzyme activity. Haematocrit and the number of other blood cells may be used as an indicator of metal pollution (Javed and Usmani 2012). If heavy metals start to accumulate in aquatic organisms, higher trophic-level consumers may also be subjected to even concentrations of these heavy metals, leading to far worse consequences. For mitigating the adverse effects of phytoremediation, metal-binding proteins like metallothionein can be used in terms of detoxification (Barwick and Maher 2003).

Organic/Inorganic Compounds: Organic halogenated compounds including polycyclic aromatic hydrocarbons (PAHs), polychlorinated biphenyls (PCB), polychlorinated dibenzofurans (PCDF), polychlorinated-*p*-dioxins (PCDD), polybrominated diphenyl ethers (PDBE), and dichlorodiphenyldichloroethylene (DDE), which may be of both anthropogenic and natural origin, can also contaminate the aquaculture environment. These compounds are widely used in the manufacture of paints, pharmaceuticals, plastics, pesticides, skin care products, and solvents. Usually, the magnitude of these types of pollution increases in their proximity to industrial areas. They are lipophilic in nature and are generally found in adipose tissues of culture organisms and can contribute toward bioaccumulation and biomagnification through aquaculture products and eventually end up in humans (Perugini et al. 2013).

Antibiotics: Antimicrobials are compounds that inhibit the growth of microorganisms in a system. Recent developments in aquaculture have led to ensuring the safety of an aquatic product; thus, to decrease the rate of risk related to the products, the use of antimicrobials is increased. The increased use of antimicrobial substances has also increased the resistance in the microbes. This has reduced the medical efficacy of medicine against the microbes (Fair and Tor 2014). Various studies have found the presence of antibiotics and antimicrobial residues in aquaculture products. Antibiotics such as oxytetracycline and florfenicol are administered for the treatment of certain bacterial diseases. They may also be found in wastewater from human and animal production systems. If they are released into the environment and reach the nearby wetlands, they may have adverse effects (Zheng et al. 2012).

Toxins: A significant percentage of aquaculture products that are harvested for human consumption need to be enriched in essential supplements. To do so, aquatic products are fed with protein-rich aquafeed. This enriches the protein level in the fishes and other crops. Commercially available aquafeed is a mixture of vegetables or animal origin (Deutsch et al. 2007; FAO 2001). The main disadvantage with the use of crop ingredients in fish is that they compose a significant volume of anti-nutrient factors that cause adverse effects on human health (Puri et al. 2019; Chakraborty et al. 2019). These compounds can affect human health in the form of toxins. In addition, marine toxins are biotoxins that are produced by certain types of algae. Some are severe in potency, while some are moderate. These algae which are consumed by other aquatic organisms accumulate the toxins within them until they reach a lethal amount and get transferred to humans and further induce toxicity in people. These include diarrhetic shellfish poisoning (DSP), paralytic shellfish poisoning (PSP), ciguatera fish poisoning (CFP), neurotoxic shellfish poisoning (NSP), amnesic shellfish poisoning (ASP), and tetrodotoxin (TTX) which is a strong neurotoxin (Theodorou et al. 2020). These mycotoxins, which are naturally occurring secondary metabolites in plants, are commonly found in filamentous fungi including Aspergillus, Penicillium, and Fusarium (Egbuta et al. 2017; Bennett and Klich 2003). Toxins produced by other organisms also have an impact on human health. Okadaic acid is one of the most common toxins that are encountered by humans produced by *Prorocentrum* sp. *Dinophysis* sp. causes DSP, which results in nausea and diarrhea, meaning that these toxins are more of a neurotoxin (Kamat et al. 2013). Domoic acid is produced by Pseudo-*nitzschia* sp. Symptoms include memory loss, headaches, and sometimes even fatal conditions caused by ASP (Tasker 2021). PSP is the consequence of consuming saxitoxin-containing shellfish which is infested by *Gymnodinium* sp. The raw consumption of these organisms without neutralizing the toxins causes poisoning. The symptoms include tingling sensation, numbness, and even cardiac arrest if consumed in higher concentration (Reboreda et al. 2010). These poisons would cause widespread death in aquaculture systems and may contaminate the entire culture. Furthermore, it could transmit through the trophic levels and intensify toxicity before finally reaching the customer.

Parasites: Food safety risks imposed by the introduction of parasites into aquaculture systems are also a major concern. Most of the fishes carry more or less specific parasites. The threats can arise during the production of aquatic products. The invasion of wild fishes or other aquatic organisms can deliver unwanted parasites to the aquaculture pond (Woo et al. 2020). These parasites may infect and propagate within the farming fishes. These unwanted parasites may cause less to severe damage to the fishes and can also be fatal for the other neighboring ponds resulting in a concerned issue over the aquaculture industry (Shinn et al. 2015). Additionally, parasitic infestations are likely to reduce the quality of the farmed animals, and their high reproductive capacity and contagious behavior may create tragedy in such systems. Translocation of fingerlings and broodstock should be done with extreme care so as not to bring any uninvited parasites into the culture pond. Interaction with wild stocks of fish, absence of proper biosecurity standards, irregular water quality parameters that further stress the fish, and changes in weather patterns can all influence the chances of parasitic infestations (Olivier 2002). *Bonamiaostreae*, a molluscan parasite spread all over Europe through the import of juvenile flat oysters, *Ostrea edulis*, which were a vastly cultured oyster variety in Spain (Cigarría and Elston 1997). Using chemotherapeutics and other biological medicines may be a farmer's first choice to control the growth of parasites into the pond but recent studies also suggest the use of other controlling methods such as biocides, mechanical methods, herbal extract, and the use of prophylaxis or vaccination to prevent the outgrowth of such harmful entities into the farming area (Buchmann 2022).

Bacteria: The aquatic pond is a home for other organisms as well as the farming fishes. Among the other living entities, the bacterial load can be harmful to the crops. They can colonize in the water body as well as in the fishes, making the fishes vulnerable to different diseases, diseases caused by bacteria re not only dependent on the disease-causing nature of bacteria but also the immunological status of the harvesting fishes as well as their exposure to the contaminating environment (Johnson and Paull 2011). Bacterial infections are common in those aquaculture systems that do not have proper water management systems and poor water quality parameters. From a food safety perspective, these can be separated into those organisms that cause disease in the animals themselves and those that can be introduced post-harvest and pose a direct risk to the consumer. Bacterial pathogens can cause significant losses in cultured animals (examples); however, most of them pose a limited direct risk to consumers as most are not zoonotic. There are some exceptions though, including *Streptococcus agalactiae*. Mycobacteria species and pathogenic Vibrio species such as *Vibrio vulnificus* and *Vibrio parahaemolyticus*. Some of the common pathogens that cause damage to freshwater fishes belong to the group of *Aeromonas*, *Flavobacterium*, *Pseudomonas*, *Acinetobacter*, *Shewanella*, and others (Austin et al. 2007; Kozinska 1999; Kozińska et al. 2014; Kozinska and Pekala 2004). Treatment of a bacterial pathogen directly affects farmed fish. Shellfish poses other potential food safety risks though it may drive the selection of antibiotic-resistant bacteria. Fecal contamination can lead to infections due to bacteria like *Salmonella* and *Escherichia coli*. Aquatic organisms cultured in these places that make use of organic manures and fertilizers from animal wastes are prone to these waterborne infections. These can have a direct impact on the cultured organisms and those who consume them which start in mild diseases which transform into serious illnesses. *Clostridium botulinum* is a spore-forming bacteria that is quite often overlooked in Hazard Analysis and Critical Control Point (HACCP), their strains can survive in high temperatures and can cause diseases in fish and people who consume these fishes (Huss et al. 1974). Alternatives of antibiotics such as using bacteriophages, quorum inhibitors, and so on can also be used in aquaculture practice wherever applicable.

Virus Infections: Viruses also put food safety at risk in the aquaculture industry. Among the severe diseases in aquatic products, viral infections are very dangerous when compared with other microbe-borne infections. These cause heavy damage to the farmed crops and are also one of the major reasons for high economic loss (Borrego et al. 2017; Rodger 2016; Lafferty et al. 2015). The invading species in the farming area carry most of the viruses, especially due to the lack of a proper barrier in an open culture where the pathogen can be brought by the invaders from long distances via ocean currents (McCallum et al. 2004). These viruses form colonies in the farming fishes and when consumed by humans, they can cause significant damage to the human immune system. One of the most prominent is Norovirus, which can affect people of all age groups and is associated with the consumption of uncooked shellfish such as oysters. It is a single-stranded RNA virus that is estimated to affect 1 in 219 people every year in the United Kingdom alone (Tam et al. 2012). The hepatitis E virus is transmitted through fecal and oral routes, which account for cases in the elderly related to the consumption of shellfish. The hepatitis A virus, which is another human pathogen, is linked to the consumption of bivalve shellfish. They are associated with about 1.5 million cases annually and are related to 300,000 people affected in 1980 making it one of the largest foodborne outbreaks of all time (Potasman et al. 2002). According to Koch's Postulates, one of the conventional ways to identify the virus that has infected crops is to diagnose the infection using cell culture and then identify the virus. These approaches are highly helpful in determining the relationship between the pathogen and the host (Rivers 1937). There are also some viruses that have been identified that are quite hard to culture in cells; PRV and PMCV are some of them. Nowadays, PCR has become the most widely acceptable tool to diagnose different types of viral infections, detecting their nucleic acids (Mugimba et al. 2021). Although various methods are present to diagnose viral infection and viral load in fishes, the proper treatment of several viral infections is yet to come.

Safety Protocol: These safety hazards are always a point of concern in any aquaculture system. Some are biological, chemical, and environmental hazards that could disrupt the whole foundation on which an aquaculture farm is built. For added safety, there lies HACCP. The principles of HACCP help in enhancing food safety to eliminate any potential risk involved in the making. A team of people from various expertise in the field of aquaculture analyze everything from scratch like feed supply, water, and site, implement various solutions to problems that pop up, and implement automation in places that require sterile conditions (Reilly and Käferstein 1997). Production of safe and sustainable food in aquaculture systems should be the top priority for farm owners in the future. As the population rises, so does the need for good-quality aquaculture products (Table 9.1).

9.3 Public Health Implications of Antimicrobial Use in Aquaculture

In aquaculture, antibiotics are used as treatments or, in certain nations, as feed additives to reduce morbidity and mortality from contagious pathogenic microorganisms. Aquaculture is the food supply industry with the fastest rate of growth and significantly contributes to global food security (Okocha et al. 2018). Prophylactic use of antibiotics is common in intensive aquaculture production systems. In many developing countries, the use of antibiotics in aquaculture is relatively unregulated, and even where laws do exist, stringent compliance is largely nonexistent (Okeke et al. 2022). Widespread and indiscriminate use of antibiotics may lead to the emergence of foodborne illnesses, affect natural microbial communities, and potentially promote the emergence of antibiotic resistance in non-target organisms. The spread of bacterial resistance genes may pose a risk to humans and animals, through the development of antibiotic resistance (Larsson and Flach 2022). Given that some

Table 9.1 Various sources and their causes in the food safety risk associated with aquaculture

Sources	Category	Causes	References
Antibiotics	Oxytetracycline	Antibiotics such as chloramphenicol and oxytetracycline are administered for treatment procedures of certain bacterial diseases that tend to accumulate in wastewater in large amounts that are then released into the environment and tend to reach nearby wetlands damaging the environment	Zheng et al. (2012)
Bacterial infections	*Salmonella*, *E. coli*, *Clostridium botulinum*, *Vibrio parahaemolyticus*	Animals cultured in these places that use organic manures and fertilizers from animal wastes are prone to these waterborne infections, these can have a direct impact on the organisms that are cultured and on the people who consume them which starts in mild diseases that transform into serious illnesses	Huss et al. (1974)
Heavy metals	Pb, Hg, As, Cd	Internal damage, pathological changes, changes in osmotic regulation and enzyme activity, hematocrit, and other blood cells as indicators of metal pollution	Javed and Usmani (2012)
Marine toxins	DSP, PSP, CFP, NSP, ASP, and TTX which is a strong neurotoxin	Mass mortality in aquaculture	Theodorou et al. (2020)
Organic/inorganic compounds	PAH, PCB, PCDF, PCDD, PDBE, and DDE	Lipophilic in nature and are generally found in adipose tissues of culture organisms they can contribute toward bioaccumulation, and biomagnification through aquaculture products and eventually end up in humans	Perugini et al. (2013)
Parasites	*Bonamiaostreae*	*Bonamiaostreae* a molluscan parasite spread all over Europe by the import of juvenile flat oysters called *Ostrea edulis* which were a vastly cultured oyster variety in Spain	Cigarría and Elston (1997)
Viral infections	Hepatitis E virus, norovirus	Transmitted through fecal and oral route A virus that is accepted as a human pathogen is linked with the consumption of bivalve shellfishes; they are associated with about 1.5 million cases annually and are related to 300,000 people affected in 1980 making it one of the largest foodborne outbreaks of all time	Tam et al. (2012), Potasman et al. (2002)

bacteria that cause disease in fish are highly related to bacteria that cause infections in humans and terrestrial animals, it is possible that there are risks of direct transfer of AMR between fish and shellfish, humans, and other animals. Antibiotic-resistant bacteria can also transfer resistance genes and the mobile elements they carry on to other species and genera of bacteria by horizontal gene transfer (Watts et al. 2017) providing an indirect method of spreading antimicrobial resistance (AMR) (Larsson and Flach 2022). This may reduce the ability of antibiotics to effectively treat infections in humans and animals. The unused metabolites of antibiotics that remain in the environment can have profound, long-lasting impacts. Nowadays, using antibiotics without causing resistance has become an intricate issue. To assure public health problems brought on by this, healthy aquaculture methods, proper and controlled use of antibiotics, and additional research to close the current methodological gaps including disease prevention and treatment approaches are required.

9.3.1 Antibiotic Uses

Antibiotics are frequently used in aquaculture although the approved and permitted drugs in aquaculture are different among countries as are their uses and governing laws. At least in some cases, bacterial diseases can be prevented by proper maintenance and by providing a hygienic environment for the farm and hatcheries. The primary rationale for implementing antibiotics in aquaculture was to combat as well as prevent diseases that have a negative impact on farmed fish and shellfish. They can be used effectively to treat animals and stop the spread of disease. Generally, they are introduced to the farm species through their food, directly into the water, or by injection. One of the most commonly used antibiotics in aquaculture, oxytetracycline, is applied to alleviate bacterial infections of fish and shellfish such as vibriosis, *aeromonosis*, *pseudomonosis*, *lactococcosis*, and *furunculosis* (Leal et al. 2019). Other commonly used antibiotics include amoxicillin and sulfadiazine-trimethoprim (Alderman and Hastings 1998). However, a variety of antibiotics with varying dose rates and withdrawal times are routinely utilized in the aquaculture industry, depending on the region and nation. Overuse of these antibiotics may result in the emergence of bacteria that are resistant to antibiotics. One of the causes of ARB in the environment has been attributed to the usage of antibiotics as feed additives in food animals (Aly and Albutti 2014). AMR results from the excessive and uncontrolled use of antibiotics, thus now impacting human and animal health globally. It is thus the primary concern with antibiotic use in aquaculture. Humans can contract antibiotic-resistant strains by direct contact with contaminated water or through cultivated organisms that have been contaminated, which have arisen as a result of the careless use of antibiotics.

Many of the antimicrobial agents that are used in aquaculture are critically important for humans. Because the occurrence of resistance to these antimicrobials in human pathogens considerably limits the therapeutic options for treating human diseases, the use of these antimicrobials in animals should be limited or avoided in order to reduce the spread of drug resistance (Heuer et al. 2009). Indirect contact with water or drinking water may result in the spread of drug-resistant bacteria to humans, and the resistant strain can result in treatment failures in people. Some of the human pathogens that include direct transfer from water include *Vibrio cholera*, *V. parahaemolyticus*, *V. vulnificus*, *Shigella* species, and *Salmonella* species, or opportunistic pathogens, such as *Aeromonas hydrophila*, *Plesiomonas shigelloides*, *Edwardsiellatarda*, *Streptococcus iniae*, and *E. coli* (Heuer et al. 2009). Additionally, if antibiotic-resistant microbes are not adequately treated, there may be challenges, which could have a detrimental impact on the country's economy given that aquaculture is the sector of the food supply that is increasing the fastest and contributes to human food security. Furthermore, there may be serious environmental consequences if antibiotic residue from aquaculture circulates into the environment.

9.3.2 Antibiotic Residues in Aquaculture

One of the possible harmful effects of antibiotic use is the accumulation of antibiotic residues in the aquatic environment. Even in low quantities, antibiotic residues in aquatic environments can cause life-threatening infections. The residues left behind in the aquatic environment have the potential to persist and settle, and they may eventually move up the food chain and ultimately reach humans.

Antimicrobial residues in the aquatic environment can change the microbiome of the environment and, as a result, the ability of the ecosystem to regulate, provide for, and support itself (Schar et al. 2020). Responses to this risk include allergy, toxicity, intestinal flora changes, and antibiotic-resistant bacteria selection. As antibiotic residues in aquaculture can gradually rise along with antibiotic use, it is possible to conclude that there is a direct relationship between each of them. Thereby, the effects caused by antibiotic residues in humans will depend on the type, toxicity, and dose of the antibiotic used. According to Paige et al. (1997), individuals are also in danger from antimicrobial agent residues in aquaculture products, in addition to the risk posed to human health by antibiotic-resistant bacteria in aquaculture. Organisms that have had antibiotic treatment may have residues in their tissues that last for a very long period and even through additional processing procedures to create edible commodities that can get into consumers. Bacanlı and Başaran (2019) has reviewed the effects of antibiotic residues—sulfamethazine, oxytetracycline, and furazolidone residues causing toxic effects like allergy, immunopathological effects, and carcinogenicity, gentamicin residues causing mutagenicity and nephropathy, and chloramphenicol residues causing hepatotoxicity, anaphylactic shock, and bone marrow toxicity in humans. Moreover, water tainted with antibiotic residues can have negative effects on the environment and eventually put human health at risk if it is used for drinking.

9.3.3 Toxicological Effects

Most of the studies have focused on the acute toxicity of antibiotics in aquaculture and limited information is available on the chronic toxicity. The gut microbiome of farmed fish might shift as a result of antibiotic use in aquaculture. The fish's fundamental wellness and development may be impacted by this upheaval, which may result in microbial community imbalances in their guts. The changed gut microbiota of these fish may also have an impact on human health by consumption. Apart from this, directly or indirectly having contact with contaminated water, especially for drinking purposes can lead to considerable health effects.

As the natural microorganisms present in the skin, intestines, and other bodily regions can act as a potential barrier to the entry of foreign substances, an imbalance resulting from the infusion of antimicrobial drugs can cause negative medical consequences to this intricate and crucial ecological system of the human body. The studies investigated so far have shown that there will be a greater imbalance in the natural microbes present in the human body due to the antimicrobial drugs, depending on the dose and exposure time, the bodily conditions, and so on. The toxicological effects of antibiotic exposure in humans, particularly in the gut, which is home to several species of bacteria, fungi, viruses, and archaea, were reviewed by Okeke et al. in (2022). There was a decrease in Enterobacteriaceae, overgrowth of *Candida* and *Enterococci* species, as well as infection with *Clostridium difficile* brought on by the antibiotic ceftriaxone; meanwhile, Beta lactams boosted the fecal sample's microbial burden and Clindamycin disrupts the microbiota in the gut. Besides, vancomycin influenced a reduction of firmicutes in the human gut. Furthermore, erythromycin, azithromycin, and clarithromycin alter the intestinal floral composition and impact the metabolism of the human gut. This includes decreased Actinobacteria, increased Bacteroidetes, and Proteobacteria, decreased bile-salt hydrolase, and increased macrolide resistance. Likewise, ciprofloxacin caused the human stomach's microbial diversity to decline, and the combination of colistin and amoxicillin shifted the gut microbiota. Early life exposure to macrolides increases the risk of developing asthma and increases the likelihood of gaining weight due to antibiotics (Korpela et al. 2016).

Most studies have focused on the gut microbes and yet a lot to discover the toxicological effects induced by the antimicrobial agents on the normal microbes in the skin and other body parts. In a study conducted by Adamsson et al. (1997), cefadroxil administration significantly reduced the aerobic microflora in the oropharynx but had no discernible effect on the anaerobic microflora and phenoxymethylpenicillin intake significantly reduced the total number of anaerobic oropharyngeal microorganisms. Another study by Christensson et al. (1991) revealed a significant decrease in *Streptococcus salivarius* and *Veillonella cocci* after receiving amoxicillin treatment, whereas enterobacteria increased in the oropharyngeal microflora of individuals receiving treatment for respiratory tract infections. Sullivan et al. (2001) thoroughly evaluated the effects of several antibiotic classes on humans, a higher chance of getting diarrhea, colonizing with yeasts and *Clostridium difficile*, and other side effects when taking cephalosporins were explained. Linezolid notably reduced the number of bifidobacteria, lactobacilli, clostridia, and Bacteroides while significantly increasing the number of *Klebsiella* strains and vastly reducing the number of enterococci.

9.4 Plant Microbial Ecology

Agriculture that is organic and free of chemicals is very important today. The use of agricultural chemicals such as DDT, lindane, and HCH is now a regular practice. In developed countries and poor nations, there is a regular practice of using cheap pesticides and it can affect the environment and even the health of human beings. The lingering residue in the meal has the potential to contaminate both the environment and the food supply. Enhancing food quality, increasing food output, and protecting the environment from persistent pesticide traces are all significant goals. The population of the planet is growing every day. An adult needs approximately 2900 kcal per day. FAO estimates that between 800 and 850 million people worldwide are undernourished. It is also important to remember that around 75% of cereals in the world have been used to feed animals. The provision of food is widely recognized as being crucial to the survival of people everywhere. The importance of better nutrition and food production increased significantly. Nowadays, improvised methods for producing high-quality and safe chemical-free food are more important. Genetically modified organisms (GMOs) are pest and disease-resistant. Improving plant variety with GMOs can reduce the utilization of harmful pesticides and chemicals because they now have an inborn ability to resist pathogens. Plant microbial ecology is influenced by several key elements, including biofilm development, physicochemical characteristics of plant surfaces, epiphytic fitness, and interactions between microbes and plants. Contamination by enteric pathogens in agricultural plant products has emerged as one of the main contributors to the spread of human food-borne diseases. This resulted in the activities surrounding the public health concern. Animal and human hosts have been linked to enteric infections. Contrary to contamination caused in the field, most of the contamination is attributed to post-harvest contamination. The compounds made from plant tissue harvesting may be able to lessen the biology and impact of intestinal infections. Controlling and halting the spread of such enteric pathogen infections is crucial in the modern day. From the spread of such intestinal pathogens, an appropriate disease management strategy must be implemented (Brandl 2006).

9.4.1 *Agricultural Plants as the Vehicles for Food-Borne Pathogens*

Agricultural plants can act to transport food-borne pathogens from agricultural fields through manures utilized during their growth and food production. The spread of zoonotic pathogens, which are often found in food, can pose a threat to the public's health. Manure is frequently employed in agricultural fields to promote the better growth and development of agricultural plants. Enteric pathogens may contaminate such manures. The dangers of microbial contamination may also be increased by using manure that has not been properly composted or by adding the excrement of domestic or wild animals that roam freely. The spread of enteric bacteria that cause foodborne illness can also be caused by field workers due to poor hygiene habits and lack of on-site sanitation. Shigellosis, an intestinal infection, received particular attention from people during the pandemic. In addition, pesticide applications made with tainted water and crop irrigation are regarded as major sources of inoculum in the field. Microbial pollution can ultimately result in problems with public health (refer to Fig. 9.2). Bioaerosols have emerged as one of the main means of transferring pollutants from feces to plant crops and ultimately to the people. The agricultural crops are being contaminated by microbes. Air currents are mainly involved in the spreading of diseases among plant microbial ecology. Pathogenic E. Coli is excreted by infected house flies and is subsequently transmitted to apples through various insect vectors, including vinegar flies (*Drosophila*

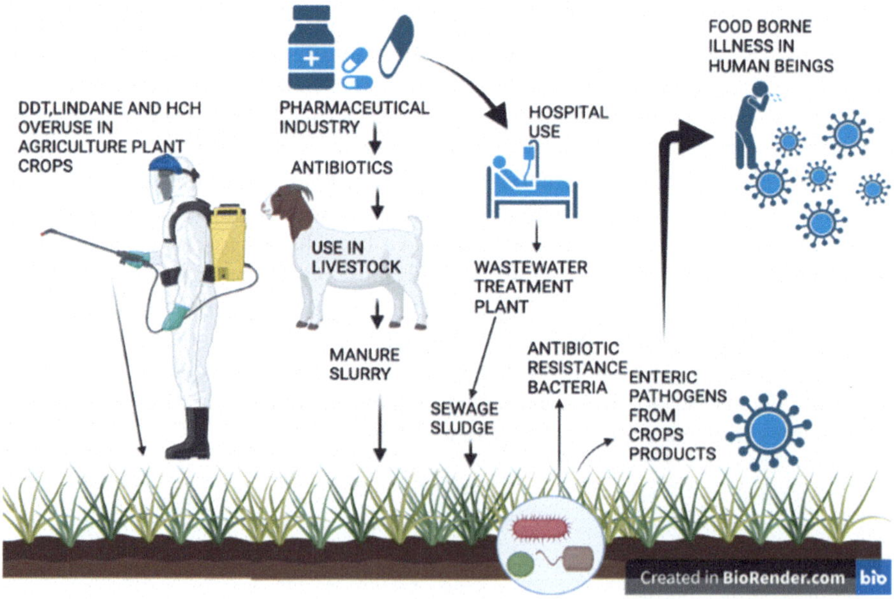

Fig. 9.2 Public health impact on antibiotic overuse and antibiotic residue. (Created with BioRender https://biorender.com/)

melanogaster) and fruit flies (*Ceratitis capitata*). It has also been documented that insects spread epiphytic bacteria on damp leaf surfaces. Air currents and insects can cause a great spread of pathogens in agricultural fields and affect public health. Foodborne illness outbreaks and epidemic events: From 1973 to 1997, the United States was plagued by foodborne illness epidemics. Between 1990 and 2003, epidemics increased by the second greatest. Major pathogens were caused by fruits and vegetables. Most commonly impacted by bacterial enteric illnesses were tomatoes, green onions, carrots, and cabbage. Juices made from unpasteurized fruits and vegetables were detected with enteric pathogens. Contamination connected to pre- and post-harvesting could have a serious impact. *Salmonella enterica*, pathogenic *E. coli*, *Shigella* spp., *Listeria monocytogenes*, *Campylobacter* spp., *Staphylococcus aureus*, *Yersinia* spp., and *Bacillus cereus* are a few of the enteric pathogens that can cause sickness. One of Japan's earlier epidemics affected 6000 people in the area due to bacterial pathogens. Associated with the enterotoxigenic serotype O157:H7, a severe foodborne pathogen that can cause hemolytic uremic syndrome and result in mortality, especially in children, the elderly, and the immunocompromised, *E. coli* strains also cause significant events. *Pseudomonas aeruginosa*, *Burkholderiacepacia (115)*, *Erwinia* spp., and *Enterococcus feacalis* are a few pathogens that frequently live on plant surfaces and can live on both plants and mammals. Humans may contract food-borne illnesses from the spread of enteric bacteria through fruits and vegetables consumed by human beings. A crucial factor in the spread of food-borne diseases across agricultural plants is fitness and stability on plant surfaces.

9.4.2 Antibiotic Uses, Residues, and Toxicological Effects

The naturally occurring chemical compound known as an antibiotic is used to kill bacteria (bactericidal) and to prevent the growth of germs (bacteriostatic). Nowadays, antibiotics are extensively used in agriculture. The use of antibiotics in farming, poultry, and managing livestock disease, as a result, has an increasing output to the environment. Antibiotics are no longer permitted in the majority of emerging nations. However, many impoverished nations continue to use antibiotics. Most disease-causing microorganisms have developed genes for antibiotic resistance. Lack of a high-quality treatment facility, incorrect use of antibiotics, and inefficient antibiotic use can all contribute to public health problems that increase morbidity and mortality. The majority of the time, the use of non-essential antibiotics has been unregulated in developing nations. The environment can undoubtedly be impacted by unregulated use. The use of animal or plant products that utilize antibiotic residue is the main cause of the antibiotic resistance gene's production. Antibiotic resistance can become a serious problem in illnesses and public health. The most crucial thing is related to its effect on the bodily functions. The use of strong medication dosages would have been crucial. Such antibiotic residue will travel through contaminated water sources, animal populations, and other factors. It is critical to regulate and keep an eye on the excessive use of antibiotics and goods with antibiotic residue intended for human consumption. The usage of antibiotics in agriculture and cattle results in their release into the environment. Antibiotic pollution is a major concern nowadays. At its most extreme, antibiotic pollution might produce genes that are resistant to antibiotics. The main effect of antibiotic misuse is the development of antibiotic resistance genes. Nowadays, using antibiotics has become a standard practice in agriculture. It has been estimated that there is an increase in crop output, which has been contaminated by antibiotic use. Antibiotic use for veterinary purposes has also grown to be a problem. Three types of poultry production, including breeding poultry, broilers, and laying hens, showed different leftover antibiotic releases. The use of non-essential antibiotics has been unregulated in poor nations.

The productivity of poultry and cattle is severely impacted by the usage of non-essential antibiotics. The sickness that spreads as a result of the use of these antibiotics must be managed and controlled. Antibiotic use has increased in developing nations in order to improve animal health. Antibiotics or antimicrobials are applied via water to farming systems (such as chicken farms) in order to promote growth and provide prophylaxis. The utilization of such a system can undoubtedly impact the environment through waste streams from livestock and veterinary practices as well as from human waste. Antibiotic residue exposure also affects other organisms in the environment. Healthy birds that had been needlessly exposed to antibiotics included some. They might also be exposed to larger doses. Some organisms may develop antibiotic-resistant genes as a result of altered ribosomal target locations (Alderman and Hastings 1998).

Antibiotic residue exposure also affects other organisms in the environment. Healthy birds that had been needlessly exposed to antibiotics included some. They might also be exposed to larger doses. Some organisms may develop antibiotic resistance as a result of altered ribosomal target locations. The demands for food production are rising daily as a result of factors including the rising standard of living and the increased global demand for animal proteins for human consumption. It has been discovered that several countries have started employing antibiotics for improved development in order to increase agricultural production. The ingestion of these antibiotic-residue-tainted meats, milk products, and eggs typically has a significant negative impact on human health.

9.5 Antibiotic Residues as Emerging Pollutants

Several antibiotics belonging to the family beta-lactam, fluoroquinolone have been studied due to their significant usage in veterinary as well as in human medicine. The emerging pollutants in the form of antibiotics are exponentially increasing in aquatic ecosystems. These pollutants have different chemical properties, and their origin seems different. In our environment, antibiotics are distributed more randomly and their presence in an aquatic ecosystem is not taken into consideration, even though it causes a greater impact on human health and the environment (Godoy and Sánchez 2020).

Different types of antibiotic residues cause harmful effects on the environment. The occurrence of this pollutant seriously affects the epigenetic system. Some of the non-antibiotic pharmaceutical chemicals promote the antibiotic-resistant gene through mutations. So, the metformin present in the aquatic system can promote antibiotic-resistant bacteria dissemination. The studies related to the exposure of *E. coli* to metformin reveal that the metformin concentration ranging from 10^{-6} to 200 mg/L resulted in the development of multi-antibiotic resistance. Further exposure to prolonged days revealed an increase in the mutation frequency of *E. coli*. The upregulation in the expression of antibiotic-resistant genes can also be detected (Wei et al. 2022).

Antibiotics are also used in food-producing animals. These residues can arise from a number of causes, including the use of antibiotics without a prescription or the contamination of feed with treated animals' excrement. For those who ingest it, this raises a number of health issues. The major problems involved are the transfer of antibiotic-resistant bacteria to humans, carcinogenicity, and toxic effects.

The studies in African countries reflect that there are no clear regulatory laws for controlling antibiotic contamination in feedstuffs. The uses of antibiotics are majorly found in the treatment of animal farming to reduce the outbreak of diseases, thereby promoting animal health. Antibiotics and their residues such as tetracyclines, cephalosporins, macrolides, chloramphenicol, nitrofurans, and beta-lactams are found in chicken meat, rabbit meat, milk, eggs, animal-derived foods, and cattle meat. Several antibiotic residues such as beta-lactams, aminoglycosides, sulfonamides, and tetracyclines cause allergic reactions. The other major effect includes the selective disruption of important bacteria living inside the intestines of humans which destroys the other incoming pathogenic bacteria that cause disease (Darwish et al. 2013).

9.6 Biodegradable Antibiotics

The massive use of antibiotics results in environmental contamination and the development of AMR (Yang et al. 2019). In aquaculture, the term "degradable antimicrobial" refers to antimicrobials that are fully degradable to very basic components, such as a single amino acid, within a short period rather than degradable to large fragments because large fragments can still have some antibacterial activity to contribute to AMR and antimicrobial selective pressure (Wu et al. 2018). The peptide polymer, α-poly-L-lysine, which is entirely made of the amino acid lysine, is well recognized to have antibacterial activity, but its use as an antimicrobial agent is highly limited because of its high cytotoxicity. To lessen the cytotoxicity of the resultant peptide polymers and the acute toxicity to zebrafish while barely affecting antibacterial activity, we replace a portion of the positively charged lysine residues with the hydrophilic serine residue. The best peptide polymers satisfy all the requirements for antibiotic substitutes in aquaculture, including excellent antibacterial activity against common pathogenic bacteria, a significant increase in zebrafish survival rates, and no induction of antibacterial resistance in bacteria.

9.7 Impact of Antibiotic Resistance Strain on Public Health

Food-borne infectious diseases are mainly formed by the ability of the pathogens to form a biofilm which leads to the outbreak of such diseases (Štefánek et al. 2022). To treat the infectious disease in fish, the usage of antibiotics has occurred (Ghimpețeanu et al. 2022). The presence of trace amounts of these antibiotics in fish and meat seriously affects public health. The studies reveal that the ability of pathogens to form biofilm results in the transferring of resistant genetic material via quorum sensing (Waters and Bassler 2005). Information about the existence of antibiotic residues in food and how they affect human health is lacking. The invention of antimicrobials led to several changes, increasing life expectancy rates, decreasing the incidence of the diversity of infectious diseases, simplifying the various operative procedures, and curing many skin cuts that may become fatal to us. These antibiotics have harmful and challenging aspects in our lives and environment. They are now recognized as an emerging threat to public health and animal health. The study (Tariq 2022) reveals that we are fighting against these pathogens in the form of diplomatic war and the side effects of these antibiotics threaten our environment too.

An aquatic ecosystem is considered to have higher mobility of organisms and genetic elements. Compounds used in veterinary medicine that are connected to the manufacturing of animal feed frequently contain antibiotics. Antibiotic use remains an inevitable part of treating bacterial illnesses. Even though several antibiotics are banned, their application in animal food leads to the appearance of their residues in the foodstuffs which may result in many health risks. In the case of human health, these residues accelerate antibiotic resistance development and induce several other pathologies such as cancer, bone marrow toxicity, nephropathy, and reproductive disorders. It also promotes the transfer of antibiotic-resistant bacteria to humans. The importance of regulation of the antibiotic mainly in animal breeding decelerates its adverse consequences (Arsène et al. 2022). Among the AMR organisms, ARB are considered an important group. It is more complex to elucidate the transmission dynamics of ARB. This complexity is mainly due to the adaptability of the bacteria to various stressors. The human-associated bacteria mainly from the sewage cause human-to-human transmissions. Mainly three health impacts are considered which include exposure, colonization, and infection. The exposure of fecal indicator bacteria (FIB) in bathing waters establishes the presence of ARB. At the same time, it is evident that the quantification of environmental transmission is more difficult (Leonard et al. 2022).

The municipal waste treatment plants act as a breeding ground and thereby easily help in the dissemination of antibiotic resistance for ARB. The unmetabolized, undigested remnants of antibiotics are transferred into the municipal waste treatment plants, thereby leading to the development of ARB and the induction of antibiotic-resistant genes. The higher rate of dissemination results in a higher impact on human health and causes potential risks to the economy (Soni et al. 2022).

9.8 Managing Antimicrobial Residues for Food Safety

The major problem with the use of antibiotics is the presence of persistent antimicrobial residues in ecosystems. While cooking, some antimicrobial residues such as sulfonamides, tetracyclines, and fluoroquinolones get converted to other forms and thus reduce the risk of ingestion, but heating does not always guarantee that these antibiotic residues will be completely broken down in animal products such as grilled meat. Quinolones and -lactams have high stability, leading to substantial issues for human health because leftover antibiotics from heat-treated milk can enter the dairy industry and be consumed by consumers (Kühne et al. 2001). Many of these animal products are consumed by people in western countries as a result of global trade. It is important to specify here the seafood and fish raised in aquaculture. The consumption of these animal products is also rising quickly due to their many advantageous traits. Therefore, it would also be important for western nations to monitor these imports as thoroughly as possible (Gaspar et al. 2019). These nations also have powerful detecting techniques at their disposal. China has developed a regulatory framework that is crucial in promoting responsible antibiotic use and ensuring food safety. A number of international and national regulatory organizations make ongoing efforts to regulate antibiotic use in accordance with international standards while taking into account the unique circumstances of each nation (Okocha et al. 2018). Antibiotics that are more likely to cause antibiotic resistance should be banned. Producers must use antibiotics only when necessary and as directed by a veterinarian. Routinely check for antibiotic residue. To ensure food safety, vegetative bacteria are often destroyed by thermal processing, although this approach may have unfavorable impacts on nutrition and quality. Antimicrobial peptides from plants and animals are of interesting efficiency against microbes like Hevein- and Knottin-like peptides, lipid transfer proteins (LTPs), and plant defensins.

9.8.1 Regulation

Antibiotic use is on the rise in aquaculture and other food industries. Because it is used indiscriminately, it poses a health risk to consumers, posing a number of risks related to residues in the tissues of treated animals and products made from them (Crawford 1985). World Health Organization (WHO) encouraged countries' veterinary, agricultural, and pharmaceutical authorities—as well as other relevant parties—to think about phasing out the use of antibiotics as growth promoters over time. Additionally, the organization suggests that animals should only be given antibiotics when prescribed by a veterinarian, and that antibiotic use in food animals should be regulated. Regulations are to be framed to manage the antimicrobial residues and currently, it is a challenging task to form a unified regulation as the use of antimicrobials varies from one country to another. A consortium of manufacturers and concerned legal organizations are required to find a pragmatic solution to antibiotic residues in food.

9.8.2 Reduced Need for Prudent Use of Antibiotics in Aquaculture

Aquaculture is the continuously growing dynamic food sector. The higher risk of disease outbreaks creates the use of different antibiotics in this field. The residues of antibiotics are retained in the farmed animals, and it gradually leads to the public health impacts by the consumption of those products derived from aquaculture. The regulation of the use of antibiotics and their consumption needs several control systems. Among the several concerns for consumers, hypersensitivity reactions, carcinogenicity, and development of antimicrobial drug resistance are the major ones. Fish production has a significant role in the main animal farming sector. The disease outbreaks occur in the population that is confined to a small area like an intensive production system. For such kinds of prevention of outbreaks and for the promotion of growth the use of antibiotics finds major application. The dependence on antimicrobials and the increased consumption of meat and aquaculture products lead to major health impacts. Environmental pollution, as well as human welfare, is one of the associated concerns (Rigos and Troisi 2005). Disease outbreak prevention by the application of antibiotics seriously affects human health. There are no such rules for some other countries and those countries have the largest aquaculture production. A subset of bacteria that infect fish are members of the same genera that infect humans. Thus, it raises the possibility that aquaculture will be the means by which antibiotic resistance spreads to humans (Heuer et al. 2009). In Nordic countries, we need the prescription of a veterinarian for getting antibiotics. Most countries have included guidelines for the prudent use of antibiotics (Wierup 2001). Around 12 tons of antibiotic discharge occurs in the surface waters of the south china sea from the Mekong River (Shimizu et al. 2013). In coastal waters, the non-resistant bacteria receive the resistance gene from other resistant bacteria, and they thereby integrate the gene along with the chromosome (De la Cruz and Davies 2000). Accurate administration of monitoring systems minimizes the impact and emergence of antibiotic-resistant genes (Chowdhury et al. 2021).

9.9 Effects of Antibiotics as the Feed Additive in the Food Animals

Antibiotics are chemical substances that have therapeutic uses for the clinical assessment of infectious agents in plants, animals, and humans. However, a sizable amount of the produced antibiotics annually are used for other purposes rather than as chemotherapeutics. In the United States alone, some 24.6 million pounds of antibiotics are used as growth promoters in animal husbandry each year (Oliver et al. 2011). Meanwhile, growth-enhancing properties of antibiotics were exploited and the introduction of antibiotics in animal feed has been globally practiced. The fundamental concept of antibiotics' growth-promoting properties is not entirely understood. It is believed that a significant amount of nutrients are used by the bacteria that live in animal feed, which causes contamination of the feed. They deteriorate the nutrient quality of the feed and microbial growth inhibits assimilation of nutrients and their absorption from the intestine. The growth-enhancing property of antibiotics is closely associated with its ability to retard these harmful microbes. Animals administered with antibiotics show certain beneficial effects on health as well as their growth. Pigs supplied with antibiotics as feed additives require less than 10–15% of the feed to reach desired growth. The major expense in animal husbandry is the cost of feed. The incorporation of antibiotics in feed helps in reducing feed expenditure. The addition of antibiotics to animal feed helps in the efficient transformation of feed into animal products. The daily growth conversion ratio of animals administered with antibiotics is found to be improved by 10% than other animals that feed on antibiotic-free diet. The quality of meat obtained from animals supplemented with antibiotic feed is improved in terms of high protein level as well as lower lipid content than animals not supplied with antibiotics (Hughes and Heritage 2002). The incorporation of penicillin and tetracycline in poultry feed helped in improving egg production and ability to hatch besides the feed conversion ratio (Gustafson and Bowen 1997). Health of such organisms that feed on antibiotic-mixed food is drastically improved. Animals administered with chlortetracycline and sulfamethazine in feed helped animals in overcoming the risk of bovine respiratory disease, morbidity, and chronic respiratory disease (Gallo and Berg 1995). China accounts for about 60% of the aquaculture fish production globally and also promotes the use of antibiotics in large quantities to enhance disease management and ensure productivity (Mo et al. 2017). Antimicrobial regimens are typically included in surface-coated or pellet feed in aquaculture, as well as through alternative techniques such as direct injection or water immersion. The use of antibiotics in food animals has shown a tremendous increase globally. Antimicrobials have not been used in aquaculture. Health issues and food safety concerns due to unregulated antibiotic use remain unaddressed in most parts of developing nations globally. Proper administration of antibiotics should not result in their presence in edible tissues of the administered animal even in very low concentrations. However, the presence of antibiotic residue in edible tissue is due to neglecting the recommended dosage and label directions, nonfulfillment of directed withdrawal periods, high dosage administration at a single site of injection, mixing errors, feeding with medicated feeds, antibiotic-contaminated equipment usage, and animal reasons such as age, pregnancy, allergy, and chemical interaction. The AMR caused due to the remains of antibiotic residue forms risk factors for human beings. For example, chloramphenicol remains show a high risk of cancer development and low concentration might trigger aplastic anemia, which is often fatal—several other toxic effects like immunopathological and carcinogenic effects of sulfamethazine and oxytetracycline, allergy caused by penicillin, mutagen effects, and nephropathy of gentamicin. In aquaculture, therapeutic levels of antibiotics are generally adapted for a short duration of time through an oral route to fishes in tanks or cages. This approval committee might set rules and regulations for antibiotic uses including withdrawal time, dosage forms, tolerance, and permission for route delivery. The common route of introducing antibiotics to aquatic animals is by mixing antibiotics along with the specific formulated feed. The efficiency of antibiotic metabolism in fish is very low and they excrete unused antibiotics through feces (Burridge et al. 2010). Aquaculture practices have paved the way for the outbreak of many bacterial diseases, which demand for the increased use of antimicrobials (Defoirdt et al. 2011)—diseases caused by pathogenic bacteria, parasitic infection, agrochemical residues, veterinary drugs, and contamination due to heavy metals as the potential hazards of products from aquaculture (Reilly et al. 1997).

9.10 Degradation of Organic Antibiotic Residues (Biotechnological Applications)

Antibiotic residues in the environment, especially in soil and water sources, can harm ecosystems and promote the emergence of antibiotic resistance. Antibiotics can be broken down via photolysis, hydrolysis, and biodegradation in the natural environment. However, most antibiotic half-lives are extended, which suggests that the degradation rates are often low and that antibiotics have traits of persistence in the environment (Zheng et al. 2022), thus giving the significance of biotechnological applications on the breakdown of organic antibiotic residues. Generally, the biotechnological application addresses using organisms mainly bacteria and fungi, by degrading or transforming the molecules. Microbial degradation has a great role in the removal mechanism of antibiotic residues. Certain microbes can metabolize or may utilize these residues for their growth and development. According to Liao et al. (2016), the important drug sulfonamide, which is used in aquaculture to treat illness in aquatic animals, degrades when it encounters a microbial community that includes Actinobacteria, Bacteroidetes, chloroflexi, firmicutes, proteobacteria, and verrucomicrobia. Additionally, a rise in temperature has been found to accelerate microbial deterioration. However, the bacterial community and richness diminished as the material degraded and firmicutes and Bacteroidetes eventually took control. This may shed light on the need for additional studies into how bacteria break down antibiotics. It is anticipated that more bacterial phyla can degrade the antibiotics making it a feasible biotechnological application to this deadly situation. Some bacteria or fungi possess enzymes that have the ability to degrade antibiotics. This is based on the principle of stimulating chemical reactions that destroy the molecular structure. These enzymes have broad-spectrum activity against numerous antibiotics or can target specific antibiotic molecules. Beta-lactamase is one of the most prevalent and significant hydrolytic enzymes that render cephalosporins, carbapenems, and monobactams (Bush 2010). Wang et al. (2021) have demonstrated the biodegradation activity of beta-lactamase from *Ochrobactrum tritici* (OtLac) toward penicillin V potassium. The piece of work is to research and reveal the enzyme behind the degradation, yet enzyme-linked biodegradation is an emerging area of research for antibiotic residues.

Correspondingly, genetic engineering techniques can be used to modify microorganisms by enhancing their ability to degrade antibiotic residues. The genetically engineered *E. coli* which recombinantly expressed by cloning the tetracycline degrading enzyme, TetX derived from *Bacteroides fragilis* that senses and degrades tetracycline antibiotic residue indicates that it is possible to combat antibiotic residues in the environment effectively and safely by employing transgenic microbes (Mu et al. 2018).

Some of the other promising antibiotic removal approaches include advanced oxidation processes (AOPs) and phytoremediation. AOPs are one of the wastewater treatment methods with a wide range of possible applications. The pedestal of the removal mechanism is in situ production of –OH, which interacts quickly with most organic molecules (aside from chlorinated alkanes) despite lacking selectivity of attack. As it has a strong non-selective oxidation ability and a high redox potential, –OH can oxidize several groups in the antibiotic's molecular structure (Zheng et al. 2022). Many AOPs mainly use combinations of strong oxidizing agents such as ozone, hydrogen peroxide, or ultraviolet light to degrade organic compounds. Lester et al.'s (2011) research on $UV/H_2O_2/O_3$ demonstrated that the amount of H_2O_2 present was a crucial factor in the two subprocesses of UV/H_2O_2 and H_2O_2/O_3, both of which could fully mineralize the antibiotics ciprofloxacin and trimethoprim. Besides, a removal efficiency of greater than 90% was observed by Östman et al. (2019) for trimethoprim and erythromycin by ozonation in a treatment plant.

The phytoremediation method uses plants to remove or break down contaminants from the environment. Several plants can absorb and destroy antibiotic residues through their metabolic mechanisms. For instance, the seedling stage of the species *Eichhornia crassipes* was proposed to be applied for phytoremediation by Yan et al. (2021). Typically, phytoremediation is facilitated through constructed wetlands, which are treatment systems for improving water quality. Thereby, harmful pollutants will be absorbed from the contaminated water by plants with their ability, to break down, and store them. It is unrealistic to predict that every contaminant will be metabolized and stored, as this will rely on the kind and toxicity of the molecule as well as the characteristics of the plant.

These methods therefore present promising means of reducing antibiotic pollution's negative effects on the ecosystem and helping to create prospective approaches to manage antibiotic-contaminated areas. The degradation of antibiotics and methods employed will differ based on the class of antibiotic residues and several other underlying factors that need to be further looked into. Additionally, more investigation is recommended to determine whether any abiotic elements together with these techniques could accelerate the total elimination of antibiotic residues from the environment. Currently, each link's impact on all aspects is thoroughly explored, which raises the possibility that knowledge gaps will progressively be addressed in the years to come.

9.11 Global Recommendations

Alexander Fleming, who discovered penicillin, already warned that bacteria could become resistant to antibacterial drugs in 1945 during his Nobel prize speech. For every antimicrobial drug discovered, the microbes will eventually develop resistance to it. Over time, the drug-resistant pathogen will acquire new mechanisms for resistance, usually through genetic changes eventually developing the AMR. Over several decades, the applications of antibacterial drugs have increased, and it has led to the misuse of these drugs in both humans and food-producing animals which ultimately resulted in the evolution of more resistant bacteria. As a result, the existing antibacterial drugs have been ineffective which accelerated a global health security emergency that rapidly outpaced the available treatment options. A comprehensive and integrated strategy approach grounded in the "one health" idea is necessary to address the cross-sectional character of the AMR epidemic, which involves various health professionals (Ferri et al. 2017).

9.11.1 Policymakers and Health Authorities

AMR endangers every aspect of public health and medical practice. These days, there is a growing problem due to the rise in mortality and health problems. An old problem is AMR. The society has been significantly threatened by AMR. It is imperative that human protection be practiced. To advance and address the AMR issue, non-governmental groups, health agencies, and professional societies are far more crucial. Strengthening the AMR research field and developing efficient approaches to address the existing AMR problem are also crucial. Reducing the amount of AMR applied to food components can help lower the risk of AMR. Much greater importance is placed on food security and its protection against AMR. Effective prevention is required to stop the unhealthily wide-ranging source of antibiotic residue.

9.11.2 Industry

AMR is a concern to society today because of its enormous effects on population, health, and society at large. It is crucial that all facets of society work together to prevent AMR-related problems from primarily harming a safe and healthy society. To address the issue of AMR, a broad approach involving the agriculture, healthcare, and pharmaceutical industries is necessary. The massive action plan for every sector is adequate to address the problem. The development of novel antimicrobial drugs and universal vaccinations is crucial. It is important to stop the abuse of antibiotics in the agriculture and aquaculture industries. This would significantly aid in preventing the overuse of antimicrobial residue development. The AMR Industry Alliance (AMRIA) is crucial in reducing AMR.

Pharmaceutical companies, particularly widely recognized global organizations, have the potential to make impactful actions against the AMR because they are equipped with the necessary employees with broad, multi-disciplinary experience to tackle and address the concerns of AMR. Also, their broad customer network can be used to facilitate the sharing of ongoing practices and to understand the various measures adopted at regional levels. Although pharmaceutical companies are not primarily responsible for the success of AMS initiatives, the skillsets found within pharmaceutical companies can complement the skillsets of other stakeholders in multisectoral AMS partnerships.

9.11.3 Research

Significant changes in research prospects during the last decade have prompted concerted initiatives at the international, national, and regional levels. The European Union helped to establish a Joint Programming Initiative on AMR (JPIAMR), which brings together 27 nations worldwide and enables them to coordinate their national research efforts and fund transnational research projects, in order to energize the actions at various levels. To prevent AMR issues, there are wide preventive measures for raising funds worldwide. Awareness and campaigning programs have taken place worldwide to prevent AMR (Rogers Van Katwyk et al. 2019).

It is crucial to make recommendations for efficient research protocols and evaluation trials in the scientific community. Open access and date options allow for a broad range of research findings. The current state of AMR policy interventions greatly depends on national and international research activities. Mitigating the threat posed by AMR will also require substantial collaboration among researchers and policymakers (Rogers Van Katwyk et al. 2019).

References

Adamsson I, Edlund C, North CE, Sjöstedt S (1997) Comparative effects of cefadroxil and phenoxymethylpenicillin on the normal oropharyngeal and intestinal microflora. Infection 25:154–158

Ahsan DA, Roth EVA (2010) Farmers' perceived risks and risk management strategies in an emerging mussel aquaculture industry in Denmark. Mar Resour Econ 25(3):309–323

Alderman DJ, Hastings TS (1998) Antibiotic use in aquaculture: development of antibiotic resistance—potential for consumer health risks. Int J Food Sci Technol 33(2):139–155

Aly SM, Albutti A (2014) Antimicrobials use in aquaculture and their public health impact. J Aquacult Res Dev 5(4):1

Arsène MMJ, Davares AKL, Viktorovna PI, Andreevna SL, Sarra S, Khelifi I, Sergeyevna DM (2022) The public health issue of antibiotic residues in food and feed: causes, consequences, and potential solutions. Vet World 15(3):662. https://doi.org/10.14202/vetworld.2022.662-671

Austin B, Austin DA, Munn CB (2007) Bacterial fish pathogens: disease of farmed and wild fish, vol 26. Springer, Dordrecht, The Netherlands, p 552

Bacanlı M, Başaran N (2019) Importance of antibiotic residues in animal food. Food Chem Toxicol 125:462–466. https://doi.org/10.1016/j.fct.2019.01.033

Barwick M, Maher W (2003) Biotransference and biomagnification of selenium copper, cadmium, zinc, arsenic and lead in a temperate seagrass ecosystem from Lake Macquarie Estuary, NSW, Australia. Mar Environ Res 56(4):471–502. https://doi.org/10.1016/S0141-1136(03)00028-X

Bennett JW, Klich M (2003) Mycotoxins. Clin Microbiol Rev 16(3):497–516

Borrego JJ, Valverde EJ, Labella AM, Castro D (2017) Lymphocystis disease virus: its importance in aquaculture. Rev Aquac 9(2):179–193

Brandl MT (2006) Fitness of human enteric pathogens on plants and implications for food safety. Annu Rev Phytopathol 44:367–392

Buchmann K (2022) Control of parasitic diseases in aquaculture. Parasitology 149(14):1985–1997

Burridge L, Weis JS, Cabello F, Pizarro J, Bostick K (2010) Chemical use in salmon aquaculture: a review of current practices and possible environmental effects. Aquaculture 306(1–4):7–23. https://doi.org/10.1016/j.aquaculture.2010.05.020

Bush K (2010) Bench-to-bedside review: the role of β-lactamases in antibiotic-resistant Gram-negative infections. Crit Care 14(3):1–8

Chakraborty P, Mallik A, Sarang N, Lingam SS (2019) A review on alternative plant protein sources available for future sustainable aqua feed production. Int J Chem Stud 7(3):1399–1404

Chao W, Zheng SS, Wang PF, Jin Q (2014) Effects of vegetations on the removal of contaminants in aquatic environments: a review. J Hydrodyn B 26(4):497–511. https://doi.org/10.1016/S1001-6058(14)60057-3

Chowdhury S, Ghosh S, Aleem MA, Parveen S, Islam MA, Rashid MM et al (2021) Antibiotic use and resistance in food animal production: what have we learned from Bangladesh? Antibiotics 10(9):1032. https://doi.org/10.3390/antibiotics10091032

Christensson B, Nilsson-Ehle I, Ljungberg B, Nömm I, Oscarsson G, Nordström L et al (1991) A randomized multicenter trial to compare the influence of cefaclor and amoxycillin on the colonization resistance of the digestive tract in patients with lower respiratory tract infection. Infection 19(4):208–215

Cigarría J, Elston R (1997) Independent introduction of Bonamiaostreae, a parasite of Ostrea edulis, to Spain. Dis Aquat Org 29:157–158

Crawford LM (1985) The impact of residues on animal food products and human health. Sci Tech Rev 4(4):669–723

Darwish WS, Eldaly EA, El-Abbasy MT, Ikenaka Y, Nakayama S, Ishizuka M (2013) Antibiotic residues in food: the African scenario. Jpn J Vet Res 61(Suppl):S13–S22

De la Cruz F, Davies J (2000) Horizontal gene transfer and the origin of species: lessons from bacteria. Trends Microbiol 8(3):128–133

Defoirdt T, Sorgeloos P, Bossier P (2011) Alternatives to antibiotics for the control of bacterial disease in aquaculture. Curr Opin Microbiol 14(3):251–258. https://doi.org/10.1016/j.mib.2011.03.004

Deutsch L, Gräslund S, Folke C, Troell M, Huitric M, Kautsky N, Lebel L (2007) Feeding aquaculture growth through globalization: exploitation of marine ecosystems for fishmeal. Glob Environ Chang 17(2):238–249

Egbuta MA, Mwanza M, Babalola OO (2017) Health risks associated with exposure to filamentous fungi. Int J Environ Res Public Health 14(7):719

Fair RJ, Tor Y (2014) Antibiotics and bacterial resistance in the 21st century. Perspect Med Chem 6:PMC-S14459

FAO (2001) Fisheries department aquaculture development. 1. Good aquaculture feed manufacturing practice. FAO, Rome, Italy

FAO Food and Agriculture Organization (2014) The state of world fisheries and aquaculture: opportunities and challenges, vol 4. Food and Agriculture Organization United Nations, Rome, pp 40–41

Ferri M, Ranucci E, Romagnoli P, Giaccone V (2017) Antimicrobial resistance: a global emerging threat to public health systems. Crit Rev Food Sci Nutr 57(13):2857–2876. https://doi.org/10.1080/10408398.2015.1077192

Gallo GF, Berg JL (1995) Efficacy of a feed-additive antibacterial combination for improving feedlot cattle performance and health. Can Vet J 36(4):223

Gaspar AF, Santos L, Rosa J, Leston S, Barbosa J, Pouca ASV et al (2019) Development and validation of a multi-residue and multi-class screening method of 44 antibiotics in salmon (Salmo salar) using ultra-high-performance liquid chromatography/time-of-flight mass spectrometry: application to farmed salmon. J Chromatogr B 1118:78–84. https://doi.org/10.1016/j.jchromb.2019.04.038

Ghimpețeanu OM, Pogurschi EN, Popa DC, Dragomir N, Drăgotoiu T, Mihai OD, Petcu CD (2022) Antibiotic use in livestock and residues in food—a public health threat: a review. Food Secur 11(10):1430. https://doi.org/10.3390/foods11101430

Godoy M, Sánchez J (2020) Antibiotics as emerging pollutants in water and its treatment. In: Antibiotic materials in healthcare. Academic, pp 221–230. https://doi.org/10.1016/B978-0-12-820054-4.00012-4

Gustafson RH, Bowen RE (1997) Antibiotic use in animal agriculture. J Appl Microbiol 83(5):531–541. https://doi.org/10.1046/j.1365-2672.1997.00280.x

Heuer OE, Kruse H, Grave K, Collignon P, Karunasagar I, Angulo FJ (2009) Human health consequences of use of antimicrobial agents in aquaculture. Clin Infect Dis 49(8):1248–1253. https://doi.org/10.1086/605667

Hughes P, Heritage J (2002) Antibiotic growth-promoters. Feed Tech 6(8):20–22

Hunt D, Boyd CE (1981) Alkalinity losses from ammonium fertilizers used in fish ponds. Trans Am Fish Soc 110(1):81–85

Huss HH, Pedersen A, Cann DC (1974) The incidence of Clostridium botulinum in Danish trout farms: I. Distribution in fish and their environment. Int J Food Sci Technol 9(4):445–450. https://doi.org/10.1111/j.1365-2621.1974.tb01793.x

Javed M, Usmani N (2012) Toxic effects of heavy metals (Cu, Ni, Fe, Co, Mn, Cr, Zn) to the haematology of Mastacembelusarmatus thriving in Harduaganj Reservoir, Aligarh, India. Glob J Med Res 12(8):1–8

Johnson PT, Paull SH (2011) The ecology and emergence of diseases in fresh waters. Freshw Biol 56(4):638–657

Kamat PK, Rai S, Nath C (2013) Okadaic acid induced neurotoxicity: an emerging tool to study Alzheimer's disease pathology. Neurotoxicology 37:163–172. https://doi.org/10.1016/j.neuro.2013.05.002

Korpela K, Salonen A, Virta LJ, Kekkonen RA, Forslund K, Bork P, De Vos WM (2016) Intestinal microbiome is related to lifetime antibiotic use in Finnish pre-school children. Nat Commun 7(1):1–8

Kozinska A (1999) Atypical cases of disorders in cyprinid wintering caused by Pseudomonas fluorescens infection. Bull Eur Assoc Fish Pathologists 19:216–220
Kozinska A, Pekala A (2004) First isolation of Shewanella putrefaciens from freshwater fish—a potential new pathogen of fish. Bull Eur Assoc Fish Pathologists 24(4):189–193
Kozińska A, Paździor E, Pękala A, Niemczuk W (2014) Acinetobacter johnsonii and Acinetobacter lwoffii—the emerging fish pathogens. J Vet Res 58(2):193–199
Kühne M, Körner U, Wenzel S (2001) Tetracycline residues in meat and bone meals. Part 2: The effect of heat treatments on bound tetracycline residues. Food Addit Contam 18(7):593–600
Lafferty KD, Harvell CD, Conrad JM, Friedman CS, Kent ML, Kuris AM, Powell EN, Rondeau D, Saksida SM (2015) Infectious diseases affect marine fisheries and aquaculture economics. Annu Rev Mar Sci 7:471–496
Larsson DJ, Flach CF (2022) Antibiotic resistance in the environment. Nat Rev Microbiol 20(5):257–269
Leal JF, Santos EB, Esteves VI (2019) Oxytetracycline in intensive aquaculture: water quality during and after its administration, environmental fate, toxicity and bacterial resistance. Rev Aquac 11(4):1176–1194
Leonard AF, Morris D, Schmitt H, Gaze WH (2022) Natural recreational waters and the risk that exposure to antibiotic resistant bacteria poses to human health. Curr Opin Microbiol 65:40–46. https://doi.org/10.1016/j.mib.2021.10.004
Lester Y, Avisar D, Gozlan I, Mamane H (2011) Removal of pharmaceuticals using combination of $UV/H_2O_2/O_3$ advanced oxidation process. Water Sci Technol 64(11):2230–2238
Liao X, Li B, Zou R, Xie S, Yuan B (2016) Antibiotic sulfanilamide biodegradation by acclimated microbial populations. Appl Microbiol Biotechnol 100:2439–2447
McCallum HI, Kuris A, Harvell CD, Lafferty KD, Smith GW, Porter J (2004) Marine ecosystems as 'open' systems. Trends Ecol Evol 11(19):585–591
Mo WY, Chen Z, Leung HM, Leung AOW (2017) Application of veterinary antibiotics in China's aquaculture industry and their potential human health risks. Environ Sci Pollut Res 24:8978–8989
Mu Z, Zou Z, Yang Y, Wang W, Xu Y, Huang J et al (2018) A genetically engineered Escherichia coli that senses and degrades tetracycline antibiotic residue. Synth Syst Biotechnol 3(3):196–203
Mugimba KK, Byarugaba DK, Mutoloki S, Evensen Ø, Munang'andu HM (2021) Challenges and solutions to viral diseases of finfish in marine aquaculture. Pathogens 10(6):673
Okeke ES, Chukwudozie KI, Nyaruaba R, Ita RE, Oladipo A, Ejeromedoghene O, Atakpa EO, Agu CV, Okoye CO (2022) Antibiotic resistance in aquaculture and aquatic organisms: a review of current nanotechnology applications for sustainable management. Environ Sci Pollut Res 29(46):69241–69274. https://doi.org/10.1007/s11356-022-22319-y
Okocha RC, Olatoye IO, Adedeji OB (2018) Food safety impacts of antimicrobial use and their residues in aquaculture. Public Health Rev 39(1). https://doi.org/10.1186/s40985-018-0099-2
Oliver SP, Murinda SE, Jayarao BM (2011) Impact of antibiotic use in adult dairy cows on antimicrobial resistance of veterinary and human pathogens: a comprehensive review. Foodborne Pathog Dis 8(3):337–355. https://doi.org/10.1089/fpd.2010.0730
Olivier G (2002) Disease interactions between wild and cultured fish—perspectives from the American Northeast (Atlantic Provinces). Bull Eur Assoc Fish Pathologists 22(2):102–109
Östman M, Björlenius B, Fick J, Tysklind M (2019) Effect of full-scale ozonation and pilot-scale granular activated carbon on the removal of biocides, antimycotics and antibiotics in a sewage treatment plant. Sci Total Environ 649:1117–1123
Paige JC, Tollefson L, Miller M (1997) Public health impact on drug residues in animal tissues. Vet Hum Toxicol 39(3):162–169
Perugini M, Manera M, Tavoloni T, Lestingi C, Pecorelli I, Piersanti A (2013) Temporal trends of PCBs in feed and dietary influence in farmed rainbow trout (Oncorhynchus mykiss). Food Chem 141(3):2321–2327. https://doi.org/10.1016/j.foodchem.2013.05.062
Potasman I, Paz A, Odeh M (2002) Infectious outbreaks associated with bivalve shellfish consumption: a worldwide perspective. Clin Infect Dis 35(8):921–928. https://doi.org/10.1086/342330

Puri S, Shingh S, Tiwari P (2019) Mycotoxins: a threat to food security and health. Int J Appl Sci Biotechnol 7(3):298–303

Reboreda A, Lago J, Chapela MJ, Vieites JM, Botana LM, Alfonso A, Cabado AG (2010) Decrease of marine toxin content in bivalves by industrial processes. Toxicon 55(2–3):235–243. https://doi.org/10.1016/j.toxicon.2009.07.029

Reilly A, Käferstein FJAR (1997) Food safety hazards and the application of the principles of the hazard analysis and critical control point (HACCP) system for their control in aquaculture production. Aquac Res 28(10):735–752. https://doi.org/10.1046/j.1365-2109.1997.00939.x

Reilly AC, Lima DS, Phillip M (1997) Food safety and products from aquaculture: an article presented at the joint FAO/NACA/WHO Study Group on Food Safety Issues with Products from Aquaculture, Bangkok, Thailand, pp 22–26

Rigos G, Troisi GM (2005) Antibacterial agents in Mediterranean finfish farming: a synopsis of drug pharmacokinetics in important euryhaline fish species and possible environmental implications. Rev Fish Biol Fish 15:53–73

Rivers TM (1937) Viruses and Koch's postulates. J Bacteriol 33(1):1–12

Rodger HD (2016) Fish vaccines. In: Fish disease causing economic impact in global aquaculture. Springer, Cham, Switzerland, pp 1–34

Rogers Van Katwyk S, Grimshaw JM, Nkangu M, Nagi R, Mendelson M, Taljaard M, Hoffman SJ (2019) Government policy interventions to reduce human antimicrobial use: a systematic review and evidence map. PLoS Med 16(6):e1002819. https://doi.org/10.1371/journal.pmed.1002819

Schar D, Klein EY, Laxminarayan R, Gilbert M, Van Boeckel TP (2020) Global trends in antimicrobial use in aquaculture. Sci Rep 10(1):21878

Shimizu A, Takada H, Koike T, Takeshita A, Saha M, Nakada N et al (2013) Ubiquitous occurrence of sulfonamides in tropical Asian waters. Sci Total Environ 452:108–115. https://doi.org/10.1016/j.scitotenv.2013.02.027

Shinn AP, Pratoomyot J, Bron JE, Paladini G, Brooker EE, Brooker AJ (2015) Economic costs of protistan and metazoan parasites to global mariculture. Parasitology 142(1):196–270

Soni K, Jyoti K, Chandra H, Chandra R (2022) Bacterial antibiotic resistance in municipal wastewater treatment plant; mechanism and its impacts on human health and economy. Bioresour Technol Rep 19:101080. https://doi.org/10.1016/j.biteb.2022.101080

Štefánek M, Wenner S, Borges V, Pinto M, Gomes JP, Rodrigues J et al (2022) Antimicrobial resistance and biofilms underlying catheter-related bloodstream coinfection by Enterobacter cloacae complex and Candida parapsilosis. Antibiotics 11(9):1245. https://doi.org/10.3390/antibiotics11091245

Sullivan Å, Edlund C, North CE (2001) Effect of antimicrobial agents on the ecological balance of human microflora. Lancet Infect Dis 1(2):101–114

Tacon AG, Metian M (2008) Aquaculture feed and food safety: the role of the food and agriculture organization and the Codex Alimentarius. Ann N Y Acad Sci 1140(1):50–59. https://doi.org/10.1196/annals.1454.003

Tam CC, Rodrigues LC, Viviani L, Dodds JP, Evans MR, Hunter PR et al (2012) Longitudinal study of infectious intestinal disease in the UK (IID2 study): incidence in the community and presenting to general practice. Gut 61(1):69–77. https://doi.org/10.1136/gut.2011.238386

Tariq KA (2022) Antimicrobial resistance, climate change, and public health. In: Climate change and microbial diversity. Apple Academic, p 109

Tasker RA (2021) Domoic acid. In: Advances in neurotoxicology, vol 6. Academic Press, pp 49–88. https://doi.org/10.1016/bs.ant.2021.03.002

Theodorou JA, Moutopoulos DK, Tzovenis I (2020) Semi-quantitative risk assessment of Mediterranean mussel (Mytilus galloprovincialis L.) harvesting bans due to harmful algal bloom (HAB) incidents in Greece. Aquac Econ Manag 24(3):273–293. https://doi.org/10.1080/13657305.2019.1708994

van Bussel CG, Schroeder JP, Mahlmann L, Schulz C (2014) Aquatic accumulation of dietary metals (Fe, Zn, Cu, Co, Mn) in recirculating aquaculture systems (RAS) changes body composition but not performance and health of juvenile turbot (Psetta maxima). Aquac Eng 61:35–42. https://doi.org/10.1016/j.aquaeng.2014.05.003

Wang P, Shen C, Cong Q, Xu K, Lu J (2021) Enzyme-catalyzed biodegradation of penicillin fermentation residues by β-lactamase OtLac from Ochrobactrumtritici. Microb Cell Factories 20:1–9

Waters CM, Bassler BL (2005) Quorum sensing: cell-to-cell communication in bacteria. Annu Rev Cell Dev Biol 21:319–346

Watts JE, Schreier HJ, Lanska L, Hale MS (2017) The rising tide of antimicrobial resistance in aquaculture: sources sinks and solutions. Marine Drugs 15(6):158. https://doi.org/10.3390/md15060158

Wei Z, Wei Y, Li H, Shi D, Yang D, Yin J et al (2022) Emerging pollutant metformin in water promotes the development of multiple-antibiotic resistance in Escherichia coli via chromosome mutagenesis. J Hazard Mater 430:128474. https://doi.org/10.1016/j.jhazmat.2022.128474

Wierup M (2001) The experience of reducing antibiotics used in animal production in the Nordic countries. Int J Antimicrob Agents 18(3):287–290. https://doi.org/10.1016/S0924-8579(01)00380-6

Woo PT, Leong JA, Buchmann K (eds) (2020) Climate change and infectious fish diseases. CABI

Wu Y, Zhang D, Ma P, Zhou R, Hua L, Liu R (2018) Lithium hexamethyldisilazide initiated superfast ring opening polymerization of alpha-amino acid N-carboxyanhydrides. Nat Commun 9(1):5297

Yan Y, Deng Y, Li W, Du W, Gu Y, Li J, Xu X (2021) Phytoremediation of antibiotic-contaminated wastewater: insight into the comparison of ciprofloxacin absorption, migration, and transformation process at different growth stages of E. crassipes. Chemosphere 283:131192

Yang C, Lou W, Zhong G, Lee A, Leong J, Chin W et al (2019) Degradable antimicrobial polycarbonates with unexpected activity and selectivity for treating multidrug-resistant Klebsiella pneumoniae lung infection in mice. Acta Biomater 94:268–280. https://doi.org/10.1016/j.actbio.2019.05.057

Zheng Q, Zhang R, Wang Y, Pan X, Tang J, Zhang G (2012) Occurrence and distribution of antibiotics in the Beibu Gulf, China: impacts of river discharge and aquaculture activities. Mar Environ Res 78:26–33. https://doi.org/10.1016/j.marenvres.2012.03.007

Zheng S, Wang Y, Chen C, Zhou X, Liu Y, Yang J et al (2022) Current progress in natural degradation and enhanced removal techniques of antibiotics in the environment: a review. Int J Environ Res Public Health 19(17):10919

Chapter 10
Phage Therapy in Aquaculture

Stavros Droubogiannis and Pantelis Katharios

Abstract Phage therapy is a promising alternative to antibiotics in aquaculture, which is facing increasing concerns over antimicrobial resistance (AMR). Bacteriophages are viruses that specifically infect and replicate within bacteria, causing bacterial cell death. They are abundant in the aquatic environment and can be isolated from a variety of sources, including fish, water, and sediment. Phage therapy has been shown to be effective against a wide range of bacterial pathogens that are responsible for causing severe disease outbreaks in aquatic fauna. Phages are highly host-specific, and therefore they can target and eliminate harmful bacteria while leaving beneficial ones unaffected. As a targeted therapy, phage therapy can be used in delicate environments like fish hatcheries where the use of broad-spectrum antimicrobials may disturb the microbial equilibrium with adverse effects on the quality of the developing larvae.

However, phage therapy in aquaculture and the aquatic environment faces several challenges. The most important challenge is the rapid development of resistance by the bacteria against phage infection. In addition, a big challenge is the lack of knowledge about the diversity and distribution of phages in aquatic environments. As such, identifying and isolating phages that are effective against specific bacterial pathogens can be time-consuming and resource-intensive, especially taking into consideration the strain diversity of several aquatic pathogens like the vibrios. Additionally, the efficacy of phage therapy can be affected by environmental factors, such as temperature and pH, and the presence of other microorganisms in the environment which can influence the host–phage interactions. Moreover, not

S. Droubogiannis
Institute of Marine Biology, Biotechnology and Aquaculture, Hellenic Centre for Marine Research, Gournes, Crete, Greece

Department of Biology, University of Crete, Iraklio, Greece

P. Katharios (✉)
Institute of Marine Biology, Biotechnology and Aquaculture, Hellenic Centre for Marine Research, Gournes, Crete, Greece
e-mail: katharios@hcmr.gr

P. Elumalai, S. Lakshmi (eds.), *Antimicrobial Resistance in Aquaculture and Aquatic Environments*, https://doi.org/10.1007/978-981-97-7320-6_10

only the regulatory framework of phage therapy applications but also the manufacturing of phage products is still far from being established.

Despite these challenges, the potential benefits of phage therapy in aquaculture are significant. It offers a sustainable and environmentally friendly alternative to antibiotics, which can reduce the risk of AMR and environmental pollution. Moreover, phage therapy can be tailored to specific bacterial pathogens, which can improve treatment efficacy and reduce the risk of non-target effects. Overall, phage therapy has the potential to revolutionize the way bacterial infections are managed in aquaculture, leading to improved health and welfare of farmed aquatic animals and sustainable aquaculture practices.

Keywords Phage therapy · Vibrio · Resistance · Fish · Hatchery

10.1 Basic Concepts Regarding Bacteriophages and Phage Therapy

Bacteriophages, also known as phages, are a type of viruses that specifically infect bacteria and are abundantly present in the world. Their immense numbers, estimated to be around 10^{31} in the biosphere, highlight their significance. The existence of phages was initially hypothesized by Ernest Hanbury Hankin, an English microbiologist, toward the end of the nineteenth century. In 1915, Frederick William Twort, an English microbiologist, published the initial scientific manuscript in the Lancet, detailing the behavior of bacteriophages. Nevertheless, credit for the discovery and nomenclature of bacteriophages is commonly attributed to Felix d'Herelle, a French microbiologist based at the Pasteur Institute. It was in 1917 that d'Herelle authored a paper delineating bacteriophages as viruses that infect bacteria (Publications Service 2007). Bacteriophages were initially employed in the treatment of bacterial infections. It is important to note that this occurred prior to the discovery of antibiotics when even minor bacterial infections posed a significant risk to human life. The application of bacteriophages for therapeutic purposes is commonly referred to as phage therapy.

Before delving into the possibilities offered by phage therapy in aquaculture, it is essential to grasp some fundamental concepts of phage microbiology. Bacteriophages serve as highly efficient "predators" of bacteria in nature, playing a crucial ecological role in controlling bacterial populations. The structure of phages consists of a protein capsid enclosing their genetic material, which can be DNA or RNA, along with a tail that attaches to the capsid. Tailed bacteriophages, which are the focus here, exhibit various morphologies, with tails that can be long or short or contractile or non-contractile. The tail which in the past played a key role in taxonomy possesses a baseplate, tail fibers, and a spike. The receptor-binding proteins at the distal end of the tail fibers interact with specific surface receptors on the bacterial host, while the spike proteins facilitate the degradation of the bacterial surface layer, allowing the phage to bind irreversibly (Fig. 10.1). Once attached, the phage injects its genetic material into the bacterial cell. Subsequently, the phage utilizes the bacterial machinery for propagation, hijacking the cell's resources.

Phages have diverse life cycles, with the two most extensively studied types being the lytic or virulent and the temperate. Upon injection of their DNA into the host cell, lytic phages initiate the production of structural proteins and genetic material, leading to the assembly of progeny virions within the host. Once assembly is complete, lytic enzymes are secreted, causing the degradation of the bacterial cell wall from within, resulting in a burst and the release of phages into the external environment. On the contrary, temperate phages may integrate their DNA into the bacterial chromosome upon infection, becoming dormant and replicating along with the bacterium until induced by damage to the host's DNA or environmental cues. Upon induction, the viral DNA excises from the host's DNA, and the phage enters the lytic cycle, releasing new virions. During the temperate phase, the phage genes integrated into the bacterial host can become functional, potentially encoding traits such as toxins and proteins associated with antibiotic resistance. For instance, the cholera toxin serves as a prime illustration, acting as the primary virulence factor of *Vibrio cholerae*. This toxin is encoded within a prophage integrated into the bacterium's chromosome. Similarly, marine pathogenic vibrios such as *V. harveyi*, *V. alginolyticus*, *V. vulnificus*, and numerous others harbor prophage-encoded toxins that enhance their virulence compared with non-phage-infected strains. This phenomenon, known as lysogenic conversion, involves the acquisition of new properties that can augment bacterial fitness, with virulence being particularly important. However, it is precisely this attribute that poses the greatest risk when utilizing phages as a therapeutic tool, as it can inadvertently transform non-virulent bacterial strains into virulent ones. Consequently, it is crucial to carefully select lytic phages while excluding temperate ones in phage therapy. Moreover, modern approaches now employ genomic analysis for a more precise selection of suitable phages.

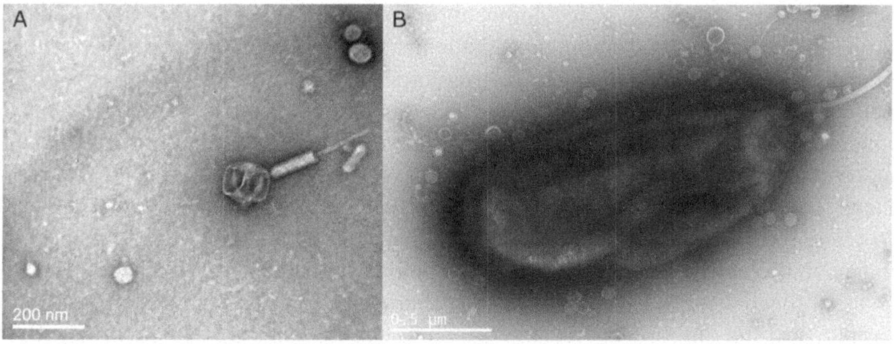

Fig. 10.1 (**a**) Vibriophage of a Myoviridae morphotype with the characteristic contractile tail. (**b**) A negatively stained *Vibrio alginolyticus* bacterial cell under bacteriophage attack. (TEM micrographs courtesy of Dr. Katharios)

Phages typically exhibit a high degree of host specificity, at times to the extent of infecting a subset of individual strains within a given species. Nevertheless, there are phages that possess a broad host range, capable of infecting different species, most commonly within the same genus. This characteristic sets phage therapy apart from other existing methods for controlling bacteria in aquaculture, providing a more focused and precise treatment strategy. The specificity of phages toward their hosts is influenced by intricate molecular interactions that occur throughout the infection cycle, although these details lie beyond the scope of this chapter. However, one crucial factor contributing to host specificity is the type and diversity of receptor-binding proteins located on the phage tail, which mediate the initial interplay between the phage and the bacterium. In Gram-negative bacteria like vibrios, these phage proteins primarily target elements of the outer membrane such as lipopolysaccharides (LPS), which are major virulence factors, as well as flagella and porins that serve as receptors for nutrient uptake from the extracellular environment (Katharios 2020).

Phages and bacteria engage in an ongoing evolutionary battle within their environmental realm. Bacteria face continuous exposure to phage "predation," prompting them to develop defensive strategies against phage infection for survival. Bacterial resistance to phages can emerge rapidly, often involving the sacrifice of receptors to which phages attach. This molecular-level process entails the down-regulation of genes encoding these receptor proteins. Resistance can also arise through genetic mutations in these proteins, leading to incompatibility between phage-binding receptors and bacterial cell receptors. Even in the face of incursion of phage genetic material, bacteria wield an array of defense mechanisms that bestow them with resistance. The CRISPR-Cas system, functioning as a bacterial "adaptive immune system," plays a pivotal role in defense against phages (Katharios 2020). Conversely, phages also adapt swiftly and develop counter-resistance mechanisms. Bacterial resistance to phage infection often involves LPS, which are important virulence determinants. In many cases, resistant strains exhibit reduced fitness or virulence compared with their wild-type counterparts, underscoring the substantial cost of resistance development for bacteria. The rapid emergence of bacterial resistance poses a significant challenge in phage therapy, a hurdle that is almost certain to arise over time. To overcome this issue, phage therapy requires meticulous design. The utilization of "phage cocktails," which involve combinations of different phages, offers a potential solution. However, the selection of appropriate phages for these cocktails demands expertise and knowledge. Previously, phage cocktails were formulated with phages exhibiting diverse host ranges. Presently, it is understood that an effective phage cocktail should ideally comprise phages that employ different receptors for initial attachment to the bacterial host. This approach mitigates the risk of bacterial resistance, as changes in multiple receptors would impose a greater cost on bacterial viability. Notably, "jumbo" phages with expansive genomes have emerged as promising candidates for phage cocktails due to their broad host range, likely attributed to a wider array of receptor-binding proteins present in their tails.

10.2 Reemergence of Phage Therapy in the Era of AMR

Despite being discovered prior to antibiotics, the clinical and therapeutic utilization of phages did not gain widespread acceptance in the Western world. Initially, their application was confined to countries in the former Eastern bloc, such as the USSR and Poland. The advent of World War II and the subsequent mass production of antibiotics played a key role in their extensive adoption, eventually overshadowing the use of phages. Antibiotics, like penicillin, could be manufactured in large quantities with consistent quality and stability, a characteristic not necessarily guaranteed with phages.

The realization of antimicrobial resistance (AMR) as a significant global crisis unfolded gradually as scientific evidence and observations accumulated over time. Even during the early stages of antibiotic use, scientists noticed that certain bacteria could persist and proliferate despite exposure to these drugs. However, it was during the period between 1990 and 2000 that the issue began to gain recognition, leading to intensified research and surveillance efforts. Esteemed organizations such as the World Health Organization (WHO) and the Centers for Disease Control and Prevention (CDC) began to prioritize AMR as a critical public health concern. In 2014, Jim O'Neill, a renowned British economist, led an international commission to investigate the impact of AMR. This initiative culminated in the publication of the Review of Antimicrobial Resistance in 2016, which stands as the most influential work on the subject and has greatly increased public awareness (O'Neill 2014). Consequently, the pursuit of alternatives has emerged as one of the scientific community's most critical research objectives.

10.3 Phage Therapy Applications in Aquaculture

Phage therapy presents a highly appealing approach to the control of pathogens in aquaculture. Bacteriophages are extensively found in aquatic environments, benefiting from the water's aqueous nature, which enhances their dispersal and increases the likelihood of encountering target bacteria. Notably, phages present in the water can traverse the gills and stomach (considering that marine fish consume water) to enter the bloodstream and potentially reach internal organs. Given that phages naturally inhabit aquatic environments, their medicinal utilization could align with organic farming practices. Unlike antibiotics, phages do not leave behind residues and pose no harm to fish, humans, or the surrounding ecosystem. Being self-replicating agents, phages offer the advantage of theoretically not requiring multiple doses in phage therapy. Furthermore, their high specificity for particular hosts renders them the most suitable option for controlling bacteria in delicate environments, such as fish hatcheries and recirculating aquaculture systems (RAS). Finally, phages can be employed during the early developmental stages of fish, where vaccination may not be feasible due to the immature state of the immune system (Katharios 2020).

Aquaculture is a fast-growing industry that includes the cultivation of aquatic organisms of different taxa, including fish, crustaceans, and mollusks. Unfortunately, aquaculture systems are vulnerable to diseases caused by a plethora of pathogens that pose significant challenges to sustainable production. Bacterial species responsible for common disease outbreaks include *Aeromonas salmonicida, Aeromonas hydrophila, Aeromonas veronii, Edwardsiella ictaluri, Edwardsiella tarda, Yersinia ruckeri, Piscirickettsia salmonis, Flabobacterium psychrophilum, Flavobacterium columnare, Pseudomonas anguilliseptica, Vibrio harveyi, Vibrio anguillarum, Vibrio alginolyticus, Vibrio parahaemolyticus, Vibrio splendidus, Photobacterium damselae* subsp. *piscicida, Moritella viscosa, Tenacibaculum maritimum, Lactococcus garvieae, Streptococcus iniae, Streptococcus parauberis, Streptococcus phocae,* and *Mycobacterium marinum* (Katharios 2020). The majority of these bacterial pathogens are naturally present in the aquatic environment, encompassing both freshwater and marine ecosystems, and have been widely used as experimental models in in vivo phage therapy applications in various aquatic organisms yielding promising results.

For the first time, in 1981, Wu et al. introduced the utilization of bacteriophages as a therapeutic approach in aquaculture (Wu et al. 1981). Subsequently, there has been significant interest in utilizing bacteriophage therapy for various aquatic animal species in controlling bacterial diseases. In parallel, evaluation of the safety, stability, and environmental impact of phage treatment in aquaculture systems has also gathered some significant attention. However, when it comes to applied phage therapy in the aquaculture setting, many things must be taken into serious consideration in order to attain the desired outcome. Phage selection, phage dosage, phage administration route, and treatment duration are key parameters that will be thoroughly discussed hereby.

10.3.1 Phage Formulation and Delivery

One of the key elements when practically applying phage therapy in aquaculture to attain the desired outcome is the optimal route of phage administration for each bacterial infection. The efficacy of bacteriophage applications relies on their ability to reach the host, which may not always be feasible. Depending on the infection's location within the body and considering factors such as penetration effectiveness and the ability to sustain high phage levels, phage preparations can be administered in diverse forms. In the aquaculture field, phages employed to control bacterial diseases can be given orally through feed, administered parenterally (via intramuscular, subcutaneous, or intraperitoneal routes), applied topically to the skin and lesions, introduced via bath treatments, or directly released into the water system.

Phage delivery through the water or orally, incorporating them into the feed, is to a great degree among the most widely used approaches. Oral administration of bacteriophages has proven to be an effective treatment for gastrointestinal infections, particularly those caused by *Vibrio* spp. Interestingly, it has been observed that orally administered bacteriophages can be absorbed into the systemic circulation, similar to the process of bacterial translocation. This suggests that oral administration may also be suitable for treating systemic infections. Several elements come into play when considering the behavior of bacteriophages within the aquatic environment. These factors encompass the concentration of bacteriophages, the existence of distinct sequences within the capsid or tail proteins that engage with receptors on enterocytes, as well as interactions with immune cells in the intestinal milieu, all contributing to the dynamics of bacteriophage transit.

The prevalent method to apply the desired substances, in our case phages, to aquaculture animals is through the water body. It can be done in different ways, ranging from a high concentration of phages with a short exposure time (immersion) to a low concentration with a long exposure time (bath). The application of bacteriophage preparations using this method is popular because it allows the bacteriophages to enter the internal organs of fish directly through the gills while providing additional benefits of environmental cleansing. Moreover, marine fish drink water continuously therefore allowing the entrance of the phages in their intestine. However, in commercial-scale aquaculture, the use of this administration route can pose challenges due to the impractically large volume of water that would be required for an effective bacteriophage treatment.

Unfortunately, there is not a universally applicable strategy for bacteriophage delivery. The appropriate approach depends on various factors and should be considered on a case-by-case basis. Administering injections, for example, may not be practical for very small fish or crustaceans and, of course, is not applicable when a large population is to be treated. Likewise, conducting bathing with a high concentration of bacteriophages can be challenging in large bodies of water, and the immersion method may vary depending on the specific environment, infection characteristics, or properties of the bacteriophages. Each administration method has its advantages and limitations, and the choice depends largely on the bacterial pathogen's nature, the animal species, and its size.

The timing of administration also plays a crucial role. Mortality rates tend to increase significantly when treatment is delayed after infection. Optimal effectiveness is observed through the prophylactic application of bacteriophages before infection and its sustained use. A study conducted by Jun et al. demonstrated a 75% increase in the survival rate of white shrimp when employing prophylactic bacteriophage immersion, along with a 50% survival rate when using phages in feed (Jun et al. 2018). Similarly, Schulz et al. noted a 16% higher survival rate in European eel and a 6% higher survival rate in rainbow trout when administering the bacteriophage preparation 1 day before infection, in comparison to treatment 24 h post-infection (Schulz et al. 2019). This improvement is likely linked to the time required for bacteriophages to multiply to an adequate concentration for effective host population control. Additionally, the prophylactic application of phages assists in regulating the concentration of pathogenic bacteria in water and/or fish, preventing them from reaching levels that could trigger disease outbreaks.

The optimal dosage for effective treatment ideally relies on data obtained from in vitro studies. The effective dose is linked to the minimum multiplicity of infection (the ratio of bacteriophages to the bacterial host) capable of significantly reducing the bacterial titer in in vitro assays. Another crucial factor influencing the dosage is the Burst size, which denotes the number of new virions produced following an infection cycle and indicates the quantity of new phages generated by a single infected bacterial cell. This factor holds significant importance as, based on our personal observations, phage therapy exhibits improved outcomes when the initial concentration of bacterial hosts is higher. This is because phages rely on these hosts for their propagation, thereby naturally adjusting the treatment dosage.

Exploring the influence of timing and the dosage of bacteriophages on the effectiveness of phage therapy through additional research may uncover intriguing findings, as a higher concentration of administration might have the potential to offset the effects of delayed treatment.

10.3.1.1 Interaction of Phages and Fish Immune System

Our understanding of the correlation between phage activity and the fish immune system is currently limited, leaving several gaps that require further investigation. Nevertheless, it is acknowledged that the method of phage administration plays a pivotal role in the established interactions between phages and the fish's immune system. In vitro studies utilizing cultured cells from mucosal surfaces, as well as in vivo experiments involving fish immersion, have provided evidence indicating that elevated levels of phages in the mucus layer play a protective role in preventing bacterial infections in the underlying epithelium (Barr et al. 2013). Upon entering the fish's body, phages initially engage with and overcome the cells of the mucosal innate immune system before aiming at their bacterial prey. The survival and presence of phages in fish organs largely rely on the specific administration route, as they encounter the host's phage clearance mechanisms and undergo sensitivity checks, such as digestive enzymes, pH changes, immune cells, and antimicrobial substances, during gut transit. The phage clearance mechanism in fish primarily involves the kidney, spleen, and intestine. This is indicated by the phage recovery results following their administration through any route within the fish's body (Dabrowska et al. 2005).

Remarkably, it has been noted that the majority of phages do not exhibit an outer lipid layer, which is usually targeted by the complement system (Carroll-Portillo and Lin 2019). This raises the question of how phages elude elements of the adaptive and innate immune system, such as antibodies and phagocytes, in order to infiltrate eukaryotic cells. One partial explanation is transcytosis, which involves either free uptake through endocytosis or crossing via a compromised gut, facilitating passive traversal through the epithelium. Another intriguing mechanism involves the phage being endocytosed by a bacterial host and hidden inside it, ensuring unrecognized access beyond epithelial cell layers. It is important to note that phages are incapable of infecting eukaryotic cells but can penetrate epithelial cell layers and spread throughout several regions inside the organism's body (Sophie et al. 2017).

Moreover, phages belonging to *Caudoviricetes* class possess the inherent capability to translocate across epithelial and endothelial cell lines (Carroll-Portillo and Lin 2019). Upon successfully penetrating the mucosal tissue of fish, a systematic screening process unfolds, encompassing both extracellular and intracellular immune recognition mechanisms. However, it is noteworthy that endocytic recognition processes commence exclusively subsequent to lysosomal degradation within the host cell, where pathogen recognition is facilitated by classical and evolutionarily conserved pattern recognition receptors (PRRs).

10.3.2 Dosage

In the field of phage therapy, two discernible approaches emerge: active and passive (Cairns et al. 2009). In active therapy, bacteriophages are administered in a dosage that can reduce the host population through multiple reproduction cycles. In passive therapy, extensive bacteriophage quantities are administered, rendering multiple reproduction cycles unnecessary, as the entire host population is lysed. It is worth noting that the passive approach is costlier; however, it can effectively circumvent bacterial defense mechanisms, including abortive infection.

The optimal ratio of bacteriophages to bacterial concentration required for successful bacteriophage therapy differs greatly, contingent upon factors such as the pathogen, fish species, and specific bacteriophage employed. Experimental investigations and field studies have utilized various doses in their assessments. For instance, in a study by Le et al., different bacteriophage amounts were used in catfish therapy, leading to survival rates that exhibited variances of up to 68% (Le et al. 2018). This highlights the importance of ascertaining the precise bacteriophage dose to ensure the success of phage therapy.

10.3.3 Monophage Therapy: Phage Cocktails

Numerous strategies for bacteriophage therapy have undergone testing and evaluation. Monophage therapy involves using a single type of bacteriophage. It is primarily used for experimental modeling of bacteriophage therapy and as a means of confirming the concept during testing of preparations. However, monophage therapy necessitates an exact match between the pathogen and the bacteriophage (Liu et al. 2022).

In the context of aquaculture, the realistic scenario often exceeds the simplicity of a "one pathogen-one disease" situation. Fish can experience concurrent infections caused by multiple strains or species of bacteria, which can impact disease outcomes and pose additional complexities for phage therapy. Consequently, polyphage therapy has become increasingly popular in the field of aquaculture (Liu et al.

2022). This approach entails the utilization of a bacteriophage cocktail comprising multiple phages that target various strains of a single bacterial species or even multiple bacterial species. One notable advantage of employing multiple phages is that they provide a more comprehensive treatment approach by effectively targeting a broad spectrum of pathogenic bacterial strains. This results in a more substantial reduction in bacterial titers and faster-acting therapeutic effects. Furthermore, using bacteriophage cocktails that target different receptors of the same bacterium may aid in mitigating the development of resistance, a significant concern in phage therapy practices.

The majority of published articles predominantly center around monophage therapy and emphasize the discovery of potentially beneficial bacteriophages. In contrast, only a limited number of articles delve into the use of cocktails or alternative combinations. In vivo experiments presented in this context illustrate the efficacy of bacteriophage therapy in the management of aquaculture-related diseases. However, the outcomes tend to vary depending on the specific combination of bacteriophages and bacteria used. Some experiments have yielded complete bacterial elimination with no recorded mortality, while others have exhibited no discernible therapeutic effect. In certain instances, disease progression was delayed, but the ultimate mortality rates did not exhibit significant deviations from those observed in the absence of bacteriophage therapy.

10.3.4 Phage-Derived Enzymes

The utilization of complete phage particles, including phage cocktails, has gained momentum in the fight against antibiotic resistance. The development of lytic enzymes derived from these viruses has garnered interest in various fields, including veterinary medicine, food science, and technology. Phage-encoded proteins with potential applications are categorized into endolysins, exolysins, and depolymerases. Endolysins primarily degrade the bacterial cell wall from within the cytoplasm of the infected bacteria during the phage lytic cycle. Conversely, exolysins or ectolysins act externally, enabling quicker access to bacteria. These enzymatic proteins have shown promise as antimicrobial agents and have demonstrated efficacy in disrupting preformed biofilms. Furthermore, the engineering of lysins has enhanced their binding and traversing capabilities through the bacterial wall. Depolymerases, another group of phage enzymes, have the potential to combat pathogens and the biofilms they form, as biofilm matrices often consist of different polymers. One significant advantage of phage lytic enzymes is that bacteria are unable to develop resistance against them, which contrasts with the current issue of phage resistance development. Although the focus of phage and lysin studies has been primarily on Gram-positive bacteria, some fish pathogens associated with disease outbreaks in fish farms, such as *Streptococcus agalactiae, Lactococcus garvieae, Renibacterium salmoninarum, Streptococcus iniae,* and *S. dysgalactiae,* have been investigated.

10.3.5 Success Stories and Commercial Applications

There are numerous success stories demonstrating the efficacy of phage therapy in combating bacterial infections. The focused approach of phage therapy enables the accurate elimination of harmful bacteria while assuring the presence of beneficial microbial populations. This characteristic significantly reduces the reliance on antibiotics and diminishes the likelihood of resistance emergence. Hereby, we will discuss and provide a summary of the most significant successful cases of applied phage therapy in the aquaculture field, highlighting the key elements that led to the desired outcome (Table 10.1).

10.3.5.1 Fish

Phage therapy has emerged as a revolutionary approach to managing bacterial infections in fish species, providing targeted and effective solutions for disease control. For instance, in the case of African catfish infected with *Pseudomonas aeruginosa*, the local application of phages resulted in a remarkable reduction of skin lesions by sevenfold (Khairnar et al. 2013). This highlights the precise eradication capabilities of phages, effectively targeting the pathogen while minimizing harm to healthy tissues. Similarly, when *Nile tilapia* were intraperitoneally injected with phages to combat *Streptococcus agalactiae* infection, a significant 3-day delay in mortality and a notable 40% reduction in overall death rates were observed (El-Araby et al. 2016). This underscores the potential of phage therapy in allowing the immune system vital time to mount a defense against the pathogen. Moreover, phage therapy trials conducted on gilthead seabream larvae setting showed that phages had a noteworthy impact on the survival of the larvae for a period of 5 days, leading to a significant reduction in mortalities (Droubogiannis and Katharios 2022; Droubogiannis et al. 2023). Another notable success story involves the administration of phages against *Pseudomonas plecoglossicida* in Aju sweetfish, where timing played a crucial role (Park and Nakai 2003). Depending on the timing of phage administration, mortality reduction ranged from 42.5% in 10 g fish to an impressive 78% in 2.4 g fish. These findings emphasize the immense potential of phage therapy in improving disease management, promoting health, and enhancing the survival rates of fish species in aquaculture settings.

10.3.5.2 Crustaceans

Applied phage therapy in crustaceans also yielded intriguing results. For example, in the case of *Vibrio harveyi* infection in brine shrimp (*Artemia franciscana*), the use of immersion-based phage treatments resulted in notable improvements in hatching success rates and enhanced survival rates (Quiroz-Guzmán et al. 2018). Hatching success reached 100% compared with approximately 50% in control

Table 10.1 Phage therapy application in aquaculture. Adapted from Schulz et al. (2022)

Pathogen	Species	Application	Outcome	References
Aeromonas hydrophila	Carp (*Cyprinus carpio*)	Intraperitoneal injection	Reduction in mortality by 100%, 60% or 50% depending on the bacteriophage or cocktail used	Yu et al. (2022)
	Cyprinid loach (*Misgurnus anguillicaudatus*)		1. Mortality drop from 39% to 0%; 2. A decrease in mortality from 100% to 43% or 17% depending on the used bacteriophage	Jun et al. (2013)
			No mortality after 7 days compared to control group (65%)	Wu et al. (1981)
		Feed	1. A decrease in mortality from 39% to 17% or 11% depending on the used bacteriophage; 2. A decrease in mortality from 96% to 47% or 27% depending on the used bacteriophage	Jun et al. (2013)
		Bath	A 47% decrease in mortality; most surviving fish showed no signs of disease	Akmal et al. (2020)
	Nile tilapia (*Oreochromis niloticus*)	Intraperitoneal injection	A 50% decrease in mortality	El-Araby et al. (2016)
		Immersion	Reduction in mortality by 37.5–55% depending on bacteriophage dose	Dien et al. (2022)
	Rainbow trout (*Oncorhynchus mykiss*)	Intraperitoneal injection	Reduction in mortality by 40% after prophylactic administration	Cao et al. (2020)
		Feed	Reduction in mortality by 70% after prophylactic administration	Cao et al. (2020)
		Bath	Reduction in mortality by 80% after prophylactic administration	Cao et al. (2020)
	Striped catfish (*Pangasianodon hypophthalmus*)	Intraperitoneal injection	Reduction in mortality by 82%, 37% or 14% depending on bacteriophage dose	Le et al. (2018)
		Feed	Reduction in mortality by 51.6–60% depending on bacteriophage dose	Dang et al. (2021)
	Zebrafish (*Danio rerio*)	Immersion	Reduction of mortality by 43.3%	Easwaran et al. (2017)

Aeromonas hydrophila and *Pseudomonas fluorescens*	European eel (*Anguilla anguilla*)	Bath	Reduction in mortality by 40%, 25% or 15% depending on time of initiation of therapy; reduction in mortality by 60% with prophylactic use	Schulz et al. (2019)
	Rainbow trout (*Oncorhynchus mykiss*)	Bath	Reduction in mortality by 25%, 15% or 10% depending on time of initiation of therapy; reduction in mortality by 36% with prophylactic use	Schulz et al. (2022)
Aeromonas salmonicida	Brook trout (*Salvelinus fontinalis*)	Immersion	Delayed disease onset by 7 days and reduced mortality from 100% to 10%	Imbeault et al. (2006)
	Rainbow trout (*Oncorhynchus mykiss*)	Intramuscular injection	Reduction in mortality from 100% to 70%	Kim et al. (2015)
	Senegalese sole (*Solea senegalensis*)	Immersion	No mortality compared to the control group (36%)	Silva et al. (2016)
Aeromonas salmonicida subsp. *salmonicida*	Atlantic salmon (*Salmo salar*)	Intraperitoneal injection	Delayed mortality; final mortality did not differ between groups	Verner-Jeffreys et al. (2007)
		Feed		
		Bath		
	Rainbow trout (*Oncorhynchus mykiss*)	Intramuscular injection	Reduction in mortality by 26.7%, no symptoms up to 14 days after bacteriophage administration	Kim et al. (2015)
Citrobacter freundii	Carp (*Cyprinus carpio*)	Intraperitoneal injection	Reduction in mortality by 100%, 45% and 0% depending on time of bacteriophage administration	Jia et al. (2020)
Citrobacter spp.	Zebrafish (*Danio rerio*)	Bath	Reduction in mortality by 17%, 23% and 26% depending on the bacteriophage or cocktail used	Royam and Nachimuthu (2020)
Edwardsiella tarda	Turbot (*Scophthalmus maximus*)	Feed	Reduction in mortality by 53%, 76% or 80% depending on bacteriophage dose	Cui et al. (2021)
	Zebrafish (*Danio rerio*)	Bath	Reduction in mortality by 50%	Nikapitiya et al. (2020)

(continued)

Table 10.1 (continued)

Pathogen	Species	Application	Outcome	References
Flavobacterium columnare	Rainbow trout (*Oncorhynchus mykiss*)	Bath	Reduction of mortality by 33–42% depending on the number of bacteriophages	Laanto et al. (2015)
	Walking catfish (*Clarias batrachu*)	Intramuscular injection	No symptoms and 100% survival	Prasad et al. (2011)
		Bath		
		Feed		
	Zebrafish (*Danio rerio*)	Immersion	Reduction in mortality by 60%	Laanto et al. (2015)
Flavobacterium psychrophilum	Atlantic salmon (*Salmo salar*)	Intraperitoneal injection	Mortality decreased from 45% to 18% and from 13% to 6% depending on the bacteriophage used	Castillo et al. (2012)
	Rainbow trout (*Oncorhynchus mykiss*)		Mortality decreased from 47% to 20% and from 80% to 47% depending on the bacteriophage used	
			23% reduction in mortality by phage administration 3 days after infection	Donati et al. (2021)
			Cocktail reduced mortality by 17–54% depending on the bacteriophage/bacterial ratio	Sundell et al. (2020)
		Feed	No significant differences in final mortality	Donati et al. (2021)
		Bath		
Lactococcus garvieae	Japanese amberjack (*Seriola quinqueradiata*)	Intraperitoneal injection	Mortality decreased from 90% to 0–50% depending on the timing of bacteriophage administration	Nakai et al. (1999)
		Feed	Mortality reduced from 65% to 10%	
	Rainbow trout (*Oncorhynchus mykiss*)		Reduction in mortality from 100% to 70% after 2 weeks	Ghasemi et al. (2014)
Photobacterium damselae subsp. Damselae	Longfin yellowtail (*Seriola rivoliana*)	Immersion	Increased egg hatch rate from 50 to 80%	Veyrand-Quirós et al. (2020)
Pseudomonas aeruginosa	African catfish (*Clarias gariepinus*)	Locally on skin lesions	A sevenfold reduction in the size of the lesions	Khairnar et al. (2013)

Pseudomonas plecoglossicida	Aju sweetfish (*Plecoglossus altivelis*)	Feed	Reduction of mortality by 42.5% when bacteriophages were administered at the time of infection in 10 g fish; in 2.4 g fish by 78% and 67% depending on time of administration	Park et al. (2000)
			1. Reduction in mortality by 40% and 73% with the cocktail; 2. Mortality reduced by 73% and 63% depending on the bacteriophage used; 3. Field infection-reduction in mortality from 18 kg/day to 6 kg after three applications of bacteriophage	Park and Nakai (2003)
Streptococcus agalactiae	Nile tilapia (*Oreochromis niloticus*)	Intraperitoneal injection	A 3-day delay and 40% reduction in mortality	Luo et al. (2018)
Streptococcus iniae	Japanese flounder (*Paralichthys olivaceus*)	Intraperitoneal injection	The decrease in mortality by 28–90% depending on the dose and time of administration	Matsuoka (2007)
Streptococcus parauberis	Japanese flounder (*Paralichthys olivaceus*)	Feed	Improved fish growth, reduced bacterial detection and improved breeding survival	Kwon et al. (2017)
Vibrio alginolyticus	Atlantic cod (*Gadus morhua*)	Immersion	Mortality delay; no statistically significant differences at the end of the experiment	Rørbo et al. (2018)
	Atlantic salmon (*Salmo salar*)		1. Reduction in mortality from 93% to 0–30% depending on the dose of bacteriophages under experimental conditions; 2. Reduction in mortality from 40% to 0% in breeding conditions	Higuera et al. (2013)
	Artemia salina		The total load of bacteria decreased by 93%	Kalatzis et al. (2016)
	Japanese sea cucumber (*Apostichopus japonicus*)	Feed	70%, 47% and 44% reduction in mortality after using a cocktail depending on the dose; no difference in survival compared to the use of antibiotics	Zhang et al. (2015)
	New Zealand rock oyster (*Saccostrea glomerata*)	Immersion	Reduction of larvae mortality by 50% after using a cocktail	Le et al. (2018)
	Turbot (*Scophthalmus maximus*)		Mortality delay; no statistically significant differences at the end of the experiment	Rørbo et al. (2018)
Vibrio anguillarum	Zebrafish (*Danio rerio*)	Immersion	Mortality reduced from 17% to 3%	Silva et al. (2014)

(continued)

Table 10.1 (continued)

Pathogen	Species	Application	Outcome	References
Vibrio campbellii	*Artemia franciscana*	Immersion	Survival of nauplii increased by 24%	Lomelí-Ortega et al. (2021)
Vibrio coralliilyticus	Pacyfic oyster (*Crassostrea gigas*)	Immersion	Reduction in larvae mortality after prophylactic use	Kim et al. (2019)
Vibro cyclitrophicus	Japanese sea cucumber (*Apostichopus japonicus*)	Feed	Mortality reduced from 81% to 18%	Li et al. (2016)
		Injection into the body cavity	Mortality reduced from 58% to 18%	
		Immersion	Mortality reduced from 63% to 18%	
Vibrio harveyi	Brine shrimp (*Artemia franciscana*)	Immersion	Bacteriophage cocktails enhanced hatching success (100%, control groups had a hatching success of around 50%) and survival rate (85–89%, control groups survival rate was 40–50%)	Quiroz-Guzmán et al. (2018)
	Artemia salina		Larval mortality decreased 24 h post-infection	Misol et al. (2020)
	Giant tiger prawn (*Penaeus monodon*)	Immersion	Larval mortality decreased by ~43%	Stalin and Srinivasan (2017)
			Reduction in mortality of larvae by 20% compared with antibiotic therapy	Karunasagar et al. (2007)
			1. In experimental infection, larvae mortality decreased by 55%; 2. With a natural outbreak, larvae mortality decreased by 69% compared to the untreated control, by 46% compared to the antibiotic treatment group	Vinod et al. (2006)
			Reduction of larvae mortality by 50%	Patil et al. (2014)
	Greenlip abalone (*Haliotis laevigata*)	Bath	Reduction in mortality of 70% compared with the control group	Wang et al. (2017)
	Turbot (*Scophthalmus maximus*)	Feed	Reduction in mortality by 58–28% depending on bacteriophage dose	Cui et al. (2021)
	Zebrafish (*Danio rerio*)	Intraperitoneal injection	1. Reduction in mortality by 27.7–33.3% depending on infectious dose with prophylactic bacteriophage application; 2. Reduction in mortality by 13.3–26.7% depending on infectious dose with therapeutic bacteriophage application	Wu et al. (2021)
	Gilthead seabream larvae (*Sparus aurata*)	Immersion	Significantly improved survival of larvae by 19–35%	Droubogiannis et al. (2023); Droubogiannis and Katharios (2022)

Vibrio parahaemolyticus	*Artemia franciscana*	Immersion	Increase in breeding success and larval survival both when using a single bacteriophage and a cocktail	Quiroz-Guzmán et al. (2018)
			Depending on the bacterial strain, larval mortality decreased by 35% or to a level comparable to the uninfected control	Makarov et al. (2019)
	Blue mussel (*Mytilus edulus*)	Immersion	Reduction of the bacteria number to undetectable levels in the tissues	Onarinde and Dixon (2018)
	Giant tiger prawn (*Penaeus monodon*)	Feed	Reduction in mortality of 40–45% by single bacteriophage and 50% by a cocktail	Alagappan et al. (2016)
	Whiteleg shrimp (*Litopenaeus vannamei*)	Immersion	Reduction in mortality of larvae in 18–21% depending on the bacteriophage; delayed therapy resulted in decreased larval survival	Lomelí-Ortega and Martínez-Díaz (2014)
			Reduction in mortality of 75 and 50% depending on time after prophylactic use; no effect of therapeutic use	Jun et al. (2018)
		Feed	Reduction in mortality of 50% for prophylactic use, no effect of therapeutic use	
			20–40% dose-dependent reduction in mortality	Ding et al. (2020)
Vibrio splendidus	Japanese sea cucumber (*Apostichopus japonicus*)	Feed	Reduction in mortality of 32–47% by a single bacteriophage, and 64% by a cocktail	Li et al. (2016)

groups, showcasing the ability of phages to effectively target and control the pathogen, consequently boosting reproductive success and overall survival of crustacean populations. Additionally, the application of phages against *Vibrio parahaemolyticus* in giant tiger prawn (*Penaeus monodon*) and whiteleg shrimp (*Litopenaeus vannamei*) has yielded significant outcomes (Jun et al. 2018; Vinod et al. 2006; Alagappan et al. 2016; Lomelí-Ortega and Martínez-Díaz 2014; Ding et al. 2020; Stalin and Srinivasan 2017; Karunasagar et al. 2007; Patil et al. 2014). Immersion-based phage treatments led to mortality reductions ranging from 18 to 21% in larvae, while prophylactic use resulted in increased larval survival (Vinod et al. 2006). These findings underscore the potential of phage therapy in combating bacterial infections in crustaceans, improving survival rates, and promoting sustainable practices in aquaculture.

10.3.5.3 Mollusks

In the realm of mollusk aquaculture, phage therapy is considered an effective strategy for combating bacterial infections. For instance, in the case of *Vibrio splendidus* infection in Japanese sea cucumber (*Apostichopus japonicus*), the application of phages through feed led to notable reductions in mortality rates, ranging from 32 to 47%, depending on the specific phage used (Zhang et al. 2015). This demonstrates the potential of phage therapy in enhancing the well-being and survival rates of mollusk populations. Additionally, studies focusing on other mollusk species, such as greenlip abalone (*Haliotis laevigata*), have reported significant reductions in mortality rates following the administration of phages through bathing (Wang et al. 2017). Mortality decreased by 70% compared with the control group, highlighting the efficacy of phage therapy in disease management within mollusk aquaculture. Similarly, in the context of *Vibrio splendidus* infection in turbot (*Scophthalmus maximus*) and Blue mussel (*Mytilus edulus*), phage therapy resulted in mortality reductions ranging from 28 to 58% and the elimination of bacteria to undetectable levels in mollusk tissues, respectively (Rørbo et al. 2018; Onarinde and Dixon 2018). These findings indicate the potential of phage therapy as an effective tool in mollusk aquaculture, offering valuable solutions for disease management and enhancing the overall health and survival rates of mollusk populations.

Another interesting application of phage therapy is for the depuration of bivalve products. Typically, depuration processes make use of the inherent filtering capability of bivalve mollusks to eliminate their intestinal contents. As a result, the likelihood of transmitting infectious agents to consumers through the consumption of contaminated mollusks is significantly reduced. In this regard, the use of phages against significant human pathogens like *E. coli* and *Vibrio parahaemolyticus* is a new promising approach (Rong et al. 2014; Pereira et al. 2017).

10.3.5.4 Phages as Biocides and Biocontrol Method in Aquaculture

The remarkable specificity of phages toward their hosts makes them an ideal approach for the management of bacteria populations in delicate environments such as fish hatcheries and RAS. Additionally, phages can be effectively utilized during the early developmental stages of fish when vaccination is not feasible due to the immaturity of the immune system. In a recently published study conducted in Finland, it was observed that phages targeting *Flavobacterium* exhibited a persistence of 14 days within the tanks of a RAS after a single administration. Notably, these phages demonstrated even longer persistence within biofilters, suggesting that the presence could be augmented by the formation of biofilms, consequently contributing to their extended viability. This discovery holds significant importance as it indicates that phages can endure the water treatment processes of RAS and can be proactively harnessed within these systems for the precise management of undesired pathogenic bacteria.

However, it is the utilization of phages as a means of controlling bacteria in hatcheries that holds the greatest potential. Despite the implementation of advanced biosecurity measures in modern marine hatcheries, pathogenic bacteria persistently find their way into fish-rearing tanks, leading to increased morbidity, mortality, and inconsistent production performance. The introduction of live feeds serves as a vehicle for bacterial entry. Disinfection of live feeds plays a crucial role in establishing healthy colonization of beneficial bacteria in the fish gut. Our research group has recently proposed a novel application of phages as a "smart disinfectant" for live feeds. We have developed and utilized phage cocktails with a wide host range that selectively target and reduce vibrios in the live feeds. Through a single administration of vibriophages during the 4-h enrichment process of live Artemia, we observed a remarkable 93% reduction in vibrio load compared with the untreated group (Kalatzis et al. 2016; Misol et al. 2020). Our ongoing efforts are dedicated to the development of phage-based disinfectants that specifically target vibrios within the Harveyi clade, such as *V. harveyi*, *V. owensii*, and *V. alginolyticus*, which are commonly found in live feeds and are associated with larval enteritis in gilthead seabream, resulting in significant losses in Mediterranean hatcheries. A notable advantage of this approach is that the treatment occurs within the batch cultures of live feeds, substantially minimizing the chances of bacterial resistance development against phages. Aquatic Biologicals, a spin-off company of the Hellenic Centre for Marine Research, is at the forefront of developing innovative aquaculture health products, including the phage "smart disinfectant."

Similar approaches have been reported in salmon hatcheries, where bacteriophages have been employed as a biocontrol agent against *Yersinia ruckeri*, the causative agent of Enteric Red Mouth disease or Yersiniosis. A Norwegian company, ACD Pharmaceuticals, has achieved significant success authorizing a commercial phage product designed for prophylactic use against *Yersinia* in salmon tanks, specifically within Norway.

10.4 Phage Therapy Compared with Antibiotics

Bacteriophage therapy offers several advantages compared with traditional antibiotic therapy. The process of isolating bacteriophages is relatively fast, easy, and cost-effective. Unlike antibiotics, bacteriophage resistance develops at a much slower rate, approximately ten times slower than resistance to antibiotics. This slower development of resistance is attributed to the ability of bacteriophages to evolve and generate new genotypes capable of re-infecting a specific bacterial strain. Bacteriophages retain infectiousness even in challenging environmental conditions and continue to replicate until there is a substantial decrease in the population density of the host bacteria. These characteristics suggest that bacteriophage therapy, unlike conventional treatments, may require fewer administrations while delivering equal or superior effectiveness.

Assessing the effectiveness of bacteriophage therapy in comparison to antibiotic therapy holds great importance. In a study conducted by Zhang et al., there were no statistically significant differences in the survival rates of sea cucumbers infected with *V. alginolyticus* when treated with a bacteriophage cocktail compared with antibiotics (Zhang et al. 2015). However, in another study by Karunasagar et al., shrimps treated with bacteriophages exhibited a 20% higher survival rate compared with those treated with antibiotics, while Vinod et al. achieved a 46% higher survival rate in shrimp after natural *V. harveyi* infection when using bacteriophages compared to antibiotics (Vinod et al. 2006; Karunasagar et al. 2007). These findings suggest that bacteriophage therapy may not only be as effective as antibiotics but potentially even more effective.

Because of the high host specificity of phages, phage therapy is more appropriate as personalized treatment. And this is how it is currently being used in human medicine. Phage therapy centers in Georgia, Belgium, and elsewhere curate biobanks of bacteriophages which are being screened against life-threatening antibiotic-resistant strains on a case-by-case basis and treatment is following the magistral preparation scheme. However, it is important to note that this personalized approach is not readily applicable in aquaculture. In the case of acute bacterial infections causing significant mortality in aquaculture production, farmers often require urgent and immediate responses. In such scenarios, a combination of antibiotics and phages is frequently seen as a potent method to combat bacterial infections. Numerous studies have explored the effects of this combination in human medicine, yielding varying results. It should be noted that antibiotics could act antagonistically to phages, either by reducing the hosts in which phage would replicate producing new virions or by altering the physiology of the bacteria affecting phage infection and replication. However, by carefully adjusting the timing of each treatment—for example, initiating phage therapy first and then introducing antibiotics—a more effective outcome might be achieved.

10.5 Challenges of Phage Therapy

Despite over a century of phage research, there are significant challenges that need to be addressed before phage therapy can be widely adopted as an industrial-scale treatment and prevention method. Recently, Culot, Grosset, and Gautier, researchers from INRA, France, have conducted an excellent review on the challenges of phage therapy in commercial aquaculture (Culot et al. 2019).

The inconsistent outcomes in phage therapy can be attributed significantly to the inadequate design of both phage therapy products and application strategies. Thorough characterization of the phages is a crucial prerequisite for effective and safe phage therapy. As mentioned earlier in this chapter, various elements of phage microbiology need to be meticulously studied in the laboratory. Comprehending factors such as burst size, host range, life cycle, and genetics of candidate phages not only establishes their suitability as therapeutic agents but also directs the optimal approach for their application, including dosage and timing.

A significant challenge lies in overcoming bacterial resistance development. As explained earlier, resistance development is a natural outcome of the coevolution between phages and bacteria. Phage cocktails offer a potential solution; however, their formulation remains a challenging task that necessitates advanced knowledge and analytical skills.

The pharmaceutical industry faces a substantial challenge in the large-scale production of phages. Cultivating phages in bioreactors is not a readily standardized process. Additionally, when phages are intended for use as therapeutic agents, stringent quality standards must be adhered to, such as ensuring the absence of endotoxins released during the lysis of Gram-negative bacterial hosts. Although technically achievable, this requirement substantially increases production costs.

However, the most substantial challenge lies in the regulatory barriers encountered in many countries, including the European Union and the United States. Obtaining licenses for phage products as pharmaceuticals can be a nightmarish process. Phages are not considered conventional pharmaceuticals, which significantly diverges their authorization process from the established norms. Moreover, the components of a phage therapy product must undergo frequent revisions and replacements to address resistance issues. As every single element of a pharmaceutical product must undergo extensive safety and efficacy testing before being approved as a new component of an authorized product, it becomes apparent that licensing phages will be impractical for the pharmaceutical industry. Moreover, producing phages as pharmaceuticals at the Good Manufacturing Practices level is highly demanding and incurs substantial costs. There is a strong advocacy effort pushing regulatory authorities to revise legislation in a manner that makes phage therapy economically viable and safe, especially in the era of AMR where alternative solutions are urgently needed. The new EU legislation has recognized the distinctions between phages and conventional drugs, and efforts are underway to update the regulations concerning phage-based veterinary medicinal products

(VMPs). On the contrary, licensing phages as biocides rather than VMPs is significantly easier and entails a considerably reduced cost for both development and production.

10.6 Future Aspects

Phage therapy is well-suited to be incorporated into multifaceted strategies aimed at enhancing fish health in aquaculture and represents an auspicious approach to combating the emergence of antibiotic-resistant bacteria. However, caution must be exercised as there is a possibility of phage resistance evolving over time. Despite the potential for resistance, the concept of phage training for therapy, involving experimental coevolution, has recently emerged and warrants further investigation. Moreover, a novel area of research, which includes the potential use of phage lytic enzymes and their interaction with the immune system of cultured aquatic organisms, is getting attention. The presence of newly discovered phages with diverse phenotypes and genotypes poses a challenge in this field. Each phage exhibits varying host specificity, efficacy, and mechanisms of action against bacteria. Consequently, it becomes vital to integrate and consolidate knowledge of the specific relationships between phages and pathogens. To tackle this challenge, it is necessary to obtain stronger evidence through in vivo studies to assess the efficacy and safety of phage lytic enzymes. These studies will offer more realistic insights into the interaction between phages, their enzymes, and the immune systems of cultured aquatic organisms, helping to make this approach a well-established biotechnology in aquaculture.

The future of phage therapy in aquaculture shows great potential as scientists continue to explore innovative approaches and advancements in the field. One significant aspect that is likely to shape the future of phage therapy is the practice of phage engineering. By utilizing genetic engineering techniques, researchers can modify phages to improve their effectiveness, expand their ability to target different hosts, and optimize their therapeutic capabilities. Phage engineering enables the deliberate manipulation of phage genomes to introduce desired traits and enhance their performance against specific bacterial pathogens. It focuses on the development of phages with broader host ranges, as naturally occurring phages often have limited effectiveness against a wide range of bacterial strains due to their narrow host range. Through genetic modifications, scientists can broaden the host range of phages, allowing them to target a wider spectrum of bacterial pathogens. This expansion of the host range holds great potential to enhance the applicability of phage therapy in aquaculture, as it enables the effective control of multiple bacterial species known to cause diseases in various aquatic organisms. Another exciting avenue in phage engineering involves enhancing the stability and persistence of phages in aquatic environments. Factors such as temperature, salinity, and UV radiation can impact the viability and survival of phages in aquaculture systems. Through phage engineering, researchers can develop phages that can withstand these environmental stressors, thereby improving their overall effectiveness as therapeutic agents. This may involve modifying phage capsid proteins or introducing protective mechanisms that enhance their stability and prolong their activity in aquatic settings.

Furthermore, phage engineering offers the potential for developing customized phage cocktails to address specific bacterial infections in aquaculture. Phage cocktails consist of mixtures of different phages that target multiple strains or species of bacteria simultaneously. By combining phages with complementary host ranges and mechanisms of action, researchers can create synergistic effects, increasing the likelihood of successful treatment outcomes. Phage cocktails provide a versatile and robust approach to combat bacterial infections, particularly in cases where bacterial pathogens exhibit high levels of diversity and resistance.

In addition, advancements in the delivery systems for phage therapy in aquaculture may shape its future. Ensuring effective delivery of phages to the infection site is crucial for achieving optimal therapeutic outcomes. Scientists are exploring various methods, such as nanoparticle-based delivery systems, biofilms, and encapsulation techniques, to improve the efficiency of phage delivery and protect them from degradation or inactivation. These delivery systems hold promise in enhancing the precision and effectiveness of phage therapy by ensuring an adequate number of phages reach the targeted pathogens, thereby maximizing their therapeutic impact.

Moreover, continuous investigation into the interactions between phages and their bacterial hosts, as well as the mechanisms underlying bacterial resistance to phages, will play a vital role in advancing the development of highly effective phage therapies. By acquiring a deeper comprehension of the intricate dynamics between phages and their bacterial hosts, researchers can pinpoint strategies to overcome bacterial resistance and bolster the long-term effectiveness of phage therapy.

In conclusion, the future of phage therapy in aquaculture shows great potential, with the important roles of phage engineering and advancements in delivery systems. Through phage engineering, researchers can enhance key properties of phages, such as host range and environmental stability. Concurrently, the development of targeted delivery systems ensures efficient and precise delivery of phages to the infection sites. Ongoing research in phage-host interactions and resistance mechanisms will further contribute to the development of safe, effective, and sustainable phage-based treatments, supporting the growth and sustainability of aquaculture practices worldwide.

Funding Information The study has been funded by European Union's Horizon Europe research and innovation program, under grant agreement No. 101084204 Cure4Aqua. The views and opinions expressed are, however, those of the authors only and do not necessarily reflect those of the European Union. Neither the European Union nor the granting authority can be held responsible for them.

References

Akmal M, Rahimi-Midani A, Hafeez-ur-Rehman M et al (2020) Isolation, characterization, and application of a bacteriophage infecting the fish pathogen Aeromonas hydrophila. Pathogens 9:215. https://doi.org/10.3390/pathogens9030215

Alagappan K, Karuppiah V, Deivasigamani B (2016) Protective effect of phages on experimental *V. parahaemolyticus* infection and immune response in shrimp (Fabricius, 1798). Aquaculture 453:86–92. https://doi.org/10.1016/j.aquaculture.2015.11.037

Barr JJ, Auro R, Furlan M et al (2013) Bacteriophage adhering to mucus provide a non–host-derived immunity. Proc Natl Acad Sci 110:10771–10776. https://doi.org/10.1073/pnas.1305923110

Cairns BJ, Timms AR, Jansen VA, Connerton IF, Payne RJ (2009) Quantitative models of in vitro bacteriophage-host dynamics and their application to phage therapy. PLoS Pathog 5(1):e1000253. https://doi.org/10.1371/journal.ppat.1000253

Cao Y, Li S, Han S et al (2020) Characterization and application of a novel Aeromonas bacteriophage as treatment for pathogenic Aeromonas hydrophila infection in rainbow trout. Aquaculture 523:735193. https://doi.org/10.1016/j.aquaculture.2020.735193

Carroll-Portillo A, Lin HC (2019) Bacteriophage and the innate immune system: access and signaling. Microorganisms 7(12):625. https://doi.org/10.3390/microorganisms7120625

Castillo D, Higuera G, Villa M et al (2012) Diversity of Flavobacterium psychrophilum and the potential use of its phages for protection against bacterial cold water disease in salmonids. J Fish Dis 35:193–201. https://doi.org/10.1111/j.1365-2761.2011.01339.x

Cui H, Cong C, Wang L et al (2021) Protective effectiveness of feeding phage cocktails in controlling Vibrio harveyi infection of turbot (Scophthalmus maximus). Aquaculture 535:736390. https://doi.org/10.1016/j.aquaculture.2021.736390

Culot A, Grosset N, Gautier M (2019) Overcoming the challenges of phage therapy for industrial aquaculture: a review. Aquaculture 513:734423. https://doi.org/10.1016/j.aquaculture.2019.734423

Dabrowska K, Switała-Jelen K, Opolski A et al (2005) Bacteriophage penetration in vertebrates. J Appl Microbiol 98:7–13. https://doi.org/10.1111/j.1365-2672.2004.02422.x

Dang THO, Xuan TTT, Duyen LTM et al (2021) Protective efficacy of phage PVN02 against haemorrhagic septicaemia in striped catfish (Pangasianodon hypophthalmus) via oral administration. J Fish Dis 44:1255–1263. https://doi.org/10.1111/jfd.13372

Dien LT, Ky LB, Huy BT et al (2022) Characterization and protective effects of lytic bacteriophage pAh6.2TG against a pathogenic multidrug-resistant Aeromonas hydrophila in Nile tilapia (Oreochromis niloticus). Transbound Emerg Dis 69:e450. https://doi.org/10.1111/tbed.14048

Ding T, Sun H, Pan Q et al (2020) Isolation and characterization of *Vibrio parahaemolyticus* bacteriophage vB_VpaS_PG07. Virus Res 286:198080. https://doi.org/10.1016/j.virusres.2020.198080

Donati VL, Dalsgaard I, Sundell K et al (2021) Phage-mediated control of Flavobacterium psychrophilum in aquaculture: In vivo experiments to compare delivery methods. Front Microbiol 12:628309. https://doi.org/10.3389/fmicb.2021.628309

Droubogiannis S, Katharios P (2022) Genomic and biological profile of a novel bacteriophage, vibrio phage virtus, which improves survival of *Sparus aurata* larvae challenged with *Vibrio harveyi*. Pathogens 11. https://doi.org/10.3390/pathogens11060630

Droubogiannis S, Pavlidi L, Skliros D et al (2023) Comprehensive characterization of a novel bacteriophage, vB_VhaS_MAG7 against a fish pathogenic strain of *Vibrio harveyi* and its *in vivo* efficacy in phage therapy trials. Int J Mol Sci 24:8200. https://doi.org/10.3390/ijms24098200

El-Araby DA, El-Didamony G, Megahed M (2016) New approach to use phage therapy against *Aeromonas hydrophila* induced motile Aeromonas septicemia in Nile tilapia. J Marine Sci Res Dev 6:2

Easwaran M, Dananjaya SHS, Park SC, Lee J, Shin H, & de Zoysa M (2017) Characterization of bacteriophage pAh-1 and its protective effects on experimental infection of *Aeromonas hydrophila* in Zebrafish (*Danio rerio*). Journal of Fish Diseases 40(6):841–846. https://doi.org/10.1111/jfd.12536

Ghasemi SM, Bouzari M, Emtiazi G (2014) Preliminary characterization of Lactococcus garvieae bacteriophage isolated from wastewater as a potential agent for biological control of lactococcosis in aquaculture. Aquac Int 22:1469–1480. https://doi.org/10.1007/s10499-014-9761-z
Higuera G, Bastías R, Tsertsvadze G et al (2013) Recently discovered Vibrio anguillarum phages can protect against experimentally induced vibriosis in Atlantic salmon, Salmo salar. Aquaculture 392–395:128–133. https://doi.org/10.1016/j.aquaculture.2013.02.013
Imbeault S, Parent S, Lagacé M et al (2006) Using bacteriophages to prevent furunculosis caused by Aeromonas salmonicida in farmed brook trout. J Aquat Anim Health 18:203–214. https://doi.org/10.1577/H05-033.1
Jia K, Yang N, Zhang X et al (2020) Genomic, morphological and functional characterization of virulent bacteriophage IME-JL8 targeting Citrobacter freundii. Front Microbiol 11:585261. https://doi.org/10.3389/fmicb.2020.585261
Jun JW, Kim JH, Shin SP et al (2013) Protective effects of the Aeromonas phages pAh1-C and pAh6-C against mass mortality of the cyprinid loach (Misgurnus anguillicaudatus) caused by Aeromonas hydrophila. Aquaculture 416:289–295. https://doi.org/10.1016/j.aquaculture.2013.09.012
Jun JW, Han JE, Giri SS et al (2018) Phage application for the protection from acute Hepatopancreatic necrosis disease (AHPND) in *Penaeus vannamei*. Indian J Microbiol 58:114–117. https://doi.org/10.1007/s12088-017-0694-9
Kalatzis PG, Bastías R, Kokkari C, Katharios P (2016) Isolation and characterization of two lytic bacteriophages, φSt2 and φGrn1; phage therapy application for biological control of *Vibrio alginolyticus* in aquaculture live feeds. PLoS One 11:e0151101
Karunasagar I, Shivu MM, Girisha SK et al (2007) Biocontrol of pathogens in shrimp hatcheries using bacteriophages. Aquaculture 268:288–292. https://doi.org/10.1016/j.aquaculture.2007.04.049
Katharios P (2020) The prospect of phage therapy in fish hatcheries. Aquac Eur 45:9–13
Khairnar K, Raut MP, Chandekar RH et al (2013) Novel bacteriophage therapy for controlling metallo-beta-lactamase producing *Pseudomonas aeruginosa* infection in catfish. BMC Vet Res 9:264. https://doi.org/10.1186/1746-6148-9-264
Kheirollahpour M, Mehrabi M, Dounighi NM, Mohammadi M, Masoudi A (2020) Nanoparticles and vaccine development pharmaceutical. Nanotechnology 8(1):6–21. https://doi.org/10.2174/2211738507666191024162042
Kim JH, Choresca CH, Shin SP et al (2015) Biological control of Aeromonas salmonicida subsp. salmonicida infection in rainbow trout (Oncorhynchus mykiss) using Aeromonas phage PAS-1. Transbound Emerg Dis 62:81–86. https://doi.org/10.1111/tbed.12075
Kim HJ, Jun JW, Giri SS, Chi C, Yun S, Kim SG, Kim SW, Kang JW, Han SJ, Kwon J, Oh WT, & Park SC (2019) Application of the bacteriophage pVco-14 to prevent Vibrio coralliilyticus infection in Pacific oyster (Crassostrea gigas) larvae. J Invertebr Pathol 167:107244. https://doi.org/10.1016/j.jip.2019.107244
Kwon AS, Kang BJ, Jun SY et al (2017) Evaluating the effectiveness of Streptococcus parauberis bacteriophage Str-PAP-1 as an environmentally friendly alternative to antibiotics for aquaculture. Aquaculture 468:464–470. https://doi.org/10.1016/j.aquaculture.2016.11.013
Laanto E, Bamford JKH, Ravantti JJ, Sundberg L-R (2015) The use of phage FCL-2 as an alternative to chemotherapy against columnaris disease in aquaculture. Front Microbiol 6:829. https://doi.org/10.3389/fmicb.2015.00829
Le TS, Nguyen TH, Vo HP et al (2018) Protective effects of bacteriophages against *Aeromonas hydrophila* causing motile Aeromonas septicemia (MAS) in striped catfish. Antibiotics 7:16
Li Z, Zhang J, Li X et al (2016) Efficiency of a bacteriophage in controlling vibrio infection in the juvenile sea cucumber Apostichopus japonicus. Aquaculture 451:345–352. https://doi.org/10.1016/j.aquaculture.2015.09.013
Liu R, Han G, Li Z et al (2022) Bacteriophage therapy in aquaculture: current status and future challenges. Folia Microbiol (Praha) 67:573–590

Lomelí-Ortega CO, Martínez-Díaz SF (2014) Phage therapy against Vibrio parahaemolyticus infection in the whiteleg shrimp (*Litopenaeus vannamei*) larvae. Aquaculture 434:208–211. https://doi.org/10.1016/j.aquaculture.2014.08.018

Lomelí-Ortega CO, Martínez-Sández AJ, Barajas-Sandoval DR et al (2021) Isolation and characterization of vibriophage vB_Vc_SrVc9: an effective agent in preventing Vibrio campbellii infections in brine shrimp nauplii (Artemia franciscana). J Appl Microbiol 131:36–49. https://doi.org/10.1111/jam.14937

Luo X, Liao G, Liu C et al (2018) Characterization of bacteriophage HN48 and its protective effects in Nile tilapia (Oreochromis niloticus) against Streptococcus agalactiae infections. J Fish Dis 41:1477–1484. https://doi.org/10.1111/jfd.12852

Makarov R, Lomelí-Ortega CO, Zermeño-Cervantes LA et al (2019) Evaluation of a cocktail of phages for the control of presumptive Vibrio parahaemolyticus strains associated to acute hepatopancreatic necrosis disease. Aquac Res 50:3107–3116. https://doi.org/10.1111/are.14252

Matsuoka S (2007) Phage therapy against β-hemolytic streptococcicosis of Japanese flounder (Paralichthys olivaceus). Fish Pathol 42:181–189. https://doi.org/10.3147/jsfp.42.181

Misol GN, Kokkari C, Katharios P (2020) Biological and genomic characterization of a novel jumbo bacteriophage, Vb_VhaM_pir03 with broad host lytic activity against *Vibrio harveyi*. Pathogens 9:1–38. https://doi.org/10.3390/pathogens9121051

Nakai T, Sugimoto R, Park K-H et al (1999) Protective effects of bacteriophage on experimental Lactococcus garvieae infection in yellowtail. Dis Aquat Org 37:33–41. https://doi.org/10.3354/dao037033

Nikapitiya C, Chandrarathna H, Dananjaya SHS et al (2020) Isolation and characterization of phage (ETP-1) specific to multidrug-resistant pathogenic Edwardsiella tarda and its in vivo biocontrol efficacy in zebrafish (Danio rerio). Biologicals 63:14–23. https://doi.org/10.1016/j.biologicals.2020.02.002

O'Neill J (2014) Antimicrobial resistance: tackling a crisis for the health and wealth of nations. Review on Antimicrobial Resistance, London

Onarinde BA, Dixon RA (2018) Prospects for biocontrol of *Vibrio parahaemolyticus* contamination in blue mussels (*Mytilus edulus*)—a year-long study. Front Microbiol 9:1043

Park SC, Nakai T (2003) Bacteriophage control of *pseudomonas plecoglossicida* infection in ayu Plecoglossus altivelis. Dis Aquat Org 53:33–39

Park SC, Shimamura I, Fukunaga M et al (2000) Isolation of bacteriophages specific to a fish pathogen, Pseudomonas plecoglossicida, as a candidate for disease control. Appl Environ Microbiol 66:1416–1422. https://doi.org/10.1128/AEM.66.4.1416-1422.2000

Patil JR, Desai SN, Roy P et al (2014) Simulated hatchery system to assess bacteriophage efficacy against *Vibrio harveyi*. Dis Aquat Org 112:113–119

Pereira C, Moreirinha C, Teles L et al (2017) Application of phage therapy during bivalve depuration improves *Escherichia coli* decontamination. Food Microbiol 61:102–112. https://doi.org/10.1016/j.fm.2016.09.003

Prasad Y, Kumar D, Sharma AK (2011) Lytic bacteriophages specific to Flavobacterium columnare rescue catfish, Clarias batrachus (Linn.) from columnaris disease. J Environ Biol 32:161–168

Publications Service (2007) On an invisible microbe antagonistic toward dysenteric bacilli: brief note by Mr. F. D'Herelle, presented by Mr. Roux. Res Microbiol 158:553–554. https://doi.org/10.1016/j.resmic.2007.07.005

Quiroz-Guzmán E, Peña-Rodriguez A, Vázquez-Juárez R et al (2018) Bacteriophage cocktails as an environmentally-friendly approach to prevent *Vibrio parahaemolyticus* and *Vibrio harveyi* infections in brine shrimp (*Artemia franciscana*) production. Aquaculture 492:273–279. https://doi.org/10.1016/j.aquaculture.2018.04.025

Rong R, Lin H, Wang J et al (2014) Reductions of *Vibrio parahaemolyticus* in oysters after bacteriophage application during depuration. Aquaculture 418–419:171–176. https://doi.org/10.1016/j.aquaculture.2013.09.028

Rørbo N, Rønneseth A, Kalatzis PG et al (2018) Exploring the effect of phage therapy in preventing *vibrio anguillarum* infections in cod and turbot larvae. Antibiotics 7:42

Royam MM, Nachimuthu R (2020) Isolation, characterization, and efficacy of bacteriophages isolated against Citrobacter spp. an in vivo approach in a zebrafish model (Danio rerio). Res Microbiol 171:341–350. https://doi.org/10.1016/j.resmic.2020.03.002

Schulz P, Robak S, Dastych J, Siwicki AK (2019) Influence of bacteriophages cocktail on European eel (*Anguilla Anguilla*) immunity and survival after experimental challenge. Fish Shellfish Immunol 84:28–37

Schulz P, Pajdak-Czaus J, Siwicki AK (2022) In vivo bacteriophages' application for the prevention and therapy of aquaculture animals—chosen aspects. Animals 12

Silva YJ, Costa L, Pereira C et al (2014) Phage therapy as an approach to prevent Vibrio anguillarum infections in fish larvae production. PLoS One 9. https://doi.org/10.1371/journal.pone.0114197

Silva YJ, Moreirinha C, Pereira C et al (2016) Biological control of Aeromonas salmonicida infection in juvenile Senegalese sole (Solea senegalensis) with Phage AS-A. Aquaculture 450:225–233. https://doi.org/10.1016/j.aquaculture.2015.07.018

Sophie N, Kristi B, Padman BS et al (2017) Bacteriophage transcytosis provides a mechanism to cross epithelial cell layers. MBio 8:e01874-17. https://doi.org/10.1128/mbio.01874-17

Stalin N, Srinivasan P (2017) Efficacy of potential phage cocktails against *Vibrio harveyi* and closely related vibrio species isolated from shrimp aquaculture environment in the south east coast of India. Vet Microbiol 207:83–96. https://doi.org/10.1016/j.vetmic.2017.06.006

Sundell K, Landor L, Castillo D et al (2020) Bacteriophages as biocontrol agents for Flavobacterium psychrophilum biofilms and rainbow trout infections. Phage 1:198–204. https://doi.org/10.1089/phage.2020.0016

Verner-Jeffreys DW, Algoet M, Pond MJ et al (2007) Furunculosis in Atlantic salmon (Salmo salar L.) is not readily controllable by bacteriophage therapy. Aquaculture 270:475–484. https://doi.org/10.1016/j.aquaculture.2007.04.032

Veyrand-Quirós B, Gómez-Gil B, Lomeli-Ortega CO et al (2020) Use of bacteriophage vB_Pd_PDCC-1 as biological control agent of Photobacterium damselae subsp. damselae during hatching of longfin yellowtail (Seriola rivoliana) eggs. J Appl Microbiol 129:1497–1510. https://doi.org/10.1111/jam.14700

Vinod MG, Shivu MM, Umesha KR et al (2006) Isolation of *Vibrio harveyi* bacteriophage with a potential for biocontrol of luminous vibriosis in hatchery environments. Aquaculture 255:117–124. https://doi.org/10.1016/j.aquaculture.2005.12.003

Wang Y, Barton M, Elliott L et al (2017) Bacteriophage therapy for the control *of Vibrio harveyi* in greenlip abalone (*Haliotis laevigata*). Aquaculture 473:251–258. https://doi.org/10.1016/j.aquaculture.2017.01.003

Wu J-L, Lin H-M, Jan L et al (1981) Biological control of fish bacterial pathogen, *Aeromonas hydrophila*, by bacteriophage AH 1. Fish Pathol 15:271–276. https://doi.org/10.3147/jsfp.15.271

Wu L, Tian Y, Pang M et al (2021) A novel vibriophage vB_VhaS_PcB-1G capable of inhibiting virulent Vibrio harveyi pathogen. Aquaculture 542:736854. https://doi.org/10.1016/j.aquaculture.2021.736854

Yu H, Zhang L, Feng C et al (2022) A phage cocktail in controlling phage resistance development in multidrug resistant Aeromonas hydrophila with great therapeutic potential. Microb Pathog 162:105374. https://doi.org/10.1016/j.micpath.2021.105374

Zhang YA, Salinas I, Li J, Parra D, Bjork S, Xu Z, LaPatra SE, Bartholomew J, Sunyer JO (2010) IgT a primitive immunoglobulin class specialized in mucosal immunity. Nat Immunol 11(9):827–835. https://doi.org/10.1038/ni.1913

Zhang J, Cao Z, Li Z et al (2015) Effect of bacteriophages on *Vibrio alginolyticus* infection in the sea cucumber, *Apostichopus japonicus* (Selenka). J World Aquac Soc 46:149–158

Chapter 11
Bacteriocin and Predatory Bacteria

Heba Mahboub, Taghreed N. Almanaa, Amer Al Ali, Mujtaba Farooq Rana, and Hiam Elabd

Abstract This chapter covers the rule of bacteriocins and predatory bacteria in Antimicrobial Resistance (AMR) of aquaculture and aquatic environments. A wide variety of bacteriocins have been obtained from bacteria (either Gram-negative or positive) described for their suggestive benefits and possible applications. Bacteriocins might be used separately or in combination against bacterial infections, targeting lowering the rate of antibiotic usage and the resulting AMR. This will require studying their types, classification, mode of action, resistance pathways, and immune responses. Immune response in the case of strains producing bacteriocins includes proteins and transporting mechanisms that differ between bacteriocins types. This chapter will broaden our knowledge about bacteriocins and predatory bacteria.

Keywords Bacteriocins · Antibiotic-resistant bacteria · Predatory bacteria · Immune response · Aquaculture

---

H. Mahboub (✉)
Department of Aquatic Animal Medicine, Faculty of Veterinary Medicine, Zagazig University, Zagazig, Sharkia, Egypt

T. N. Almanaa
Department of Botany and Microbiology, College of Science, King Saud University, Riyadh, Saudi Arabia

A. Al Ali
Department of Medical Laboratory Sciences, College of Applied Medical Sciences, University of Bisha, Bisha, Saudi Arabia

M. F. Rana
Department of Biochemistry, College of Medicine, Taif University, Al-Taif, Saudi Arabia

H. Elabd
Aquatic Animals' Medicine Department, Faculty of Veterinary Medicine, Benha University, Toukh, Egypt

P. Elumalai, S. Lakshmi (eds.), *Antimicrobial Resistance in Aquaculture and Aquatic Environments*, https://doi.org/10.1007/978-981-97-7320-6_11

11.1 Introduction

Based on the rules of World Health Organization (WHO), the broad expansion of antibiotic-resistant bacteria (ABR) and sluggish innovation of recent types of antibiotics are measured as major public health problems (Alzahrani et al. 2023; Aslam et al. 2021; Hernando-Amado et al. 2019; Tartor et al. 2018). In 66 countries, there is a declaration from the Global Antibiotic Resistance and Use Surveillance System regarding elevated degrees of antibiotic resistance (Hu et al. 2021). And the hazards induced by antibiotics and their resistance are increasing and cannot be miscalculated (Sun et al. 2020). The overuse of antibiotics for extended time has permitted infectious agents to acclimatize to antibiotics, accordingly lessens their efficacy. The extent of resistant strains of bacteria from one person to another or in the environment from human to non-human source, such as from animal that is used for feeding human and received antibiotics like those operated to cure human diseases, is a crucial factor in the emergence of antibiotic resistance (Wee et al. 2020). Because of emerging antimicrobial resistance, current researchers are performed to recognize and try natural new therapeutic agents and strategies which are friendly environmental and available products and have many privileges on public health (Abdel Rahman et al. 2022; Mahboub et al. 2022a, b, c, 2024; Mansour et al. 2022). For this reason, to reduce residues of antibiotic plus expanding resistance of genes, it is essential to discover alternatives for antibiotics and biochemical supplements (Elabd et al. 2022a, b; El-Bouhy et al. 2021a, b; Hamed et al. 2022; Rahman et al. 2023).

Innovations in identifying bacteriocins and their classification have increased attention in utilizing these substances either for therapy or new feed supplements. These are identified as proteins or antimicrobial peptides and are formed by bacteria either Gram-negative or positive (Benítez-Chao et al. 2021). The efficiency of bacteriocins and their releasing strains prove great success in suppressing spoilage of many food and virulent bacteria in diverse food products (Yu et al. 2023). The efficiency of numerous bacteriocins with their importance to cure infections in humans and animals has also been proven a great success (Huang et al. 2021). Many bacteriocins are described to be active in confronting different virulent bacteria (Pulingam et al. 2022).

11.2 Bacteriocins

11.2.1 Origin, Composition, Synergistic Interaction, and Mode of Action

Bacteriocins were originally exploited from 100 years and have been discovered to be formed by various species of archaea and bacteria (Hernando-Amado et al. 2019; Besse et al. 2015; Riley and Wertz 2002). Various bacteriocins, originate from lactic acid bacteria and diverse strains of lactic acid bacteria, associated with food groups could release antibacterial proteins or bacteriocins which are strongly efficient against foodborne pathogens including *Pseudomonas aeruginosa, P. fluorescens, Staphylococcus aureus*, *Salmonella typhi*, *Escherichia coli* O157:H7, *Shigella flexneri*, *Clostridium botulinum*, and *Listeria monocytogenes* (Darbandi et al. 2022).

They are composed of a heterogeneous family of minute protein molecules synthesized by ribosomes which are considered non-biologically active precursor peptides with potent antimicrobial action at accurate doses (Chikindas et al. 2018). These peptides are characterized by bacteriostatic or bactericidal activity against bacteria (Hatakka and Saxelin 2008), or rarely, in contradiction of a diverse array of distinct bacteria (Draper et al. 2008).

Bacteriocins are considered peptides that have antimicrobial activity, a number of them have numerous privileges like anticancer, antiviral, and antibiofilm. Such characteristics can be enhanced via synergistic interfaces of these bacteriocins with other particles which are bioactive like phages, antibiotics, essential oils, and nanoparticles. Several studies are revealing the impacts of these mixtures as recent and promoting therapeutic protocols, as they may endow numerous privileges over current treatments. Especially, bacteriocins can provide advantage from mixing with nanoparticles which in turn can promote their solubility, and stability, plus guard them from degradation of enzymes, minimize their interactions with other particles and enhance their bioavailability. Moreover, joining bacteriocins with other antimicrobials is predicted to lessen the antibiotic resistance' growth because of the contribution of many mechanisms of actions. In addition, another related incentive for this synergism including lessening the dose of every antimicrobial constituent, thus minimizing the negative impacts like toxicity. Additionally, such combination could prolong the bacteriocins' efficacy when they act as anticancer and antiviral substances. Based on these outcomes, synergistic impacts of bacteriocin mixtures enable them to be utilized like medicines, as well as in other different applications involving antispoilage, against viruses, antibiofilms and for combating cancer (Zgheib et al. 2020).

Bacteriocins are detected in the normal environment at a lesser dose that is adequate for suppressing other challenging strains. Worth mentioning, the basic role of bacteriocins is initially associated with self-signals and repels abilities. In progressive doses than normal ratios, bacteriocins exhibit potent activity as antimicrobials and could make pore formation with further membrane' disorders, or obstructing division of cell (Chikindas et al. 2018). Because of the nano-molecular size of bacteriocins, and their thermal stability, they are considered an appropriate therapeutic option (Huang et al. 2021).

Such compounds comprise of a wide array of biochemical and molecular features, as well as a broad series of activities with different mechanisms of actions (Negash and Tsehai 2020).

Bacteriocins prevent the growing of target organisms via working mainly on the envelope of the cell and via influencing expression of gene and production of protein inside the cells. Bacteriocins can exert their antimicrobial action at nano-molar concentrations, signifying that they are responsive at very little concentrations related to antimicrobial peptides (AMPs) formed by eukaryotic cells (Benítez-Chao et al. 2021).

A recent study verifies the potential activity of combining essential oils and bacteriocin confronting 18 decomposing microorganisms (involving nine strains of bacteria plus equal strains of yeasts). Strains of yeast include *Pichia* aff. *fermentans*, *Hanseniaspora opuntiae*, *Pichia kundriavzevii*, *Candida metapsilosis*, and *Kodamaea ohmeri*. The strains of spoilage bacteria involve *Klebsiella pneumoniae*, *K. variicola, Serratia marcescens*, and *Lactococcus lactis* subsp. *lactis*. The findings show that mixing enterocin KT2W2G and cinnamon exhibit a broad synergistic activity in contradiction of the nominated putrefying microorganisms. Of note, the usage of only enterocin KT2W2G doesn't demonstrate activity (Issouffou et al. 2018).

Bacteriocin-producing cells commonly develop mechanism of action to protect themselves from being engulfed by their self-bacteriocins, either by using efflux pumps or releasing self-immune proteins, or both of them (Bountra et al. 2017). Bacteriocins can enter cell membranes and induce escape of cell contents, such as inorganic phosphate and K+, or produce a bacteriostatic impact via obstructing the biosynthesis of RNA, proteins, and DNA (Baptista et al. 2020; Wang et al. 2019).

11.2.2 An Advanced Classification of Bacteriocins (Figure 1)

Bacteriocins have a wide variety (Choi et al. 2023). To capture this diverse group, it has been innovated a combined database entitled BACTIBASE which carries functional and mechanical data associated with >200 bacteriocins (Soltani et al. 2021). Particular information involves specific and sequence characteristics, such as physicochemical, structural, structural, and antimicrobial properties. Years ago, various ordering schemes are assigned for bacteriocins. Such bacteriocins are originated from lactic acid bacteria that are numerous and are categorized into different methods (dos Nascimento et al. 2012), from two (Soltani et al. 2021) to four subclasses. These categories have been updated numerous times depending on recent changes in terms of modes of action and structures of bacteriocins (Yu et al. 2023). Following the classification determined by Soltani et al. (2021), which has the privilege of being clear; bacteriocins were primarily classified into two main classes according to the existence or disappearance of the altered motifs which are detected post-translation. Furthermore, this grouping not including antimicrobial proteins and regarded only antimicrobial peptides, like colicins originated from *Escherichia coli*. Depending on such guidelines plus considering current techniques on ribosomally synthesized and post-translationally modified peptides (RiPPs) (Ahearne et al. 2021), a recent study updates the bacteriocins' grouping, which arise from the bacteria (Gram-negative and Gram-positive), into two main classes representing class I (modified) and class II (unmodified) bacteriocins (Soltani et al. 2021).

Class I represents peptides carry molecular the multitude less than 5 kDa which carry PTMs verified via enzymes coded in the bacteriocin gene group. This class I bacteriocins are associated with the RiPPs' family. Class II of bacteriocins represents unmodified peptides with molecular mass of 6–10 kDa neither involving nor fixed disulfide bridges. The alterations allow bacteriocins (class I) tolerating higher temperatures, degrading enzymes or risky pHs compared to class II, even if disulfide bridges augment the stability of peptide (class II). Further, class I has been sub-classified based on the RiPP nomenclature (Norris and Patchett 2016) which includes circular peptides, lanthipeptides, glycocins, and sactipeptides from Gram-positive bacteria (Draper et al. 2008), siderophore and nucleotide peptides (Gram-negative type), and finally lasso peptides (created from Gram-negative plus Gram-positive bacteria) (Norris and Patchett 2016). Additionally, linaridins (Claesen and Bibb 2010) and thiopeptides (Ashouri et al. 2020) basically formed by cyanobactins and actinobacteria from various cyanobacteria (Martins et al. 2018), related to this class. Lantibiotics (peptide antibiotics) have been heavily considered for years comprise 19–>50 amino acids (Doan et al. 2020; Kapinga et al. 2019) and are well-known by the existence of uncommon amino acids, defined as β-methyllanthionine and lanthionine, which occur due to dehydration of threonines and serines yielding residues of didehydrobutyrine and didehydroalanine. The thioether linkages are connecting lantibiotics which are located in-between two β-carbons (Claessen et al. 2014; Waite et al. 2020). The balance of sactibiotics is identified via the occurrence of intramolecular sulfur to α-carbon bonds including cysteine and additional deposit in place of the thioether (β-carbon) bonds which are characteristics of lanthipeptides. Further, bacteriocins (class II) are subcategorized into three classes: bacteriocins (non-pediocin-like) that lack this characteristic motif, bacteriocins (pediocin-like) that carry the sequence of YGNGV, and two-peptide bacteriocins (Soltani et al. 2021).

There is another suggestion for classifying bacteriocins that are derived from Gram-negative bacteria into two key classes (Rebuffat 2011). The first class involves great molecular volume modular proteins of 30–80 kDa named colicins-like bacteriocins; meanwhile, the other one includes few peptides of molecular mass from 1 to 10 kDa named microcins-like bacteriocins. Currently, a new classification including bacteriocins of Gram-negative bacteria plus Gram-positive bacteria is mentioned by Soltani et al. (2021) and includes two basic classes. The first one is RiPPs, while the other one carries unmodified peptides. The LAB-bacteriocins are classified into three principal classes. Class I is composed of bacteriocins with low molecular weight (≤10 kD), with post-translational modification distinguished by the existence of lanthionine (uncommon amino acids); class II is composed of different bacteriocins with little molecular mass (≤10 kD) without post-translational modification, and the last class is characterized by bacteriocins with high molecular weight (10 kD) (Alvarez-Sieiro et al. 2016; Pérez-Ramos et al. 2021) (Fig. 11.1).

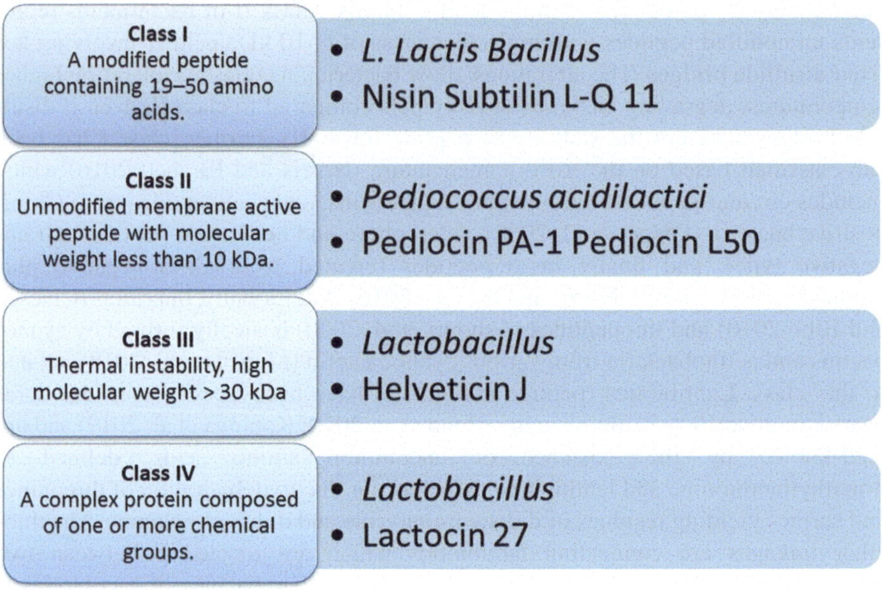

Fig. 11.1 Updated classification of bacteriocins

11.2.3 Immune Response for Different Classes of Bacteriocins (Figure 2)

11.2.3.1 Class Ia: Lantibiotic Immunity

Such system comprises of either ABC transporter [consisting subunits 2 or 3 LanEF (G)] or immune protein (LanI). Each of LanEF (G) and LanI could either work individually or as one unit interacting with converse autoimmunity. In some circumstances, an auxiliary protein named LanH is included in such own-defense system. The guideline for synthesis of numerous lantibiotics is controlled by a two-constituent signal transduction scheme, the histidine kinase (LanK) and the response regulator (LanR). LanK transmits its phosphate bond onto LanR that in turn stimulates the genes transcription related to clusters of bacteriocin (Alvarez-Sieiro et al. 2016). Under certain conditions, the individual stimulator P_{nisI} is located, which has a role in promoting the immunological function in the cell for a week, and accordingly defending it throughout its primary come across with nisin (Draper et al. 2008).

Lactococcus lactis nisin-releasing strains have the ability to avert the harmful impacts of the private nisin via exhausting two different mechanisms. The former one is associated with expression of lipoprotein protein (specific NisI immunity), which can join nisin, and then avert it from getting into its targeted lipid II (Hacker et al. 2015).

The other mechanism includes the ATP-binding cassette transporter NisFEG that ejects nisin from the membrane before formation of hole (Bierbaum and Sahl 2009). Based on this, the full immunity can only be detected in the incidence of both proteins (Ra et al. 1999). Expression of genes in case of nisin immunity via direct or indirect interfaces plays also a major role via localizing -transposon within the chromosomal *feuD* gene in *L. lactis* N8 delivering an imitative strain *L. lactis* L58 that has an improved immunity for nisin (Zhu et al. 2016).

11.2.3.2 Class Ib or Circular Bacteriocins

The gene expression for the production, passage, and immune function of such circular bacteriocins has been gathered either in single or numerous operons and are positioned either on the chromosome or plasmids. A group of ten genes, *48ABCC$_I$DD$_I$EFGH*, found on the pheromone-response plasmid pMB2 of 68 kbp, is pronounced for production of AS-48 (Martínez-Bueno et al. 1994). Northern blotting research reveals that they act like two synchronized operons (*as-48ABC* and *as-48C$_I$DD$_I$EFGH*, respectively). Both of them are substantial for the production of bacteriocin and immunity (Diaz et al. 2003).

Currently, there are promoters for the expression of operons and there is an extra promoter included in the *as-48D1EFGH* expression (Adımcılar et al. 2019). D1 was primarily designated as an immune protein owing to it providing a particular level of bacteriocin resistance (Martínez-Bueno et al. 1998). The *as-48EFGH* genes encrypt ancillary ABC transporters which are also included in self-protection. When AS-48 was added into the medium, it provided higher resistance. Such ABC transporter promotes the release of bacteriocin via ejection from the membrane (Diaz et al. 2003). Considering this data, the mutual appearance of the membrane proteins C and D$_I$ and complementary ABC transporter are needed for providing a whole immunity against AS-48.

11.2.3.3 Immunity of Class IIa Bacteriocins

The immune protein works inside the inner cell and cannot cooperate with the bacterial membrane by itself (Pérez-Ramos et al. 2021). The establishment of a potent complex includes an immune protein, bacteriocin, and the mannose PTS system. Such a complex is shaped in the case of bacteriocin found in the extracellular place. For this reason, the interaction between bacteriocin and receptors can happen for pairing immunity proteins, and then preventing the death of the cell. This finding opines the straight interface between bacteriocin and immune protein, and the starring role of the immune protein to block the pore made by the bacteriocin (Pérez-Ramos et al. 2021).

11.2.3.4 Immunity of Class IIb Bacteriocins

LagC is considered the immune protein that belongs to lactococcin G. LagC identifies its private bacteriocin by the C-terminal portion of lactococcin Gβ and the N-terminal portion of lactococcin Gα. Also, it is found that the promoting action of the EntI and LagC proteins is based on the targeted cell membranes' structures, suggesting the interface of the immune proteins by the receptors of their cognate bacteriocin (Oppegård et al. 2010). Furthermore, there are specific immunity proteins like PlnI, SkkI, and PlnL responsible for self-protection contrary to their cognate bacteriocins, plantaricin EF, sakacin 23K, and plantaricin JK (Kjos et al. 2010). Hence, it is assumed that such proteins destroy their cognate bacteriocins via releasing proteolysis enzymes.

11.2.3.5 Immunity of Class IIc, Leaderless Bacteriocins

Leaderless bacteriocins have belonged to a heterogeneous group of bacteriocins and the immune systems are poorly understood. For self-protection of the aureocin A53, the ABC transporter is included, while for getting full immunity, there must be one more protein (dos Nascimento et al. 2012). Two co-expressed proteins, AucIB and AucIA, are included. An AucIA bonds specific characteristics with other defined immune proteins, like being connected to the membrane via three alleged transmembrane regions and being a cationic protein (dos Nascimento et al. 2012).

11.2.3.6 Class IId Bacteriocins Plus Non-pediocin Bacteriocins

Class IId is composed of heterogeneous bacteriocins with basic distinctive which is linear single-peptides, for example, lactococcin 972, lactococcin A, and enterocin B (Alvarez-Sieiro et al. 2016).

Lactococcin A is a type of bacteriocin that originated from *L. lactis* subsp. LMG 213 has an action confronting strain that are adjacent to the releasing strain. It is formed as a 75-amino acid precursor plus an extension of a 21-amino acid N-terminal. The biosynthesis genes for lactococcin A include the immunity gene (*lciA*), structural gene (*lcnA*), and other genes (*lcnD and lcnC*) that encrypt the devoted ABC transport system and its auxiliary protein, are passed on a 55 kb plasmid. The mode of action in the case of lactococcin 972, a bacteriocin created by IPLA972 *L. lactis*, is based on suppressing biosynthesis of septum in the goal strain via conjugating to the precursor of the cell wall, lipid II, as it has a deficiency in lanthionine ring that can compose intramolecular bridges (Martínez-Bueno et al. 1998).

L. lactis strain creates an immune protein named LciA owning α-helix arrangement for defending itself against its own bacteriocin. Another study demonstrates a model of immunity by which the LciA immune protein acts together with the receptor (man-PTS of the *L. lactis* strain), accordingly inhibiting bacteriocin from the formation of the pore (Venema et al. 1994).

11.2.3.7 Immune Mechanism for Class III Bacteriocins

Little researches have been carried out on the self-immunity of bacteriolysin releasing strains to their precise bacteriocin. A previous study reports the confrontation to lysostaphin in *Staphylococcus simulans*. Such resistance of lysostaphin is owing to the addition of serines swapping particular glycines throughout synthesis of peptidoglycan by the immune protein (Gargis et al. 2010). Another report demonstrates that the immunity in the producer cells to the bacteriocin zoocin A is returned to the existence of the zoocin A immune factor (Zif) (Gargis et al. 2009).

11.2.4 Applications of Bacteriocins in Aquaculture

Aquaculture is a rapidly growing industrial sector as fish is regarded as indispensable and highly nutritious food essential for human food security and is a great source of protein. There is a continuous increasing demand for fish to meet the uprising population, which leads to pressure on fish captures and presence of problems related to diseases and high stocking densities (Tacon 2020). Utilizing of antibiotics for competing diseases threats mostly like sulfonamide, streptomycin, tetracycline, and chloramphenicol inducing development of antibiotic resistance (AMR) in aquaculture. This elucidated the necessity to figure out safer approaches and with the limitation of vaccine applications in aquaculture practices, bacteriocins application raised as a promising suggested solution. Bacteriocins are safe and degradable bioactive compounds produced by bacteria (Pérez-Ramos et al. 2021).

Bacteriocins are bactericidal peptides of small molecular weight, synthesized via ribosome, most likely with a length of around 25–60 amino acids. They are present in all the bacteria (Gram-negative and Gram-positive) and have antibacterial activities (Belguesmia et al. 2020). They have the power to suppress or destroy virulent bacteria and could be eliminated through the digestive system (Wang et al. 2019).

There are almost 230 various bacteriocins, 206 are formed by Gram-positive bacteria and 19 are produced via Gram-negative bacteria. They possess the antibacterial activity at minimum concentration and biodegrade leaving no risks to fish or human. They remain unchanged at various degrees of temperature and pH (Darbandi et al. 2022).

Gut isolated *Enterococcus faecium* isolated from *Mugil* spp. release broad-spectrum bacteriocins against Gram-negative and Gram-positive bacteria (Nayak et al. 2022). *Enterococcus thailandicus* secretes bacteriocin-like inhibitory substances (BLIS) against pathogenic *Lactococcus garvieae* in fish aquaculture (Lin et al. 2013). *Lactobacillus* strain produces bacteriocin against *Listeria monocytogenes* and *Salmonella enterica*. Nisin (type 1 bacteriocin) is released from *L. lactis* against *L. garvieae* (Sequeiros et al. 2015). *Bacillus subtilis* also produces bacteriocin antagonizing the fish pathogens, *Aeromonas salmonicida, A. hydrophila*, and *B. mycoides* (Darbandi et al. 2022).

Probiotic bacteria is regarded as the most applied method to present bacteriocins in aquaculture practices, compared to using bacteriocins separately or in combinations (Nayak et al. 2022).

Bacteriocins are used in aquaculture to overcome the resistance of antibiotics in the culture conditions. Probiotics are commonly safe for antagonizing diseases. *Bacillus* has a wide range of antibacterial activities. It secretes bacteriocins as thuricin, subpeptin, mersacidin, and sublancin against fish pathogens. Lactic acid bacteria (LAB) produce bacteriocin-like substances (BLIS) isolated from gut is active in combating the infection by *A. hydrophila* in freshwater crayfish (Kuebutornye et al. 2020).

Pyocin pigment secreted from *Pseudomonas aeruginosa* of aquatic origin suppress *A. hydrophila*, *Pseudomonas* sp., and *Vibrio* sp. isolated from marine algae (Nayak et al. 2022).

The use of bacteriocins from various sources involving fresh and marine water origin has been utilized in different systems of fish cultures. In aquaculture, surfactin, an antiviral component released by *A. calcoaceticus, P. aeruginosa*, and *B. subtilis*, is apparently used to control viral and bacterial infections (Nayak et al. 2022).

Bacteriocins could be used as a preservative as LAB can defend the fish and their products against infection and pathogenic contaminations and increasing the shelf life. *Enterococcus faecalis* produces sodium alginate-coated bacteriocin used as preservative for salmon fillets. Utilizing bioactive like peptides of low molecular weight and bacteriocins is considered the best option for mitigating diseases and in preservation purposes (Shan et al. 2020) (Fig. 11.2).

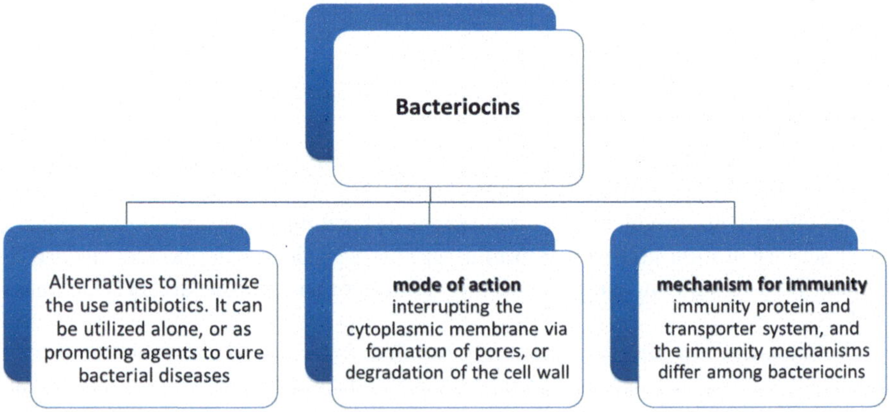

Fig. 11.2 Bacteriocins: definition, mode of action, and mechanism of immunity

11.3 Predatory Bacteria

11.3.1 Definition, Mode of Action, and Microbial Predatory Interactions

Predatory bacteria belongs to prokaryotes group that can dynamically kill and ingest other bacteria, altering the prey bacterial diversity and in turn modifies the general construction of the microbial community (Kamada et al. 2023).

Bacterial predation has been suggested as one of the discerning abilities depending on the bacterial size and shape. Predatory bacterium can attack in groups causing destruction to the prey which is commonly parallel in size to them. They can release an astonishing selection of metabolites and catalytic enzymes for lysing other organisms in the extracellular area (Pérez et al. 2016).

Predation in all aquatic environments is an extensive mode of action among living organisms. Water recycling systems appeared as a consequence of the massive water as operational requirements for aquaculture practices, which need water purification to secure safety (Kandel et al. 2014).

Microbial predatory interactions are among the ecological procedures that can alter the function of water purification such as predation by viruses. Microbial biomass in aquaculture practices might be used up at almost same level of production and microbial predation influences the development and the acclimation of prey and predators and alters the function and structure of microbial biomasses (Gallet et al. 2009).

11.3.2 Predatory Bacteria in Aquaculture

In aquaculture, some bacteria appear to protect against aquatic bacterial pathogens and could be used as bio-control alternative therapeutics to antibiotics and thus overcome the dangers of antibiotic resistance. Antibiotics are effective, for example, tetracycline and oxytetracyline, but stimulate bacterial modification, causing dissemination of antibiotic-resistant bacteria that will in turn build up in food system (Ezzedine et al. 2022; Han et al. 2015).

Among these are *Bdellovibrio* and like organisms (BALOs) are minute, motile, and Gram-negative bacteria grouping that attack other Gram-negative bacteria for preying. It comprises of families *Bacteriovoracaceae, Peredibacteraceae*, *Halobacteriovoraceae*, and *Bdellovibrionaceae*. BALOs are regarded as control practices against bacterial communities in aquaculture practices. BALOs have a physical benefit over other bacteriovorous/predatory microorganisms as protozoans which are larger and less motile and bacteriophages that are lesser but non-motile (Koval et al. 2015). They attach with flagella to prey cells and penetrate into the prey cells' periplasm. BALOs have periplasmic or epibiotic modes of predation:

periplasmic predators split by synchronous septation inside prey while epibiotic predators split by binary fission while fixed to prey (Kandel et al. 2014; Ooi et al. 2021).

BALOs are present in the gut of snakehead fish and sturgeon and protect against pathogens induced by the aquatic bacterial infections. The BALOs taken out from water and soil samples are mandatory bacterial predators. Hence, their upscaling process is harder than facultative bacterial predators like *Saprospira* and *Myxococcus* which can be developed in nutritional media. These are successful in diminishing the *Vibrio parahaemolyticus* VpAHPND cells which resulted in almost 100% mortality rates of black tiger shrimp, *Penaeus monodon*, *Penaeus vannamei*, and Pacific white shrimp post-larvae within 22–30 days in most farms (Han et al. 2015; Yeoh et al. 2021).

Saprospiraceae bacteria family belongs to "Ixotrophy" predation which means nourishing on prey trapped on sticky substances. Such bacteria occupy aqueous environments and are gliding, multicellular, filamentous, and facultative predators that might predate other bacteria, such as cyanobacteria. *Saprospira* can capture the prey by its multicellular filamentous cells, form 3D reticular structures, and kill its prey via straight interaction and lysis. *Saprospira* spp. in marine environments predates on the diatom *Chaetoceros ceratosporum* (Furusawa et al. 2003). It could be used as a bio-control predator against *Vibrio* spp., and it is effective in controlling vibriosis in shrimp aquaculture (Ooi et al. 2021).

Myxococcales, which are named as "group attacks" because of their gliding movement, lytic enzymes secretion, and antibiotics release (Pérez et al. 2016). The "predatome" is also a widely used definition that includes the protein families with the same phenotypes of predatory bacteria and the examination of the predatory-associated proteins plus the encrypting genes, predatory characters are expected as unknown in the phyla *Myxococcota* and *Bdellovibrionota* (Waite et al. 2020). Furthermore, myxobacteria can be highlighted for the detection of unknown natural yields depending on the relationship between predation and antibiotic biosynthesis (Ahearne et al. 2021).

Streptomyces is one of the predatory bacterial species; its mode of action depends on what is called "far range metabolites" as it doesn't need direct contact with prey, but can perform lysis at a distance. They attack forming groups and are capable of producing a wide variety and quantity of antibiotic substances and tributary metabolites. They don't require direct contact with the prey and their growth was convoyed by the vanishing of living prey cells (Claessen et al. 2014; Waite et al. 2020; Kumbhar et al. 2014).

Herpetosiphon and Myxobacteria exhibit similar hunting approaches and are described as micro-predators. Predation of *Myxococcus xanthus* is a common procedure, using gliding motility to look for prey and create a broad range of predatory yields that destroy and make analysis for the prey cells depending on density and need to be close to the prey cells to stimulate predation, followed by penetration and lysing of neighboring cells in the presence of the action of some lytic enzymes and secondary metabolites (Claessen et al. 2014; Kumbhar et al. 2014). The normal feeding habit of *M. xanthus* is compliant predation as a high cell density of predators is needed for effective predation. *M. xanthus* cells progressively pierce the prey colony inducing lysis of the cells directly and exhibit different patterns of predation when they prey on strains that release great contents of the exopolysaccharide galactoglucan (Pérez-Ramos et al. 2021; Pérez et al. 2016).

In general, several hunting strategies have mutual structures. For example, various predators need means of motility to run into their prey. Many predators like *Flavobacterium,* myxobacteria, *Saprospira, Cytophaga,* and *Herpetosiphon* depend on gliding motion to transfer dynamically over surfaces, via a smooth cell movement. Currently, it has been documented that *Flavobacterium johnsoniae* and *M. xanthus* can glide using proteins that transfer in helical paths, with notably various motility tools, opining that these ways may have progressed individually (Kamada et al. 2023).

Candidate phyla radiation (CPR) predatory bacteria denote ultra-minor predators. The presence of small predators as pathogens, parasites, or symbionts proves the suggestion that the extra-small size of CPR bacteria cells is linked to the minor sizes of genome, rendering these cells rely on other complete or greater functional prokaryotes genomes and greater cells. Epibiotic predator/sucker Ca. *Vampirococcus lugosii* of anoxic-photosynthetic bacteria accompanies flat and (diameter: 500–600 nm; height: 200–250 nm) stacked cells. Bacteria and facultative epibiotic predator Ca. *Velamenicoccus archaeovorus* consist of 200–300 nm coccoid cells. The human oral *Actinomyces odontolyticus* epibiont predator Ca. *Nanosynbacter lyticus* is associated with 200–300 nm diameters coccoid cells. The relationship between *N. lyticus* and *A. odontolyticus* is being recently suggested not to be predatory, but symbiotic (Tian et al. 2020; Moreira et al. 2021; Kizina et al. 2022; Bor et al. 2016; He et al. 2015).

11.3.3 Phylogenetic Tree for the Predatory Bacteria

The phylogenetic tree includes an overall of 136 sequences including predatory bacterial (16S rRNA genes) obtained from 12 phyla (involving phyla candidate) of Bdellovibrionota, Bacteroidota, Actinobacteriota, Desulfobacterota, Cyanobacteria, Myxococcota, Ca. Patescibacteria (CPR), Ca. Omnitrophica (OP3), or Ca. Absconditabacteria (SR1), Pseudomonadota, and Planctomycetota. The accessible gene sequence of "Ca. *Vampirococcus lugosii*" (with accession number of MW286273 and a molecular mass of 1071 bp) was the most shortest one among the obtained sequences, plus the phylogenetic trees without and with "Ca. *V. lugosii*" are created (Moreira et al. 2021). A recent study reveals the phylogenetic tree depending on the sequences of around 1.6 kb following position apart from the straight sequence of "Ca. *V. lugosii*" (1071 bp) (Kamada et al. 2023).

11.3.4 Main Taxa of Predatory Bacteria

There are two main taxa of predatory bacteria including the phyla Myxococcota and Bdellovibrionota, besides Bradymonadales order which is considered the third category of predation. The order Bradymonadales was previously credited to Proteobacteria phylum, class Deltaproteobacteria, but is recently allied with the Desulfobacteriota phylum (Waite et al. 2020).

The bacteria found in this order, or Bradymonabacteria, is characterized by a distinctive sort of predation that is variant from the so-named facultative and compel (Pérez et al. 2016), which are described as totally prey-reliant and prey-independent (Mu et al. 2020).

The kind of predation in Bradymonabacteria is distinguished by facultative prey-reliant and named recently as facultative (Mu et al. 2020), and the earlier description "facultative" (Pérez et al. 2016) was recently substituted with "opportunistic" (Pérez et al. 2016).

Currently, there are two recent strains that are predatory and identified from a freshwater aquarium including Myxococcus sp. MH1 (phylum Myxococcota) and Bacteriovorax stolpii HI3 (phylum Bdellovibrionota). They exhibit a broad array of preys involving *E. coli* (HB101) and fifty-two eco-friendly strains comprising of 8 strains of Gram-positive allied with the phyla Bacillota and Actinomycetota plus 44 strains of Gram-negative bacteria associated with the phyla Pseudomonadota and Bacteroidota. Meanwhile, *Myxococcus* sp. MH1 can prey on a total of 53 prey strains, the strain, Bacteriovorax stolpii HI3, can prey on *E. coli* HB101 and the 25 environmental Gram-negative strains (Inoue et al. 2022).

11.4 Conclusion

Bacteriocins can serve as alternatives to minimize the use of traditional antibiotics. It can be utilized alone, or as promoting agents to cure bacterial diseases. Bacteriocins are considered a desirable and reasonable therapeutic approach, it is therefore essential to investigate their mechanisms of action plus the resistance that arises by the releasing bacteria to their self bacteriocins. The majority of bacteriocins work through interrupting the cytoplasmic membrane via the formation of pores or degradation of the cell walls. However, part of these peptides still have unidentified mechanisms of action, particularly the group that actively works against Gram-negative bacteria. Considering immune mechanism, major bacteriocin-releasing strains own a mechanism for immunity including an immune protein and a devoted transporter system, but the immunity mechanisms differ among bacteriocins.

References

Abdel Rahman AN, Mansour DA, Abd El-Rahman GI, Elseddawy NM, Zaglool AW, Khamis T, Mahmoud SF, Mahboub HH (2022) Imidacloprid toxicity in *Clarias gariepinus*: protective role of dietary *Hyphaene thebaica* against biochemical and histopathological disruption, oxidative stress, immune genes expressions, and *Aeromonas sobria* infection. Aquaculture 555:738170. https://doi.org/10.1016/j.aquaculture.2022.738170

Adımcılar V, Kalaycıoğlu Z, Aydoğdu N, Dirmenci T, Kahraman A, Erim FB (2019) Rosmarinic and carnosic acid contents and correlated antioxidant and antidiabetic activities of 14 salvia species from Anatolia. J Pharm Biomed Anal 175:112763. https://doi.org/10.1016/j.jpba.2019.07.011

Ahearne A, Albataineh H, Dowd SE, Stevens DC (2021) Assessment of evolutionary relationships for prioritization of myxobacteria for natural product discovery. Microorganisms 9(7):1376. https://doi.org/10.3390/microorganisms9071376

Alvarez-Sieiro P, Montalbán-López M, Mu D, Kuipers OP (2016) Bacteriocins of lactic acid bacteria: extending the family. Appl Microbiol Biotechnol 100(7):2939–2951. https://doi.org/10.1007/s00253-016-7343-9

Alzahrani OM, Elumalai P, Nada HS, Ahmed SAA, Zaglool AW, Shawky SM, Alkafafy M, Mahboub HH (2023) *Pseudomonas putida*: sensitivity to various antibiotics, genetic diversity, virulence, and role of formic acid to modulate the immune-antioxidant status of the challenged *Nile tilapia* compared to Carvacrol oil. Fishes 8(1):6. https://doi.org/10.3390/fishes8010006

Ashouri G, Mahboobi Soofiani N, Hoseinifar SH, Jalali SAH, Morshedi V, Valinassab T, Bagheri D, Van Doan H, Torfi Mozanzadeh M, Carnevali O (2020) Influence of dietary sodium alginate and *Pediococcus acidilactici* on liver antioxidant status, intestinal lysozyme gene expression, histomorphology, microbiota, and digestive enzymes activity, in Asian Sea Bass (*Lates calcarifer*) juveniles. Aquaculture 518:734638. https://doi.org/10.1016/j.aquaculture.2019.734638

Aslam B, Khurshid M, Arshad MI, Muzammil S, Rasool M, Yasmeen N, Shah T, Chaudhry TH, Rasool MH, Shahid A et al (2021) Antibiotic resistance: one health one world outlook. Front Cell Infect Microbiol 11:771510. https://doi.org/10.3389/fcimb.2021.771510

Baptista RC, Horita CN, Sant'Ana AS (2020) Natural products with preservative properties for enhancing the microbiological safety and extending the shelf-life of seafood: a review. Food Res Int 127:108762. https://doi.org/10.1016/j.foodres.2019.108762

Belguesmia Y, Bendjeddou K, Kempf I, Boukherroub R, Drider D (2020) Heterologous biosynthesis of five new class II bacteriocins from *Lactobacillus paracasei* CNCM I-5369 with antagonistic activity against pathogenic *Escherichia coli* strains. Front Microbiol 11:1198

Benítez-Chao DF, León-Buitimea A, Lerma-Escalera JA, Morones-Ramírez JR (2021) Bacteriocins: an overview of antimicrobial, toxicity, and biosafety assessment by in vivo models. Front Microbiol 12:630695

Besse A, Peduzzi J, Rebuffat S, Carré-Mlouka A (2015) Antimicrobial peptides and proteins in the face of extremes: lessons from archaeocins. Biochimie 118:344–355. https://doi.org/10.1016/j.biochi.2015.06.004

Bierbaum G, Sahl H-G (2009) Lantibiotics: mode of action, biosynthesis and bioengineering. Curr Pharm Biotechnol 10(1):2–18. https://doi.org/10.2174/138920109787048616

Bor B, Poweleit N, Bois JS, Cen L, Bedree JK, Zhou ZH, Gunsalus RP, Lux R, McLean JS, He X et al (2016) Phenotypic and physiological characterization of the epibiotic interaction between TM7x and its basibiont actinomyces. Microb Ecol 71(1):243–255. https://doi.org/10.1007/s00248-015-0711-7

Bountra K, Hagelueken G, Choudhury HG, Corradi V, El Omari K, Wagner A, Mathavan I, Zirah S, Yuan Wahlgren W, Tieleman DP et al (2017) Structural basis for antibacterial peptide self-immunity by the bacterial ABC transporter McjD. EMBO J 36(20):3062–3079. https://doi.org/10.15252/embj.201797278

Chikindas ML, Weeks R, Drider D, Chistyakov VA, Dicks LM (2018) Functions and emerging applications of bacteriocins. Curr Opin Biotechnol 49:23–28. https://doi.org/10.1016/j.copbio.2017.07.011

Choi G-H, Holzapfel WH, Todorov SD (2023) Diversity of the bacteriocins, their classification and potential applications in combat of antibiotic resistant and clinically relevant pathogens. Crit Rev Microbiol 49(5):578–597. https://doi.org/10.1080/1040841X.2022.2090227

Claesen J, Bibb M (2010) Genome mining and genetic analysis of cypemycin biosynthesis reveal an unusual class of posttranslationally modified peptides. Proc Natl Acad Sci 107(37):16297–16302. https://doi.org/10.1073/pnas.1008608107

Claessen D, Rozen DE, Kuipers OP, Søgaard-Andersen L, Van Wezel GP (2014) Bacterial solutions to multicellularity: a tale of biofilms, filaments and fruiting bodies. Nat Rev Microbiol 12(2):3178. https://doi.org/10.1038/nrmicro3178

Darbandi A, Asadi A, Mahdizade Ari M, Ohadi E, Talebi M, Halaj Zadeh M, Darb Emamie A, Ghanavati R, Kakanj M (2022) Bacteriocins: properties and potential use as antimicrobials. J Clin Lab Anal 36(1):e24093. https://doi.org/10.1002/jcla.24093

Diaz M, Valdivia E, Martínez-Bueno M, Fernández M, Soler-González AS, Ramírez-Rodrigo H, Maqueda M (2003) Characterization of a new operon, as-48EFGH, from the as-48 gene cluster involved in immunity to Enterocin AS-48. Appl Environ Microbiol 69(2):1229–1236. https://doi.org/10.1128/AEM.69.2.1229-1236.2003

Doan HV, Soltani E, Ingelbrecht J, Soltani M (2020) Medicinal herbs and plants: potential treatment of monogenean infections in fish. Rev Fish Sci Aquac 28(2):260–282. https://doi.org/10.1080/23308249.2020.1712325

dos Nascimento JS, Coelho MLV, Ceotto H, Potter A, Fleming LR, Salehian Z, Nes IF, do de Bastos MCF (2012) Genes involved in immunity to and secretion of aureocin A53, an atypical class II bacteriocin produced by *Staphylococcus aureus* A53. J Bacteriol 194(4):875–883. https://doi.org/10.1128/JB.06203-11

Draper LA, Ross RP, Hill C, Cotter PD (2008) Lantibiotic immunity. Curr Protein Pept Sci 9(1):39–49. https://doi.org/10.2174/138920308783565750

Elabd H, Faggio C, Mahboub HH, Emam MA, Kamel S, El Kammar R, Abdelnaeim NS, Shaheen A, Tresnakova N, Matter A (2022a) *Mucuna pruriens* seeds extract boosts growth, immunity, testicular histology, and expression of immune-related genes of mono-sex *Nile tilapia* (*Oreochromis niloticus*). Fish Shellfish Immunol 127:672–680. https://doi.org/10.1016/j.fsi.2022.06.055

Elabd H, Youssuf H, Mahboub HH, Salem SMR, Husseiny WA, Khalid A, El-Desouky HS, Faggio C (2022b) Growth, hemato-biochemical, immune-antioxidant response, and gene expression in *Nile tilapia* (*Oreochromis niloticus*) received nano iron oxide-incorporated diets. Fish Shellfish Immunol 128:574. https://doi.org/10.1016/j.fsi.2022.07.051

El-Bouhy ZM, Reda RM, Mahboub HH, Gomaa FN (2021a) Bioremediation effect of pomegranate peel on subchronic mercury immunotoxicity on African catfish (*Clarias gariepinus*). Environ Sci Pollut Res 28(2):2219–2235. https://doi.org/10.1007/s11356-020-10599-1

El-Bouhy ZM, Reda RM, Mahboub HH, Gomaa FN (2021b) Chelation of mercury intoxication and testing different protective aspects of *Lactococcus lactis* probiotic in African catfish. Aquac Res 52(8):3815–3828. https://doi.org/10.1111/are.15227

Ezzedine JA, Scheifler M, Desdevises Y, Jacquet S (2022) A comparative study of the dynamics and diversity of Bdellovibrio and like organisms in lakes Annecy and Geneva. Microorganisms 10(10):1960. https://doi.org/10.3390/microorganisms10101960

Furusawa G, Yoshikawa T, Yasuda A, Sakata T (2003) Algicidal activity and gliding motility of Saprospira sp. SS98-5. Can J Microbiol 49(2):92–100. https://doi.org/10.1139/w03-017

Gallet R, Tully T, Evans MEK (2009) Ecological conditions affect evolutionary trajectory in a predator-prey system. Evolution 63(3):641–651. https://doi.org/10.1111/j.1558-5646.2008.00559.x

Gargis SR, Gargis AS, Heath HE, Heath LS, LeBlanc PA, Senn MM, Berger-Bächi B, Simmonds RS, Sloan GL (2009) Zif, the Zoocin A immunity factor, is a FemABX-like immunity protein with a novel mode of action. Appl Environ Microbiol 75(19):6205–6210. https://doi.org/10.1128/AEM.01011-09

Gargis SR, Heath HE, LeBlanc PA, Dekker L, Simmonds RS, Sloan GL (2010) Inhibition of the activity of both domains of Lysostaphin through peptidoglycan modification by the Lysostaphin immunity protein. Appl Environ Microbiol 76(20):6944–6946. https://doi.org/10.1128/AEM.01066-10

Hacker C, Christ NA, Duchardt-Ferner E, Korn S, Göbl C, Berninger L, Düsterhus S, Hellmich UA, Madl T, Kötter P et al (2015) The solution structure of the lantibiotic immunity protein NisI and its interactions with Nisin. J Biol Chem 290(48):28869–28886. https://doi.org/10.1074/jbc.M115.679969

Hamed HS, Amen RM, Elelemi AH, Mahboub HH, Elabd H, Abdelfattah AM, Moniem HA, El-Beltagy MA, Alkafafy M, Yassin EMM et al (2022) Effect of dietary *Moringa oleifera* leaves nanoparticles on growth performance, physiological, immunological responses, and liver antioxidant biomarkers in *Nile tilapia* (*Oreochromis niloticus*) against zinc oxide nanoparticles toxicity. Fishes 7(6):360. https://doi.org/10.3390/fishes7060360

Han JE, Mohney LL, Tang KFJ, Pantoja CR, Lightner DV (2015) Plasmid mediated tetracycline resistance of *Vibrio parahaemolyticus* associated with acute hepatopancreatic necrosis disease (AHPND) in shrimps. Aquac Rep 2:17–21. https://doi.org/10.1016/j.aqrep.2015.04.003

Hatakka K, Saxelin M (2008) Probiotics in intestinal and non-intestinal infectious diseases—clinical evidence. Curr Pharm Des 14(14):1351–1367. https://doi.org/10.2174/138161208784480162

He X, McLean JS, Edlund A, Yooseph S, Hall AP, Liu S-Y, Dorrestein PC, Esquenazi E, Hunter RC, Cheng G et al (2015) Cultivation of a human-associated TM7 phylotype reveals a reduced genome and epibiotic parasitic lifestyle. Proc Natl Acad Sci 112(1):244–249. https://doi.org/10.1073/pnas.1419038112

Hernando-Amado S, Coque TM, Baquero F, Martínez JL (2019) Defining and combating antibiotic resistance from one health and global health perspectives. Nat Microbiol 4(9):1432–1442. https://doi.org/10.1038/s41564-019-0503-9

Hu Y, Jin L, Zhao Y, Jiang L, Yao S, Zhou W, Lin K, Cui C (2021) Annual trends and health risks of antibiotics and antibiotic resistance genes in a drinking water source in East China. Sci Total Environ 791:148152. https://doi.org/10.1016/j.scitotenv.2021.148152

Huang F, Teng K, Liu Y, Cao Y, Wang T, Ma C, Zhang J, Zhong J (2021) Bacteriocins: potential for human health. Oxidative Med Cell Longev 2021:e5518825. https://doi.org/10.1155/2021/5518825

Inoue D, Hiroshima N, Nakamura S, Ishizawa H, Ike M (2022) Characterization of two novel predatory bacteria, *Bacteriovorax stolpii* HI3 and Myxococcus sp. MH1, isolated from a freshwater pond: prey range, and predatory dynamics and efficiency. Microorganisms 10(9):1816. https://doi.org/10.3390/microorganisms10091816

Issouffou C, Suwansri S, Salaipeth L, Domig KJ, Hwanhlem N (2018) Synergistic effect of essential oils and enterocin KT2W2G on the growth of spoilage microorganisms isolated from spoiled banana peel. Food Control 89:260–269. https://doi.org/10.1016/j.foodcont.2018.02.019

Kamada S, Wakabayashi R, Naganuma T (2023) Phylogenetic revisit to a review on predatory bacteria. Microorganisms 11(7):1673. https://doi.org/10.3390/microorganisms11071673

Kandel PP, Pasternak Z, van Rijn J, Nahum O, Jurkevitch E (2014) Abundance, diversity and seasonal dynamics of predatory bacteria in aquaculture zero discharge systems. FEMS Microbiol Ecol 89(1):149–161. https://doi.org/10.1111/1574-6941.12342

Kapinga IB, Limbu SM, Madalla NA, Kimaro WH, Mabiki FP, Lamtane HA, Tamatamah RA (2019) Dietary *Aspilia mossambicensis* and *Azadirachta indica* supplementation alter gonadal characteristics and histology of juvenile *Nile tilapia* (*Oreochromis niloticus*). Aquac Res 50(2):573–580. https://doi.org/10.1111/are.13931

Kizina J, Jordan SFA, Martens GA, Lonsing A, Probian C, Kolovou A, Santarella-Mellwig R, Rhiel E, Littmann S, Markert S et al (2022) Methanosaeta and "Candidatus *Velamenicoccus archaeovorus*". Appl Environ Microbiol 88(7):e02407–e02421. https://doi.org/10.1128/aem.02407-21

Kjos M, Snipen L, Salehian Z, Nes IF, Diep DB (2010) The Abi proteins and their involvement in bacteriocin self-immunity. J Bacteriol 192(8):2068–2076. https://doi.org/10.1128/JB.01553-09

Koval SF, Williams HN, Stine OC (2015) Reclassification of Bacteriovorax Marinus as Halobacteriovorax Marinus Gen. Nov., Comb. Nov. and Bacteriovorax Litoralis as Halobacteriovorax Litoralis Comb. Nov.; Description of Halobacteriovoraceae Fam. Nov. in the class Deltaproteobacteria. Int J Syst Evol Microbiol 65(Pt 2):593–597. https://doi.org/10.1099/ijs.0.070201-0

Kuebutornye FKA, Abarike ED, Lu Y, Hlordzi V, Sakyi ME, Afriyie G, Wang Z, Li Y, Xie CX (2020) Mechanisms and the role of probiotic bacillus in mitigating fish pathogens in aquaculture. Fish Physiol Biochem 46(3):819–841. https://doi.org/10.1007/s10695-019-00754-y

Kumbhar C, Mudliar P, Bhatia L, Kshirsagar A, Watve M (2014) Widespread predatory abilities in the genus *Streptomyces*. Arch Microbiol 196(4):235–248. https://doi.org/10.1007/s00203-014-0961-7

Lin Y-H, Chen Y-S, Wu H-C, Pan S-F, Yu B, Chiang C-M, Chiu C-M, Yanagida F (2013) Screening and characterization of LAB-produced bacteriocin-like substances from the intestine of grey mullet (*Mugil cephalus* L.) as potential biocontrol agents in aquaculture. J Appl Microbiol 114(2):299–307. https://doi.org/10.1111/jam.12041

Mahboub HH, Rashidian G, Hoseinifar SH, Kamel S, Zare M, Ghafarifarsani H, Algharib SA, Moonmanee T, Van Doan H (2022a) Protective effects of *Allium hirtifolium* extract against foodborne toxicity of zinc oxide nanoparticles in common carp (*Cyprinus carpio*). Comp Biochem Physiol Part C Toxicol Pharmacol 257:109345. https://doi.org/10.1016/j.cbpc.2022.109345

Mahboub HH, Faggio C, Hendam BM, Algharib SA, Alkafafy M, Abo Hashem M, Mahmoud YK, Khamis T, Abdel-Ghany HM, Masoud SR et al (2022b) Immune-antioxidant trait, *Aeromonas veronii* resistance, growth, intestinal architecture, and splenic cytokines expression of *Cyprinus carpio* Fed *Prunus armeniaca* kernel-enriched diets. Fish Shellfish Immunol 124:182–191. https://doi.org/10.1016/j.fsi.2022.03.048

Mahboub HH, Elsheshtawy HM, Sheraiba NI, Fahmy EM, Masoud SR, Mohamed EAA, Abdelnaeim NS, Mohamed DI, Ismail TA, Ahmed SAA (2022c) Dietary black cumin (*Nigella sativa*) improved hemato-biochemical, oxidative stress, gene expression, and immunological response of *Nile tilapia* (*Oreochromis niloticus*) infected by *Burkholderia cepacia*. Aquac Rep 22:100943. https://doi.org/10.1016/j.aqrep.2021.100943

Mahboub HH, Eltanahy A, Omran A, Mansour AT, Safhi FA, Alwutayd KM, Khamis T, Husseiny WA, Ismail SH, Yousefi M et al (2024) Chitosan nanogel aqueous treatment improved blood biochemicals, antioxidant capacity, immune response, immune-related gene expression and infection resistance of *Nile tilapia*. Comp Biochem Physiol B Biochem Mol Biol 269:110876. https://doi.org/10.1016/j.cbpb.2023.110876

Mansour AT, Hamed HS, El-Beltagi HS, Mohamed WF (2022) Modulatory effect of papaya extract against Chlorpyrifos-induced oxidative stress, immune suppression, endocrine disruption, and DNA damage in female *Clarias gariepinus*. Int J Environ Res Public Health 19(8):4640. https://doi.org/10.3390/ijerph19084640

Martínez-Bueno M, Maqueda M, Gálvez A, Samyn B, Van Beeumen J, Coyette J, Valdivia E (1994) Determination of the gene sequence and the molecular structure of the Enterococcal peptide antibiotic AS-48. J Bacteriol 176(20):6334–6339

Martínez-Bueno M, Valdivia E, Gálvez A, Coyette J, Maqueda M (1998) Analysis of the gene cluster involved in production and immunity of the peptide antibiotic AS-48 in *Enterococcus faecalis*. Mol Microbiol 27(2):347–358. https://doi.org/10.1046/j.1365-2958.1998.00682.x

Martins J, Leikoski N, Wahlsten M, Azevedo J, Antunes J, Jokela J, Sivonen K, Vasconcelos V, Fewer DP, Leão PN (2018) Sphaerocyclamide, a prenylated cyanobactin from the cyanobacterium *Sphaerospermopsis* sp. LEGE 00249. Sci Rep 8:14537. https://doi.org/10.1038/s41598-018-32618-5

Moreira D, Zivanovic Y, López-Archilla AI, Iniesto M, López-García P (2021) Reductive evolution and unique predatory mode in the CPR bacterium *Vampirococcus lugosii*. Nat Commun 12(1):2454. https://doi.org/10.1038/s41467-021-22762-4

Mu D-S, Wang S, Liang Q-Y, Du Z-Z, Tian R, Ouyang Y, Wang X-P, Zhou A, Gong Y, Chen G-J et al (2020) Bradymonabacteria, a novel bacterial predator group with versatile survival strategies in saline environments. Microbiome 8(1):126. https://doi.org/10.1186/s40168-020-00902-0

Nayak A, Karunasagar I, Chakraborty A, Maiti B (2022) Potential application of bacteriocins for sustainable aquaculture. Rev Aquac 14(3):1234–1248. https://doi.org/10.1111/raq.12647

Negash AW, Tsehai BA (2020) Current applications of bacteriocin. Int J Microbiol 2020:4374891. https://doi.org/10.1155/2020/4374891

Norris GE, Patchett ML (2016) The glycocins: in a class of their own. Curr Opin Struct Biol 40:112–119. https://doi.org/10.1016/j.sbi.2016.09.003

Ooi MC, Goulden EF, Smith GG, Bridle AR (2021) Predatory bacteria in the haemolymph of the cultured spiny lobster *Panulirus ornatus*. Microbiology 167(11):001113. https://doi.org/10.1099/mic.0.001113

Oppegård C, Emanuelsen L, Thorbek L, Fimland G, Nissen-Meyer J (2010) The Lactococcin G immunity protein recognizes specific regions in both peptides constituting the two-peptide bacteriocin Lactococcin G. Appl Environ Microbiol 76(4):1267–1273. https://doi.org/10.1128/AEM.02600-09

Pérez J, Moraleda-Muñoz A, Marcos-Torres FJ, Muñoz-Dorado J (2016) Bacterial predation: 75 years and counting! Environ Microbiol 18(3):766–779. https://doi.org/10.1111/1462-2920.13171

Pérez-Ramos A, Madi-Moussa D, Coucheney F, Drider D (2021) Current knowledge of the mode of action and immunity mechanisms of LAB-bacteriocins. Microorganisms 9(10):2107. https://doi.org/10.3390/microorganisms9102107

Pulingam T, Parumasivam T, Gazzali AM, Sulaiman AM, Chee JY, Lakshmanan M, Chin CF, Sudesh K (2022) Antimicrobial resistance: prevalence, economic burden, mechanisms of resistance and strategies to overcome. Eur J Pharm Sci 170:106103. https://doi.org/10.1016/j.ejps.2021.106103

Ra R, Beerthuyzen MM, de Vos WM, Saris PEJ, Kuipers OP (1999) Effects of gene disruptions in the Nisin gene cluster of *Lactococcus lactis* on Nisin production and producer immunity. Microbiology 145(Pt 5):1227–1233. https://doi.org/10.1099/13500872-145-5-1227

Rahman ANA, Elkhadrawy BA, Mansour AT, Abdel-Ghany HM, Yassin EMM, Elsayyad A, Alwutayd KM, Ismail SH, Mahboub HH (2023) Alleviating effect of a magnetite (Fe_3O_4) nanogel against waterborne-lead-induced physiological disturbances, histopathological changes, and lead bioaccumulation in African catfish. Gels 9(8):641. https://doi.org/10.3390/gels9080641

Rebuffat S (2011) Bacteriocins from Gram-Negative bacteria: a classification? In: Drider D, Rebuffat S (eds) Prokaryotic antimicrobial peptides: from genes to applications. Springer, New York, pp 55–72. https://doi.org/10.1007/978-1-4419-7692-5_4

Riley MA, Wertz JE (2002) Bacteriocins: evolution, ecology, and application. Ann Rev Microbiol 56:117–137. https://doi.org/10.1146/annurev.micro.56.012302.161024

Sequeiros C, Garcés ME, Vallejo M, Marguet ER, Olivera NL (2015) Potential aquaculture probiont *Lactococcus lactis* TW34 produces Nisin Z and inhibits the fish pathogen *Lactococcus garvieae*. Arch Microbiol 197(3):449–458. https://doi.org/10.1007/s00203-014-1076-x

Shan C, Wu H, Zhou J, Yan W, Zhang J, Liu X (2020) Synergistic effects of bacteriocin from *Lactobacillus panis* C-M2 combined with dielectric barrier discharged non-thermal plasma (DBD-NTP) on *Morganella* sp. in aquatic foods. Antibiotics 9(9):593. https://doi.org/10.3390/antibiotics9090593

Soltani S, Hammami R, Cotter PD, Rebuffat S, Said LB, Gaudreau H, Bédard F, Biron E, Drider D, Fliss I (2021) Bacteriocins as a new generation of antimicrobials: toxicity aspects and regulations. FEMS Microbiol Rev 45(1):fuaa039. https://doi.org/10.1093/femsre/fuaa039

Sun S, Korheina DKA, Fu H, Ge X (2020) Chronic exposure to dietary antibiotics affects intestinal health and antibiotic resistance gene abundance in oriental river prawn (*Macrobrachium nipponense*), and provokes human health risk. Sci Total Environ 720:137478. https://doi.org/10.1016/j.scitotenv.2020.137478

Tacon AGJ (2020) Trends in global aquaculture and Aquafeed production: 2000–2017. Rev Fish Sci Aquac 28(1):43–56. https://doi.org/10.1080/23308249.2019.1649634

Tartor Y, Taha M, Mahboub H, El Ghamery M (2018) Yeast species associated with diseased fish: occurrence, identification, experimental challenges and antifungal susceptibility testing. Aquaculture 488:134–144. https://doi.org/10.1016/j.aquaculture.2018.01.033

Tian R, Ning D, He Z, Zhang P, Spencer SJ, Gao S, Shi W, Wu L, Zhang Y, Yang Y et al (2020) Small and mighty: adaptation of Superphylum patescibacteria to groundwater environment drives their genome simplicity. Microbiome 8(1):51. https://doi.org/10.1186/s40168-020-00825-w

Venema K, Haverkort RE, Abee T, Haandrikman AJ, Leenhouts KJ, de Leij L, Venema G, Kok J (1994) Mode of action of LciA, the Lactococcin A immunity protein. Mol Microbiol 14(3):521–532. https://doi.org/10.1111/j.1365-2958.1994.tb02186.x

Waite DW, Chuvochina M, Pelikan C, Parks DH, Yilmaz P, Wagner M, Loy A, Naganuma T, Nakai R, Whitman WB et al (2020) Proposal to reclassify the Proteobacterial classes Deltaproteobacteria and Oligoflexia, and the phylum Thermodesulfobacteria into four phyla reflecting major functional capabilities. Int J Syst Evol Microbiol 70(11):5972–6016. https://doi.org/10.1099/ijsem.0.004213

Wang J, Zhang S, Ouyang Y, Li R (2019) Current developments of bacteriocins, screening methods and their application in aquaculture and aquatic products. Biocatal Agric Biotechnol 22:101395. https://doi.org/10.1016/j.bcab.2019.101395

Wee BA, Muloi DM, van Bunnik BAD (2020) Quantifying the transmission of antimicrobial resistance at the human and livestock interface with genomics. Clin Microbiol Infect 26(12):1612–1616. https://doi.org/10.1016/j.cmi.2020.09.019

Yeoh HI, Izzatty R, Furusawa G, Amirul A-AA, Shu-Chien AC, Sung YY (2021) The vibrio-predatory filamentous bacteria effectively removed acute hepatopancreatic necrosis disease (AHPND) causative *Vibrio parahaemolyticus* in vitro. Aquac Rep 21:100910. https://doi.org/10.1016/j.aqrep.2021.100910

Yu W, Guo J, Liu Y, Xue X, Wang X, Wei L, Ma J (2023) Potential impact of combined inhibition by bacteriocins and chemical substances of foodborne pathogenic and spoilage bacteria: a review. Foods 12(16):3128. https://doi.org/10.3390/foods12163128

Zgheib H, Drider D, Belguesmia Y (2020) Broadening and enhancing bacteriocins activities by association with bioactive substances. Int J Environ Res Public Health 17(21):7835. https://doi.org/10.3390/ijerph17217835

Zhu D, Li R, Liu F, Xu H, Li B, Yuan Y, Saris PEJ, Qiao M (2016) Mu insertion in feuD triggers the increase in Nisin immunity in *Lactococcus lactis* subsp. lactis N8. J Appl Microbiol 120(2):402–412. https://doi.org/10.1111/jam.13015

Chapter 12
Application of Probiotics, Prebiotics, and Synbiotics in Aquaculture

P. G. Preena, J. C. Anjana, and V. J. Rejish Kumar

Abstract Sustainable aquaculture production is made possible by ameliorating stress, mitigating microbial infections, and preventing water quality deterioration. Diseases are one of the major concerns in intensive aquaculture systems. Frequent application of chemotherapeutics such as antimicrobials can exacerbate antibiotic resistance and will hamper environmental sustainability. Rational aquaculture advocates alternative strategies for improving the growth and immunity of cultured organisms. This includes the application of probiotics, prebiotics, immunostimulants, vaccines, phage and photodynamic therapy, and various herbal products. Among these, probiotics are found to have multiple applications in aquaculture. The application of probiotics in aquaculture improves the health of cultured organisms and the water quality of the culture system. Probiotics administered through feed (gut probiotics) are proven to promote digestion and absorption of the host organism. Furthermore, they contribute to the increased synthesis of essential nutrients such as vitamins, fatty acids, and carbohydrates, thereby augmenting the immune system of the host. Probiotics play a role in managing aquatic pathogens by generating inhibitory compounds, competing for nutrients and energy resources, contending for mucosal surface colonization, and disrupting quorum sensing. Besides, the water quality can also be improved by the administration of probiotics directly (pond probiotics) as bioaugmentors in the culture system. Microbes such as *Bacillus* sp., *Lactobacillus* sp., *Lactococcus* sp., *Carnobacterium* sp., *Bifidobacterium* sp., *Streptomyces*, *Saccharomyces cerevisiae*, and so on are the dominant groups reported as probiotics in aquaculture. The application of prebiotics such as inulin, oligofructose, galactooligosaccharides, mannan oligosaccharides, and so on along with probiotics is known as synbiotics and improves the activity and survival of probiotic bacteria. This chapter summarizes the diversity of probiotic and prebiotic

P. G. Preena · J. C. Anjana
Faculty of Ocean Science and Technology, Kerala University of Fisheries and Ocean Studies, Panangad, Kochi, Kerala, India

V. J. Rejish Kumar (✉)
Department of Aquaculture, Kerala University of Fisheries and Ocean Studies, Panangad, Kochi, Kerala, India

P. Elumalai, S. Lakshmi (eds.), *Antimicrobial Resistance in Aquaculture and Aquatic Environments*, https://doi.org/10.1007/978-981-97-7320-6_12

compounds, their selection strategies, mode of action, method of administration, and individual and combined application of probiotics and prebiotics as synbiotics in aquaculture. The article also addresses the limitations, challenges, safety concerns, and prospects of probiotics and synbiotics for augmenting sustainable and eco-friendly aquaculture production.

Keywords Probiotics · Prebiotics · Synbiotics · Competitive exclusion · Aquaculture · *Lactobacillus*

12.1 Introduction

The major goal of the aquaculture industry is to maximize production efficiency along with economic and environmental sustainability. The intensification of production can be achieved by regulating stressful conditions and bacterial and viral infections and improving water quality. As a part of aquatic disease management, traditional strategies, viz., usage of veterinary medicines were adopted. However, the frequent application of chemical additives such as antimicrobials can lead to severe antimicrobial resistance (AMR) and a deteriorated ecosystem. This constraint affects the successful management of aquatic diseases and results in massive economic losses. Hence, in sustainable aquaculture production, alternative curative measures were practiced to improve the growth and immunity of the aquatic organisms. These strategies include various immune-stimulants, vaccines, growth promoters, herbal products, probiotics, prebiotics, phage therapy, photodynamic therapy, and a combination of different immune-modulators. Among these, probiotics are found to be a promising alternative with their multiple applications in aquaculture to improve the health of the cultured species. The application of probiotics, either as live or dead forms, in combination with prebiotics as synbiotics seems to offer better performance in cultured systems. This chapter emphasizes different kinds of organisms from various sources having probiotic potential, their selection criteria, diverse prebiotic compounds, synbiotics, and their mode of action of probiotics individually and in combination with prebiotics, their mode of delivery and application in aquaculture. A flowchart showing the description and beneficial action of probiotics, prebiotics, and synbiotics employed in aquaculture is given in Fig. 12.1.

12.2 Probiotics

In aquaculture, bacterial diseases pose a significant threat, leading to substantial mortality among cultured species. The utilization of probiotics in aquaculture has garnered increased attention recently, thanks to their varied capabilities and minimal adverse effects in combating microbial pathogens. The term "probiotic" is

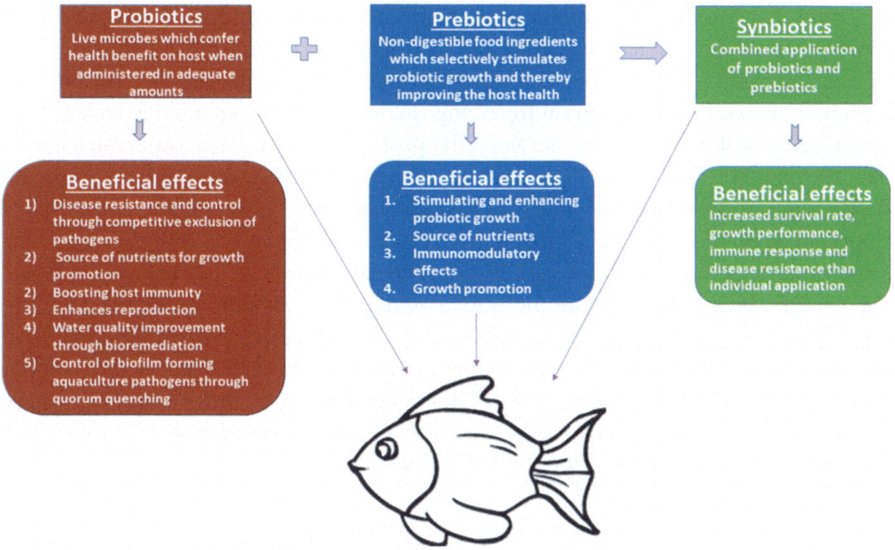

Fig. 12.1 A flowchart showing the description and beneficial effects of probiotics, prebiotics, and synbiotics used in aquaculture

derived from the Latin word *pro* and the Greek word *bios* indicating "for life" (Schrezenmeir and De Vrese 2001). According to FAO/WHO (2001), probiotics could be defined as the live microbes that boost up the overall health of the host when supplied in adequate amounts and hence can be used in controlling diseases. A live, dead, or microbial cell component can be administered as a probiotic through either feed or rearing water, thereby improving the microbial balance, health status, disease resistance, feed intake, growth performance, and reduced stress response of aquatic animals (Hai 2015). It is reported that diverse Gram-positive and Gram-negative bacteria, yeasts, and algae were found to possess probiotic features and were extensively used in aquaculture. The microbial groups belonging to lactic acid bacteria (LAB), Bacillaceae, Vibrionaceae, pseudomonads, and yeasts are the dominant groups proposed to be used as probiotics.

Probiotic microbes are well known to promote absorption, digestion by producing enzymes such as proteases, amylases, and so on, and inhibition of bacterial pathogens through competitive binding with the epithelial cells of the gastrointestinal tract (GIT) (Akther et al. 2015). In addition, they also enhance the production of nutrients such as vitamins, fatty acids, carbohydrates, and so on, thereby boosting the immunity of the host. They were reported to possess better performance in aquaculture acclimatized to 0–30 ppt salinity (Rahiman et al. 2010) as well as bringing out their wide application. The application of probiotics in aquaculture rearing salmonids, shrimps, and so on was found to be successful in combating fish pathogens for the last few decades. Unlike terrestrial living beings, the intestinal microbiota of aquatic species do not simply exist as entities, however, are in constant interaction with the aquatic environment and the host (Holzapfel et al. 1998). The microbes present in the ambient environment are continuously ingested along with the feed or drinking water and can also colonize the axenic egg surface. This indicates that the bacteria inhabiting the surrounding environment have a great impact on the gut microbial composition, thereby influencing the health of the aquatic organism. It is also stated that the interactions between the probiotics and host are not restricted to the intestinal tract but are also active in the gills, skin, and surrounding environment. Hence, the cultured environment, not necessarily from feed, plays a vital role in supplying probiotics to the aquatic host (Fuller 1989). Hence, the definition of "Probiotic" was later modified as "live microbial adjunct which has a beneficial effect on the host by modifying the host-associated or ambient microbial community, by ensuring improved use of the feed or enhancing its nutritional value, by enhancing the host response toward disease, or by improving the quality of its ambient environment" (Verschuere et al. 2000). This illustrates the additional role of probiotics as biocontrol agents. As Wang et al. (2008) suggested, probiotics are often referred to as "friendly" or "healthy" bacteria.

12.2.1 Strategies for the Selection of Probiotic

Numerous factors need to be considered for the effective application of probiotics. Appropriate selection of bacterial strains and administration in adequate doses is crucial. The strains recovered from any source such as microbial culture collections, cultured environment, rearing system, uncontaminated seawater, gut of healthy fishes, and so on (Gram and Hjelm 2002; Planas et al. 2006; Abidi 2003; Ringo et al. 2007) can be explored for probiotic development. The probiotic strains isolated from various natural sources like internal organs of the host, viz., gut, kidney, gills, gonads, and so on are well known as putative probiotics in contrast to non-putative which are commercially developed from other sources (Chauhan and Singh 2019). The strains that can be used as probiotic feed supplements and biocontrol agents should satisfy some important criteria. Several steps have to be taken to ensure the safety and performance of the probiotics. The general rules to be followed for the appropriate selection of probiotics are discussed below (Gram et al. 1999; Merrifield et al. 2010; Kumar et al. 2016; Chauhan and Singh 2019).

1. Preliminarily, a promising probiotic candidate should be non-pathogenic and harmless to the host and humans.
2. They should be devoid of virulence genes and plasmid-encoded AMR genes, thereby diminishing the pathogenicity and rapid spread of AMR through horizontal gene transfer.
3. They should also have the capacity to maintain the hereditary traits and possess beneficial growth characteristics like short generation time and lag phase.
4. It is highly encouraged to have the capability to withstand low pH and bile salts, resistance to gastric juice, and the ability to colonize the intestinal epithelial surface and adhere to intestinal mucus.

5. Those that have antagonistic activity against aquaculture pathogens such as *Aeromonas salmonicida*, *Vibrio anguillarum*, and so on are preferred. In vitro antagonism studies are encouraged wherein various aquaculture pathogens will be exposed to probiotic candidates; however, a positive in-vivo effect is not predictable.
6. It is highly preferable to select the probiotic strains isolated from the host itself to ensure a high survival rate in the GIT of the host.
7. An ideal probiotic should also have the capability to produce significant digestive enzymes and vitamins, thereby enhancing the nutritional benefits and immune response.
8. They should be viable under normal storage conditions (in feed) and freeze-drying (industrial use).

Bacteria isolated from various sources can be subjected to multiple screening steps to measure their potential as efficient probiotics. The procedures include biochemical indexing, in vitro analysis of antagonistic potential against aquatic pathogens, tolerance study against acid, bile, and gastric juice, adhesion assays, drug susceptibility, and biofilm formation. This is followed by the identification of bacteria through phenotypic and genotypic analysis. The strains that are selected as probiotics are advisable to confirm with in-vivo studies for measuring antagonism and pathogenicity before the field application. "Gnotobiotics" (bacteria-free) are significant as model organisms to determine the exact action of pathogen/probiont because normal flora are absent from the gut to mask the test results (Gomez-Gil et al. 2011). Those candidates that have satisfied most of the listed criteria can be successfully applied in aquaculture.

12.2.2 Promising Probiotic Candidates in Aquaculture

During the earlier period, the concept of probiotic, the intestinal microbiota, was mainly limited to Gram-positive bacteria especially of the genera *Lactobacillus*, *Bifidobacterium*, and *Streptococcus* (Verschuere et al. 2000). Meanwhile, due to the frequent environmental water flow through the digestive tract, major microbes inhabiting the intestine are transient. There are several probiotic bacteria along with other microbes identified in the GI tract of aquatic animals even though several potentially pathogenic microbes, viz., *Salmonella*, *E. coli*, and so on were also reported. These probiotic groups include Gram-positive bacteria such as *Lactobacillus*, *Bacillus*, *Bifidobacterium*, *Streptomyces*, *Carnobacterium*, *Streptococcus*, *Enterococcus*, *Clostridium*, *Lactococcus*, *Micrococcus*, *Microbacterium*, *Arthrobacter*, *Paenibacillus*, and so on and Gram-negative bacteria such as *Pseudomonas*, *Vibrio*, *Aeromonas*, *Roseobacter*, *Alteromonas*, *Pseudoalteromonas*, *Rhodosporidium*, *Roseobacter*, *Shewanella*, and so on along with the yeast *Saccharomyces*, fungi *Aspergillus*, and algae *Spirulina* (Martınez Cruz et al. 2012; Ringø 2020).

12.2.2.1 Gram-Positive Bacteria

LAB belonging to the order Lactobacillales are one of the most popular probiotics used in the aquaculture industry. Several species of LAB have been explored because they represent the major microflora of the fish intestines and hence are more capable of binding the epithelial cell surface by attaching to intestinal mucus (Ringo and Gatesoupe 1998). *L. bugaricus*, *L. acidophilus*, *L. plantarum*, *L. casei*, *L. paracasei*, *L. gasseri*, *L. fermentum*, *L. salivarius*, *L. reuteri*, *L. rhamnosus*, *L. helveticus*, *L. johnsonii*, and so on are the predominant LAB groups used in aquaculture (Nwanna 2015). Around 75% of the microbial taxa found in the gut of healthy fishes are Firmicutes group, which comprise LAB and Bacillales. LAB were found to possess multiple functions such as inhibition of colonization of pathogens to intestinal mucus, antagonistic effect through the production of acetic acid and lactic acid, boosting the immunity, and so on (Gomez-Gil et al. 2011). *L. plantarum*, later known as *Carnobacterium divergens*, was found to protect turbot against pathogenic *Vibrio splendidus* and *V. pelagius* (Ringo 1999). *Carnobacterium* isolated from salmon showed antagonism against *Aeromonas salmonicida*, *A. hydrophila*, *Phtobacterium damselae*, *Vibrio* sp., and *Flavobacterium psychrophilum* when administered to Atlantic salmon and rainbow trout (Robertson et al. 2000). Different species of LAB such as *L. acidophilus*, *L. lactis*, *L. plantarum*, and *L. sporogenes* were proved to show protective effects against vibriosis in *Carassius auratus* (Sun et al. 2018). An overall enhanced growth performance, nutrient utilization, disease resistance, and fish health were the beneficial effects shown by LAB in different cultured fishes. Bacillus groups are also the most dominant groups in exerting probiontic potential by increased survival rate and reduced disease susceptibility (James et al. 2023). *Bacillus subtilis*, *B. licheniformis*, *B. coagulans*, *B. pumilus*, *B. vireti*, and so on were found to exhibit improved digestion and immune response and increased disease resistance against fish pathogens in various fishes such as tilapia, shrimp, prawn, goldfish, koi carp, and so on (Rohani et al. 2022). The mixture of *B. subtilis* and *B.megaterium* increased the yield of *Litopenaeus vannamei* and improved water quality (Olmos et al. 2011). *Bifidobacterium* is also another widely used probiotic strain in aquaculture with *Bifidobacterium bifidum*, *B. lactis*, *B. breve*, *B. longhum*, and *B. infantis* as the major species (Nwanna 2015). As Lactobacilli and Bifidobacteria are the normal inhabitants of fish intestine, they lack virulence and pathogenicity and hence most preferably selected as probiotics (Martınez Cruz et al. 2012). *B. bifidum*, *B. longhum*, and *B. thermophilum* together with *Bacillus* and *Lactobcaillus* increase the overall immunity in tilapia and sturgeon fish (Rohani et al. 2022). The combination of *Bifidobacterium*, *Bacillus*, and *Lactobacillus* was also exploited in recirculating aquaculture systems (RASs) for promoting water quality and feed efficiency (Zibiene and Zibas 2019).

Actinobacteria groups are also effective in utilizing as probiotics in aquaculture, among which *Streptomyces* play a vital role. They were found to enhance disease control and overall immunity by acting as gut probiotics and bioaugmentor in aquaculture by producing bacteriocin and antimicrobial compounds, competitive exclusion, and quorum quenching (QQ) activity (James et al. 2023). *Streptococcus phocae* is another important strain having probiotic potential demonstrated in shrimps by increased phagocytic activities and protection against *Vibrio parahaemolyticus* (Pattukumar et al. 2013). *Streptococcus* species together with *Lactobacillus*, *Bifidobacterium*, and fungus have shown improved growth and survival rates in *Litopenaeus vannamei* (Javadi and Khatibi 2017). *Enterococcus faecalis* is also reported to combat vibriosis while administered as a probiotic in rainbow trout (Rodriguez-Estrada et al. 2009). Improvement in the reproductive performance was noted in swordtail with the administration of co-mixture probiotics of *Enterococcus*, *Bifidobacterium*, and *Lactobacillus* (Abasali and Mohamad 2011). *Lactococcus lactis* was also utilized as a feed probiotic (as a vaccine) and pond probiotic (as a biocontrol) for improving the immunity in fishes (Beck et al. 2017). In addition, *Micrococcus luteus*, *Microbacterium*, *Brevibacterium*, and so on are the other well-known Gram-positive groups that provide nutritional sources like vitamins and improve the feed conversion ratio (FCR) in cultured fishes like tilapia (Chauhan and Singh 2019). *Clostridium butyricum* is the only species of clostridium reported to be used as probiotics in shrimp aquaculture as it does in humans (Zhao et al. 2017).

12.2.2.2 Gram-Negative Bacteria

The most widely used Gram-negative probionts in aquaculture include vibrios, aeromonads, and pseudomonads. *Vibrio* and *Pseudomonas* normally exist as indigenous microbiota in marine fish, while aeromonads constitute the gut of freshwater fishes (Balcázar et al. 2006). Vibriosis is one of the major diseases affecting cultured fishes; however, several species of vibrios can still be employed to mitigate the pathogenic vibrios. *Vibrio Pelagius* and *V. mediterranei* were found to be effective colonizers of fish gut compared with true probionts, thereby eliminating the attachment of pathogens to the epithelial cell surface (Gomez-Gil et al. 2011). Like vibrios, aeromonads were also exploited as probiotics to fight against similar genera of pathogens such as *Aeromonas hydrophila*, *A. veronii*, *A. sobria*, and so on. For example, feed supplemented with *A. hydrophila* offered protection against *A. salmonicida* in goldfish (Kumar et al. 2016). A combination of *Pseudomonas aeruginosa* and *P. synxantha* was also found to be an enriched feed for *Penaeus latisulcatus* (Van Hai et al. 2010). The antagonistic property of *Pseudomonas fluorescens* was also proved by inhibitory activity against major fish pathogens like *A. salmonicida* and *V. anguillarum* through competition for limited iron (Gram et al. 1999). *Roseobacter* belonging to Rhodobacterales was significant in hatcheries to protect Scallop and turbot larvae from the attack of pathogens like vibrios (Balcázar et al. 2006). *Psychrobacter* is found to be an efficient probiotic as it exhibits antagonism against *V. harveyi* and *Aeromonas* sp. in infected shrimps through extracellular antimicrobial production (Franco et al. 2017).

12.2.2.3 Other Probiotic Candidates

Besides bacteria, yeast, fungi, and algae also demonstrate various probiotic properties. Yeasts constitute a major portion of intestinal microbiota in fishes and are not affected by antibiotic treatment. Because of their diverse metabolic potential and immunostimulant properties, they help in improving the immunity and digestion of the host organism. In addition, they also increase the survival rate, feed efficiency, blood biochemistry, and growth in challenged fishes. Their capability to secrete polyamine molecules and better adhesion to intestinal mucus makes them an efficient probiont (Perez-Sanchez et al. 2013). *Saccharomyces cerevisiae* is the most popular dietary probiotic yeast in modulating the immune response of various fishes such as trout, tilapia, and so on against popular fish pathogens (Yilmaz et al. 2022). *Debaryomyces hansenii*, a halotolerant yeast proved its efficiency as a probiotic in aquaculture (Navarrete and Tovar-Ramírez 2014). Other than yeasts, fungi like *Aspergillus niger* modulate the growth performance, hematology, and immunity in ornamental fishes (Jasim et al. 2022). Among microalgae, *Tetraselmis suecica* and *Chaetoceros* were observed to diminish pathogenic bacteria while promoting the proliferation of beneficial bacteria in the GIT of white shrimps (Nagappan et al. 2021). The intestinal microbiota of tilapia was also improved by the inclusion of *Schizochytrium* sp. in the diet (Souza et al. 2020).

Even though probiotics are efficient in exerting antibacterial effects, their antiviral and antifungal activities have also gained attention. However, the mechanism of antiviral and antifungal activity is not clearly understood. In most cases, inhibition of viruses is by the action of bacterial extracellular enzymes. The antiviral effect has been displayed with the combined action of *Vibrio* sp., *Aeromonas* sp., and *Pseudomonas* sp., in salmons, against infectious hematopoietic necrosis virus (Kamei et al. 1988). Several species of vibrios like *V. alginolyticus* reduced the incidence of WSSV syndrome by modulating the immune response in shrimps (Rodriguez et al. 2007). While aantifungal activity of the probiotics is rarely reported. *Aeromonas media* was shown to give protection in eel and silver perch against *Saprolegnia parasitica* (Jahangiri and Esteban 2018). *Lactobacillus plantarum* and *Pseudomonas* had also shown antifungal activity to fight against saprolegniasis in catfish and trout, respectively (Chauhan and Singh 2019).

12.2.3 Beneficial Effects of Probiotics in Host: Mode of Action

The mode of action of probiotics is mainly associated with intestinal microbiota. As already discussed in the previous sections, probiotics stimulate the immunity and improve the growth performance in fishes by producing digestive enzymes and essential nutrients. It also exhibits an antagonistic effect against pathogens by producing inhibitory substances and competing with pathogens through competitive exclusion. Probiotics could also improve water quality by acting as bioaugmentors in the aquaculture production system.

The possible mechanism by which probiotic protects the fish is categorized as follows.

12.2.3.1 Competitive Exclusion of Pathogens

The occurrence of intestinal pathogens can be eliminated by modifying the intestinal microbiota through the ingestion of probiotics. The action through which probiotic strains fight/compete for the space (adhesion site), nutrients, energy, chemicals, and oxygen is often termed "competitive exclusion," which improves the feed utility, immune response, growth, and survival of the host. Two competitive strategies such as exploitation and interference competition have been considered to bring out competitive exclusion (Knipe et al. 2021).

Exploitation Competition

Exploitation competition involves the action in which space and nutrients are captured and utilized, thereby eliminating the survival of pathogens. This competition is indirect and restricts the growth of competitors through the immediate consumption of limited resources.

Competition for Space/Adhesion Sites

As already discussed, one of the desirable criteria in selecting efficient probiont is their successful colonization in the gut and getting adhered to the epithelial cell surface, thereby inhibiting the pathogen attachment to the intestinal mucus. The probiotic strains diminish the capability of pathogens in adherence through competition for adhesion receptors, thereby antagonizing the pathogens (Hai 2015). The electrostatic and hydrophobic interactions, steric and passive forces, and lipoteichoic acids are the responsible factors that influence adhesion (Hoseinifar et al. 2018). The attachment can be either specific because of the binding of probiotic strains on the receptors on epithelial cells/surface of adherent bacteria or non-specific due to the physiochemical agents (Lazado et al. 2015). This is the first probiotic effect because the pathogenic infection is initiated through adhesion to the cell surface (Hasan and Banerjee 2020). The colonizing capacity of probiotics is not restricted to the intestinal tract whereas active on fish skin, gills, and so on (Perez-Sanchez et al. 2013). *Flavobacterium psychrophilum* was found to be excluded from the infection site by the competitive colonization of probiotic strains on fish skin (Burbank et al. 2011). This competitive exclusion of potential pathogens exerts an additional advantage of reducing the need for antibiotic therapy in intensive aquaculture systems (Balcázar et al. 2006). The selection of indigenous microbes as probiotics has a great possibility for competition because they are already adapted to the environment (Lalloo et al. 2010). This is exemplified by the inhibition of adhesion of intestinal pathogens such as *P. aeruginosa*, *Klebsiella*, and *E. coli* by the indigenous *Lactobacillus* (Nwanna 2015). *Aeromonas hydrophila*, a normal intestinal flora, is found to cause huge mortalities in various fishes such as catfish, tilapia, carp, and so on; however, feed with *A. hydrophila* offered protection against *A. salmonicida* in goldfish (Kumar et al. 2016). This indicates that the opportunistic pathogens can also be modified to make them suitable for regulating diseases in aquaculture. The attachment is critical in biofilm formation and hence the pathogenicity of the microbes. Thus, the adhesive probiotic strains play a vital role in eliminating non-adhesive cells of social "cheaters" from the population (Schluter et al. 2015). There also exists a strong competition for biofilm formation by producing certain molecules like rhamnolipid for dissipating the pathogens (Knipe et al. 2021). The biofilm-forming property of *V. parahaemolyticus* could also inhibited by *Pseudomonas aeruginosa*, thereby affecting the structural integrity and making them more susceptible to antimicrobials (Vinoj et al. 2015).

Competition for Nutrients/Energy

The capability to compete for available nutrients and energy with others in the same ecological niche at the fastest rate determines the survival rate of the microbial population. A good probiotic candidate can outcompete the enemies by consuming all the existing nutrients that would have been otherwise used up by the pathogens. Thus, by emptying the available nutrients in the intestinal tract, the survival of pathogens is negatively affected. Siderophore (iron-chelating agent)-producing bacteria are usually employed as probiotics because of their potency to sequester ferric iron under iron-limited conditions and deprive of iron to the pathogens (Gram et al. 1999). In spite of siderophores acting as public goods, they are usually regarded as competitive phenotypes since "private" siderophore-producing probiotic strains have greater iron-binding affinity than lower-affinity siderophore-forming pathogens (Joshi et al. 2006). In an experiment demonstrated by Brunt and Austin (2005), it was shown that antibiosis via the formation of inhibitory substances was not the reason behind the inactivation of pathogenic *Streptococcus iniae* and *Lactobacillus garviae* by *Bacillus* spp. The siderophores produced by the *Bacillus* spp. could outcompete the pathogens by making the iron unavailable to them under restricted iron resources (Wuertz et al. 2021).

Interference Competition

Production of Inhibitory Compounds

Interference competition explains the production of inhibitory compounds, mainly antimicrobials that encompass strain-specific bacteriocins to wide-spectrum antibiotics (Knipe et al. 2021). Antagonism is a natural phenomenon where microbial interactions play a key role in maintaining equilibrium between healthy and pathogenic microbial populations (Pandiyan et al. 2013). The inhibitory substances exert negative impacts on pathogens through bactericidal or bacteriostatic activity.

Diverse compounds, viz., organic acids, enzymes like lysozymes and proteases, volatile fatty acids like acetic acid, lactic acid, butyric acid, and propionic acid, hydrogen peroxide, siderophores, and so on other than antibiotics and bacteriocins were found to possess antimicrobial action (Chauhan and Singh 2019). Their production in the host intestine provides a barrier against the propagation of pathogens and induces pH alteration due to organic acid production. Rames et al. (2015) demonstrated the antimicrobial action of *Bacillus licheniformis* and *B. pumilus* and could withstand low pH and high concentrations of bile. Probiotics such as *Bacillus cereus*, *Paenibacillus polymyxa*, and *Pseudomonas* sp. exhibited antagonism against pathogenic vibrios in shrimps (Kumar et al. 2016). Probiotic strains recovered from the gut of ornamental fishes were shown inhibitory action against pathogens like *V. alginolyticus* and *A. hydrophila* through the production of antimicrobial metabolites (Martınez Cruz et al. 2012). Besides antagonistic activity, they were found to colonize the GI tract and enhance the growth of healthy microflora in the gut. *Lactobacillus* spp. is one of the most significant probiotic strains known for producing bactericidal proteins, organic acids, hydrogen peroxide, and diacetyl, which could have inhibited aquatic pathogens and improved the immune response against the infectious agents (Hai 2015). The tropodithietic acid produced by the probiotic *Phaeobacter gallaeciensis* could effectively combat several fish pathogens (D'Alvise et al. 2012). The siderophores recovered from the probiotic strain *Pseudomonas fluorescens* were found to inhibit several Vibrio species (Hasan and Banerjee 2020).

Bacteriocins have gained more attention nowadays as antimicrobial compounds. They are the widely accepted strategy to control antibiotic resistance in aquaculture and to improve the health of aquatic species. These heterogeneous groups of antimicrobial peptides are being synthesized by diverse species and possess a wide spectrum of action (Pereira et al. 2022). Bacteriocins are usually secreted by probiotics at the site of action itself. Endogenous LAB, viz., *Lactobacillus*, *Lactococcus*, *Carnobacterium*, and *Enterococcus* are well known for producing bacteriocins to fight against pathogens (Perez-Sanchez et al. 2013). Divercins and piscicocins are the two significant class II bacteriocins produced by *Carnobacterium* from fish intestines (Metivier et al. 1998). The bacteriocin produced by the probiotic strain *Bacillus subtilis* showed antimicrobial activity against multiple pathogens such as *A. salmonicida*, *A. hydrophila*, *P. flourescens*, and *B. mycoides* (Banerjee et al. 2017). The probiotic strain, *Bacillus velezensis*, could also produce lytic enzymes like ß-1,3-glucanase in addition to four bacteriocins (Kuebutornye et al. 2020) for killing enemies.

The energy investment involved in exploitation and interference competition strategies varies. The energy expenditure for antimicrobial production leads to slower growth and a low reproduction rate (Little et al. 2008). While in exploitation competition, the energy is mainly utilized for growth, and hence, they are rapid growers. This could be made clear with the help of an experiment with two co-existing marine actinomycetes where *Salinispora arenicola* consumed the fuel in interference competition through the production of antibiotics and *Salinispora tropica* in exploitation competition for growth promotion (Knipe et al. 2021). Species diversity and biodiversity are mainly improved by interference competition (Shank et al. 2011).

12.2.3.2 Source of Nutrients for Growth Promotion and Enzymes for Better Digestion and Feed Utilization

The growth rate of aquatic animals is significantly influenced by the action of probiotics through the supply of essential nutrients or supporting nutrient uptake and effective digestive enzyme processes. In several reports, enhancement of growth was observed in the fishes fed with probiotics. It is highly essential to understand whether the taste of probiotics is desirable to aquatic animals and is not being clearly studied so far. The nutrient absorption was stimulated by probiotics by elevating the surface area of GIT in the host (Hasan and Banerjee 2020). Probiotic strains can colonize the GI tract very rapidly and permanently because of their fast multiplication efficacy compared with their lower expulsion rate (Rohani et al. 2022). Increased appetite and weight gain rate, FCR, specific growth rate, protein efficiency ratio, protein retention efficiency, and lipid retention efficiency are the prominent features encountered in case studies dealing with probiotic diet (Goh et al. 2022). These features are the results of the action of probiotics in improvising the gut flora leading to the secretion of necessary nutrients through improved digestive enzyme actions (Rohani et al. 2022). Nourishment of vitamins and fatty acids in the fish was observed through the application of Bacteroides and *Clostridium* in aquaculture (Kumar et al. 2016). LAB are one of the main sources of nutrients mainly vitamins (Wuertz et al. 2021). As suggested by Ringø (2020), *Pseudomonas* sp., *Agrobacterium* sp., *Microbacterium* spp., *Brevibacterium* sp., and *Staphylococcus* sp. are some of the important strains in contributing to nutrient enrichment in fishes. Proteins and lipids were significantly increased with an evident weight gain in commercially important Nile tilapia supplemented with the probiotics *Streptococcus faecium*, *Saccharomyces cerevisiae*, and *Lactobacillus acidophilus* (Lara et al. 2003). The biogenic metabolites such as vitamins, amino acids, short-chain fatty acids (SCFAs), and polyamines provided by *Bacillus* sp. could act as growth factors and stimulate appetite and thus amend feeding (Goh et al. 2022). The nutrient-enriched photosynthetic machinery in phytoplankton like microalgae is utilized for probiotic development as it supplies abundant polyunsaturated fatty acids and vitamins that are not synthesized by higher animals themselves (Kumar et al. 2016). A diatom, *Chaetoceros muelleri* together with *V. alginolyticus*, reached a sufficient density and provided good results when fed to shrimps (Gomez et al. 2000). The addition of LAB such as *Lactobacillus lactis*, *Lactococcus casei*, and *Pediococcus acidilactici* to the rotifer *Brachionus plicatilis* promoted the growth rate to a large extend in shrimps (Planas and Vazquez 2004).

Probiotics also participate in enhancing digestion by modifying the intestinal microbiota and improving the enzymatic activity which results in better food utilization. The beneficial effect on the digestive processes could be due to the secretion of extracellular enzymes such as amylases, proteases, lipases, chitinases, phytases, cellulases, trypsin, lysozyme, and so on (Wang et al. 2019). The enzymatic activity not only promotes feed conversion but also neutralizes the toxic effects of pathogenic metabolites (Rohani et al. 2022). The probiotic yeast, *Debaryomyces hansenii*, was found to secrete trypsin and amylase which aids digestion and synthesizes polyamines, viz., spermine and spermidine that supported the differentiation and maturation of the GI tract in aquatic animals (Tovar et al. 2002). The apparent digestion of crude protein, dry matter, and phosphorous was made possible by *Bacillus* sp. in white shrimps (Martınez Cruz et al. 2012). Through enhancing the digestion of certain compounds, the lipid levels in the blood get controlled and also mitigate the issues due to intolerance to several compounds (Ringø et al. 2018). The feed supplemented with whole and treated *Saccharomyces cerevisiae* to rainbow trout improves the growth rate due to their proper nutrient digestibility (Perez-Sanchez et al. 2013). Plant extracts are supplied in large amounts in modern aquafeed, and hence, cellulose, chitin, and non-starch hydrocarbon degrading probiotics are significant because those substrates are indigestible in the host system (Wuertz et al. 2021).

12.2.3.3 Boosting Immune Response

Shrimps, fish larvae, and other invertebrates rely mainly on non-specific immune responses. Modulation of the non-specific immune system is one of the most beneficial effects of probiotics; meanwhile, the immune modulatory function is influenced by the type, dose, source, and duration of supplementation (Liu et al. 2012). Probiotics generally stimulate innate immune functions by reducing the release of inflammatory mediators and increasing the secretion of anti-inflammatory mediators (Hancz 2022). Several commercial probiotic strains were found to boost the humoral and cellular components of the innate immune system in both fish and shellfish. Most of the immunostimulants of probiotics are derived from their cell wall like lipopolysaccharides, glucans, and muramyl dipeptide (Perez-Sanchez et al. 2013). Probiotics enhance the immune response by elevating the number of lymphocytes, leukocytes, granulocytes, erythrocytes, monocytes, macrophages, natural killer activity, neutrophil adherence and migration, plasma bactericidal activity, cytotoxicity, complement activity, systemic antibody production, phagocytic and superoxide dismutases, globulin and albumin levels, and respiratory burst activities (Hai 2015). The immune response in the host is activated by the production of interleukins, cytokines, chemokines, and other effector molecules through a signaling cascade induced by the recognition of probiotic strains (Hoseinifar et al. 2018). The structural component in the cell wall of yeast probiotic, β-glucan, comprises receptors specific for phagocytic cells, and their binding releases signaling molecules responsible for the release of new WBCs, which contributes to improved immune status and disease resistance in fishes (Rohani et al. 2022). The TLR2 signaling system is the one that participated in the recognition of *Psychrobacter* and thereby induced an immune response in grouper fish (Sun et al. 2014). The potent immunostimulant, SCFA produced by probiotics could stimulate the overall immunity in fish (Montalban-Arques et al. 2015).

A significant increase in acidophilic granulocytes, intraepithelial lymphocytes, and serum lysozyme and bactericidal activity was shown in Nile tilapia through the heterogeneous expression of selected probiotics (Kumar et al. 2016). Mixed cultures of probiotics, viz., *Lactobacillus acidophilus*, *Bacillus subtilis*, *Clostridium butyricum*, and *Saccharomyces cerevisiae* improved the nonspecific immune response via migration of neutrophils, lysozyme activity, and plasma bactericidal activity in Nile Tilapia thereby combating *E. tarda* infections (Martınez Cruz et al. 2012). The probiotic strain *Clostridium butyricum* supported rainbow trout to resist vibriosis through phagocytosis of leukocytes (Das et al. 2017). The application of LAB is also significant; for example, *Lactobacillus rhamnosus* was found to stimulate the respiratory burst activity in rainbow trout (Balcázar et al. 2006). Immunomodulation also helps the aquatic species attain stress tolerance. The plasma cortisol level was found to decrease after evoked stresses in probiotic-treated fishes like Nile tilapia and sea bream with *Saccharomyces cerevisiae* and *Lactobacillus* respectively (Wuertz et al. 2021). The low cortisol level prevents the immunosuppressive effect on antibody production and thus improves stress resistance and health status. An enhanced hepatic tolerance was also detected in probiotic-treated Zebrafish due to the decreased oxidative stress levels (Gioacchini et al. 2014).

12.2.3.4 Effects on Reproduction

Numerous studies have documented the positive action of probiotics on the reproductive performance of fish. The demand for high nutrients in breeding cultured species could be satisfied by the application of probiotics. Large amounts of proteins, fatty acids, vitamins, and carotenoids are contributed by probionts for enhancing the reproductive performance in broodstock fishes (Martınez Cruz et al. 2012). The egg production, fertilization rate, and embryonic development are found to be greatly enhanced in probiotic-supplemented fishes as compared with the untreated ones (Gioacchini et al. 2014). Proteins and fatty acids form the essential part of the yolk and play a vital role in oocyte formation and maturation, maintaining reproductive physiology and spawning and post-spawning activities and increasing the vitellogenesis rate (Rohani et al. 2022). Probiotic strains such as *Lactobacillus acidophilus*, *Lactobacillus casei*, and *B. subtilis* were found to improve reproductive performance in fishes (Hasan and Banerjee 2020). A profound increase in gonadosomatic index, fecundity, viability, and production of fry was observed in ornamental fishes supplemented with *B. subtilis* (Ghosh et al. 2007). The administration of lactobacillus to fertilized zebrafish eggs could enhance larval performance in growth and survival (Padenia 2021).

12.2.3.5 Water Quality Improvement

In addition to the beneficial effects within the host, probiotics can also improve the water quality by converting the complex substances in cultured systems to simpler forms. Gram-positive bacteria like *Bacillus* sp. are more efficient in converting organic matter to carbon dioxide in contrast to Gram-negative bacteria, where the organic components get assimilated to biomass (Rohani et al. 2022). It is good to introduce probiotics in sufficient amounts in production ponds to eliminate the deposition of particulate and dissolved organic carbon generated during the culture. These "water probiotics" could employ diverse mechanisms like bioremediation and biocontrol for the removal of toxic organic wastes and to control infectious agents. Well-known bioremediators such as *Bacillus*, *Acinetobacter*, *Pseudomonas*, *Rhodopseudomonas*, *Cellulomonas*, *Nitrobacter*, *Nitrosomonas*, and so on can be exploited as probiotics by direct application in aquaculture systems to ameliorate water quality (Kumar et al. 2016). Nitrogenous compounds like ammonia and sulfur-containing wastes like hydrogen sulfide persist in aquaculture through fish excretion and unused feed gets converted to non-toxic forms using nitrifying and sulfur-oxidizing bacteria, respectively. A commercial probiotic product comprising the mixture of *Nitrosomonas* sp., *Nitrobacter* sp., *Saccharomyces cereviae*, and *Bacillus* sp. was found to have positive impacts on beneficial microbiota of shrimps and considerably reduced the concentration of inorganic nitrogen and phosphates related to untreated control groups (Jiqiu et al. 2006). Minimal accumulation of organic carbon could minimize the overgrowth of phytoplankton. In addition, several probiotic strains possess algicidal activities on microalgae for healthy aquaculture production (Balcázar et al. 2006). A combination of two different commercial probiotics comprising *Bacillus*, *Pediococcus*, and *Enterococcus* was found to effectively degrade ammonia and reduce total *Vibrio* count in a recent study demonstrated by Hassan et al. (2022). The routine application of probiotics in cultured systems could improve the pH, temperature, dissolved oxygen, and ammonia and maintain a healthy environment for shrimp cultivation (Banerjee et al. 2010). Lactobacilli bacteria were found to influence these parameters by increasing the digestion efficacy and thereby reducing the generation of waste products in shrimp cultured systems (Aguirre-Guzman et al. 2012).

The application of probiotics in two emerging systems, viz., RAS and biofloc technology (BFT) gained more attention considering their efficacy in improving water quality parameters and combating pathogens. By improving the water quality, probiotic supplementation in RAS enhances feed utilization, stress control, immune response, and growth rate of cultured fishes (Zibiene and Zibas 2019). The treatment of penaeid culture with *B. pumilus* improved water quality and helped to protect against the marine pathogen, *Vibrio* (Banerjee et al. 2010). The probiotic application in BFT also brings out the abundance of beneficial bacteria in both the cultured environment and fish gut, thereby inhibiting the pathogens, improving the water quality, and enhancing the immunity, growth, and survival of fishes (Rohani et al. 2022). The introduction of probiotic *B. subtilis* to the biofloc system could enhance the water quality along with the abundance of beneficial intestinal flora in

Macrobrachium rosenbergii (Qiu et al. 2023). Thus, probiotics can be successfully employed in ponds and production systems as water additives better than feed additives for reducing water pollution.

12.2.3.6 Quorum Quenching

Quorum sensing (QS) is the bacterial regulatory mechanism that governs the expression of virulence genes in a density-dependent manner by producing signal molecules known as auto-inducers, predominantly N-acyl homoserine lactones (AHLs). Thus, the disruption of QS has been taken as an important anti-pathogenic strategy by using probiotic strains as QQ bacteria. *Bacillus* sp. was identified as one of the most significant quorum quenchers on biofilm formation and virulence factors production of aquatic pathogens like *Aeromonas hydrophila* (Chu et al. 2014). Several species of *Bacillus* such as *B. cereus*, *B. thuringiensis*, *B. subtilis*, *B. licheniformis*, *B. circulans*, *B. methylotrophicus*, and so on possessed QQ activity toward Gram-negative aquatic pathogens by secreting AHL lactonases (James et al. 2020). The autoinducer antagonists produced by *Lactobacillus*, *B. cereus*, and *Bifidobacterium* could degrade the signal molecules formed by pathogens (Chauhan and Singh 2019). New QQ probiotics such as *B. thuringiensis* and *B. cereus* were demonstrated recently to control biofilm-forming *V. harveyi* and *V. alginolyticus* (vibriosis) in sea bass in addition to having suitable probiotic features such as resistance to pH and bile salts, better adherence on mucus, and production of exoenzymes (Ghanei-Motlagh et al. 2020). Marine red algae (*Delisea pulchra*) also exhibited QS antagonistic activity for protecting rainbow trout, Artemia, and Brachionus from vibriosis (Kumar et al. 2016). Thus, the AHL-degrading probiotic strains are found to be the better substitute for antibiotics in aquaculture for the biocontrol of aquatic pathogens.

12.2.4 Mode of Application

Probiotics can be administered into the feed pellets under suitable conditions or directly in to water mainly for bioaugmentation. Majority of the probiotic strains are processed in such a way as to be applied in feed additives for better performance. However, in the case of penaeid cultured systems, the most common way of application is through water to improve the water quality (Amiin et al. 2023). Here, we discuss the mode of delivery of probiotics as feed additives and water additives (pond probiotics).

12.2.4.1 Probiotics as a Feed Additive

The administration of probiotics to the feed generally stimulates the colonization of the strains on the surface of the intestinal tract. This enhances the competitive advantage over the pathogens as discussed in the previous sections. Mohapatra et al. (2013) demonstrated the use of feed probiotics comprising *L. lactis*, *B. subtilis*, and *S. cerevisiae* in *Labeo rohita* fingerlings, which showed excellent growth, nutrient retention, and digestibility. The application of *Lactobacillus rhamnosus* to the feed was noticed to enhance the fecundity of *Danio rerio* (Gioacchini et al. 2014). The probiotics were found to be applied as either a single culture or a mixed culture. Even though earlier studies were focused on the individual application, mixed strains are more efficient in defending against multiple pathogens in the host (Kesarcodi-Watson et al. 2012). A combination of *L. plantarum* and *L. lactis* was found to be more successful in modulating immunity against *S. iniae* in Japanese flounder than sole culture (Beck et al. 2015). It is also essential to maintain the viability of probiotics in feed pellets and the number of applications should be increased if it is least viable on frequent checking. Hence, most probably the probiotic cultures are applied in the form of live spray or freeze-dried. Viable probiotic-fed species have shown better immunity and disease resistance in most cases as compared with heat-killed strains (Hai 2015). The viable culture of *B. pumilus* and *B. licheniformis* stimulated the respiratory burst and lysozyme activity in rohu than the dead cells (Rames et al. 2015). However, the inactivated probiotic strains were also observed to induce the immune system and suggested as a better alternative to live probiotics where there arises a safety issue (Salinas et al. 2006). In addition to the live intact cells, the subcellular components and extracellular products also possessed the same positive impacts in a host as live cells (Brunt and Austin 2005). The probiotics can also be supplied in spore form to withstand hostile conditions like reduced moisture content in dry feed and low pH in the stomach (Wuertz et al. 2021). The endospore-forming lactic acid-producing bacteria belonging to the Bacillaceae family like *B. subtilis* and *B. coagulans* are utilized for probiotic application in aquaculture (Kumar et al. 2016).

Microencapsulation is another way of administering probiotics to the feed, thereby protecting the microbes from low pH and digestive enzymes to deliver the strains safely to the intestine. Emulsion, starch adhesion, spray drying, and extrusion are the methods employed where concentrated microbial cells are encapsulated using alginate, carboxymethylcellulose, and chitosan (Rokka and Rantamäki 2010). The probiotics can be given as encapsulations by enriching with live feed such as Artemia, rotifer, microalgae, and copepods. This has the selective advantage of keeping the probiotic viable by proliferation within the live feed (Hai 2015). The encapsulated form of *L. planatrum* and *B. subtilis* to *Artemia nauplii* possessed good activity when challenged with vibrios (Esteban 2012).

The probiotic composition in the correct dose is also an inevitable factor to ensure positive results. An elevated dose does not assure additional protection in the host and may bring out the suppression of the non-specific immune system in the host. A probiotic feed containing *B. subtilis* at a concentration of 0.1% could promote the growth rate, better feed conversion, and immune responses at the same efficacy as 1% in fishes (Kumar et al. 2016). The probiotic density is found to be appropriate at 10^5 CFU mL^{-1} and better cellular immune responses were observed at 10^7 CFU mL^{-1}. However, probiotic cells of *L. brevis* at the rate of 10^9 CFU mL^{-1} were needed to protect hybrid tilapia from *A. hydrophila* (Hai 2015). Like dosage, the duration and frequency of application are also significant in obtaining accurate results. The period of probiotic administration was found to be as short as 5 days to 8 months, and the prolonged application can diminish the immune response (Chauhan and Singh 2019). Short-term benefits are only possible with probiotic feed supplementation and data on long-term effects are not reported. It is reported that probiotics should be constantly added in Atlantic cod larvae as the probiotic colonization is transient (Skjermo et al. 2015). Thus, the optimal dose and period of administration are mainly determined by the probiont, type of host, rearing conditions, health status, and objective of the applications. For better performance, probiotics together with other growth promoters like prebiotics, collectively termed synbiotics, can be exploited and are discussed in the following sections.

12.2.4.2 Probiotics as a Water Additive

The probiotics can be administered directly in the live form to culture water for enhancing fish health by improving the water quality parameters. In contrast to the feed probiotics where the action is targeted to the fish intestine, the action of pond probiotics is to keep the surrounding environment healthy. Here, the water probiotics mainly act as bioaugmentors stimulating the conversion of complex toxic products to simplest nontoxic metabolites. The water probiotics play a lead role in modifying the microbial composition of the water column and sediments and tend to outgrow the pathogens. The water probiotics can be directly incorporated into ponds/tanks as water additives, bacterial suspension, or by immersion methods (Hai 2015). In addition to the influence on water quality, direct supplementation to rearing water is advisable for all ages of fish to improve the intestinal microbiota as well. Because of the immature digestive system in fish larvae, feed probiotics via feeding and injection have limited applications. The growth promotion genes were upregulated, and stress-related proteins were downregulated in cod larvae when commercial probiotics were incorporated as such into rearing water and around 70% of the fish intestine comprised the probiotic population of *L. plantarum* even though feed probiotics were not provided (Jahangiri and Esteban 2018). Hence, the mixture of water probiotics and enriched live feed is highly recommended as the accurate way of probiotic supplementation.

12.3 Prebiotics

Gibson and Roberfroid were the first to propose the notion of prebiotics in 1995. Prebiotics are dietary compounds resistant to digestion that confer health benefits to the host by stimulating the growth and/or activity of specific bacteria in the colon. According to DeVrese and Schrezenmeir (2008), prebiotics, which are predominantly fibers and indigestible dietary components, selectively increase the growth and/or activity of certain genera of microorganisms in the colon, most notably lactobacilli and bifidobacteria. Prebiotics have been found to promote the health of hosts' gut microbiota and have the potential to increase feed utilization (Dawood and Koshio 2016). Rendering to various literature reviews, diverse prebiotic combinations successfully encourage the growth of different aquatic organisms. According to Ringo et al. (2010), the pioneer report on prebiotics in aquaculture is by Hanley et al. (1995).

12.3.1 Significance of Prebiotics

The effectiveness of prebiotics in altering the composition of the gut microbiota has been established, as they facilitate the proliferation of beneficial bacteria while concurrently inhibiting the growth of undesired (pathogenic) bacteria. In the field of aquaculture, there is growing interest in probiotics as a substitute for antibiotics. Alternatively, a more feasible approach in modulating the GIT microbiota in aquatic organisms involves the utilization of prebiotics, which creates an environment conducive to the proliferation of specific bacterial species. This strategy holds the potential to enhance fish growth efficiency while concurrently reducing the host organism's vulnerability to diseases. Prebiotics are concentrated fish-specific foods that have been linked to enhancements in growth, feed efficiency, gut shape, intestinal enzyme activity, immunological position, disease resistance, and stress responses. When prebiotics are combined with probiotics, they elicit comprehensive immunological responses. The existence of probiotics depends on the intestine's appropriate pH, which is maintained by prebiotics. They strengthen the growth of probiotics, which boosts the immune system. In addition, they promote the development and reproduction of the beneficial microflora. The probiotics consume these prebiotics as food. Selective fermentation plays a pivotal role in determining the potential prebiotic properties of dietary substrates upon their arrival in the large intestine. Prebiotics exhibit targeted effects on the microbiota, thereby conferring health benefits to the host. When prebiotics undergo fermentation by the indigenous flora at the specific site where they manifest their prebiotic effects, like other low-digestible carbohydrates, they enhance the production of intestinal gases and exert an osmotic impact on the GIT.

To be classified as a prebiotic, a substance should possess the following attributes (Gibson et al. 2004):

1. Resilience against gastric acidity and gastrointestinal (GI) absorption.
2. Susceptibility to fermentation by the intestinal microbiota.
3. Selective stimulation of the growth and activity of intestinal bacteria associated with promoting health and overall well-being refers to the targeted enhancement of specific beneficial bacteria in the gut for improving the overall health of fish.

12.3.2 Types of Prebiotics Employed in Aquaculture

The contemporary delineation of prebiotics suggests the utilization of specific agents to foster the proliferation of endogenously defensive gut microbiota. A prebiotic that falls within the broad group of carbohydrates is dietary fiber. Burkitt et al. (1972) defined dietary fiber as "the sum of polysaccharides and lignin that are not digested by the endogenous secretions of the human gastrointestinal tract." Prebiotics can be categorized into soluble types, such as inulin and oligofructose, insoluble types, such as cellulose, and mixed types, including bran. Extensive research has shown that dietary fibers, including prebiotics, undergo fermentation by the anaerobic microbiota inhabiting the large intestine. This fermentation process leads to the generation of lactic acid, short-chain fatty acids (acetate and propionate), and gases (H_2, CO_2, and CH_4), which are subsequently utilized by the host organism (Roberfroid 1993).

Inulin, fructooligosaccharides (FOS), short-chain FOS, mannan oligosaccharides (MOS), galacto oligosaccharides, xylo oligosaccharides, arabino xylooligosaccharides, iso maltooligosaccharides, and GroBiotic-A are some of the common prebiotics proven to be effective in fish. Various prebiotics used in aquaculture and their beneficial action in different fishes are listed in Table 12.1.

12.3.2.1 Inulin

The impact of inulin on the gut microbiota, especially within the colon of endothermic organisms, has been observed to yield advantageous outcomes (Possemiers et al. 2009). Inulin, a composite blend of fructose polymers, is characterized by its heterogeneity. It is discerned from FOS, a subset of inulin, based on the variations in the degree of polymerization. The inulin has prospective benefits in aquaculture for stimulating favorable gut microbiome, suppressing infections, and inflating immune response, though it is not a natural fiber. Olsen et al. (2001) were the first to perform a study using inulin in salmonids where they fed high-level (15%) dietary inulin to on-growing Arctic charr. The outcomes reported by Bakke-McKellep et al. (2007) after providing dietary inulin to Atlantic salmon are consistent with those seen in Arctic charr by influencing digestive parameters and gut microbiota (Ringo et al. 2006).

Table 12.1 Commonly used prebiotics in aquaculture

Sl. No.	Prebiotics	Fish species	Observations	References
1.	Inulin	Sturgeon	No change in RBC count, increase in total WBC count, decrease in dose-dependent alkaline phosphatase	Ahmdifar et al. (2011)
		Gilthead seabream	Increase in serum complement activity, increase in leukocyte phagocytic activity	Cerezuela et al. (2008); Cerezuela et al. (2012)
		Hybrid surubim	Increase in lactic acid bacteria, decrease in growth of *Vibrio* spp. and *Pseudomonas* spp., increase in total Ig (restricted to synbiotic treatment), no change in serum protein or lysozyme	Mourino et al. (2012)
		Nile tilapia	Increase in superoxide activity, lysozyme activity, and protection from *A. hydrophila* challenge	Ibrahem et al. (2010)
2.	β-glucan	Atlantic salmon and Rainbrow trout	Stimulates the recovery potential of immunosuppressed cells in teleost fishes	Petit and Wiegertjes (2016)
		Common carp	Immunomodulation by β-glucan through the production of SCFAs in the GIT	Hadiuzzaman et al. (2022)
		Rohu fish	Increase in lysozyme activity, oxidative burst, and phagocytosis	Misra et al. (2006)
		Atlantic cod, flounder fish, sea bream, catfish	Increase in growth and survival rate	Hadiuzzaman et al. (2022)

(continued)

Table 12.1 (continued)

Sl. No.	Prebiotics	Fish species	Observations	References
3.	Arabinoxylan oligosaccharide (AXOS)	Siberian sturgeon	Increase in phagocytic activity and respiratory burst activity	Geralylou et al. (2012); Rurangwa et al. (2008)
		African catfish	Increase in acetate, propionate, and total SCFA production	Rurangwa et al. (2008)
4.	Mannanoligosaccharide (MOS)	Atlantic salmon	Increase in protein and energy concentration in the whole body	Grisdale-Helland et al. (2008);
		Rainbow trout	Increase in absorptive surface in the posterior gut region, increase in microvilli density and microvilli length	Staykov et al. (2007); Dimitroglou et al. (2008)
		European lobster	Early survival and morphological development of early juvenile stage	Daniels et al. (2006)
		European sea bass	Increase in growth and feed conversion, decrease in lipid vacuolization	Torrecillas et al. (2007)
		Tilapia	No effect on growth performance, feed conversion, and gross gastrointestinal morphology but increase in survival rate	He et al. (2003)
		Tiger shrimp	Increase in growth, feed conversion, and survival	Genc et al. (2007)
		Chanel catfish	No effect on growth performance, hematology or immune function	Welker et al. (2007)
5.	Galactooligosaccharide (GOS)	Atlantic salmon	Increase in nitrogenous and energy losses in the non-fecal nitrogen excretion	Grisdale-Helland et al. (2008)
		Red drum	Increase in protein and organic ADC values and decrease in lipid ADC values	Burr et al. (2008)

(continued)

Table 12.1 (continued)

Sl. No.	Prebiotics	Fish species	Observations	References
6.	Fructooligosaccharide (FOS)	Atlantic salmon	Higher feed efficiency and energy retention, no effect on whole blood neutrophil oxidative radical production, and serum lysozyme activity	Grisdale-Helland et al. (2008)
		Caspian roach	Increase in serum Ig, lysozyme activity, and resistance to salinity challenge	Song et al. (2006); Soleimani et al. (2012)
		Japanese flounder	Lysozyme activity, phagocytic percentage, and index	Ye et al. (2011)
		Red swamp cray fish	Increase in expression of immune-related genes (crustin1, lysozyme, SOD, and proPO), increase in phagocytic activity and survival against *A. hydrophila*	Dong and Wang (2013)
		Sea cucumber	Increase in phagocytosis, but no significant immunological improvement observed	Zhang et al. (2010)
		Stellate sturgeon	Increase in serum lysozyme activity	Akrami et al. (2013)
		Yellow croaker	No statistically significant innate immunity improvement observed	Qinghui et al. (2011)
7.	Isomaltooligosaccharide (IMO)	Pacific white shrimp	No effect on microbial population, immune responses, and resistance	Li et al. (2009)
8.	Xylooligosaccharide (XOS)	Crucian carp	Increase in growth and no effect on survival, increased enzymatic activity	Xu et al. (2009)

12.3.2.2 Fructooligosaccharides

FOS are defined as chains of β-D-fructans, characterized by the presence of short- and medium-length fructosyl units. These units are connected through β-(2-1) glycosidic linkages and are attached to a terminal glucose unit, thus forming FOS. Mammalian digestive systems cannot hydrolyze the β(2-1) glycosidic linkages because they lack β-fructosidases. Certain bacteria that express this enzyme, including *Lactobacillus* and *Bifidobacteria*, can ferment FOS (Sghir et al. 1998; Manning and Gibson 2004). Consequently, the incorporation of FOS in the diet will specifically promote the growth and viability of these bacterial species within the

GIT of fishes. Certain fishes, including Atlantic salmon, hybrid tilapia, and turbot larvae, were demonstrated to grow faster when FOS is added to their diets. The supplementation of short-chain fructooligosaccharides has been proven to have positive effects on nutritional utilization, growth promotion, and reduced susceptibility to diseases in diverse aquatic species, owing to the modulation of GI microbiota (Respondek et al. 2008).

12.3.2.3 Mannanoligosacharides (MOS)

MOS are glucomannoprotein complexes formed from the cell wall of yeast *Saccharomyces cerevisiae*, as stated by Sohn et al. (2000). MOS may be advantageous as feed additions for fish and shellfish because mannose-containing compounds produce intracellular signaling that may boost the production of proinflammatory cytokines.

12.3.2.4 Galactooligosaccharides

Galactooligosaccharides (GOS), a type of prebiotic, are composed of 2–20 units of galactose and glucose. In a study conducted by Burr et al. (2008), it was observed that red drum fish fed a basal diet exhibited significantly higher apparent digestibility coefficients (ADC) for protein and organic matter, while displaying significantly lower ADC values for lipids. In contrast to these findings, Grisdale-Helland et al. (2008) found that GOS had no impact on the Atlantic salmon's ability to digest its food, as well as its ability to consume it or grow.

12.3.2.5 Arabinoxylooligosacharides

The primary non-starch polysaccharides contained in many cereal grains are called arabinoxylans (AXE), and they are a type of dietary fiber (Swennen et al. 2006). They are composed of arabinofuranosyl moieties bound to D-xylopyranosyl residues with β-(1, 4) links. The effects of arabinoxylooligosacharides are not well studied in aquatic organisms.

12.3.2.6 Prebiotic Mixtures

The commercial product GroBiotic-A is a blend of dry fermentation products and dairy ingredient components which is partially autolyzed brewer's yeast. Insoluble β-glucans are one of the several polysaccharides that make up the yeast membrane. It is well known that both yeast and β-glucans from other sources can cause fish to develop immune reactions (Dalmo and Bøgwald 2008). Inulin, FOS, and GOS in *Litopenaeus vannamei* post larvae (Oktaviana and Yuhana 2014) and β-glucan, inulin, and MOS in Chinese mitten crab (Lu et al. 2019) demonstrated beneficial effects to the host. Prebiotic blends that are commercially available, such as Immunogen (prebiotic combo of β-glucan and MOS), have been shown to promote growth in common carp (Ebrahimi et al. 2012). The prebiotic β-glucan has been found to work well with a variety of other prebiotics to promote the growth of aquatic animals. The combination of MOS and β-glucan has demonstrated the potential to significantly enhance the growth performance of various aquatic species, including Nile tilapia, snakehead, Caspian trout, Chinese mitten crab, and Caspian trout (Selim and Reda 2015; Ismail et al. 2019).

12.3.3 The Impact of Incorporating Prebiotics into Aquaculture Diets

Prebiotics play a crucial role as dietary additives in aquaculture, exerting multifaceted effects on growth, microbial dynamics within the GIT, immune enhancement, and stress resistance improvement.

12.3.3.1 Immunomodulatory Effects

The immune systems can be strengthened by prebiotics, which are complex carbohydrates that can serve as an energy source for gut bacteria (Mohammadi et al. 2022). Immunosaccharides are any prebiotics that can increase an organism's immunological response (Nawaz et al. 2018). These immunosaccharides can be used in aquaculture as an alternative to antibiotics in the control of aquatic animal health (Vazirzadeh et al. 2020). Thus prebiotics can boost an aquatic animal's immune response in one of two ways: by encouraging the growth of its gut microbiota or by stimulating its innate immune system (Song et al. 2014). Prebiotics enhance non-specific immune responses, and according to Bailey et al. (1991), prebiotics can modify microbes of the GI tract by increasing immune responses. Utilizing immunostimulants can boost non-specific defense mechanisms, leading to heightened resistance to infectious diseases. This enhancement occurs by reinforcing both innate humoral and cellular defense mechanisms, ultimately promoting the increased growth of aquatic fishes. The immunomodulatory characteristics of prebiotics are intricately linked to pattern recognition receptors expressed in macrophages, including beta (β)-glucan receptors and dentin-1 receptors. Prebiotics function as growth factors for specific commensal bacteria, inhibiting the adhesion and invasion of pathogenic microorganisms in the colon epithelium. They achieve this by competing for the same glycoconjugates present on the surface of epithelial cells. In addition, prebiotics contribute to altering the pH of the colon, modifying barrier function, enhancing mucus production, promoting the production of short-chain fatty acids, and inducing cytokine production. This direct effect of prebiotics enhances innate immunity by improving the phagocytosis process, as highlighted by Abbas et al. in 2012. The resulting higher internal acidity activates protease, leading to increased microbicidal activity (Abbas et al. 2012). Prebiotics have a beneficial effect on gut-associated lymphoid tissue. Inulin, FOS, mannanooligosaccharides, GOS, and arabinogalactans are included as the nutritional therapy products. Utilized to promote optimal bowel function, they favor the growth of normal bacterial flora while inhibiting the development of pathogens. These immunosaccharides play a significant role in enhancing the functionality of phagocytic cells, leading to heightened bactericidal activities. Furthermore, they elicit a stimulation of natural killer cells, complement system, lysozyme activity, and antibody response within the host organism (Ringo et al. 2010).

12.3.3.2 Growth Performance

The gut microbiota of a host can thrive with the aid of prebiotics (Dawood and Koshio 2016). This could promote the growth of aquatic organisms. Prebiotics can aid in the utilization of feed. As a result, FCR is lowered and growth can be enhanced (Ganguly et al. 2013). A review of the literature reveals that diverse prebiotic combinations can successfully stimulate the growth of different aquatic species (Wee et al. 2002).

12.3.3.3 Disease Resistance

Prebiotic combinations, including β-glucan and MOS, have been shown to boost the disease resistance of a variety of aquatic animals. According to the report of Gu et al. (2011), the prebiotic mixture increased the resistance to infection in sea cucumbers against *Vibrio splendidus*, Atlantic cod against *Vibrio anguillarum*, and common carp against *Aeromonas hydrophila.* As noted by Gultepe et al. in 2012, MOS alone did not demonstrate the ability to enhance fish immune systems and disease resistance. In contrast, β-glucan has been extensively documented to independently improve the growth, immune systems, and disease resistance of aquatic animals. Recent research has shown that prebiotic combinations of β-glucan and MOS are the best to use in aquaculture species. In addition, the correct prebiotic combination dosage must be established to prevent the side effects of immunosuppression.

12.3.3.4 Stress Mitigation

Aquaculture productivity is significantly impacted by abiotic variables (Mugwanya et al. 2022). Abiotic factor fluctuations that are outside of the ideal range might cause rapid demise and disease epidemics in aquaculture. The main objective of intensifying the aquaculture sector is to generate affordable and cost-effective products. However, this may impose stress on the aquaculture species within the farming system (Das et al. 2017). Studies have shown that prebiotics by themselves can reduce stress in aquaculture animals. For instance, Dawood et al. (2016) observed that β-glucan alone reduced stress in red sea bream (*Pagrus major*). According to Hoseinifar et al. (2014), FOS alone has been able to ameliorate stress in common carp. Some prebiotics can work properly in an ideal environment. The biological activity of prebiotics was regulated by abiotic factors such as temperature. Research conducted by Hoseinifar et al. (2017) has demonstrated the potential of prebiotic combinations, including GOS, FOS, and inulin, in alleviating stress among aquaculture species.

12.4 Synbiotics

The co-administration of probiotics and prebiotics is collectively known as "synbiotics." The term synbiotic was coined by Gibson and Roberfroid (1995). The combined use of probiotics and prebiotics results in a synergistic effect, wherein the prebiotic fosters the growth and activity of specific probiotic bacteria residing in the gut. Probiotic action is mainly based on the positive effects on the host, while prebiotics are targeted to probiotics for their better performance. The efficacy of the probiotics such as viability, intestinal balance, pathogen inhibition, and so on has been overall improved with the simultaneous application of probiotics and prebiotics (Goh et al. 2022). Besides, synbiotic diet promotes intestinal enzyme functions, feed utilization, and food digestion by secretion of digestive enzymes thereby making the essential nutrients available to the host (Rohani et al. 2022). Synbiotic feeding was found to provide significantly better results compared with probiotics or prebiotics applied alone in aquaculture for fish health. Through the combined application, prebiotics provide the fermentable energy source to probiont over the other competitive intestinal microflora (Merrifield et al. 2010). Besides favorable to probiotics, prebiotics promote the growth of other beneficial bacteria, which support host health (Huynh et al. 2019). There is also an additional advantage that prebiotics could bring out the selective enhancement of probiotics only and not the pathogens. Prebiotics developed from king oysters improved the survival of probiotic *L. plantarum* while not the virulent strains like *V. alginolyticus* and *V. harveyi* during synbiotic application (Prabawati et al. 2021). The immunostimulatory action and the associated resistance against pathogens of synbiotic treatment are also appreciable. During a synbiotic application of probiotic *Bifidobacteria* along with the prebiotic oligosaccharides, the former enhanced the secretion of immunostimulants in the fish intestines through the fermentation of the latter, thereby invading the pathogens (Mussatto and Mancilha 2007). The hemato-biochemical parameters such as Hb, RBC, WBC, and so on could also be improved through synbiotic application (Rohani et al. 2022). Above all, in addition to the stimulation effect, prebiotics also act as protective agents thereby reducing the probiotic inactivation during storage and processing. During the drying process, the prebiotic carbohydrates stabilize the protein structural integrity of probiotic cells (Goderska 2012).

The application of synbiotics in aquaculture was reported for the first time in rainbow trout (Panigrahi et al. 2005) and white shrimps (Li et al. 2009). Thereafter, the synbiotics were administered to various commercial fishes such as cobia, yellow croaker, Japanese flounder, lobster, shrimps, and so on (Huynh et al. 2017). It is reported that prebiotic β-glucans applied in synbiotics could have direct effects on the immune system and also supply substrates for the growth, survival, and colonization of probiotics (Huynh et al. 2017). Synbiotic diet could improve the fish's health through the release of exoenzymes and bioactive molecules from synbiotic metabolic reactions, thereby augmenting feed utilization through better nutrient absorption (Yilmaz et al. 2022). They could also enhance the immunity by triggering the secretion of cytokines and thus better phagocytosis, respiratory burst activity, and nitric oxide production (Yilmaz et al. 2022). A dietary supplement of synbiotic mixture containing 1%, *Enterococcus faecalis*, 1% polyhydroxyl butyrate, and 0.4% MOS considerably enhanced the hemato-immunological response and growth performance and also decreased the cumulative mortality rate in rainbow trout (Rodriguez-Estrada et al. 2009). Lee et al. (2018) noticed a higher survival rate of Japanese eel against *V. anguillarum* by showing improved growth and immunity after being treated with synbiotic mixture containing *B. subtilis* strain WB60 and MOS. Enhanced protection against bacterial diseases with consistent patterns was observed in the synbiotic application when compared with the individual administration of probiotics and prebiotics. This can be confirmed by a case study in which 95% of the survival rate of shrimps was observed when treated with synbiotic combination of *Pseudoalteromonas piscicida* and MOS followed by 87% with prebiotic and 84% with probiotic (Ramadhani et al. 2019).

12.4.1 Considerations for Synbiotic Application

While developing the synbiotic formulation, proper selection, correct dose, and proportion of probiotics and prebiotics are vital. An inappropriate combination and dosage of probiotics and prebiotics could retard the proper growth of fish. Inaccurate selection of prebiotic substrates could hinder the fermentation rate which leads to prebiotic deposition and thereby inhibits the host's digestive functions (Goh et al. 2022). Hence, to achieve the huge benefits from synbiotics, the metabolizing capacity of the prebiotics by the corresponding probiotic strains needs to be well studied. An under- and over-dosage of synbiotic could also have negative impacts on the host. For example, a proportion of *Bacillus megaterium*/g of feed at 10^8 CFU with 0.2% prebiotic IMO had given significant results than with a higher concentration of 10^{10} CFU (Li et al. 2009). The frequency of administration and type of aquatic species are also important factors in synbiotic selection. The daily regimen of synbiotic treatment is advisable compared with weekly or monthly administration (Munaeni et al. 2014). Because of the disparity of the intestinal flora of different aquatic individuals, the efficacy of the synbiotic composition varies (Hoseinifar et al. 2018). Hence, for the application in various fishes, proper standardization is essential for sustainable function. Thus, it can be concluded that multiple features need to be taken into account with both in-vitro and in-vivo studies prior to the application of synbiotics in aquaculture.

12.5 Conclusion, Research Gaps, and Future Perspectives

A wide range of microorganisms having probiont potential for aquaculture application have been studied so far. These probiotics in combination with indigestible compounds called prebiotics possess multiple beneficial effects such as intestinal microbial balance, disease resistance through competitive exclusion, enhanced digestion and immune response, water quality improvement, and QQ for better fish cultivation. As a substitutive strategy to chemotherapeutics, probiotics have nowadays been selected as the better alternative for combating the major pathogens in aquaculture. However, the existence of AMR genes in probiotic strains and the probability of subsequent transfer to other intestinal pathogens raise a major concern (Preena et al. 2020). Egervärn et al. (2010) reported the horizontal gene transfer of AMR genes from *L. plantarum* to *Listeria monocytogenes* in the GI tract. Hence, while selecting the probiotic candidates, utmost care needs to be taken to ensure the absence of AMR and virulence genes to ensure the safety of fishes, thereby reducing the economic losses. More studies are needed to determine the best way of mode of delivery of probiotics to reach the destination. In addition to the above-mentioned techniques, immobilized probiotics can be used by cell entrapment in gel matrices wherein reduced contamination, increased stability and protection, and better nutrient supplementation compared with free cell systems. Meanwhile, this technology is least exploited in aquaculture, and recently, probiotics were immobilized with oyster and walnut shells for better bioremediation in aquaculture sediments (Yi et al. 2021).

Probiotic screening methods should be improvised by unraveling the exploitation and interference-based mechanism underlying the probiont–pathogen–host interactions. Special efforts and considerations need to be undertaken to check the antagonistic potential of probiotics against pathogens with minimal effects on intestinal beneficial microbiota. For this, in addition to in vitro studies, in vivo studies with experimental challenge assay should be designed to understand the competitive life strategy of probiotic candidates. This could be of great significance in measuring the safety and suitability of probiotic species on cultured species meant for human consumption. Besides, species-specific probiotic efficacy needs to be understood prior to application in novel aquatic species. Thus, the development of novel probiotics along with high-purity prebiotics for better aquaculture production is the need of the hour.

Molecular techniques such as recombinant technology, mutagenesis, and strain improvement can be further utilized to increase the efficacy of probiotics to be species-specific. The effects of probiotic action against viruses, fungi, and parasites can also be studied in detail to mitigate wide-spectrum infections in fishes. Due to the increasing demand for probiotics in aquaculture, advanced "omic"-based molecular tools such as whole-genome sequencing, metagenomics, proteomics, and metatranscriptomics coupled with high-performance liquid chromatography and nuclear magnetic resonance-based metabolome analysis allow better investigation and selection of appropriate probiotics. The generated data could explore the exact functioning and diverse combination of synbiotic mixtures in accurate dose, proportion, and duration. The state of the art of probiotics, prebiotics, and synbiotics could bring out the paradigm shift in the aquaculture industry which leads to the emergence of good-quality fishes, consumer health assurance, and enhanced sustainability and economic benefits.

Acknowledgments This work was supported by the Centre of Excellence in Sustainable Aquaculture and Aquatic Health Management (CAAHM), Plan fund 2023-24, Kerala University of Fisheries and Ocean Studies.

Conflict of InterestThe authors declare that they have no conflict of interest.

References

Abasali H, Mohamad S (2011) Dietary prebiotic immunogen supplementation in reproductive performance of platy (*Xiphophorus maculatus*). Vet Res 4:66–70

Abbas AK, Lichtman AH, Pillai S (2012) Cellular and Molecular Immunology, seventh ed., Saunders Elsevier Philadelphia, PA, USA, 2012

Abidi R (2003) Use of probiotics in larval rearing of new candidate species. Aquaculture Asia 8:15–16

Aguirre-Guzman G, Lara-Flores M, Sanchez-Martínez JG, Campa-Cordova AI, Luna-González A (2012) The use of probiotics in aquatic organisms: a review. Afr J Microbiol Res 6:4845–4857

Ahmdifar E, Akrami R, Ghelichi A et al (2011) Effects of different dietary prebiotic inulin levels on blood serum enzymes, hematologic, and biochemical parameters of great sturgeon (*Huso huso*) juveniles. Comp Clin Pathol 20:447–451

Akrami R, Iri Y, Khoshbavar H et al (2013) Effect of dietary supplementation of fructooligosaccharide (FOS) on growth performance, survival, lactobacillus bacterial population and hematoimmunological parameters of stellate sturge juvenile. Fish Shellfish Immunol 35:1235–1123

Akther N, Wu B, Memon AM, Mohsin M (2015) Probiotics and prebiotics associated with aquaculture: a review. Fish Shellfish Immunol 45:733–741

Amiin MK, Lahay AF, Putriani RB, Reza M, Putri SME, Sumon MAA, Jamal MT, Santanumurti MB (2023) The role of probiotics in vannamei shrimp aquaculture performance—a review. Vet World 16:638–649

Bailey J, Blankenship L, Cox N (1991) Effect of fructooligosaccharide on salmonella colonization of the chicken intestine. Poult Sci 70:2433–2438

Bakke-McKellep AM, Penn MH, Salas PM, Refstie S, Sperstad S, Landsverk T, Ringø E, Krogdahl Å (2007) Effects of dietary soybean meal, inulin and oxytetracycline on gastrointestinal histological characteristics, distal intestine cell proliferation and intestinal microbiota in Atlantic salmon (*Salmo salar* L.). Brit J Nutr 97:699–713

Balcázar JL, de Blas I, Ruiz-Zarzeula I, Cunningham D, Vendrell D, Múzquiz JL (2006) The role of probiotics in aquaculture. Vet Microbiol 114:173–186

Banerjee S, Khatoon H, Sharif M, Yusof FM (2010) Enhancement of Penaeus monodon shrimp postlarvae growth and survival without water exchange using marine *Bacillus pumilus* and periphytic microalgae. Fish Sci 76:481–487

Banerjee G, Nandi A, Ray AK (2017) Assessment of hemolytic activity, enzyme production and bacteriocin characterization of *Bacillus subtilis* LR1 isolated from the gastrointestinal tract of fish. Arch Microbiol 199:115–124

Beck BR, Kim D, Jeon J, Lee SM, Kim HK, Kim OJ, Holzapfel WH (2015) The effects of combined dietary probiotics *Lactococcus lactis* BFE920 and *lactobacillus plantarum* FGL0001 on innate immunity and disease resistance in olive flounder (*Paralichthys olivaceus*). Fish Shellfish Immunol 42:177–183

Beck BR, Lee SH, Kim D, Park JH, Lee HK, Kwon S-S, Song SK (2017) A *Lactococcus lactis* BFE920 feed vaccine expressing a fusion protein composed of the OmpA and FlgD antigens from *Edwardsiella tarda* was significantly better at protecting olive flounder (*Paralichthys olivaceus*) from edwardsiellosis than single antigen vaccines. Fish Shellfish Immunol 68:19–28

Brunt J, Austin B (2005) Use of a probiotic to control lactococcosis and streptococcosis in rainbow trout, *Oncorhynchus mykiss* (Walbaum). J Fish Dis 28:693–701

Burbank DR, Shah DH, La Patra SE, Fornshell G, Cain KD (2011) Enhanced resistance to coldwater disease following feeding of probiotic bacterial strains to rainbow trout (*Oncorhynchus mykiss*). Aquaculture 321:185–190

Burkitt DP, Walker ARP, Painter NS (1972) Effect of dietary fibre on stools and transit-times, and its role in the causation of disease. Lancet 300:1408–1411

Burr G, Hume M, Neill WH, Gatlin DM III (2008) Effects of prebiotics on nutrient digestibility of a soybeanmeal-based diet by red drum *Sciaenops ocellatus* (*Linnaeus*). Aquac Res 39:1680–1686

Cerezuela RA, Cuesta J, Meseguer MA (2008) Esteban, effects of inulin on gilthead seabream (*Sparus aurata* L.) innate immune parameters. Fish Shellfish Immunol 24(2008):663–668

Cerezuela R, Guardiola FA, Meseguer J, Esteban MA (2012) Increases in immune parameters by inulin and *Bacillus subtilis* dietary administration to gilthead seabream (*Sparus aurata* L.) did not correlate with disease resistance to *Photobacterium damselae*. Fish Shellfish Immunol 32:1032–1040

Chauhan A, Singh R (2019) Probiotics in aquaculture: a promising emerging alternative approach. Symbiosis 77:99–113

Chu W, Zhou S, Zhu W, Zhuang X (2014) Quorum quenching bacteria *bacillus* sp. QSI-1 protect zebrafish (*Danio rerio*) from *Aeromonas hydrophila* infection. Sci Rep 4:5446

D'Alvise PW, Lillebø S, Prol-Garcia MJ, Wergeland HI, Nielsen K, Bergh Ø et al (2012) *Phaeobacter gallaeciensis* reduces *vibrio anguillarum* in cultures of microalgae and rotifers, and prevents vibriosis in cod larvae. PLoS One 7:e43996

Dalmo RA, Bøgwald J (2008) β-Glucans as conductors of immune symphonies. Fish Shellfish Immunol 25:384–396

Daniels C, Boothroyd D, Davies S, Pryor R, Taylor D, Wells C (2006) Bio-MOS improves growth and survival of cultures lobsters. Shellfish News 21:23–25

Das S, Mondal K, Haque S (2017) A review on application of probiotic, prebiotic and synbiotic for sustainable development of aquaculture. J Entomol Zool Stud 5:422–429

Dawood MA, Koshio S (2016) Recent advances in the role of probiotics and prebiotics in carp aquaculture: a review. Aquaculture 454:243–251

Dawood MA, Koshio S, Ishikawa M, Yokoyama S, El Basuini MF, Hossain MS, Nhu TH, Dossou S, Moss AS (2016) Effects of dietary supplementation of lactobacillus rhamnosus or/and *Lactococcus lactis* on the growth, gut microbiota and immune responses of red sea bream, Pagrus major. Fish Shellfish Immunol 49:275–285

DeVrese M, Schrezenmeir J (2008) Probiotics, prebiotics, and synbiotics. In: Food biotechnology. Springer, Berlin Heidelberg, pp 1–66

Dimitroglou A, Davies S, Sweetman J (2008) The effects of dietary mannan oligosaccharides on the intestinal histology of rainbow trout (Oncorhynchus mykiss). Abstracts/Comp. Biochem Physiol A 150:S63

Dong C, Wang J (2013) Immunostimulatory effects of dietary fructooligosaccharides on red swamp crayfish, *Procambarus clarkii* (Girard). Aquac Res 44:1416–1424

Ebrahimi G, Ouraji H, Khalesi M, Sudagar M, Barari A, Zarei Dangesaraki M, Jani Khalili K (2012) Effects of a prebiotic, immunogen®, on feed utilization, body composition, immunity and resistance to *Aeromonas hydrophila* infection in the common carp *Cyprinus carpio* (*Linnaeus*) fingerlings. J Anim Physiol Anim Nutr 96:591–599

Egervärn M, Lindmark H, Olsson J, Roos S (2010) Transferability of a tetracycline resistance gene from probiotic *lactobacillus reuteri* to bacteria in the gastrointestinal tract of humans. Antonie Leeuwenhoek 97:189–200

Esteban MA (2012) An overview of the immunological defences in fish skin. ISRN Immunol Article ID 853470, p 29

FAO/WHO (2001) Health and nutritional properties of probiotics in food including powder milk with liver lactic acid bacteria. Food and Agriculture Organization and World Health Organization Joint report

Franco R, Martın L, Arenal A, Santiesteban D, Sotolongo J, Cabrera H et al (2017) Evaluation of two probiotics used during farm production of white shrimp *Litopenaeus vannamei* (Crustacea: Decapoda). Aquac Res 48:1936–1950

Fuller R (1989) A review: probiotics in man and animals. J Appl Bacteriol 66:365–378

Ganguly S, Dora KC, Sarkar S, Chowdhury S (2013) Supplementation of prebiotics in fish feed: a review. Rev Fish Biol Fish 23:195–199

Genc MA, Aktas M, Genc E, Yilmaz E (2007) Effects of dietary mannan oligosaccharides on growth, body composition and hepatopancreas histology of Penaeus semisulcatus (de Haan 1844). Aquac Nutr 13:156–161

Geralylou JZC, Souffreau E, Rurangwa S, D'Hondt L, Callewaert CM (2012) Courtin, effects of arabionxylan-oligosaccharides (AXOS) on juvenile Siberian sturgeon (*Acipenser baerii*) performance, immune responses and gastrointestinal microbial community. Fish Shellfish Immunol 33:718–724

Ghanei-Motlagh R, Mohammadian T, Gharibi D et al (2020) Quorum quenching properties and probiotic potentials of intestinal associated bacteria in Asian sea bass *Lates calcarifer*. Mar Drugs 18:23

Ghosh S, Sinha A, Sahu C (2007) Effect of probiotic on reproductive performance in female live-bearing ornamental fish. Aquac Res 38:518–526

Gibson GR, Roberfroid MB (1995) Dietary modulation of the human colonic microbiota: introducing the concept of prebiotics. J Nutr 125:1401–1412

Gibson GR, Probert HM, Loo JV, Rastall RA, Roberfroid MB (2004) Dietary modulation of the human colonic microbiota: updating the concept of prebiotics. Nutr Res Rev 17:259–275

Gioacchini G, Giorgini E, Olivotto I, Maradonna F, Merrifield DL, Carnevali O (2014) The influence of probiotics on zebrafish *Danio Rerio* innate immunity and hepatic stress. Zebrafish 11:98–106

Goderska K (2012) Different methods of probiotics stabilization. IntechOpen

Goh JXH, Tan LTH, Law JW-F, Ser H-L, Khaw K-Y, Letchumanan V, Lee L-H, Goh B-H (2022) Harnessing the potentialities of probiotics, prebiotics, synbiotics, paraprobiotics, and postbiotics for shrimp farming. Rev Aquac 14:1478–1557

Gomez GB, Roque A, Turnbull JF (2000) The use and selection of probiotic bacteria for use in the culture of larval aquatic organisms. Aquaculture 191:259–270

Gomez-Gil B, Roque A, Soto-Rodriguez S (2011) Probiotics in the larval culture of aquatic organisms. In: Montet D, Ray RC (eds) Aquaculture microbiology and biotechnolgy, vol 2. Taylor and Francis CRC Press, Science Publishers, pp 31–46

Gram L, Hjelm M (2002) The selection and potential use of aquatic bacteria as fish probiotics. ICES Council Meeting Documents, ICES, Copenhagen, Denmark

Gram L, Melchiorsen J, Spanggaard B, Huber I, Nielsen T (1999) Inhibition of *Vibrio anguillarum* by *Pseudomonas fluorescens* strain AH2, a possible probiotic treatment of fish. Appl Environ Microbiol 65:969–973

Grisdale-Helland JB, Helland SJ, Gatlin DM (2008) The effects of dietary supplementation with mannanoligosaccharide, fructooligosaccharide or galactooligosaccharide on the growth and feed utilization of Atlantic salmon (*Salmo salar* L.). Aquaculture 283:163–167

Gu M, Ma H, Mai K, Zhang W, Bai N, Wang X (2011) Effects of dietary β-glucan, mannan oligosaccharide and their combinations on growth performance, immunity and resistance against *Vibrio splendidus* of sea cucumber, *Apostichopus japonicas*. Fish Shellfish Immunol 31:303–309

Gultepe N, Hisar O, Salnur S, Hossu B, Tanrikul TT, Aydın S (2012) Preliminary assessment of dietary mannanoligosaccharides on growth performance and health status of gilthead seabream *Sparus auratus*. J Aquat Anim Health 24:37–42

Hadiuzzaman M, Moniruzzaman M, Shahjahan M, Bai SC, Min T, Hossain Z (2022) β-glucan: mode of action and its uses in fish immunomodulation. Front Mar Sci 9:1–15

Hai NV (2015) The use of probiotics in aquaculture. J Appl Microbiol 119:917–935

Hancz C (2022) Application of probiotics for environmentally friendly and sustainable aquaculture: a review. Sustain MDPI 14:15479

Hanley F, Brown H, Carbery J (1995) First observations on the effects of mannan oligosaccharide added to hatchery diets for warmwater hybrid red tilapia. In: Poster at the 11th annual symposium on biotechnology in the feed industry, Lexington, KY

Hasan KN, Banerjee G (2020) Recent studies on probiotics as beneficial mediator in aquaculture: a review. JoBAZ 81:53. https://doi.org/10.1186/s41936-020-00190-y

Hassan MA, Fathallah MA, Elzoghby MA et al (2022) Influence of probiotics on water quality in intensified Litopenaeus vannamei ponds under minimum-water exchange. AMB Expr 12:22

He S, Xu G, Wu Y, Weng H, Xie H (2003) Effects of IMO and FOS on the growth performance and non-specific immunity in hybrid tilapia. Chinese Feed 23:14–15

Holzapfel WH, Haberer P, Snel J, Schillinger U, Huis in't Veld J (1998) Overview of gut flora and probiotics. Int J Food Microbiol 41:85–101

Hoseinifar SH, Soleimani N, Ringø E (2014) Effects of dietary fructooligosaccharide supplementation on the growth performance, haematoimmunological parameters, gut microbiota and stress resistance of common carp (*Cyprinus carpio*) fry. Br J Nutr 112:1296–1302

Hoseinifar SH, Ahmadi A, Khalili M, Raeisi M, Van Doan H, Caipang CM (2017) The study of antioxidant enzymes and immune-related genes expression in common carp (Cyprinus carpio) fingerlings fed different prebiotics. Aquac Res 48:5447–5454

Hoseinifar SH, Sun Y-Z, Wang A, Zhou Z (2018) Probiotics as means of diseases control in aquaculture, a review of current knowledge and future perspectives. Front Microbiol 9:2429

Huynh T-G, Shiu Y-L, Nguyen T-P, Truong Q-P, Chen J-C, Liu C-H (2017) Current applications, selection, and possible mechanisms of actions of synbiotics in improving the growth and health status in aquaculture: a review. Fish Shellfish Immunol 64:367–382

Huynh TG, Hu SY, Chiu CS, Truong QP, Liu CH (2019) Bacterial population in intestines of white shrimp, *Litopenaeus vannamei* fed a synbiotic containing Lactobacillus plantarum and galactooligosaccharide. Aquac Res 50:807–817

Ibrahem JMD, Fathi M, Meslhy S, El-Aty AMA (2010) Effect of dietary supplementation of inulin and vitamin C on the growth hematology, innate immunity and resistance of Nile tilapia (*Oreochromis niloticus*). Fish Shellfish Immunol 29:241–246

Ismail M, Wahdan A, Yusuf MS, Metwally E, Mabrok M (2019) Effect of dietary supplementation with a synbiotic (Lacto Forte) on growth performance, haematological and histological profiles, the innate immune response and resistance to bacterial disease in *Oreochromis niloticus*. Aquac Res 50:2545–2562

Jahangiri L, Esteban MA (2018) Administration of probiotics in the water in finfish aquaculture systems: a review. Fishes MDPI 3:33. https://doi.org/10.3390/fishes3030033

James G, Das BC, Jose S, Kumar VJR (2020) *Bacillus* as an aquaculture friendly microbe. Aquac Int 29:323–353. https://doi.org/10.1007/s10499-020-00630-0
James G, Geetha PP, Puthiyedathu ST, Jayadradhan RKV (2023) Applications of actinobacteria in aquaculture: prospects and challenges. 3 Biotech 13:42
Jasim SA, Abdelbasset BK, Shichiyakh RA, Al-Shawi SG et al (2022) Probiotic effects of the fungi, *Aspergillus Niger* on growth, immunity, haematology, intestine fungal load and digestive enzymes of the common carp, *Cyprinus carpio*. Aquac Res 53:3828–3840
Javadi A, Khatibi SA (2017) Effect of commercial probiotic (Protexin®) on growth, survival and microbial quality of shrimp (*Litopenaeus vannamei*). Nutr Food Sci 47:204–216
Jiqiu L, Beiping T, Kangsen M (2006) Comparative study between probiotic bacterium *Arthrobacter* XE-7 and chloramphenicol on protection of *Penaeus chinensis* postlarvae from pathogenic Vibrios. Aquaculture 253:140–147
Joshi F, Archana G, Desai A (2006) Siderophore cross-utilization amongst rhizospheric bacteria and the role of their differential affinities for Fe3+ on growth stimulation under iron-limited conditions. Current Microbiol 53:141–147
Kamei Y, Yoshimizu M, Ezura Y, Kimura T (1988) Screening of bacteria with antiviral activity from fresh water salmonid hatcheries. Microbiol Immunol 32:67–73
Kesarcodi-Watson A, Kaspar H, Lategan MJ, Gibson L (2012) Performance of single and multi-strain probiotics during hatchery production of Greenshell™ mussel larvae, Perna canaliculus. Aquaculture 354:56–63
Knipe H, Temperton B, Lange A, Bass D, Tyler CR (2021) Probiotics and competitive exclusion of pathogens in shrimp aquaculture. Rev Aquac 13:324–352
Kuebutornye FKA, Abarike ED, Lu Y, Hlordzi V, Sakyi ME, Afriyie G, Wang Z, Li Y, Xie CX (2020) Mechanisms and the role of probiotic *Bacillus* in mitigating fish pathogens in aquaculture. Fish Physiol Biochem 46:819–841
Kumar V, Roy S, Meena DK, Sarkar UK (2016) Application of probiotics in shrimp aquaculture: importance, mechanism of action and methods of administration. Rev Fisheries Sci Aquacult 24:342–368
Lalloo R, Moonsamy G, Ramchuran S, Gorgens J, Gardiner N (2010) Competitive exclusion as a mode of action of a novel *Bacillus cereus* aquaculture biological agent. Lett Appl Microbiol 50:563–570
Lara F, Olvera N, Guzman M, López M (2003) Use of the bacteria *Streptococcus faecium* and *Lactobacillus acidophilus*, and the yeast *Saccharomyces cerevisiae* as growth promoters in Nile tilapia (*Oreochromis niloticus*). Aquaculture 216:193–201
Lazado CC, Caipang CMA, Estante EG (2015) Prospects of host associated microorganisms in fish and penaeids as probiotics with immunomodulatory functions. Fish Shellfish Immunol 45:2–12i
Lee S, Katya K, Hamidoghli A, Hong J, Kim D-J, Bai SC (2018) Synergistic effects of dietary supplementation of *Bacillus subtilis* WB60 and mannanoligosaccharide (MOS) on growth performance, immunity and disease resistance in Japanese EEL, *Anguilla japonica*. Fish Shellfish Immunol 83:283–291
Li J, Tan B, Mai K (2009) Dietary probiotic *Bacillus* OJ and isomaltooligosaccharides influence the intestine microbial populations, immune responses and resistance to white spot syndrome virus in shrimp (*Litopenaeus vannamei*). Aquaculture 291:35–40
Little AEF, Robinson CJ, Peterson SB, Raffa KF, Handelsman J (2008) Rules of engagement: interspecies interactions that regulate microbial communities. Annual Rev Microbiol 62:375–401
Liu CH, Chiu CH, Wang SW, Cheng W (2012) Dietary administration of the probiotic, *Bacillus subtilis* E20, enhances the growth, innate immune responses, and disease resistance of the grouper, *Epinephelus coioides*. Fish Shellfish Immunol 33:699–706
Lu J, Bu X, Xiao S, Lin Z, Wang X, Jia Y, Wang X, Qin JG, Chen L (2019) Effect of single and combined immunostimulants on growth, anti-oxidation activity, non-specific immunity and resistance to *Aeromonas hydrophila* in Chinese mitten crab (*Eriocheir sinensis*). Fish Shellfish Immunol 93:732–742

Manning TS, Gibson GR (2004) Prebiotics. Best Pract Res Clin Gastroenterol 18:287–298

Martınez Cruz P, Ibanez AL, Hermosillo OAM, Saad HCR (2012) Use of probiotics in aquaculture. ISRN Microbiol 916845:1–13

Merrifield DL, Dimitroglou A, Foey A, Davies SJ, Baker RTM, Bøgwald J, Castex M, Ringø E (2010) The current status and future focus of probiotic and prebiotic applications for salmonids. Aquaculture 302:1–18

Metivier A, Pilet MF, Dousset X, Sorokine O, Anglade P, Zagorec M et al (1998) Divercin V41, a new bacteriocin with two disulphide bonds produced by *Carnobacterium divergens* V41: primary structure and genomic organization. Microbiology 144:2837–2844

Misra CK, Das BK, Mukherjee SC, Pattnaik P (2006) Effect of long term administration of dietary β-glucan on immunity, growth and survival of *Labeo Rohita* fingerlings. Aquaculture 255:82–94

Mohammadi G, Hafezieh M, Karimi AA, Azra MN, Doan HV, Tapingkae W, Abdelrahman HA, Dawood MAO (2022) The synergistic effects of plant polysaccharide and *Pediococcus acidilactici* as a synbiotic additive on growth, antioxidant status, immune response, and resistance of Nile tilapia (*Oreochromis niloticus*) against *Aeromonas hydrophila*. Fish Shellfish Immunol 120:304–313

Mohapatra S, Chakraborty T, Kumar V, De Boeck G, Mohanta KN (2013) Aquaculture and stress management: a review of probiotic intervention. J Anim Physiol Anim Nutr 97:405–430

Montalban-Arques A, De Schryver P, Bossier P, Gorkiewicz G, Mulero V, Gatlin DMI, Galindo-Villegas J (2015) Selective manipulation of the gut microbiota improves immune status in vertebrates. Front Immunol 6:512

Mourino JLP, Do Nascimento VF, Jatob AB, Da Silva BC, Jesus GFA, Seiffert WQ (2012) Effect of dietary supplementation of inulin and *W. Cibaria* on haemato-immunological parameters of hybrid Surubim (*Pseudoplatystoma* sp). Aquac Nutr 18:73–80

Mugwanya M, Dawood MA, Kimera F, Sewilam H (2022) Anthropogenic temperature fluctuations and their effect on aquaculture: a comprehensive review. Aquac Fish 7:223–243

Munaeni W, Yuhana M, Widanarni W (2014) Effect of micro-encapsulated synbiotic at different frequencies for luminous vibriosis control in white shrimp (*Litopenaeus vannamei*). Microbiol Indonesia 8:73–80

Mussatto SI, Mancilha IM (2007) Non-digestible oligosaccharides: a review. Carbohydr Polym 68:587–597

Nagappan S, Das P, AbdulQuadir M, Thaher M, Khan S, Mahata C, Al-Jabri H, Vatland AK, Kumar G (2021) Potential of microalgae as a sustainable feed ingredient for aquaculture. J Biotechnol 341:1–20

Navarrete P, Tovar-Ramírez D (2014) [Chapter 5] Use of yeasts as probiotics in fish aquaculture. In: Hernandez-Vergara M, Perez-Rostro C (eds) Sustainable aquaculture techniques. IntechOpen Limited, London, UK, pp 135–172. https://doi.org/10.1016/B978-0-12-801238-3.00014-3

Nawaz A, Irshad S, Hoseinifar SH, Xiong H (2018) The functionality of prebiotics as immunostimulant: evidences from trials on terrestrial and aquatic animals. Fish Shellfish Immunol 76:272–278

Nwanna LC (2015) Use of probiotics in aquaculture. Appl Trop Agr J 15:76–83

Oktaviana A, Yuhana M (2014) The use of synbiotics to prevent imnv and *Vibrio harveyi* coinfection in *Litopenaeus vannamei*. HAYATI Journal of Biosciences 21:12

Olmos J, Ochoa L, Paniagua-Michel J, Rosalia C (2011) Functional feed assessment on *Litopenaeus vannamei* using 100% fish meal replacement by soybean meal, high levels of complex carbohydrates and *Bacillus* probiotic strains. Mar Drugs 9:1119–1132

Olsen RE, Myklebust R, Kryvi H, Mayhew TM, Ringø E (2001) Damaging effect of dietary inulin on intestinal enterocytes in *Arctic charr* (*Salvelinus alpinus* L.). Aquac Res 32:931–934

Padenia U (2021) Manipulating the microbiome of larval Zebrafish, *Danio rerio*. World Fisheries University, Pukyong National University, Busan, S. Korea

Pandiyan P, Balaraman D et al (2013) Probiotics in aquaculture. Drug Invent Today 5:55–59

Panigrahi A, Kiron V, Puangkaew J, Kobayashi T, Satoh S, Sugita H (2005) The viability of probiotic bacteria as a factor influencing the immune response in rainbow trout *Oncorhynchus mykiss*. Aquaculture 243:241–254

Pattukumar V, Kanmani P, Kumar RS, Yuvaraj N, Paari A, Arul V (2013) Improved resistance to *Vibrio parahaemolyticus* in black tiger shrimp *Penaeus monodon* treated with *Streptococcus phocae* PI80 and *Bacillus subtilis*. Isr J Aquac Bamidgeh IJA 10

Pereira WA, Piazentin ACM, de Oliveira RC et al. (2022) Bacteriocinogenic probiotic bacteria isolated from an aquatic environment inhibit the growth of food and fish pathogens. Sci Rep 12:5530. https://doi.org/10.1038/s41598-022-09263-0

Perez-Sanchez T, Ruiz-Zarzeula I, de Blas I, Balcazar JL (2013) Probiotics in aquaculture: a current assessment. Rev Aquac 5:1–14

Petit J, Wiegertjes GF (2016) Long-lived effects of administering β-glucans: indications for trained immunity in fish. J Dev Comp Immunol 64:93–102

Planas M, Vazquez J, Marqu'es J, P'erez R, Gonz'alez M, Murado M (2004) Enhancement of rotifer (*Brachionus plicatilis*) growth by using terrestrial lactic acid bacteria. Aquaculture 240:313–329

Planas M, Perez-Lorenzo M, Hjelm M, Gram L, Uglenes Fiksdal I, Bergh O, Pintado J (2006) Probiotic effect in vivo of *Roseobacter* strain 27-4 against *Vibrio (Listonella) anguillarum* infections Iin turbot (*Scophthalmus maximus* L.) larvae. Aquaculture 255:323–333

Possemiers S, Grootaert C, Vermeiren J, Gross G, Marzorati M, Verstraete W, Van de Wiele T (2009) The intestinal environment in health and disease—recent insights on the potential of intestinal bacteria to influence human health. Curr Pharmaceut Des 15:2051–2065

Prabawati E, Hu S-Y, Chiu S-T, Balantyne R, Risjani Y, Liu C-H (2021) A synbiotic containing prebiotic prepared from a by-product of king oyster mushroom, *Pleurotus eryngii* and probiotic, *Lactobacillus plantarum* incorporated in diet to improve the growth performance and health status of white shrimp, *Litopenaeus vannamei*. Fish Shellfish Immunol 120:155–165

Preena PG, Swaminathan TR, Kumar VJR, Singh ISB (2020) Antimicrobial resistance in aquaculture: a crisis for concern. Biologia 75:1497–1517. https://doi.org/10.2478/s11756-020-00456-4

Qinghui Ai, Houguo Xu, Kangsen Mai, Wei Xu, Jun Wang, Wenbing Zhang (2011) Effects of dietary supplementation of Bacillus subtilis and fructooligosaccharide on growth performance, survival, no-specific immune response and disease resistance of juvenile large yellow croaker, Larimichthys crocea, Aquaculture 317:155–161

Qiu Z, Xu Q, Li S, Zheng D, Zhang R, Zhao J, Wang T (2023) Effects of probiotics on the water quality, growth performance, immunity, digestion, and intestinal flora of giant freshwater prawn (*Macrobrachium rosenbergii*) in the biofloc culture system. Water 15:1211. https://doi.org/10.3390/w15061211

Rahiman KM, Jesmi Y, Thomas A, Hatha AAM (2010) Probiotic effect of *Bacillus* NL110 and *Vibrio* NE17 on the survival, growth performance and immune response of *Macrobrachium rosenbergii* (de Man) School of Environment Sciences. Mahatma Gandhi University, Kottayam, India

Ramadhani DE, Widanarni W, Sukenda S (2019) Microencapsulation of probiotics and its applications with prebiotic in Pacific white shrimp larvae through *Artemia* sp. J Akuak Indonesia 18:130–140

Rames D, Vinothkanna A, Rai AK, Vignesh VS (2015) Isolation of potential probiotic *Bacillus* spp. and assessment of their subcellular components to induce immune responses in *Labeo rohita* against *Aeromonas hydrophila*. Fish Shellfish Immunol 45:268–276

Respondek F, Goachet AG, Julliand V (2008) Effects of dietary short-chain fructooligosaccharides on the intestinal microflora of horses subjected to a sudden change in diet. J Anim Sci 86:316–323

Ringo E (1999) Does *Carnobacterium divergens* isolated from Atlantic Salmon, *Salmo salar* L., colonize the gut of early developing turbot, *Scophthalmus maximus* L., larvae? Aquac Res 30:229–232

Ringø E (2020) Probiotics in shellfish aquaculture. Aquac Fish 5:1–27

Ringo E, Gatesoupe FJ (1998) Lactic acid bacteria in fish: a review. Aquaculture 160:177–203
Ringø E, Hoseinifar SH, Ghosh K, Van Doan H, Beck BR, Song SK (2018) Lactic acid bacteria in finfish – an update. Front Microbiol 9:1818
Ringo E, Sperstad S, Myklebust R, Mayhew TM, Olsen RE (2006) The effect of dietary inulin on the aerobic bacteria associated with hindgut of Arctic charr (*Salvelinus alpinus* L.). Aquac Res 37:891–897
Ringo E, Salinas L, Olsen RE, Nyhaug A, Myklebust R, Mayhew TM (2007) Histological changes in intestine of Atlantic Salmon (*Salmo salar* L.) following in vitro exposure to pathogenic and probiotic bacterial strains. Cell Tissue Res 328:109–116
Ringo E, Olsen RE, Gifstad TO, Dalmo RA, Amlund H, Hemre GI, Bakke AM (2010) Prebiotics in aquaculture: a review. Aquac Nutr 16:117–136
Roberfroid MB (1993) Dietary fiber, inulin, and oligofructose—a review comparing their physiological effects. Crit Rev Food Sci Nutr 33:103–148
Robertson PAW, O'Dowd C, Burrells C, Williams P, Austin B (2000) Use of *Carnobacterium* sp. as a probiotic for Atlantic salmon (*Salmo salar* L.) and rainbow trout (*Oncorhynchus mykiss*, Walbaum). Aquaculture 185:235–243
Rodriguez J, Espinosa Y, Echeverria F, Cardenas G, Roman R, Stern S (2007) Exposure to probiotics and and [beta]-1,3/1,6-glucans in larviculture modifies the immune response of *Penaeus vannamei* juveniles and both the survival to white spot syndrome virus challenge and pond culture. Aquaculture 273:405–415
Rodriguez-Estrada U, Satoh S, Haga Y, Fushimi H, Sweetman J (2009) Effects of single and combined supplementation of *Enterococcus faecalis*, mannan oligosaccharide and polyhydroxybutyrate acid on growth performance and immune response of rainbow trout *Oncorhynchus mykiss*. Aquac Sci 57:609–617
Rohani MF, Islam SMM, Hossain MK, Ferdous Z, Siddik MAB, Nuruzzaman M, Padeniya U, Brown C, Shahjahan M (2022) Probiotics, prebiotics and synbiotics improved the functionality of aquafeed: upgrading growth, reproduction, immunity and disease resistance in fish. Fish Shellfish Immunol 120:569–589
Rokka S, Rantamäki P (2010) Protecting probiotic bacteria by microencapsulation: challenges for industrial applications. Eur Food Res Technol 231:1–12
Rurangwa E, Delaedt Y, Geraylou Z, Van De Wiele T, Courtin CM, Delcour JA, Ollevier F (2008) Dietary effect of arabinoxylan oligosaccharides on zootechnical performance and hindgut microbial fermentation. In: Siberian sturgeon and African catfish. Aquaculture Europe, Krakow, Poland, September 15–18, 2008, pp 569–570
Salinas I, Dıaz-Rosales P, Cuesta A, Meseguer J, Chabrillon M, Morinigo M-A, Esteban MA (2006) Effect of heat-inactivated fish and non-fish derived probiotics on the innate immune parameters of a teleost fish (*Sparus aurata* L.). Vet Immunol Immunopathol 111:279–286
Schluter J, Nadell CD, Bassler BL, Foster KR (2015) Adhesion as a weapon in microbial competition. ISME J 9:139–149
Schrezenmeir J, De Vrese M (2001) Probiotics, prebiotics, and synbiotics—approaching a definition. Am J Clin Nutr 73:361S–364S
Selim KM, Reda RM (2015) Beta-glucans and mannan oligosaccharides enhance growth and immunity in Nile tilapia. N Am J Aquac 77:22–30
Sghir A, Chow JM, Mackie RI (1998) Continuous culture selection of bifidobacteria and lactobacilli from human fecal samples using fructooligosaccharides as selective substrate. J Appl Microbiol 85:769–777
Shank EA, Klepac-Ceraj V, Collado-Torres L, Powers GE, Losick R, Kolter R (2011) Interspecies interactions that result in *Bacillus subtilis* forming biofilms are mediated mainly by members of its own genus. Proc Natl Acad Sci 108:E1236–E1243
Skjermo J, Bakke I, Dahle SW, Vadstein O (2015) Probiotic strains introduced through live feed and rearing water have low colonizing success in developing Atlantic cod larvae. Aquaculture 438:17–23

Sohn KS, Kim MK, Kim JD, Han IK (2000) The role Immunostimulants in monogastric animal and fish—review. Asian-Australian J Anim Sci 13:1178–1187

Soleimani N, Hoseinifar SH, Merrifield DL, Barati M, Abadi ZH (2012) Dietary supplementation of fructooligosaccharide (FOS) improves the innate immune response, stress resistance, digestive enzyme activities and growth performance of Caspian roach (*Rutilus rutilus*) fry. Fish Shellfish Immunol 32:316–321

Song Z, Wu T, Cai L, Zhang L, Zheng X (2006) Effects of dietary supplementation with *Clostridium butyricum* on the growth performance and humoral immune response in *Miichthys miiuy*. J Zhejiang Univ Sci B 7:596–602

Song SK, Beck BR, Kim D, Park J, Kim J, Kim HD, Ringø E (2014) Prebiotics as immunostimulants in aquaculture: a review. Fish Shellfish Immunol 40:40–48

Souza FP, Lima ECS, Urrea-Rojas AM, Suphoronski SA, Facimoto CT, Bezerra Júnior JDS, Oliveira TES, Pereira UP, Santis GWD, Oliveira CAL, Lopera-Barrero NM (2020) Effects of dietary supplementation with a microalga (*Schizochytrium* sp.) on the hemato-immunological, and intestinal histological parameters and gut microbiota of Nile tilapia in net cages. PLoS One 15:e0226977

Staykov Y, Spring P, Denev S, Sweetman J (2007) Effect of a mannan oligosaccharide on the growth performance and immune status of rainbow trout (*Oncorhynchus mykiss*). Aquac Int 15:153–161

Sun YZ, Xia HQ, Yang HL, Wang YL, Zou WC (2014) TLR2 signaling may play a key role in the probiotic modulation of intestinal microbiota in grouper *Epinephelus coioides*. Aquaculture 430:50–56. https://doi.org/10.1016/j.aquaculture.2014.03.042

Sun Y, He M, Cao Z, Xie Z, Liu C, Wang S, Zhou Y (2018) Effects of dietary administration of *Lactococcus lactis* HNL12 on growth, innate immune response, and disease resistance of humpback grouper (*Cromileptes altivelis*). Fish Shellfish Immunol 82:296–303

Swennen K, Courtin CM Lindemans GC, Delcour JA (2006) Large-scale production and characterisation of wheat bran arabinoxylooligosaccharides. J Sci Food Agric 86:1722–1731

Torrecillas S, Makol A, Caballero MJ, Montero D, Robaina L, Real F, Sweetman J (2007) Immune stimulation and improved infection resistance in European sea bass (*Dicentrarchus labrax*) fed mannan oligosaccharides. Fish Shellfish Immunol 23:969–981

Tovar D, Zambonino J, Cahu C, Gatesoupe F, Vazquez R, Lesel R (2002) Effect of live yeast incorporation in compound diet on digestive enzyme activity in sea bass (*Dicentrarchus labrax*) larvae. Aquaculture 204:113–123

Van Hai N, Buller N, Fotedar R (2010) Encapsulation capacity of *Artemia nauplii* with customized probiotics for use in the cultivation of western king prawns (*Penaeus latisulcatus* Kishinouye, 1896). Aquac Res 41:893–903

Vazirzadeh A, Marhamati A, Rabiee R, Faggio C (2020) Immunomodulation, antioxidant enhancement and immune genes up-regulation in rainbow trout (*Oncorhynchus mykiss*) fed on seaweeds included diets. Fish Shellfish Immunol 106:852–858

Verschuere L, Rombaut G, Sorgeloos P, Verstraete W (2000) Probiotic bacteria as biological control agents in aquaculture. Microbiol Mol Biol Rev 64:655–671

Vinoj G, Jayakumar R, Chen JC, Withyachumnarnkul B, Shanthi S, Vaseeharan B (2015) N-hexanoyl-L-homoserine lactone-degrading *Pseudomonas aeruginosa* PsDAHP1 protects zebrafish against *Vibrio parahaemolyticus* infection. Fish Shellfish Immunol 42:204–212

Wang YB, Li JR, Lin J (2008) Probiotics in aquaculture: challenges and outlook. Aquaculture 281:1–4

Wang A, Ran C, Wang Y, Zhang Z, Ding Q, Yang Y, Olsen RE, Ringø E, Bindelle J, Zhou Z (2019) Use of probiotics in aquaculture of China—a review of the past decade. Fish Shellfish Immunol 86:734–755

Wee W, Hamid NKA, Mat K, Khalif RIAK, Rusli ND, Rahman MM, Kabir MA, Wei LS (2002) The effects of mixed prebiotics in aquaculture: a review. Aquac Fish

Welker TL, Lim C, Yildirim-Aksoy M, Shelby R, Klesius PH (2007) Immune response and resistance to stress and Edwardsiella ictaluri challenge in channel catfish, *Ictalurus punctatus*, fed diets containing commercial whole-cell yeast or yeast subcomponents. J World Aquac Soc 38:24–35

Wuertz S, Schroeder A, Wanka KM (2021) Probiotics in fish nutrition—long-standing household remedy or native nutraceuticals? Water 13:1348. https://doi.org/10.3390/w13101348

Xu B, Wang Y, Li J, Lin Q (2009) Effects of prebiotic xylooligosaccharides on growth performance and digestive activities of allogynogenetic crucian carp. Fish Physiol Biochem 35:351–357

Ye JDK, Wang FD, Li YZ (2011) Single or combined effects of fructo- and mannan oligosaccharide supplements and *Bacillus clausii* on the growth, feed utilization, body composition, digestive enzyme activity, innate immune response and lipid metabolism of Japanese flounder *Paralichthys olivaceus*. Aquac Nutr 17:902–911

Yi M, Wang C, Wang H, Zhu X, Liu Z, Gao F, Ke X, Cao J, Wang M et al (2021) The *in situ* remediation of aquaculture water and sediment by commercial probiotics immobilized on different carriers. Water Reuse 11:572–585

Yilmaz S, Yilmaz E, Dawood MAO, Ringo E, Ahmadifar E, Abdel-Latif KM-R (2022) Probiotics, prebiotics, and synbiotics used to control vibriosis in fish: a review. Aquaculture 547:737517

Zhang Q, Ma H, Mai K, Zhang W, Liufu Z, Xu W (2010) Interaction of dietary Bacillus subtilis and fructooligosaccharide on the growth performance, non-specific immunity of sea cucumber, *Apostichopus japonicus*. Fish Shellfish Immunol 29:204–211

Zhao X, Yang J, Wang L, Lin H, Sun S (2017) Protection mechanism of *Clostridium butyricum* against *Salmonella enteritidis* infection in broilers. Front Microbiol 8:1523

Zibiene G, Zibas A (2019) Impact of commercial probiotics on growth parameters of European catfish (*Silurus glanis*) and water quality in recirculating aquaculture systems. Aquuacult Int 27:1751–1766

Chapter 13
Bacteriocins as an Antimicrobial Agent for the Treatment of AMR in Aquaculture

Sweta Patel, Aneri Joshi, and Sriram Seshadri

Abstract Aquaculture is a large source of food worldwide. Excessive use of antibiotics and antimicrobial agents has led to the development of antibiotic resistance in aquatic systems. This resistance can be transferred to humans via the food chain, and it can contribute to the development of antibiotic resistance in human pathogens through the transfer of resistance genes from aquatic bacteria to human pathogens. There are various researches going on to combat this issue. There are various researches going on to combat this issue using plant and microbe-based remedies against aqua pathogens. Antimicrobial peptides, more specifically, bacteriocins, are small proteinaceous molecules that are synthesized and secreted by bacteria. Bacteriocins are considered a promising alternative to conventional antibiotics due to their high specificity, low toxicity, and ability to be produced by natural microorganisms. Addition of bacteriocin or bacteriocin-secreting bacteria to aquaculture acts as an alternative to control morbidity and is considered safe as it does not have any adverse effect. The use of bacteriocins as antimicrobial agents can help control bacterial infections and reduce the selective pressure that drives the emergence of antibiotic resistance.

Keywords Aquaculture · Antimicrobial resistance (AMR) · Antibiotic resistance · Bacteriocins · Antimicrobial proteins

---

S. Patel
Institute of Science, Nirma University, Ahmedabad, Gujarat, India

Iwana Consultancy Solution Pvt Ltd, Mumbai, Maharashtra, India

A. Joshi · S. Seshadri (✉)
Institute of Science, Nirma University, Ahmedabad, Gujarat, India
e-mail: sriram.seshadri@nirmauni.ac.in

P. Elumalai, S. Lakshmi (eds.), *Antimicrobial Resistance in Aquaculture and Aquatic Environments*, https://doi.org/10.1007/978-981-97-7320-6_13

Abbreviations

AMP	Antimicrobial protein
AMR	Antimicrobial resistance
ARB	Antimicrobial-resistant bacteria
ARGs	AMR genes
QQ	Quorum quenching
QS	Quorum sensing

13.1 Introduction

13.1.1 Aquaculture and Infection

Fish cultures are usually done in ponds which are either outdoor or earthen (clay lined). Over time, they increase with organic load due to feed, fish growth, and excretions. High organic load favors bacterial and viral growth in water, eventually infecting aquatic animals. Open oceans, large lakes, and other natural open systems facilitate the dilution of organic load and dissolve waste, but it facilitates the entry of pathogens directly or indirectly into the open system (Yanong et al. 2022). Zoonotic pathogens and disease spread due to them cause considerable problem in fishery industries. These pathogens are Gram-positive (Mycobacteriaceae, Streptococcaceae, and Erysipelotrichaceae families) and Gram-negative (Aeromonadaceae, Vibrionaceae, Pseudomonadaceae, Enterobacteriaceae, and Hafniaceae families) bacteria, parasite (cestodes, trematodes, nematodes) (Ziarati et al. 2022), virus (viral hemorrhagic septicemia virus, infectious hematopoietic necrosis virus, infectious salmon anemia virus, *Piscine orthoreovirus*, tilapia lake virus, covert mortality nodavirus, shrimp hemocyte iridescent virus, and abalone herpesvirus) (Kibenge 2019), and fungi.

To respond to diseases in aquatic environments as we exit a worldwide pandemic of zoonotic origin, we must concentrate on prompt diagnosis to establish the etiology. The known diversity of RNA viruses is doubled by metagenomic investigation of the aquatic virome. The findings show a sizable RNA virome that contains numerous aquatic virus groups that were previously undiscovered. Aside from demonstrating the enormous diversity of aquatic viruses, metagenomics has also demonstrated that these viruses have a considerably larger impact on host communities than was previously thought (Hutson et al. 2023; Zhang et al. 2022). Apart from metagenome, environmental DNA/RNA (eDNA/eRNA) technique is also widely used to identify bacterial and fungal contaminant in aquaculture (Bohara et al. 2022).

13.2 Why There Is a Concern for AMR in Aquaculture

Antimicrobial resistance (AMR) is a growing concern in aquaculture because the use of antibiotics and other antimicrobial agents in aquaculture can lead to the development and spread of antibiotic-resistant bacteria in aquatic environments (Sahoo et al. 2016). Aquaculture is an important source of food globally; therefore it can also contribute to the development of antibiotic resistance in human pathogens through the transfer of resistance genes from aquatic bacteria to human pathogens (Preena et al. 2020). This can occur through the food chain, direct contact with aquatic environments, or the use of aquaculture products such as fish meal or fish oil in animal feed.

To enhance aqua production, the use of antibiotics has increased worldwide. But overuse of antibiotics has increased the risk of AMR. Also, it has been found that compared to the natural habitat fishes, cultured fishes have high antimicrobial resistance to pathogens because of the preventive strategy. For example, the Gram-negative bacteria *Aeromonas hydrophila* and *Plesiomonas shigelloides* showed increased antimicrobial resistance in cultured catfish than in the natural ecosystem catfish through plasmid-mediated (R-plasmid) AMR. Zoonotic pathogens, such as *Streptococcus iniae*, *Aeromonas hydrophila*, *Vibrio vulnificus*, *Photobacterium damselae*, and *Mycobacterium marinum*, carry extended-spectrum beta-lactamases and other AMR genes (ARGs) that spread through the food web. Certain antibacterial genes like tel, sul, and qnr are transported by several site-specific recombination events (Preena et al. 2020). The most frequent antimicrobial-resistant bacteria (ARB) investigated belonged to *Vibrio* spp. (25%) and *Aeromonas* spp. (18%), followed by the Enterobacteriaceae *Escherichia coli* (7%) (Caputo et al. 2023).

The emergence and spread of antibiotic-resistant bacteria can limit the effectiveness of antibiotics in treating bacterial infections in humans and animals and pose a significant public health threat. The World Health Organization (WHO) has identified AMR as one of the top global public health threats, and the responsible use of antibiotics in aquaculture is an important step toward addressing this threat (Thornber et al. 2022).

13.2.1 Available Treatment Against AMR

There are several available treatments and therapeutics for AMR (Fig. 13.1) in aquaculture, including:

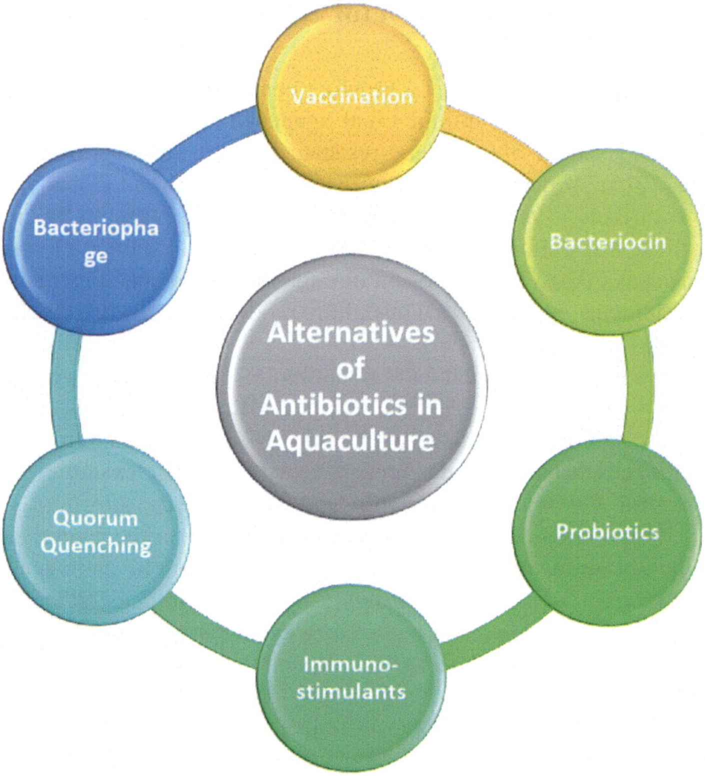

Fig. 13.1 Various available options for antibiotics in aquaculture

13.2.1.1 Vaccination

Vaccines based on inactive bacterial pathogen are more successful in aquaculture (Sommerset et al. 2005), and some vaccines based on live attenuated bacteria are also available in market for aquaculture (Ma et al. 2019). The first fish vaccination program was initiated in 1942 against *Aeromonas salmonicida* in cutthroat trout (*Oncorhynchus clarkii*) (Bondad-Reantaso et al. 2023). With advancement in biotechnology, many vaccines are commercialized. Vaccination against most common diseases affecting aquaculture species, such as furunculosis, which is caused by the bacterium *Aeromonas salmonicida*, and *vibriosis*, has been found to help in reduction of antibiotic usage in aquaculture (Caputo et al. 2023). For large-scale aquafarming, oral mode of vaccination is preferred compared to injecting individual fish (Du et al. 2022).

However, difficulties persist because the majority of marketed vaccinations are intraperitoneal injections of killed whole cell pathogen preparations. The conventional method is less efficient and requires booster dose. New technology can be approached for fish vaccination, but it is expensive. Due to lack of efficient adjuvants and a lack of understanding of the immune response, progress in the creation of mucosal vaccines has been hampered. In some nations, the price of injecting fish may be too high to treat the disease (Adams 2019; Ma et al. 2019; Nayak 2020).

13.2.1.2 Immune Stimulants

They are chemical compounds that activate innate immune system. They are bio-friendly agent such as probiotic, vitamins, polysaccharides, herbs, and biological factors added in to feed formulation, which can control and kill pathogenic bacteria (Wang et al. 2017). Fishes have a strong innate immune system to cope with various infections. Various pathogen-associated molecules can be ingested to elevate defense system against pathogen (Wangkahart et al. 2019). They are shown to improve disease resistance via nonspecific immunity. Levamisole, a chemical compound, is increasing phagocytic activity and resistance against *Aeromonas salmonicida* infection. FK 656, a synthetic peptide, provides protection against *Aeromonas salmonicida*. Lipopolysaccharide and muramyl dipeptide stimulate B cells. Certain herbs like *Ocimum tenuiflorum*, *Zingiber officinale*, and *Allium cepa are also immunostimulants due to their anti-stress and anti-oxidative properties. The disadvantages of such products include limited efficacy, high cost, the risk of overdose leading to immunosuppression, and a lack of broad-spectrum effectiveness* (Kela 2022).

13.2.1.3 Quorum Quenching

Quorum sensing (QS) signals produced by heavy bacterial growth produce virulence factors which are capable of causing massive death in aquaculture (Priya et al. 2022). They communicate with each other using small molecules called autoinducers which are detected on the cell membrane by receptors and eventually regulate certain gene expression of bacteria (Defoirdt et al. 2008). To prevent QS, quorum quenching (QQ) is the strategy used in aquaculture. One of the strategies is to use another bacteria like Bacillus spp. as QQ against Vibrio spp. These species are commensal microflora of fish gut and act as bio-controlling agents against pathogenic Vibrio spp. via reducing its C6-HSL, which is responsible for pigment production and quorum sensing (Müller et al. 2023). In silico and in vitro studies have shown resistance development to QQ by microorganism either by reducing gene expression to activate QS signalling or by mutation in gene related to efflux pump in P. aeruginosa (Krzyżek 2019).

13.2.1.4 Bacteriophage

Viruses which can invade bacteria and cause lysis are called bacteriophage. Phage therapy showed prominent results when given in combination against common bacterial infection. Low-cost and eco-friendly nature makes using them feasible (Liu et al. 2022). Host-phage relationship is also complex and their host-specific nature makes them useful against limited disease/bacteria. Bacteria have gain protection

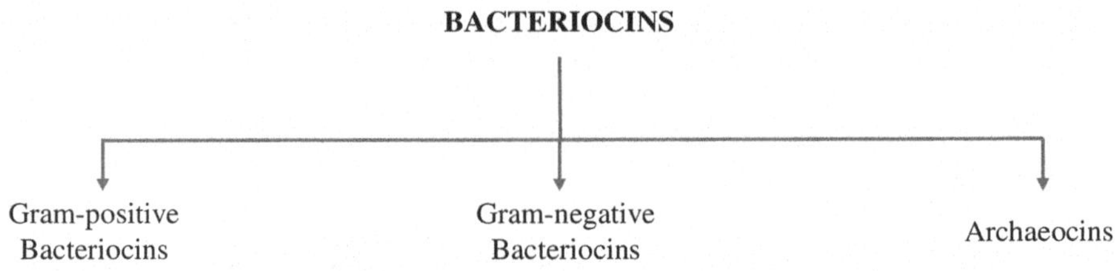

Fig. 13.2 Bacteriocin classification

system against phage such as adsorption inhibition, CRISPR-Cas mechanism, and mutation in genes (Bondad-Reantaso et al. 2023; Ninawe et al. 2020).

13.2.1.5 Probiotics

Probiotics are live beneficial bacteria that are ingested through food or water and help maintain a healthy balance of microorganisms inside the body by producing bacteriocins, siderophores, lysozymes, proteases, and hydrogen peroxides which can inhibit the growth of harmful pathogens. Probiotics could be used effectively in aquaculture diets to reduce antibiotic adverse effects. By promoting these fish feed additives, it may increase the productive performance and feed utilization of the fish, which would increase fish production and protect public health (El-Saadony et al. 2021).

13.3 Bacteriocin

All living systems produce antimicrobial protein (AMP), also called antimicrobial peptide, due to their relatively small molecular size. They are believed to be the primary line of defense in the host system that protect against intruders. There are two types of AMPs produced by bacteria. The first one is called bacteriocin which is ribosomally synthesized peptide, and the other is produced non-ribosomally (non-coded gene for AMPs), like ε-poly-L-lysine (Chikindas et al. 2018). Bacteriocins, a type of AMP, are only toxic to bacteria which are closely related to producing strain (Arthur et al. 2014). Bacteriocin is composed of toxin gene, immunity gene, and lysis gene. Here, toxic gene acts as bacteriocin; an immunity gene provides self-protection against toxin; and a lysis gene helps in releasing of toxin from producing cell (Suphan et al. 2012).

Colicin produced from *Escherichia coli* was the first discovered bacteriocin in 1928 for its antimicrobial activity against some strains of *E. coli*. First classification and nomenclature were done by Fredericq in 1957 (Suphan et al. 2012) (Fig. 13.2).

Usually, they are named based on producer species (Table 13.1). Further, they are divided based on receptor specificity and immunity they have. For example, colicin

Table 13.1 Naming of bacteriocin based on producer species

Name of bacteria	Bacteriocin
Escherichia coli	Colicins
Pseudomonas aeruginosa	Pyocins
Enterobacter cloacae	Cloacins
Bacillus cereus	Cerecins
Yersinia pestis	Pesticins
Pediococcus spp.	Pediocin
Lactococcus lactis	Nisin and lacticin
Leuconostoc spp.	Leucocin
Enterococcus spp.	Enterocin
Bacillus megaterium	Megacins
Carnobacterium spp.	Carnocyclin
Klebsiella pneumoniae	Klebicin
Lactobacillus helveticus	Helveticin I

is classified as per its receptor specificity in 17 subtypes such as colicin A, B, C, and so on. Each subtype is further denoted as A1, A2, etc. depending on their different immunity pattern. (Fredericq 1957). Gram-p ositive bacteria produce a diverse range of bacteriocins which are broadly classified as bacteriocins and termed Gram-positive bacteriocins. The Gram-negative bacteria mostly produce colicins and microcins, which are termed Gram-negative bacteriocins (Lagos 2013). There is another class of bacteriocins which are produced by archaea commonly known as archaeocins (Besse et al. 2015).

13.3.1 Gram-Positive Bacteriocins

Klaenhammer classified Gram-positive bacteriocins into four subclasses based on molecular mass, thermostability, posttranslational modifications, amino acids, enzymatic activity, and mode of action (Oscáriz and Pisabarro 2001) (Fig. 13.3).

13.3.1.1 Class I Bacteriocins

Class I bacteriocins are commonly known as lantibiotics. Lantibiotics are heat-stable peptides (<5 kb) and possess posttranslationally modified amino acids like lanthionine and B-methyllanthionine and some dehydrated amino acids. Lantibiotics are further classified into type A and type B, based on their biological activity, structure, and charge (Bharti et al. 2015; Negash and Tsehai 2020).

Type A lantibiotics have nisin and lacticin 3147, secreted from Gram-positive *Lactococcus* species. They have a positive charge and are linear in structure (screw-shaped), with 21–38 amino acids in size. They act on the target cells by forming

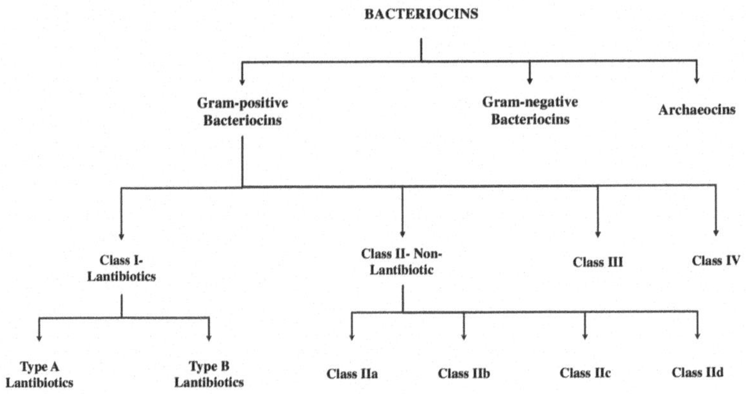

Fig. 13.3 Classification of Gram-positive bacteriocins

pores and depolarizing the cytoplasmic membrane (de Arauz et al. 2009; Ryan et al. 1996).

Type B lantibiotics have mersacidin secreted by *Bacillus* spp. They have a globular, secondary structure and are comparatively smaller in size, having 19 amino acids, and do not possess any net charge. They interfere with cellular enzymatic reactions, such as cell wall synthesis, and act on leader proteins cleaving by an ABC transporter (Negash and Tsehai 2020) (Fig. 13.4).

13.3.1.2 Class II Bacteriocins

Class II bacteriocins are heat stable, non-lanthionine, not posttranslationally modified, having 30–60 amino acids, and positively charged at neutral pH (Negash and Tsehai 2020). An amphiphilic helical structure allows these class II bacteriocins to insert into the membrane target, depolarizing the cell membrane (Bharti et al. 2015). Class II bacteriocins are further classified into four subclasses; IIa, IIb, IIC, and IId.

Class IIa has podocin-like bacteriocin with an anti-listerial effect. The members of this family are characterized by a conserved amino acid sequence (YGNGVXaaC). Some examples of class IIa include pediocin PA-1 and pediocin AcH produced by *Pediococcus acidilactici*, *sakacins* produced by *Lactobacillus sakei*, leucocin A produced by *Leuconostoc gelidum*, *enterocins* A and P produced by *Enterococcus faecium, and carnobacteriocin* produced by *Carnobacterium* sp. These class IIa bacteriocins are active against other *Lactobacillus* species but are particularly effective against *Listeria monocytogenes*. The best-known bacteriocin is pediocin PA-1, produced by *Pediococcus acidilactici* (Negash and Tsehai 2020).

Class IIb bacteriocin activity depends on the two components, in which two distinct peptides, type E (enhanced) and type S (synergistic), act synergistically to generate an antimicrobial effect. This class includes lacticin F and lactococcin G, secreted by *Lactobacillus* strain. The class IIb bacteriocins of this class form pores

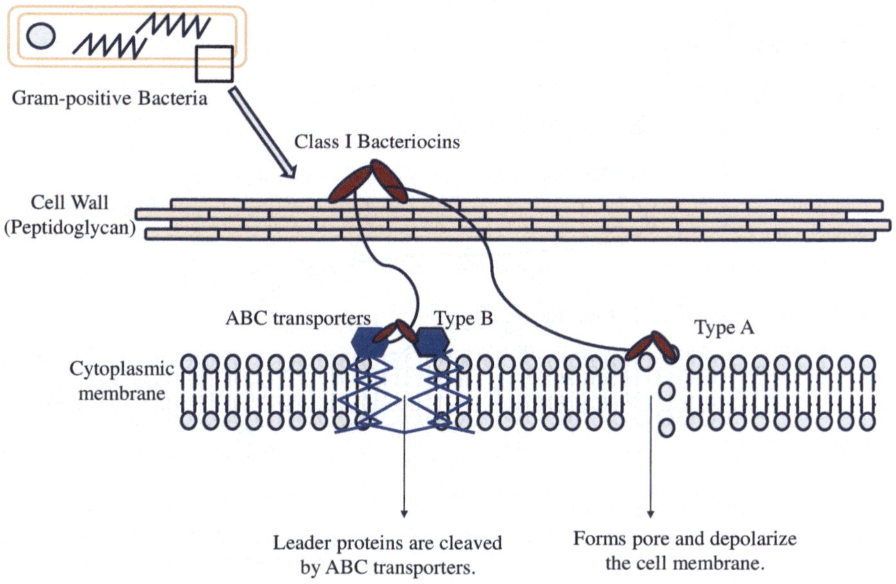

Fig. 13.4 Mechanism of class I bacteriocins. (Adapted from Bharti et al. 2015; Negash and Tsehai 2020)

in the membrane of target cells, which release various ions by disrupting the transmembrane electric potential, and pH gradient, eventually dissipating the proton motive force, eventually resulting into decrease of the ATP concentration, and causing cell death (Bharti et al. 2015) (Fig. 13.5).

Class IIc bacteriocins are cyclic/circular with their N- and C-terminal covalently linked. They are hydrophobic, 3400–7300 Da in size, and cationic in nature. The cyclic structure of these bacteriocins helps permeabilize the membrane by disrupting the proton motive force, resulting in cell death. Due to this cyclic/circular structure, they have an advantage in antimicrobial activity (Negash and Tsehai 2020). Class IIc bacteriocins include gassericin A produced by *Lactobacillus gasseri* LA39, circularin A produced by *Clostridium beijerinckii*, and carnocyclin A produced by *Carnobacterium* spp. (Kawai et al. 1998, 2004).

Class IId includes other bacteriocins, such as lactocin A and B, which require lipid or carbohydrate moieties for their activity. This class does not have much information (Negash and Tsehai 2020).

13.3.1.3 Class III Bacteriocins

Class III bacteriocins are heat-labile, with high molecular weight. They have a catalytic domain at the N-terminus, while C-terminus represents the target recognition site. This class includes colicins, megacins from *Bacillus megaterium*, klebicin from *Klebsiella pneumoniae*, helveticin I from *Lactobacillus helveticus*, and

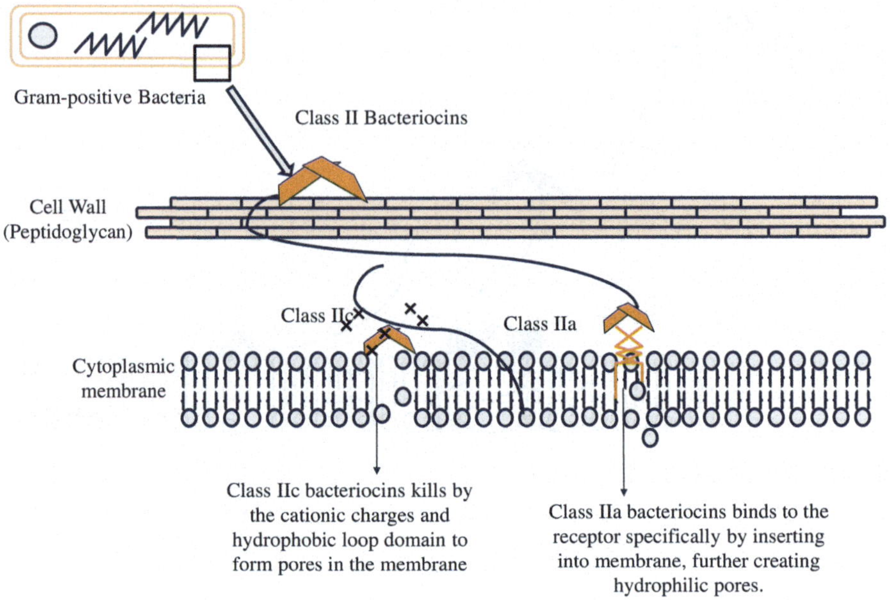

Fig. 13.5 Mechanism of class II bacteriocins. (Adapted from Oscáriz and Pisabarro 2001)

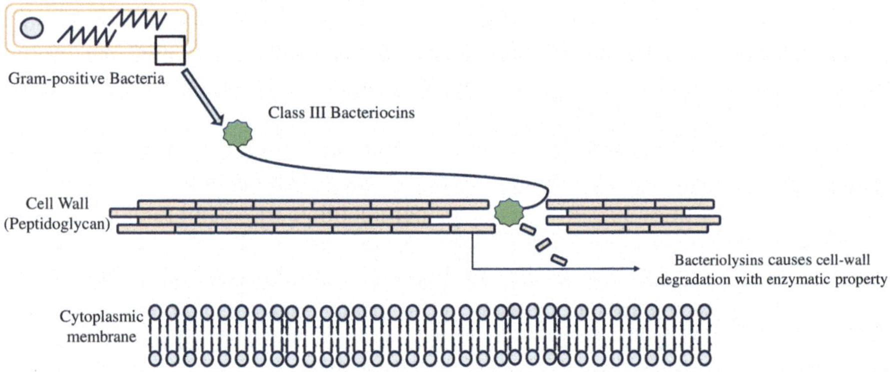

Fig. 13.6 Mechanism of class III bacteriocin

enterolysin produced from *Enterococcus faecalis* (Chikindas et al. 2018). Bacteriolysins are a group of bacteriocin with proteins with an enzymatic property that causes cell wall degradation (Sahoo et al. 2016) (Fig. 13.6).

13.3.1.4 Class IV Bacteriocins

This class IV of bacteriocins are posttranslationally modified into circular head-to-tail-like structures, having a complex synthesizing machinery (Sahoo et al. 2016). They are cyclic peptides containing either a glycoprotein (lactocin 27) or lipoprotein (lacstrepcins) (Bharti et al. 2015; Oscáriz and Pisabarro 2001).

13.3.2 Gram-Negative Bacteriocins

Kolter and Moreno have classified the Gram-negative bacteriocins, which were distinguished from the Gram-positive bacteriocins. The Gram-negative bacteriocin have included colicins and microcins having molecular weight of 10 kDa and belonging to the Enterobacteriaceae family (Oscáriz and Pisabarro 2001). Gram-negative bacteria family Enterobacteriaceae also produces peptides and proteins like molecules which can be classified as bacteriocins. Gram-negative bacteriocins include two major groups: colicins and microcins (Bharti et al. 2015) (Fig. 13.7).

Colicins are produced by *Escherichia coli* and have a colicinogenic plasmid named pCol. They have a high molecular weight, around 30–80 kDa. These colicins are further subdivided into two types depending on their size, copy number, amplification ability, and ability to be transferred by conjugation. Type 1 colicins have small plasmids, which occur in large copy numbers. It includes pColE1 with a 6.6 kb size. Type II colicins have large plasmids, occurring in one or two per cell. It

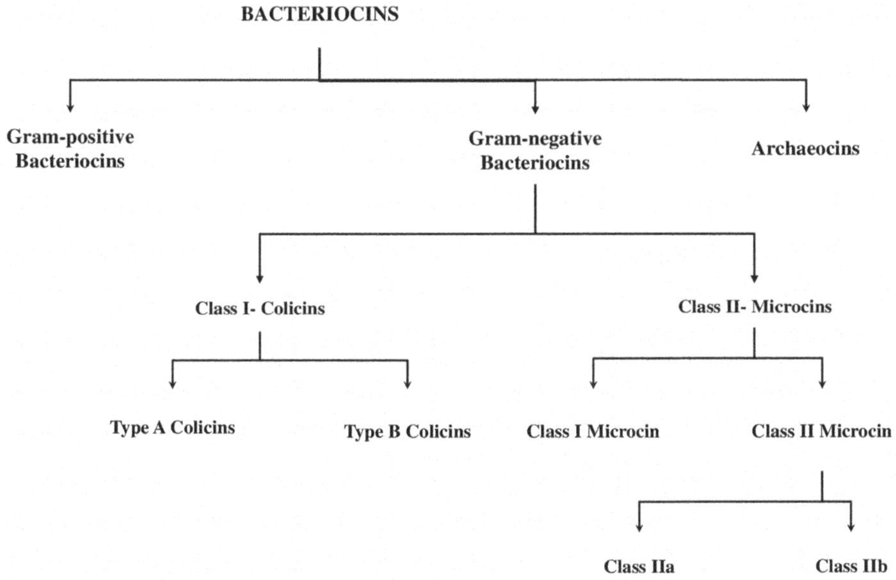

Fig. 13.7 Classification of Gram-negative bacteriocins

includes pColH1 with a 94 kb size. Colicins act on target cells by making voltage-dependent channels in the inner membrane by interfering with their nuclease activity (DNase, 16s rRNase, and tRNase activity) or inhibiting their peptidoglycan synthesis (Marković et al. 2022).

Enteric bacteria produce microcins with low molecular weight (1–10 kDa). Microcins are classified into two classes, depending on their location, posttranslational modification, gene organization, and leader peptide sequence (Negash and Tsehai 2020). Class I microcins include B17, C7–C51, and J25, with low molecular mass peptide <5 kDa. They are plasmid-encoded and undergo posttranslational modifications (Marković et al. 2022).

Class II microcins are subgrouped into class IIa and IIb. They have a high molecular weight of around 5–10 kDa. Class IIa includes microcin L, V, and N, which are plasmid-encoded and do not undergo posttranslational modification. Class IIa includes microcin L, V, and N, which are plasmid-encoded and do not undergo posttranslational modification, while the class IIb includes microcin E492, M, and H47, which undergo posttranslational modification-C-terminal siderophore (Marković et al. 2022).

13.3.2.1 Mode of Action of Gram-Negative Bacteriocin Colicin

The Gram-negative bacteriocin acts on the target cell by various bactericidal mechanisms. Pore-forming colicins can form pores in the target cell's cell membrane, eventually leading to cell membrane leakage, change in ion gradient, electrochemical disbalance, or cell death. The nuclease activity of nuclease colicins will contain DNase and RNase, which will target the DNA or RNA of the target cell. Peptidoglycan is a very important component of the cell wall of bacteria, and the peptidoglycanase colicin will inhibit the synthesis of peptidoglycan, which will lead to bacterial cell wall destruction and, eventually, bacterial death (Darbandi et al. 2022) (Fig. 13.8).

13.3.3 Archaeocins

Archaeocins are peptides from archaea which are ribosomally produced or can be stepwise multienzyme-synthesized. They are further classified as halocins, produced by *Halobacteria*, and sulfolobicins, produced by *Crenarchaeota*. These archaeocins are synthesized mainly in the stationary phase (Besse et al. 2015; Liu et al. 2022; O'Connor and Shand 2002) (Fig. 13.9).

Halocins are 4–5 kDa in size. The most common halocins include H1, produced from *Haloferax mediterranei*, and H6, produced from *Haloferax gibbonsii*. Sulfolobicins, another type of archaeocins, are completely different from halocins as they are predominantly associated with cells. Sulfolobicins are likely to show similar characteristic features to classical bacteriocins, as they are proteinaceous and directed against the strain (O'Connor and Shand 2002) (Table 13.2).

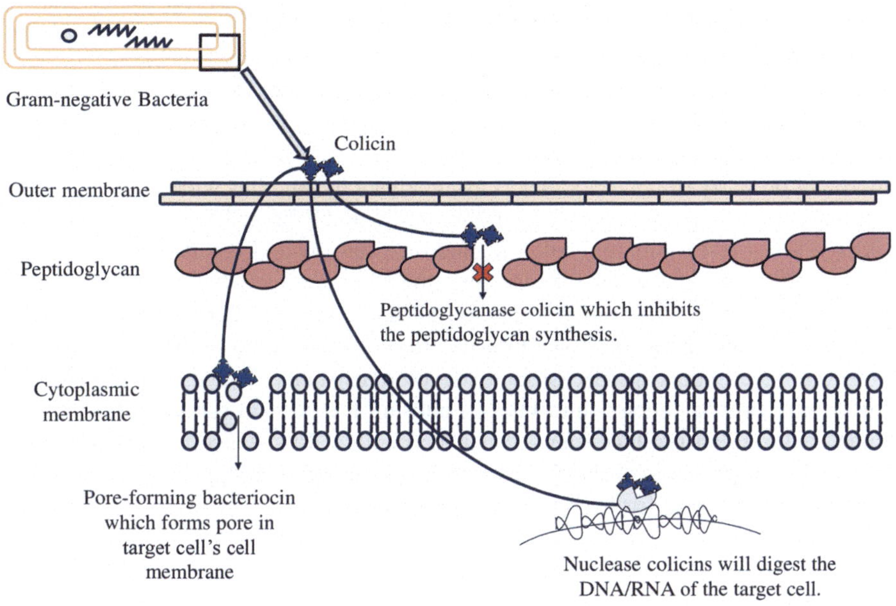

Fig. 13.8 Mode of action of colicins

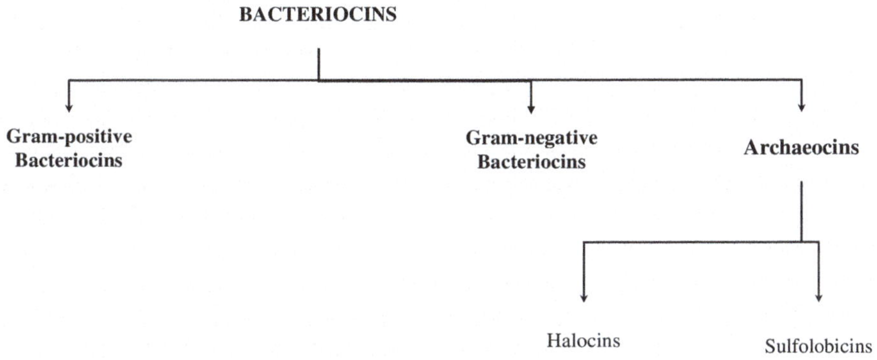

Fig. 13.9 Classification of archaeocins

Table 13.2 Naming of archaeocin based on producer species

Name of bacteria	Archaeocin
Haloferax mediterranei strain M2a (previously Xia3)	HalH1
Haloferax gibbonsii Ma2.39	HalH6
Halobacterium sp. GN101	HalR1
Uncharacterized halophilic strain	HalA4
Haloferax mediterranei Sech7a	HalSech7A
Sulfolobus islandicus HEN2/2	Sulfolobicins

13.4 Bacteriocins: Advantages and Limitations

Initially, bacteriocin was used as a food preservative. Later on, based on its potential, it is used in human and veterinary health. Bacteriocin has proven potency against clinically important pathogens and multidrug resistance bacteria. The antimicrobial property of bacteriocin acts as an advantage and has wide applications in pharmaceutics, cosmetics, and food industry (Pircalabioru et al. 2021). Lower cytotoxicity, lower environmental half-life, and their narrow and broad-spectrum specificity make them very useful. Despite their numerous advantages, marketed bacteriocin is lower in number compared to identified one. This is because of the low yield and expensive process of purification of bacteriocin (Schofs et al. 2020). However, bacteriocins are also prone to develop resistance. Resistance to ward lantibiotic subclass has been found via cell altering mechanism (Draper et al. 2015). Some bacteria can produce nisin resistance peptide. Resistance toward pediocin-like class IIa bacteriocin is also found to have downregulated Man-PTS gene expression which is receptor for bacteriocin (Kjos et al. 2011).

13.5 Future Aspects

There is an increase of bacterial pathogens in freshwater systems, which leads to economic loss in the fishery sector. In the aquaculture systems, there has been an outbreak of bacterial disease which has increased the use of antibiotics. One way to overcome fish mortality is by bacteriocins, which can be used as a safe and alternative reason for other strategies (Nayak et al. 2022). But this use of antibiotics can have several drawbacks, including the production loss of fish farming. The vaccination and antibiotic strategies are the most ancient but tedious processes; thus, there has been research about the new treatment strategy as an alternative. To avoid this, new therapeutics for fish farming and infection control strategies have been introduced, such as bacteriocins (Pereira et al. 2022).

Bacteriocin has numerous qualities which make them suitable to use as an alternative to antibiotics, as they are nontoxic to eukaryotic cells and can be used as a safe alternative to traditional methods. In addition to that, they are primary metabolites as they are synthesized ribosomally compared to antibiotics which are secondary metabolites (Darbandi et al. 2022). These bacteriocins have stability at higher saline concentrations of up to 10%, which is quite a remarkable quality. As bacteriocin has a narrow killing spectrum of bacteria, it can reduce the incidence of drug resistance pathogens by limiting the selective pressure to evolve in bacteria (Suphan et al. 2012). Bacteriocin can also be widely used as genetic engineering techniques can easily evolve and manipulate it (Benítez-Chao et al. 2021). Also, it can be formulated or encapsulated to prevent its enzymatic degradation in gut (Mathur et al. 2017). Thus, as an antimicrobial property, this bacteriocin can target bacterial infection and enhance the immune structure of aquatic species, as they have a wide range of producers, varied mode of action, and biochemical properties (Pereira et al. 2022). M ore research on the mode of action of many bacteriocins is required such as class IId and class IV bacteriocin, and synergistic effect of different bacteriocins may help to combat resistance development.

References

Adams A (2019) Progress, challenges and opportunities in fish vaccine development. Fish Shellfish Immunol 90:210–214. https://doi.org/10.1016/j.fsi.2019.04.066

Arthur TD, Cavera VL, Chikindas ML (2014) On bacteriocin delivery systems and potential applications. Future Microbiol 9(2):235–248. https://doi.org/10.2217/fmb.13.148

Benítez-Chao DF, León-Buitimea A, Lerma-Escalera JA, Morones-Ramírez JR (2021) Bacteriocins: an overview of antimicrobial, toxicity, and biosafety assessment by in vivo models. Front Microbiol 12:630695

Besse A, Peduzzi J, Rebuffat S, Carré-Mlouka A (2015) Antimicrobial peptides and proteins in the face of extremes: lessons from archaeocins. Biochimie 118:344–355. https://doi.org/10.1016/j.biochi.2015.06.004

Bharti V, Mehta A, Singh S, Jain N, Ahirwal L, Mehta S (2015) Bacteriocin: a novel approach for preservation of food. Int J Pharm Pharm Sci 7:20–29

Bohara K, Yadav AK, Joshi P (2022) Detection of fish pathogens in freshwater aquaculture using EDNA methods. Diversity 14(12):1015. https://doi.org/10.3390/d14121015

Bondad-Reantaso MG, MacKinnon B, Karunasagar I, Fridman S, Alday-Sanz V, Brun E, Le Groumellec M, Li A, Surachetpong W, Karunasagar I, Hao B, Dall'Occo A, Urbani R, Caputo A (2023) Review of alternatives to antibiotic use in aquaculture. Rev Aquac:1–31. https://doi.org/10.1111/raq.12786

Caputo A, Bondad-Reantaso MG, Karunasagar I, Hao B, Gaunt P, Verner-Jeffreys D, Fridman S, Dorado-Garcia A (2023) Antimicrobial resistance in aquaculture: a global analysis of literature and National Action Plans. Rev Aquac 15(2):568–578. https://doi.org/10.1111/raq.12741

Chikindas ML, Weeks R, Drider D, Chistyakov VA, Dicks LMT (2018) Functions and emerging applications of Bacteriocins. Curr Opin Biotechnol 49:23–28. https://doi.org/10.1016/j.copbio.2017.07.011

Darbandi A, Asadi A, Ari MM, Ohadi E, Talebi M, Zadeh MH, Emamie AD, Ghanavati R, Kakanj M (2022) Bacteriocins: properties and potential use as antimicrobials. J Clin Lab Anal 36(1):e24093. https://doi.org/10.1002/jcla.24093

de Arauz L, Juncioni AF, Jozala PG, Mazzola, and Thereza Christina Vessoni Penna. (2009) Nisin biotechnological production and application: a review. Trends Food Sci Technol 20(3):146–154. https://doi.org/10.1016/j.tifs.2009.01.056

Defoirdt T, Boon N, Sorgeloos P, Verstraete W, Bossier P (2008) Quorum sensing and quorum quenching in vibrio Harveyi: lessons learned from in vivo work. ISME J 2(1):19–26. https://doi.org/10.1038/ismej.2007.92

Draper LA, Cotter PD, Hill C, Paul Ross R (2015) Lantibiotic resistance. Microbiol Mol Biol Rev 79(2):171–191. https://doi.org/10.1128/MMBR.00051-14

Du Y, Hu X, Miao L, Chen J (2022) Current status and development prospects of aquatic vaccines. Front Immunol 13(2022). https://doi.org/10.3389/fimmu.2022.1040336

El-Saadony MT, Alagawany M, Patra AK, Kar I, Tiwari R, Dawood MAO, Dhama K, Abdel-Latif HMR (2021) The functionality of probiotics in aquaculture: an overview. Fish Shellfish Immunol 117:36–52. https://doi.org/10.1016/j.fsi.2021.07.007

Fredericq P (1957) Colicins. Ann Rev Microbiol 11:7–22. https://doi.org/10.1146/annurev.mi.11.100157.000255

Hutson KS, Davidson IC, Bennett J, Poulin R, Cahill PL (2023) Assigning cause for emerging diseases of aquatic organisms. Trends Microbiol 31:681. https://doi.org/10.1016/j.tim.2023.01.012
Kawai Y, Saito T, Kitazawa H, Itoh T (1998) Gassericin a; an uncommon cyclic bacteriocin produced by lactobacillus Gasseri LA39 linked at N- and C-terminal ends. Biosci Biotechnol Biochem 62(12):2438–2440. https://doi.org/10.1271/bbb.62.2438
Kawai Y, Kemperman R, Kok J, Saito T (2004) The circular bacteriocins gassericin A and circularin A. Curr Protein Peptide Sci 5(5):393–398. https://doi.org/10.2174/1389203043379549
Kela E (2022) Significance of Immunostimulants in aquaculture: a review. Aquac Fish 6(1):1–4. https://doi.org/10.24966/AAF-5523/100046
Kibenge FSB (2019) Emerging viruses in aquaculture. Curr Opin Virol 34:97–103. https://doi.org/10.1016/j.coviro.2018.12.008
Kjos M, Nes IF, Diep DB (2011) Mechanisms of resistance to bacteriocins targeting the mannose phosphotransferase system. Appl Environ Microbiol 77(10):3335–3342. https://doi.org/10.1128/AEM.02602-10
Krzyżek P (2019) Challenges and limitations of anti-quorum sensing therapies. Front Microbiol 10:2473. https://doi.org/10.3389/fmicb.2019.02473
Lagos R (2013) Bacteriocins. In: Maloy S, Hughes K (eds) Brenner's encyclopedia of genetics, 2nd edn. Academic Press, San Diego, pp 277–279
Liu R, Han G, Li Z, Cun S, Hao B, Zhang J, Liu X (2022) Bacteriophage therapy in aquaculture: current status and future challenges. Folia Microbiol 67(4):573–590. https://doi.org/10.1007/s12223-022-00965-6
Ma J, Bruce TJ, Jones EM, Cain KD (2019) A review of fish vaccine development strategies: conventional methods and modern biotechnological approaches. Microorganisms 7(11). https://doi.org/10.3390/microorganisms7110569
Marković KG, Grujović MŽ, Koraćević MG, Nikodijević DD, Milutinović MG, Semedo-Lemsaddek T, Djilas MD (2022) Colicins and Microcins produced by enterobacteriaceae: characterization, mode of action, and putative applications. Int J Environ Res Public Health 19(18):11825. https://doi.org/10.3390/ijerph191811825
Mathur H, Field D, Rea MC, Cotter PD, Hill C, Paul Ross R (2017) Bacteriocin-antimicrobial synergy: a medical and food perspective. Front Microbiol 8:1205. https://doi.org/10.3389/fmicb.2017.01205
Müller M, Spiers AJ, Tan A, Mujahid A (2023) Investigating quorum-quenching marine bacilli as potential biocontrol agents for protection of shrimps against early mortality syndrome (EMS). Sci Rep 13(1):4095. https://doi.org/10.1038/s41598-023-31197-4
Nayak SK (2020) Current prospects and challenges in fish vaccine development in India with special reference to Aeromonas Hydrophila vaccine. Fish Shellfish Immunol 100:283–299. https://doi.org/10.1016/j.fsi.2020.01.064
Nayak A, Karunasagar I, Chakraborty A, Maiti B (2022) Potential application of bacteriocins for sustainable aquaculture. Rev Aquac 14(3):1234–1248. https://doi.org/10.1111/raq.12647
Negash AW, Tsehai BA (2020) Current applications of bacteriocin. Int J Microbiol 2020:e4374891. https://doi.org/10.1155/2020/4374891
Ninawe AS, Sivasankari S, Ramasamy P, Seghal Kiran G, Selvin J (2020) Bacteriophages for aquaculture disease control. Aquac Int 28(5):1925–1938. https://doi.org/10.1007/s10499-020-00567-4
O'Connor EM, Shand RF (2002) Halocins and sulfolobicins: the emerging story of archaeal protein and peptide antibiotics. J Ind Microbiol Biotechnol 28(1):23–31. https://doi.org/10.1038/sj/jim/7000190
Oscáriz JC, Pisabarro AG (2001) Classification and mode of action of membrane-active Bacteriocins produced by gram-positive bacteria. Int Microbiol 4(1):13–19. https://doi.org/10.1007/s101230100003
Pereira WA, Carlos MN, Mendonça AV, Urquiza VÞ, Marteinsson JG, LeBlanc PD, Cotter EF, Villalobos JR, Oliveira RPS (2022) Use of probiotic bacteria and bacteriocins as an alter-

native to antibiotics in aquaculture. Microorganisms 10(9):1705. https://doi.org/10.3390/microorganisms10091705

Pircalabioru G, Gratiela LI, Popa LM, Gheorghe I, Popa M, Barbu IC, Cristescu R, Chifiriuc M-C (2021) Bacteriocins in the era of antibiotic resistance: rising to the challenge. Pharmaceutics 13(2):196. https://doi.org/10.3390/pharmaceutics13020196

Preena PG, Swaminathan TR, Kumar VJR, Singh ISB (2020) Antimicrobial resistance in aquaculture: a crisis for concern. Biologia 75(9):1497–1517. https://doi.org/10.2478/s11756-020-00456-4

Priya PS, Boopathi S, Raghul Murugan B, Haridevamuthu AA, Arockiaraj J (2022) Quorum sensing signals: aquaculture risk factor. Rev Aquac 15(3):1294–1310. https://doi.org/10.1111/raq.12774

Ryan MP, Rea MC, Hill C, Ross RP (1996) An application in Cheddar cheese manufacture for a strain of Lactococcus Lactis producing a novel broad-Spectrum Bacteriocin, Lacticin 3147. Appl Environ Microbiol 62(2):612–619. https://doi.org/10.1128/aem.62.2.612-619.1996

Sahoo TK, Jena PK, Patel AK, Seshadri S (2016) Bacteriocins and their applications for the treatment of bacterial diseases in aquaculture: a review. Aquac Res 47(4):1013–1027. https://doi.org/10.1111/are.12556

Schofs L, Sparo MD, Sánchez SF, Bruni. (2020) Gram-positive bacteriocins: usage as antimicrobial agents in veterinary medicine. Vet Res Commun 44(3):89–100. https://doi.org/10.1007/s11259-020-09776-x

Sommerset I, Krossøy B, Biering E, Frost P (2005) Vaccines for fish in aquaculture. Expert Rev Vaccines 4(1):89–101. https://doi.org/10.1586/14760584.4.1.89

Suphan B, Riley MA, Robinson SM (2012) Bacteriocins of aquatic microorganisms and their potential applications in the seafood industry. (https://scholar.googleusercontent.com/scholar?q=cache:ZsC-B2KAQtEJ:scholar.google.com/+b%5BPDF%5D+Bacteriocins+of+aquatic+microorganisms+and+their+potential+applications+in+the+seafood+industry&hl=en&as_sdt=0,5. Accessed 22 May 2023

Thornber K, Abul Bashar M, Ahmed S, Bell A, Trew J, Hasan M, Hasan NA, Alam MM, Chaput DL, Haque MM, Tyler CR (2022) Antimicrobial resistance in aquaculture environments: unravelling the complexity and connectivity of the underlying societal drivers. Environ Sci Technol 56(21):14891–14903. https://doi.org/10.1021/acs.est.2c00799

Wang W, Sun J, Liu C, Xue Z (2017) Application of Immunostimulants in aquaculture: current knowledge and future perspectives. Aquac Res 48(1):1–23. https://doi.org/10.1111/are.13161

Wangkahart E, Secombes CJ, Wang T (2019) Studies on the use of flagellin as an immunostimulant and vaccine adjuvant in fish aquaculture. Front Immunol 9:3054. https://doi.org/10.3389/fimmu.2018.03054

Yanong RPE, Francis-Floyd R, Petty BD (2022) Infectious diseases in aquaculture—exotic and laboratory animals. MSD veterinary manual. https://www.msdvetmanual.com/exotic-and-laboratory-animals/aquaculture/infectious-diseases-in-aquaculture. Accessed 9 May 2023

Zhang Q-Y, Ke F, Gui L, Zhao Z (2022) Recent insights into aquatic viruses: emerging and reemerging pathogens, molecular features, biological effects, and novel investigative approaches. Water Biology and Security 1(4):100062. https://doi.org/10.1016/j.watbs.2022.100062

Ziarati M, Zorriehzahra MJ, Hassantabar F, Mehrabi Z, Dhawan M, Sharun K, Emran TB, Dhama K, Chaicumpa W, Shamsi S (2022) Zoonotic diseases of fish and their prevention and control. Vet Q 42(1):95–118. https://doi.org/10.1080/01652176.2022.2080298

Chapter 14
Biosecurity in Aquaculture

B. L. Chandhana, Ardhra Vijayan, and V. J. Rejish Kumar

Abstract Aquaculture stands out as one of the rapidly expanding domains within the food industry with a global annual growth rate of around 11%, supplying over 16% of the world's animal protein, mostly for human consumption. It is an indispensable part of the economy of most developing countries. The growth of aquaculture and the consecutive increase in the transboundary movement of live and processed aquatic organisms have been attributed to an increase in the spread of diseases. In the realm of animal husbandry, the term biosecurity is commonly employed to describe the preventative measures implemented to safeguard against infectious diseases. It can be defined as "a set of standard scientific measures adopted to exclude pathogens (prevent the establishment and spread of pathogens) from the host and the rearing environment". Appropriate biosecurity measures can curb diseases and monetary losses in aquaculture. Biosecurity measures in farms and hatcheries should be maintained at the entry, internal as well as exit levels to ensure complete control over the health management. Standard biosecurity procedure involves key steps such as hazard identification, risk assessment and risk management. In general, the programme involves practices and procedures involving preventive methods such as quarantine, certification and vaccination, monitoring of organisms as well as the system, disinfection and disease treatment. Overall sustainability of the aquaculture depends on the biosecurity measures and best management practices followed by hatchery/grow-out operators and exporters. This chapter discusses various methods adopted to ensure biosecurity at different levels in farms

B. L. Chandhana
Department of Aquaculture, Kerala University of Fisheries and Ocean Studies, Panangad, Kochi, Kerala, India

PAGR Centre, National Bureau of Fish Genetic Resources, Kochi, Kerala, India

A. Vijayan
Department of Aquatic Animal Health Management, Kerala University of Fisheries and Ocean Studies, Panangad, Kochi, Kerala, India

V. J. Rejish Kumar (✉)
Department of Aquaculture, Kerala University of Fisheries and Ocean Studies, Panangad, Kochi, Kerala, India

P. Elumalai, S. Lakshmi (eds.), *Antimicrobial Resistance in Aquaculture and Aquatic Environments*, https://doi.org/10.1007/978-981-97-7320-6_14

and hatcheries such as quarantine, health certifications including SPF and SPR, vaccination, disease surveillance, system management, effluent treatment and various legislative measures developed by different agencies, organizations or nations.

Keywords Biosecurity · Aquaculture · Aquatic animal health management · Quarantine · SPR

14.1 Introduction

Fisheries and aquaculture have been extensively acknowledged as the future of global food and nutritional security. The total fish production is 178 million metric tons and its 88.5% is intended for human consumption (FAO 2022). Output from the capture sector has been in a stagnating phase during the previous few years for various reasons such as declining resources, over-exploitation, inefficient regulations, mismanaged fishing practices, etc. (National Fisheries Policy 2020). At the same time, aquaculture sector has been exponentially growing over the past decade with a current global annual growth rate of around 11%, supplying over 16% of the world's animal protein (FAO 2022). It has already become an indispensable part of the economies of most developing countries (Vijayan and Sanil 2010).

The phenomenal advancements in the aquaculture sector and the successive growth in the export of live and processed aquatic products have increased the spread of diseases. Recurrent disease outbreaks due to viral, bacterial, fungal and parasitic pathogens have damaged the growth and sustainability of the sector (Stentiford et al. 2012). It is estimated that a loss of more than 6.0 billion US$ per annum occurs in the sector due to diseases alone making it a critical element in successful aquaculture operation (Mishra et al. 2018). There exist certain pathogens and diseases which are regarded as important and listed by the World Organisation for Animal Health (previously OIE) for their potential to severely impact the environment or economy (Table 14.1). The recent increase in the emergence and spread of transboundary aquatic animal diseases (TAADs) significantly affects the ornamental trade also. It has thus been acting as a hindrance in achieving a higher expected annual growth rate in the sector. This will also affect the socioeconomic status of developing countries for which aquaculture is an integral part of the economy (Adam and Gunn 2017).

Adapting proper procedures to prevent disease outbreaks in the aquaculture systems is the need of the hour to prevent losses in the sector. This is because most of the aquatic pathogens are opportunistic in nature and hence they will cause the diseases if the animal is under stress. Any change in the equilibrium between fish (the host), environment and pathogen will result in a disease condition. But most of the standard veterinary methods used for prevention, control as well as eradication of transmittable diseases, which are well-advanced in terrestrial systems, are still in the infancy regarding aquaculture (Scarfe and Palić 2020). Thus, the age-old axiom "prevention is better than cure" appertains in the case of aquatic animal diseases

Table 14.1 Aquatic animal diseases listed by OIE

Finfish	Shellfish (crustaceans and molluscs)	Amphibian
Aphanomyces invadans (EUS)	Acute hepatopancreatic necrosis disease (AHPND)	*Batrachochytrium dendrobatidis*
Epizootic haematopoietic necrosis virus (EHNV)	*Aphanomyces astaci* (crayfish plague)	*Batrachochytrium salamandrivorans*
Gyrodactylus salaris	Decapod iridescent virus 1	*Ranavirus* species
HPR-deleted or HPR0 infectious salmon anaemia virus (ISAV)	*Hepatobacter penaei* (necrotizing hepatopancreatitis)	
Infectious haematopoietic necrosis virus (IHNV)	Infectious hypodermal and haematopoietic necrosis virus (IHHNV)	
Koi herpesvirus (KHV)	Infectious myonecrosis virus (IMNV)	
Red Sea bream iridovirus (RSIV)	*Macrobrachium rosenbergii* nodavirus (white tail disease)	
Salmonid alphavirus (RSIV)	Taura syndrome virus (TSV)	
Spring viraemia of carp virus (SVCV)	White spot syndrome virus (WSSV)	
Tilapia lake virus (TiLV)	Yellow head virus genotype 1	
Viral haemorrhagic septicaemia virus (VHSV)	*Abalone herpesvirus*	
	Bonamia ostreae	
	Bonamia exitiosa	
	Marteilia refringens	
	Perkinsus marinus	
	Perkinsus olseni	
	Xenohaliotis californiensis	

(Sitjà-Bobadilla and Oidtmann 2017). The concept and approaches of biosecurity in aquaculture that can effectively eliminate infectious diseases will be helpful for farms and other systems to be certified as pathogen-free (Cameron 2002, 2004; OIE 2019). The chapter discusses biosecurity principles, plans and their various components and risk analysis procedures for planning biosecurity measures.

14.2 What Is Biosecurity?

Biosecurity or biological security, in an extensive sense, can be defined as a "strategic and integrated approach, encompassing the policy and regulatory frameworks that analyze and manage risks in the sectors of food safety, animal life and health, and plant life and health, including associated environmental risk" (Burnette et al. 2013). It "is a set of management and physical measures which, when used together, cumulatively reduce the risk of infection in aquatic animal populations within an aquaculture establishment" (OIE 2022). The latitude of biosecurity concerned with aquaculture has yet to be described in an ample manner. Biosecurity is a shared responsibility, where every party involved in various stages of the process of animal production has a diverse but pivotal role in the employment of the entire programme. Any failure or omission in the chain of biosecurity will vitiate the overall exertion to establish and maintain biosecurity (Pruder 2004). The risks and cost-effective measures should be identified for planning and employing biosecurity in an aquaculture establishment to accomplish the objectives within the plan. The measures required will differ based on factors such as the type of pathogens, the species cultured, type of culture system, management practices, surrounding conditions, location, etc. (OIE 2022). Suitable biosecurity measures at entry, internal and exit levels can prevent evolving health issues and lessen impressions of disease by applying the principle of preventing diseases rather than curative response (Bera et al. 2018). Some general on-farm biosecurity measures to overcome the problems associated with aquaculture systems are listed in Table 14.2.

14.3 Biosecurity Plan

As per the definition of OIE, a biosecurity plan means a document that identifies potential pathways for the introduction of pathogenic agents into, or spread within, or release from a zone, compartment or aquaculture establishment and describes the measures applied to mitigate the identified risk, in accordance with the recommendations in the Aquatic Code (OIE 2022). A plan can help you decide what steps to take to create a thorough biosecurity programme that will best safeguard your system (Palić et al. 2015) (Fig. 14.1).

All partners, staff and any government official should be involved in developing the plan, which must be documented. This guarantees that the biosecurity plan complies with all applicable governmental regulations and that farm workers can carry it out. The plan should be examined and revised annually. Everyone must be able to comprehend, use and enforce the plan for it to be effective, which means that every stakeholder should be covered by the plan every day. Plans for biosecurity must be practical, affordable and balanced between practicality, affordability and regulatory requirements. In the end, the suggested biosecurity measures ought to boost farm's biological, operational and financial performance. The finest biosecurity programmes are those that are created specifically for a unit and the diseases that may or may not be found there. It is advisable to incorporate the methods indicated in the OIE Aquatic Code and Manual (OIE 2022), as well as the details in the current national or local legislation, for biosecurity strategies to comply with international standards and national or local regulations. A farm or other unit will need to establish a minimal set of processes to comply with these standards, but nothing prevents developing procedures that are specific to the needs of the farm or unit (Scarfe and Palić 2020). Before establishing and executing a biosecurity strategy, gather as many details as possible about how a farm operates, as all future activities will

Table 14.2 On-farm biosecurity practices for better management of aquaculture system

Category	Biosecurity practices
Pre-stock management	• Sludge removal and disposal • Use of good water source with good quality • Water filtration and sterilization • Water conditioning for 10–15 days • Maintain proper water temperature • Frequent water quality checking • Boundary fences, protected dikes and driveways
Seed selection and stocking	• Healthy animals (eggs, larvae or broodstock) from dependable provider • Quality test • Quarantine and health certification for importing and transferring sp. • Stocking disease-free shrimp (SPF/SPR) • Disinfection of egg/larvae/broodstock • Proper stocking density
Post-stock management	• Weekly checking of pond bottom • Regular health check of the animal • No draining or abandoning of disease-affected stocks
Disease management	• Seed and feed quality • Pathogen screening of broodstock, larvae and live feed • Disease surveillance (active and passive) • Early diagnosis • Removal and proper disposal of dead and moribund animal • Hygienic handling of animal/broodstock/larvae/live feed • Prudent and responsible use of veterinary medicines • Use of vaccines/probiotics/immunostimulants • Supplying balanced diet to crops
Hygiene management	• Use of foot/hand dip/wheel bath at entry points • Dressing room and shower near main entrance • Use of disinfected clothing and boots • Limiting movement of visitors • Specific place for dining • Prohibition of smoking • Cleaning of hatchery/production units • Disinfection of equipment and utensils • Pest control
Total control of the farm	• Visitor policy • Staff awareness • Disposal of waste and effluent treatment system • Record-keeping and auditing

depend on this knowledge. This can be accomplished by conversing with the owner, inspecting the farm or filling out a set of questions regarding the farming operation, how it is handled, the diseases observed, how they have been controlled, etc. (Palić and Scarfe 2018). The questionnaire can be completed before the farm visit and supplemented during it. Record-keeping and auditing them from time to time should be practiced at every farm or production facility to have a better understanding of the operations as well as to track down any suspected source of infection. It is

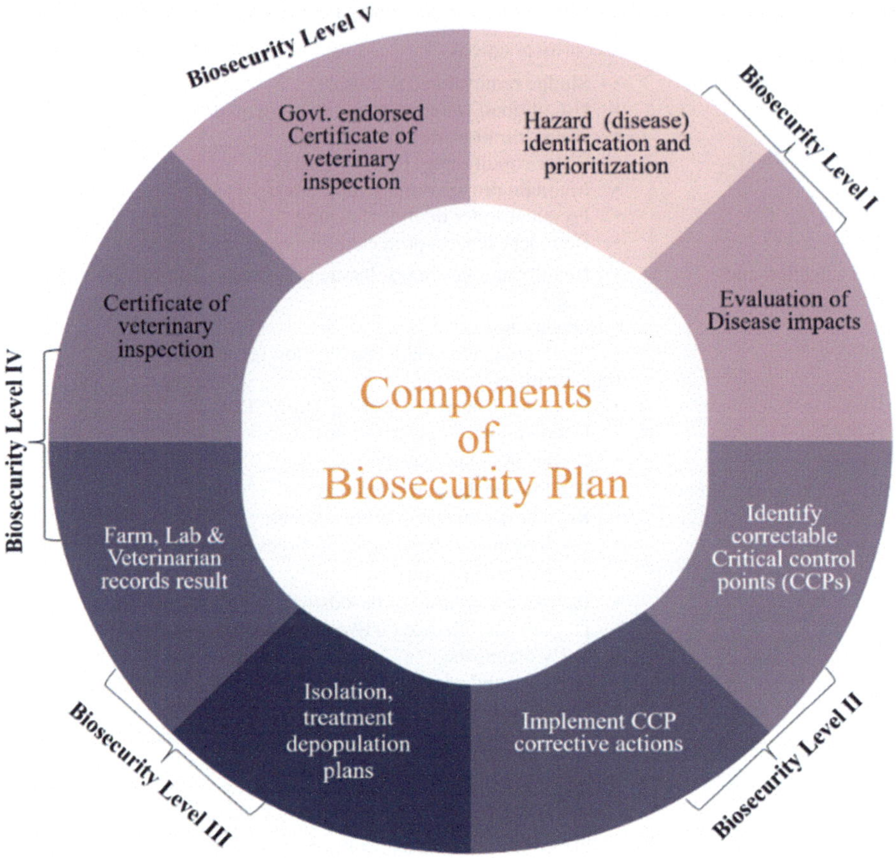

Fig. 14.1 Components of biosecurity plan

advisable to maintain records such as, cleaning and disinfection record, training records, records for livestock producers, stock received and inspection records, dispatch records, visitor records, etc.

14.4 Biosecurity Management

The biosecure management of an aquaculture system demands the proper identification of the routes of transmission of diseases such as into the system, within the system and from the system. In other words, there should be a thorough understanding of the entry, internal and exit levels of the system. Animals, people and pathogens are the major routes of transmission at all these levels. Therefore, biosecurity management may be designed to address mainly animal, people and pathogen management (Fig. 14.2). Other entry-level routes include water, feed, vehicles, vessels and other equipment. Influent water is one of the major routes of the pathogen into the farm (Moss et al. 2012). Considering the routes within the system, many production and processing divisions are in varying states of health. Therefore, there is a high danger of disease transmission between these units unless proper methods are adapted to prevent cross-contamination. The local wild life population, water supplies and neighbouring farms may all be negatively impacted by the farm's transmission pathways at exit level. Proper treatment of the effluent should also be taken care of at this level. Another way of looking at biosecurity management is at conceptual, structural and operational levels. Conceptual biosecurity involves planning of the location and setting up of the facility, while at structural level this deals with the material components such as fencing, cleaning and disinfecting facilities, testing and storage facilities, etc. Operational biosecurity consists of dynamic procedures such as effective disease monitoring through repetitive procedures, such as sterilization of touch areas, screening at the visitor's entrance point and thorough verification of transmitters and sources of stocks that must be implemented in order to prevent pathogen and its horizontal transmission (biosecurity management.tnau.ac.in). Thus, in a farm, every facility should have some general safeguard measures to aid disease prevention and control activities, the workers should be aware of the practices that may increase or decrease the risk of health hazards, and there must be proper management of the critical elements like animals, people and pathogens.

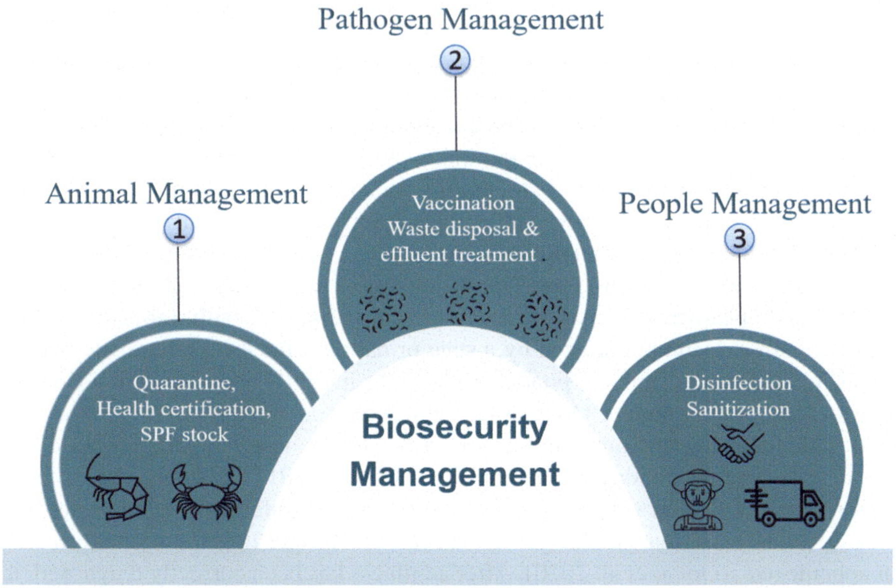

Fig. 14.2 Major elements of biosecurity management system

14.5 Animal Management

It's crucial to get healthy animals (eggs, larvae or broodstock) from a dependable provider. Determine the health indicators and illnesses of concern for each species in consultation with a fish health specialist. Learn as much about an animal's health before purchasing it from a third party, including where it originated from, whether it has ever experienced medical problems and whether it has undergone any health checks, disease tests or treatment. It is desirable to use animals from suppliers that have tested negative for a particular infection. Fish should be checked as soon as they are received if the information on animal health is lacking or ambiguous (Yanong and Erlacher-Reid 2012).

Disease outbreaks at the farm level are not usually caused by the sheer existence of a pathogen (the required cause). In addition to the main cause, other elements (also known as risk factors) are necessary for the illness to manifest. Effective risk management strategies and good husbandry practices can lessen the effects of an epidemic (Mohan et al. 2008). Diseases may also be evaded with proper feeding, handling techniques and water quality/chemistry. Quarantine; regular monitoring; checking specific pathogen-free (SPF), specific pathogen-resistant (SPR) and specific pathogen-tolerant (SPT) statuses of the stock (shrimp); vaccination; and the use of prebiotics, probiotics and immunostimulants are all good prophylactic measures (Yanong and Erlacher-Reid 2012).

14.5.1 Quarantine

According to OIE (2022), quarantine "means maintaining a group of aquatic animals in isolation with no direct or indirect contact with other aquatic animals, in order to undergo observation for a specified length of time and, if appropriate, testing and treatment, including proper treatment of the effluent waters". The principles of quarantine appertain to new fish entering a facility, fish being transferred from one area or facility to another within the system and inhabitant fish that turns infected (Yanong and Erlacher-Reid 2012).

Typically, any new aquatic stock must undergo a mandatory quarantine period, referred to as the quarantine period, as outlined in an import health standard or other legally binding document created by a state or nation. This is a prerequisite before releasing the stock to farms or ponds, ensuring it can safely integrate with the existing stock and environment. Quarantine might last from 15 days to 3 months (Jia et al. 2017). During this time, animals must be closely observed, and proper diagnostic tests must be conducted (Hadfield and Clayton 2011). Fish are checked for the presence of fin and skin erosion as well as signs of infection including loss of balance, anorexia, lethargy, changes in skin colour, changes in respiration rates, loss of appetite, etc. (Opiyo et al. 2020). After an illness has been correctly diagnosed, it should be treated for the proper amount of time using effective medications (Assefa and Abunna 2018). The construction and operation of the quarantine facility must prioritize a robust security framework, guaranteeing the isolation of imported aquatic animals. This measure aims to prevent the inadvertent release of the animals, as well as any potential pathogens, pests or other living organisms present in their transport waters. No living aquatic creatures, tools or materials should be taken out of the quarantine facility while it is under quarantine (Arthur et al. 2008). To stop possible infections from entering a production facility through diseased broodstock, juvenile or larvae, quarantine is required and may be successfully used at grow-out facility and individual hatchery level. A risk analysis should be used to assess whether quarantine or other biosecurity measures are necessary in each specific situation (Arthur et al. 2004; Murray et al. 2004; OIE 2006).

Physical barriers that effectively segregate arriving fish from the rest of the farm are the quarantine systems. Additionally, the water used in quarantine systems needs to be kept apart from the water used in the main farm, and discharges need to be managed properly. Quarantining new animals properly provides them time to adjust to water, feed and management and recuperate from handling and prevents the spread of infections to the established populations. It has been demonstrated that handling and transportation weaken disease resistance, and recovery may take weeks. Use a sign or alternative method to alert staff about the presence of illness in the population. If there are sick fish within the general population, consider segregating them in tanks within the same system or room as their healthy counterparts (Yanong and Erlacher-Reid 2012).

The following are the prerequisites for effective quarantine:

1. Physical infrastructure that is adequate for containment needed, including holding tanks, safe water input and discharge, sanitation practices and prevention of the movement of people, materials and objects
2. Operating procedures that have been established and facilities that are suitable for carrying out traceability, pathogen diagnosis, containment measures, etc.
3. Skilled workers to complete the aforementioned task without error

When a transfer or introduction is determined to have a low risk of introducing a disease, minimum quarantine rules are imposed. For situations when there is a larger danger of disease transmission, further safeguards will be needed. (http://ecoursesonline.iasri.res.in/mod/page/view.php?id=94357). When animals, water or equipment are mixed between batches, those batches must be treated as one batch for disease screening. The length of quarantine evaluation or confinement cannot be predetermined since it will depend on the candidate species and the dangers involved in its migration. Animals that are going to be transferred should be put in a quarantine facility for the necessary illness testing, certification and health checks. Any medication taken must be disclosed to the importing nation's competent authority (CA) (Arthur et al. 2008). The quarantine containment facilities should be near to or inside of current fisheries or animal health facilities, and it would be ideal if there were always someone on duty. Access should be limited to authorized employees, and the facility should be lockable. Many diseases may be treated using easily accessible medications (such as salt baths, fresh water and formalin), notably the common disorders brought on by external parasites. Other approved medications could be accessible, but they might need to be prescribed or administered by an expert. It should be emphasized that improper use of chemical therapies might result in additional health issues, such as the emergence of bacterial strains that are resistant to antibiotics. Therefore, chemical treatment should be done with great caution and professional guidance. External parasite outbreaks are particularly likely to affect wild populations. This can be avoided by giving animals that are entering a quarantine facility an initial treatment, as well as by careful observation and husbandry modifications (such as lowering the temperature, altering the food schedule or increasing holding density) (Leszczynski et al. 2017).

Programmes for health certification and quarantine are crucial tools in the battle against illnesses brought on by the introduction or movement of exotic fish and shellfish. Therefore, there is a critical need to create a set of guidelines for the transportation of aquatic animals between nations as well as within nations from one site to another to stop the spread of illness.

14.5.2 Health Certification

Any stock of aquatic animals that is exported or imported must undergo health certification before it may be released by trained staff from a quarantine facility. Additionally, the status of the stock regarding certain infections as well as the stock's overall health as determined by the numerous biological markers that are indications of the animal's health state might be included. The size of the samples taken for health monitoring should allow effective pathogen identification and screening. The quantity of fish to be sampled for health certification depends on the frequency of illness in the community. To provide an accurate representative of the vulnerable age and size groups of the animals, samples should preferably be gathered using stratified random sampling. Additionally, samples for surveillance purposes must be done during the months when the illness is most commonly seen (http://ecoursesonline.iasri.res.in/mod/page/view.php?id=94357).

Certification indicates that the person who signed the document accepts responsibility for the veracity of the claims contained within. This is crucial when, according to a pre-existing legislative framework, the certificate is required to provide a transfer licence. In other words, the signing authority has a moral and legal duty to guarantee that the representations made in the certificate are true to the best of his or her knowledge. Thus, to determine whether to sign a health certificate, the signing authority must have first-hand knowledge of or control over the personnel who offer the scientific advice. There is minimal assurance against disease spread from certificates signed by persons with insufficient training and experience. Both the importer and the exporter are affected by these certifications. Additionally, it should be highlighted that many aquatic animal infections cannot be detected by border inspections for obvious diseases, which are the present foundation for health certificate issuance in many nations. Certificates should be precise, and they should explicitly address the importing country's needs. Thus, pre-communication between responsible authorities of both exporting or importing nations is required (OIE 2022).

14.5.3 Specific Pathogen-Free (SPF) Animals

To a large extent, illness outbreaks in shrimp can be minimized by utilizing SPF shrimp. But it only refers to freedom from certain infections, not freedom from all the pathogens. SPF (specific pathogen-free) animal is cultivated in meticulously biosecure facilities, featuring proper water treatment and an enclosed environment. It originates from a population that has consistently tested negative for specific pathogens over a minimum of 2 years, necessitating the implementation of a surveillance programme. Additionally, the animal undergoes biosecure breeding practices, is provided with biosecure feeds and is subject to stringent management measures. As a result, the SPF categorization only considers the animal's past hygienic state as well as the facility and cultural circumstances in which it was grown and maintained. Regular audits are required for transparency while claiming SPF certification. An SPF animal need not be tolerant to infections and genetically diverse or perform better or worse in terms of growth. It simply relates to its hygienic state, which is not inherited. The SPF status is lost as soon as an SPF animal is exposed to possible infections (Citarasu et al. 2022).

The infections that SPF animals must be free of will vary time to time. In certain cases, only diseases on the OIE list are considered. Presently SPF shrimp populations are free from some OIE listed diseases such as WSSV, YHV, IHHNV, TSV and IMNV (Moss et al. 2012). The SPF technique is not a novel idea; rather, it is standard procedure in the case of terrestrial animals. In fact, from the perspective of a veterinarian, the concept of stocking (i.e. investing) in diseased animals is hard to comprehend. Additionally, since hygienic factors are eliminated, using SPF animals offers a good starting point for developing breeding plans. They are essential for research, international trade and stopping the transboundary spread of diseases (Alday-Sanz et al. 2020).

Specific pathogen resistance (SPR) denotes the ability to resist infection from a particular pathogen, representing a qualitative attribute where an animal is either infected or not. SPR stocks are generated by subjecting sub-lots of families to a specific pathogen (or a combination of pathogens) challenge, followed by the selection of the most resistant families as broodstock for the subsequent generation. Typically, these strains exhibit resistance specifically against the targeted pathogen. The resistance status (SPR) to a specific pathogen remains unaffected by management practices but may be lost in subsequent generations if poor animal breeding and selection strategies are employed. On the other hand, specific pathogen tolerance (SPT) indicates the ability to endure a specific illness, where an animal may get infected without necessarily developing the disease, or the manifestation of the disease is reduced. Seeds with SPT are capable of tolerating a specific pathogen at a particular level of exposure. Animals with specific pathogen tolerance (SPT) either remain disease-free or experience a milder impact compared to normal animals when infected with the designated pathogen for which they hold a tolerant status. While SPT animals are SPF (specific pathogen-free), if they have never been exposed to the pathogen, their susceptibility to infection means they cannot be classified as specific pathogen-resistant (SPR) for the same pathogen. Unless these animals also possess SPF status, they pose a potential risk as carriers of pathogens, thereby increasing the likelihood of disease spread (Sanz 2018).

14.6 Pathogen Management

Infectious agents differ in terms of their regulatory importance, capacity to persist, pathogenicity, diagnostic capabilities and management. The environment and the host, particularly, and the organism that is being attacked and its immunological status will ultimately decide whether fish become ill (Yanong and Erlacher-Reid 2012). Understanding prevalent illnesses in a specific species, possible disease reservoirs and pertinent diagnostic and management methods are crucial. While some pathogens, like the bacteria *Aeromonas hydrophila* and the ciliated protist *Trichodina*, are widespread and opportunistic, others, like *Streptococcus* spp. or *Amyloodinium*, may cause significant morbidity and mortality or have regulatory repercussions. It will be easier to detect the presence of certain infections if fresh fish are quarantined at temperatures conducive to serious illnesses. Pathogen management techniques include isolating sick individuals, using equipment designed for that population and sanitizing and disinfecting systems and equipment (Yanong et al. 2022).

In case of suspicion or occurrence of any disease in the facility, the concerned authorities must be notified. Working in couple with fish health professionals will help to review the chances of susceptibility to such diseases and to implement adequate diagnostic and management practices. Within a facility, there are "reservoirs" where pathogens may live and grow. Reservoirs that aren't living things include water, machinery, furnishings, flooring and walls. The aquaculture animals themselves, as well as other species (including frogs and birds), vegetation and live (or frozen) food all come under living reservoirs (Yanong and Erlacher-Reid 2012). Biofouling occurring at the various culture systems despite being recorded to be one of the best reservoirs to harbour various pathogens had received only little attention (Sitjà-Bobadilla and Oidtmann 2017). It's crucial to comprehend the biology of infections, especially the mechanisms that allow them to persist in reservoirs and the ease with which simple disinfectants may eliminate them. Sediments, biofilms and high animal populations can all help concentrate microorganisms. In the system, uneaten feeds and other organic debris promote disease survival and dissemination (Yanong and Erlacher-Reid 2012).

The source of water is a significant risk factor for the spread of illness. Deep-well water or municipal supplies that have been cleaned and disinfected are two examples of protected water sources. Water sources that are protected are preferred over unprotected sources, such as lake or river surface waters (Yanong et al. 2022). Unprotected incoming sources can be treated by filtering, UV sterilization, chlorination and/or ozonation to make them safer (Hedrick et al. 2000). By improving water quality and lowering pathogen loads, efforts to maintain a clean environment, such as preventing the build-up of organic debris, properly disinfecting nets and equipment and thoroughly disinfecting fish-holding units between groups of fish, help to reduce disease outbreaks (Yanong et al. 2022).

Diagnostics are critical for preventing and treating illness in aquatic animals. The diagnostic techniques progressed from microscopic to molecular characterization. Molecular tools, in addition to microscopy, histological, microbiological and immunological methods, provide early diagnosis and prevent serious deaths and economic losses in aquaculture (Kumar et al. 2021). It is common to believe that a disease's confirmation diagnosis is difficult and expensive, which may be accurate in the case of some emerging diseases. Until the host is ready to be consumed, disease diagnostics should be accessible throughout the whole life cycle of the host. For the identification of diseases at various stages, there are several modern diagnostics accessible in aquaculture. Healthy animals are screened using several diagnostic techniques to make sure they are clear of infections caused by pathogens at subclinical levels (Ananda Raja and Jithendran 2015). For better management of pathogens, it is important to have proper diagnostic tools and surveillance methods (active and passive) for the early detection of diseases.

Some infectious disorders can be effectively treated with medication, adequate cleanliness and disinfection. Other infections, such as mycobacteria, several parasites and many viruses, are difficult or impossible to cure (Yanong and Erlacher-Reid 2012). Aquaculture still uses antibiotics despite their negative consequences on the emergence of drug-resistant bacteria. There is widespread usage of biological and chemical disease control methods, including the use of probiotics, prebiotics and medicinal plants (Assefa and Abunna 2018). It is strongly advised to practice effective farm management to reduce the usage of pesticides and antibiotics. Probiotic, biocontrol and bioremediation applications appear promising; nevertheless, more study is still required to ensure the safety of both food and the environment (Chinabut and Puttinaowarat 2005). Treating numerous bacterial infections in aquatic animals solely with antimicrobials is not feasible (Dadar et al. 2016). Thus, the prevention of a wide range of bacterial and viral infections has been credited to vaccination as an effective therapy strategy (Ma et al. 2019).

14.6.1 Vaccination

A vaccination is described as a biologically based preparation created to boost immunity against a particular illness or a set of illnesses (Mondal and Thomas 2022). Vaccines are regarded as biological agents that trigger an immune response to a specific antigen derived from an infectious virus that causes illness (Czochor and Turchick 2014). One of the most crucial—and perhaps the most urgent—strategies for the prevention and management of infectious fish sickness is the advancement of vaccination. Advancements in fish vaccination over the past decade involve immunizing large stocks simultaneously and the creation of multivalent vaccines (Plant and LaPatra 2011). Use of vaccinations to protect carp against illness by immunizing them with the bacteria *Aeromonas punctata* by Snieszko et al. (1938) was the first report on protective immunity. The immune system of the fish is either exposed to the complete virus or just a portion of it throughout the vaccination procedure. After a given amount of time, immunity begins to develop (Mondal and Thomas 2022). Before administering vaccine to fish, there are a few crucial factors that need to be taken into mind: the type of fish to be immunized, the fish's immune system's condition, the aquaculture system's production cycle and life history, the diseases that need to be controlled there and the financial benefits (Adams 2016).

14.6.2 Waste Disposal and Effluent Treatment

The waste materials and effluent water from hatcheries and farms can harbour potential pathogens which if left untreated will pose threat to nearby facilities or the natural systems. There should be a proper system installed at every hatchery and farm for proper treatment and disposal of the effluent water. Solid particles if any should be removed from the effluent before treating it for pathogens. Methods such as chlorination or UV treatment should be done to ensure that no harmful pathogens get released from the system. Along with pathogen reduction, the process also helps in bringing down the amount of pollution caused by aquaculture practices (https://www.fao.org/3/x9199e/X9199E05).

14.7 People Management

Only if the management, workers and visitors are aware of and adhere to the appropriate procedures will effective biosecurity be ensured. The danger of sickness will grow if staff members and guests don't adhere to set procedures. Only authorized employees should be allowed access to critical locations. Visitors that arrive at your site just after going to another aquaculture plant must be taken seriously as a risk. Based on the life stage, species and sickness state of the fish, people should be assigned to certain work areas if the workforce is sizable enough. Fish in quarantine should be handled last if there aren't enough workers to handle and clean them—healthy fish first and then fish that may be sick. Age is another consideration. As younger fish are thought to be more prone, handling should be done from youngest to oldest fish (Yanong and Erlacher-Reid 2012).

Following the implementation of a biosecurity programme, staff should have access to continuing education opportunities. Periodic reviews of the biosecurity programme, personnel compliance, illnesses and productivity are recommended. Create biosecurity initiatives with an educational component in collaboration with a specialist in aquatic animal health. Employ disinfectant foot baths, handwashing stations, alcohol spray bottles, net disinfection stations, showers and vehicle disinfection stations to help prevent the spread of infections. To maximize their use and ease of use, they should be positioned in key areas (such as at entrances and between systems). Many germs can be eliminated by simply washing your hands with soap and water. Using an antiseptic will cut hazards even more. After usage, nets should be sanitized to stop the transfer of infections from one species of fish to another. Net dips need to be large enough and filled with enough water to completely immerse the net and, if feasible, the part of the handle that could be submerged while being used. Showers are a common feature in aquaculture facilities, mainly used to clean up after outdoor operations (Yanong and Erlacher-Reid 2012).

When transitioning between sections within a facility that have different biosecurity levels, especially when moving from areas with quarantined fish, diseased populations or older life stages to those with more vulnerable or younger life stages, employees can mitigate the risk of disease transmission by taking showers, changing clothes and removing boots. If an employee works at numerous different facilities, they should also take a shower. Pathogens may be transported on the wheels of moving vehicles that move between facilities. This risk can be decreased by having disinfectant-filled drive-through spaces at the entrances and exits (Yanong and Erlacher-Reid 2012).

14.8 Risk Analysis

Risk analysis is an assessment for determining the likelihood and effects of unfavourable events. Even though the animal health industry became acquainted with risk analysis from the last decade of the previous century itself, it hasn't been applied in aquaculture until the onset of the twenty-first century (Rodgers 2001). A biosecurity strategy may be tailored to the greatest threats by conducting a risk analysis, and any suggested biosecurity measures can be tested for their capacity to lower risk to a manageable level. Establishing the scope of a risk analysis is crucial before starting. The four stages of risk analysis consist of hazard identification, risk assessment, risk management and finally risk communication (OIE 2022). The process of hazard identification entails identifying the specific illnesses that the epidemiology unit is concerned about. Calculating each hazard's degree of risk entails calculating how likely it is for it to occur and what unfavourable outcomes could follow. To manage the risks that have been analysed, strategies must be identified and put into action. Risk analysis involves constant risk communication. It involves disclosing details on the dangers and risks that are currently in place as well as those that may arise in the future, as well as how the nation has evaluated these dangers and risks, what steps have been taken (or will be taken) to manage the risks identified and how their efficacy will be continuously evaluated. All concerned parties should be in contact with each other, and the country's trading partners, as well as with the business community and the public (Hastein et al. 2008).

Risk analysis has the potential to be applied in a variety of aquatic animal health-related fields, such as analysis of (a) disease transmission between domesticated and wild populations; (b) potential pathogen transmission via composted or ensiled fish waste; and (c) risk of disease introduction at the farm, regional or national level (Peeler 2005). In summary, the components of a biosecurity strategy that are easiest to identify are often the risk management techniques. The primary regulatory tool overseeing biosecurity, the Sanitary and Phytosanitary (SPS) Measures Agreement by the World Trade Organization (WTO), underscores the importance of utilizing risk analysis as the foundation for implementing any SPS measures (WTO 1994).

In the area of aquatic animal health, risk analysis techniques, or import risk analysis (IRA), have mostly been used to weigh the dangers of bringing exotic illnesses into a nation (Peeler et al. 2007). It is crucial to pinpoint potential entry points and conditions that can encourage the spread of the illness at the farm level using epidemiological data from observational studies, such as the risk factors studies mentioned above. Epidemiological data may be combined with additional data for risk analysis, such as pathogen characteristics, the volume of aquatic animal movements and other movements (such as those of humans and vehicles) on and off the farm (Peeler 2005). The OIE's Aquatic Animal Health Code (OIE 2022) describes the fundamental stages that must be taken in the risk analysis process, although individual member nations are free to decide on the method's specifics. A national aquatic animal health programme includes several different elements, risk analysis being only one of them (FAO/NACA 2000; Arthur et al. 2004).

14.8.1 Steps in Risk Analysis

The primary elements of the risk analysis process include hazard identification/hard analysis, risk assessment, risk management and risk communication.

1. Hazard analysis: It should be able to detect, characterize and prioritize dangers at the hazard identification step. Risk analysis uses a variety of methods and procedures to detect risks (Hayes 2002, 2003). Identification of hazards entails locating the illnesses that may have a negative impact on the welfare of aquatic animals and aquaculture output. The risks might include harmful pathogens that need to be kept away from the farm. Other infections that are known to exist in the facility and that must be controlled to minimize their effects on production (commonly referred to as production illnesses) may also be included. It may be necessary to examine "unknown hazards" (like various types of infections) that might influence on farm output because new illnesses frequently appear in aquaculture. Potential dangers that were addressed or found but weren't given the same attention or priority as others and weren't further investigated are crucial to keeping an analysis transparent (Lind et al. 2015).

2. Risk assessment: Many approaches for assessing risk within an environmental risk analysis framework have been discussed (Vose 2008). These procedures can include highly statistical quantitative methods as well as qualitative and semi-quantitative methods that draw on more arbitrary, opinion-based judgements. The likelihood of an event occurring on the farm and the consequences it would have should be determined to rate the risk associated with that event. The likelihood may be calculated by considering the routes or pathways needed for a disease to enter and for your animals to get exposed to it. The consequences can be calculated by considering how an illness would affect the output of the farm. Multiple elements of the effects such as mortality, diminished growth or food conversion, product quality, market access and treatment costs should be considered (https://www.agriculture.gov.au/fisheries/aquaculture-farm-biosecurity-plan).
3. Risk management: The process stage of risk management is when ways to lower the amount of risk are found, chosen and put into action (Bondad-Reantaso and Arthur 2008). The stage of entrance, dissemination and/or establishment of the organism in the receiving environment will determine the management goals (for instance, one goal would be to determine whether a certain action will accelerate the spread of an established organism) (Lind et al. 2015). Risk management entails determining ways to lower the identified risks to a manageable level. Extreme, moderate and medium dangers ought to be viewed as unacceptable. The agricultural biosecurity strategy would include risk management strategies to lower these hazards to manageable levels. Low risks might not necessitate a lot of mitigation work, but they can still call for some continuing monitoring to see whether the risk profile shifts over time. To bring risks down to a manageable level, there may be a wide variety of risk management strategies accessible. The feasibility, efficacy and expense of the favoured choice should all be taken into consideration. Options for risk management can lessen consequences or likelihood, or perhaps both. For instance, immunization could greatly lessen outcomes yet does not affect a pathogen's chance of entrance (https://www.agriculture.gov.au/fisheries/aquaculture-farm-biosecurity-plan).
4. Risk communication: This is the procedure through which information and viewpoints regarding hazards and risks are collected from parties who may be impacted or have an interest during the risk analysis. It also involves informing decision-makers and interested parties of the findings of the risk assessment and any recommended risk management measures. Ideally starting at the beginning of the risk analysis and continuing throughout the entire process, risk communication is a multifaceted and iterative process, the level where overall system integrity is provided (Bondad-Reantaso and Arthur 2008). Risk communication can involve large-scale activities like making public announcements about the results of a risk analysis, or it can involve asking key stakeholders for highly targeted input to help develop elements like assessment end points, conceptual models or a risk assessment (Lind et al. 2015).

14.9 Legislation and Regulation

To promote international trade in aquatic animals and aquatic animal products without posing unacceptable dangers to the health of people and aquatic animals, several variables should be taken into consideration. To maintain uniformity and to ensure the effectiveness and sustainability of the biosecurity plans and procedures, there are several national and international legal frameworks and standards in existence. These measures help all the concerned parties to be part of the biosecurity strategy as well as for its efficacious implementation. A variety of challenges, including contingency planning, new species introductions, domestic transportation of live fish and their products, fish inspection and quarantine and disease surveillance, management and eradication, should be addressed by legislation and regulation. Additionally, the task is to create legislation that supports and directs both the public and commercial sectors engaged in initiatives linked to fish health. It should encourage efficient coordination, power-sharing and communication amongst all those engaged and identify the responsibilities of the different authorities involved in concerns of fish health at the national, provincial and district levels. Fish producers should be given the opportunity to participate in decision-making processes related to the execution of fish health management programmes as sources of data and reports (Vijayan and Sanil 2010).

14.9.1 International Biosecurity Framework

- *The Sanitary and Phytosanitary (SPS) Measures Agreement*: This applies to all sanitary (concerning the health of people and animals) and phytosanitary (concerning the health of plants) actions that might have a direct or indirect effect on global trade. The SPS Agreement states that sanitary or phytosanitary measures should cover all pertinent laws, orders, regulations and specifications; testing, inspection, certification and approval procedures; quarantine treatments, including pertinent requirements for animal transportation; pertinent statistical methods, sampling procedures and risk assessment methodologies; and packaging and labelling specifications specifically related to food safety. The SPS Agreement intends to establish a set of guidelines that will prevent nations from using SPS measures to create protectionist trade barriers. However, if more extreme actions can be scientifically proven to be required to safeguard human, animal or plant life or health, these precautions must be used regularly. In situations where there is a lack of scientific evidence, a member may provisionally implement SPS measures based on the data at hand. Members must still gather the extra data required for a more complete risk assessment and reassess the sanitary or phytosanitary action if appropriate within a reasonable amount of time (Oidtmann et al. 2011).

- *OIE Standards*: The OIE is an intergovernmental organization with the goal of enhancing transparency and global collaboration in the management of animal illnesses. To carry out its obligations under the SPS Agreement, it has created two worldwide standards for aquatic animal health: the Manual of Diagnostic Tests for Aquatic Animals (Diagnostic Manual) and the Aquatic Animal Health Code (Aquatic Code) (OIE 2022). Over time, ideas on how the OIE addresses biosecurity for aquaculture have changed. OIE only advised nations to establish "general biosecurity". Aquaculture-specific biosecurity protocols have only recently been addressed by the OIE Aquatic Animal Health Standards Commission (AAHSC), which has done so by clarifying several concepts, including "biosecurity", "biosecurity plan" and "basic biosecurity conditions" (Scarfe and Palić 2020).
- Other organizations have established international guidelines associated with fisheries, aquaculture and biodiversity, for example, the Code of Conduct for Responsible Fisheries (Food and Agriculture Organization (FAO 1995), the Technical Guidelines for Responsible Fisheries (FAO 1997), the international treaty on biodiversity (Convention on Biological Diversity, 5 June 1992) and ICES guidelines on introduction and transfers of marine organisms (International Council for the Exploration of the Sea (ICES) 2005; Oidtmann et al. 2011).

14.9.2 National Biosecurity Framework

National programmes may also be developed and implemented by industry, although they are often started by the "competent authority" in a nation and approved by the national legislative body. The directive for national rules to be issued by the competent authority, i.e. the government body in charge of, or with legislative jurisdiction over, animal health and disease, is normally provided by the enactment of decisions taken by legislative bodies. Typically, this is a ministry or agency for animal husbandry (OIE 2022). However, in other nations, aquaculture-related legislation and regulations are complicated by the belief that it somehow contributes to the traditional harvest of wild fisheries or that it belongs under the purview of organizations responsible for managing natural resources or the environment. In general, implementing biosecurity strategies for specific epidemiological units, like a land-based farm, is more straightforward compared to broader epidemiological units that can be challenging to delineate—such as a natural resource (like an entire water catchment) or an open body of water (such as an ocean)—where there is minimal control over animal movement or the transmission of infectious diseases (Hastein et al. 2008).

The creation of national biosecurity plans, however, is not necessarily dependent on laws and regulations. Instead, they may be started and carried out by a sector of the economy without the involvement of the government, and in certain situations, the sector of the economy develops a position to direct the promulgation of rules (Mitchell and Stoskopf 1999).

14.10 Recent Advancements in Biosecurity

As aquatic biosecurity continues to be a significant concern, researchers are increasingly turning to artificial intelligence (AI) to bolster biosecurity efforts. Aquatic pest and disease detection and prevention might be improved by AI, which has the potential to revolutionize aquatic biosecurity. The use of automated surveillance systems is one of the main ways AI may aid in aquatic biosecurity. Water bodies may be monitored for indicators of pathogens and illnesses, alerting authorities to any potential problems, using AI-enabled cameras and sensors. This can provide a quicker reaction by reducing the amount of time needed to identify and address biosecurity concerns. AI may also be utilized to aid in the creation of aquatic biosecurity management plans that are more efficient. AI-enabled systems may analyse data from many sources, including population, weather and water quality data, to create prediction models that can guide biosecurity management choices. Finally, the public may be educated about aquatic biosecurity with the use of AI. AI-enabled systems may be used to develop engaging instructional materials that teach users how to prevent and deal with biosecurity hazards as well as improve awareness of related issues. Overall, artificial intelligence (AI) has the potential to completely transform aquatic biosecurity by enabling more effective and precise detection and control of aquatic illnesses and pests. Researchers and authorities may collaborate to build a safer aquatic environment by utilizing the potential of AI (Frąckiewicz 2023).

Another technological advancement in disease diagnosis is the use of aptamer. The aptamer technology which is otherwise called as "synthetic antibodies" is competing with antibodies in both diagnostic and therapeutic applications of both established and newly developing illnesses (Jayasena 1999). Through signal amplification, it can be employed to create sensitive biosensors for rapid detection of aquatic pathogens. One of the key qualities of aptamers to be employed as probes in pathogen detection for illness diagnosis is specificity and sensitivity towards their target molecules/cells. The special qualities of aptamers, such as target specificity, low immunogenicity and non-toxicity, point to their potential for the prevention and/or treatment of illnesses. Despite aptamers' modest size, which is equivalent to small molecule therapy and promotes systemic clearance, the target specificity of aptamers implies functional benefit over small molecule therapy (Nimjee et al. 2005). These kinds of advancements in disease diagnosis and therapeutics can help to enhance the biosecurity management.

14.11 Conclusion

The rapidly growing aquaculture sector is constrained mostly by disease outbreaks and transmission. Control of disease in the aquatic system is more challenging than other systems because the water body itself accelerates the disease spread and it is often difficult to detect and characterize. Other challenges include lack of knowledge, surveillance, proper medicine or application, immunization and experienced manpower, limitations in diagnostic tools, etc. Thus, on-farm biosecurity practices will be more helpful for effective disease management and control. In every aquaculture setting, there should be some standard operating procedures that provide standard rules for biosecurity measures. Hazard identification and risk analysis are also equally important for a better biosecurity management system. Moreover, there is an urgent need for technological development for early detection or rapid diagnosis for improving the growth of the aquaculture sector.

Acknowledgements This work was supported by the Centre of Excellence in Sustainable Aquaculture and Aquatic Health Management (CAAHM), Plan fund 2023-24, Kerala University of Fisheries and Ocean Studies.

Conflict of InterestThe authors declare that they have no conflict of interest.

References

Adam KE, Gunn GJ (2017) Social and economic aspects of aquatic animal health. Rev Sci Tech 36(1):323–329. https://doi.org/10.20506/rst.36.1.2632

Adams A (2016) Fish vaccines, 1st edn. Springer, Basel, Switzerland

Alday-Sanz V, Brock J, Flegel TW, McIntosh R, Bondad-Reantaso MG, Salazar M, Subasinghe R (2020) Facts, truths and myths about SPF shrimp in aquaculture. Rev Aquac 12(1):76–84

Ananda Raja R, Jithendran KP (2015) Aquaculture disease diagnosis and health management. In: Perumal S, Thirunavukkarasu AR, Pachiappan P (eds) Advances in marine and brackishwater aquaculture. Springer, pp 247–255. https://doi.org/10.1007/978-81-322-2271-2_23

Arthur JR, Bondad-Reantaso MG, Baldock FC, Rodgers CJ, Edgerton BF (2004) Manual on risk analysis for the safe movement of aquatic animals (FWG/01/2002). APEC/DoF/NACA/FAO. APEC Publ. No. APEC #203-FS-03.1, p 59

Arthur JR, Bondad-Reantaso MG, Subasinghe RP (2008) Procedures for the quarantine of live aquatic animals: a manual. Food and Agriculture Organization of the United Nations, Rome

Assefa A, Abunna F (2018) Maintenance of fish health in aquaculture: review of epidemiological approaches for prevention and control of infectious disease of fish. Vet Med Int 2018:5432497. https://doi.org/10.1155/2018/5432497

Bera KK, Karmakar S, Jana P, Das S, Purkait S, Pal S, Haque R (2018) Biosecurity in aquaculture: an overview. Aqua Int 12:1–4. https://srac.tamu.edu

Bondad-Reantaso MG, Arthur JR (2008) Pathogen risk analysis for aquaculture production. In: Bondad-Reantaso MG, Arthur JR, Subasinghe RP (eds) Understanding and applying risk analysis in aquaculture. FAO fisheries and aquaculture technical paper. No. 519. FAO, Rome, pp 27–46

Burnette RN, Hess JE, Kozlovac JP, Richmon JY (2013) Chapter 1. Defining biosecurity and related concepts. In: Burnette RN (ed) Biosecurity: understanding, assessing, and preventing the threat. Wiley, pp 1–16. https://doi.org/10.1002/9781118769119.ch1

Cameron AR (2002) Survey toolbox for aquatic animal diseases—a practical manual and software package. Australian Centre for International Agricultural Research (ACIAR), Monograph No. 94, p 375

Cameron AR (2004) Principles for the design and conduct of surveys to show presence or absence of infectious disease in aquatic animals. Australia Department of Agriculture and Water Resources, National Aquatic Animal Health Technical Working Group Policy Document, p 37

Chinabut S, Puttinaowarat S (2005) The choice of disease control strategies to secure international market access for aquaculture products. Dev Biol (Basel) 121:255–261
Citarasu T, Babu MM, Yilmaz E (2022) Alternative medications in shrimp health management for improved production. Aquaculture 6:738695
Czochor J, Turchick A (2014) Introduction. Vaccines. Yale J Biol Med 87(4):401–402
Dadar M, Dhama K, Vakharia VN, Hoseinifar SH, Karthik K, Tiwari R, Khandia R, Munjal A, SalgadoMiranda C, Joshi SK (2016) Advances in aquaculture vaccines against fish pathogens: global status and current trends. Rev Fish Sci Aquac 25(3):184–217. https://doi.org/10.1080/23308249.2016.1261277
FAO (1995) Code of Conduct for Responsible Fisheries Rome, p 41.
FAO (1997) Technical Guidelines for Responsible Fisheries. No. 4. Rome. p 82.
FAO (2022) The state of world fisheries and aquaculture 2022. Towards blue transformation. FAO, Rome. https://doi.org/10.4060/cc0461en
FAO/NACA (2000) The Asia regional technical guidelines on health management for the responsible movement of live aquatic animals and the Beijing consensus and implementation strategy. FAO Fisheries Technical Paper No. 402. Rome, FAO, p 53
Frąckiewicz M (2023) Artificial intelligence, TS2 space. https://ts2.space/en/ai-and-sustainable-aquatic-biosecurity-ensuring-safe-and-sustainable-aquatic-trade-with-intelligent-machines-3/
Hadfield CA, Clayton LA (2011) Fish quarantine: current practices in public zoos and aquaria. J Zoo Wildl Med 42(4):641–650
Hastein T, Binde M, Hine M, Johnsen S, Lillehaug A, Olesen NJ, Purvis N, Scarfe AD, Wright B (2008) National biosecurity approaches, plans and programmes in response to diseases in farmed aquatic animals: evolution, effectiveness and the way forward. Rev Sci Tech 27(1):125–145
Hayes KR (2002) Identifying hazards in complex ecological systems. Part 1: fault-tree analysis for biological invasions. Biol Invasions 4(3):235–249
Hayes KR (2003) Biodiversity sector ecological implications of GMOs. Robust methodologies for ecological risk assessment final report: inductive hazard analysis for GMOs
Hedrick RP, McDowell TS, Marty GD, Mukkatira K, Antonio DB, Andree KB, Bukhari Z, Clancy T (2000) Ultraviolet irradiation inactivates the waterborne infective stages of Myxobolus cerebralis: a treatment for hatchery water supplies. Dis Aquat Org 42:53–59
ICES. (2005). ICES Code of Practice on the Introductions and Transfers of Marine Organisms, pp 30.
Jayasena SD (1999) Aptamers: an emerging class of molecules that rival antibodies in diagnostics. Clin Chem 45(9):1628–1650
Jia B, St-Hilaire S, Singh K, Gardner I, A. (2017) Biosecurity knowledge, attitudes and practices of farmers culturing yellow catfish (Pelteobagrus fulvidraco) in Guangdong and Zhejiang provinces, China. Aquaculture 471:146–156
Kumar V, Roy S, Behera BK, Das BK (2021) Disease diagnostic tools for health management in aquaculture. In: Pandey PK, Parhi J (eds) Advances in fisheries biotechnology. Springer, pp 363–382. https://doi.org/10.1007/978-981-16-3215-0_21
Leszczynski JK, Tackett J, Wallace-Fields M (2017) Basic animal facility management. In: Management of animal care and use Programs in research, education, and testing. pp 599–638
Lind CE, Dana GV, Perera RP, Phillips MJ (2015) Risk analysis in aquaculture: a step-by-step introduction with worked examples. WorldFish, Penang, Malaysia
Ma J, Bruce TJ, Jones EM, Cain KD (2019) A review of fish vaccine development strategies: conventional methods and modern biotechnological approaches. Microorganisms 7(11):569. https://doi.org/10.3390/microorganisms7110569
Mishra S, Das R, Swain P (2018) Status of fish diseases in aquaculture and assessment of economic loss due to disease. In: Rao PN, Pandey BN, Pandey PN, Joshi BD (eds) Contemporary trends in fisheries and aquaculture. Today & Tomorrow's Printers and publishers, pp 183–199
Mitchell H, Stoskopf M (1999) Guidelines for development and application of aquatic animal health regulations and control programs. J Am Vet Med Assoc 214(12):1786–1789

Mohan CV, Phillips MJ, Bhat BV, Umesh NR, Padiyar PA (2008) Farm-level plans and husbandry measures for aquatic animal disease emergencies. Rev Sci Tech 27(1):161–173

Mondal H, Thomas J (2022) A review on the recent advances and application of vaccines against fish pathogens in aquaculture. Aquac Int 30:1971–2000. https://doi.org/10.1007/s10499-022-00884-w

Moss SM, Moss DR, Arce SM, Lightner DV, Lotz JM (2012) The role of selective breeding and biosecurity in the prevention of disease in penaeid shrimp aquaculture. J Inverteb Pathol 110:247–250. https://doi.org/10.1016/j.jip.2012.01.013

Murray N, MacDiarmid SC, Wooldridge M, Gummow B, Morley RS, Weber SE, Giovannini A, Wilson D (2004) Introduction and qualitative risk analysis. In: Handbook on import risk analysis for animals and animal products, vol 1, p 59

National Fisheries Policy (2020) Department of Fisheries, Government of India. https://dof.gov.in/sites/default/files/2020-12/Policy_0.pdf

Nimjee SM, Rusconi CP, Sullenger BA (2005) Aptamers: an emerging class of therapeutics. Annu Rev Med 56:555–583

Oidtmann BC, Thrush MA, Denham KL, Peeler EJ (2011) International and national biosecurity strategies in aquatic animal health. Aquaculture 320(1–2):22–33. https://doi.org/10.1016/j.aquaculture.2011.07.032

Opiyo M, Mziri V, Musa S, Kyule D, Hinzano S, Wainaina M, Magondu E, Werimo K, Ombwa V (2020) Fish disease management and biosecurity systems. In: State of Aquaculture in Kenya 2020: towards nutrition-sensitive fish food systems, pp 97–126

Palić D, Scarfe AD (2018) Biosecurity in aquaculture: practical veterinary approaches for aquatic animal disease prevention, control, and possible eradication. In: Dewulf J, Van Immerseel F (eds) Biosecurity in animal production and veterinary medicine—from principles to practice. ACCO Publishing House, Leuven, Belgium, pp 498–523

Palić D, Scarfe AD, Walster CI (2015) A standardized approach for meeting national and international aquaculture biosecurity requirements for preventing, controlling, and eradicating infectious diseases. J Appl Aquac 27(3):185–219

Peeler E (2005) The role of risk analysis and epidemiology in the development of biosecurity for aquaculture. In: Diseases in Asian aquaculture, proceedings of the fifth symposium on diseases in Asian aquaculture, Queensland, Australia, 24–28 November 2002. Asian Fisheries Society, Manila, Philippines. pp 35–46

Peeler EJ, Murray AG, Thebault A, Brun E, Giovaninni A, Thrush MA (2007) The application of risk analysis in aquatic animal health management. Prev Vet Med 81(1–3):3–20

Plant KP, LaPatra SE (2011) Advances in fish vaccine delivery. Dev Comp Immunol 35(12):1256–1262. https://doi.org/10.1016/j.dci.2011.03.007

Pruder GD (2004) Biosecurity: application in aquaculture. Aquac Eng 32:3–10. https://doi.org/10.1016/j.aquaeng.2004.05.002

Rodgers CJ (ed) (2001) Risk analysis in aquatic animal health. In: Proceedings of an international conference held in Paris, France, 8–10 February 2000. World Organisation for Animal Health, Paris. p 346

Sanz VA (2018) Specific pathogen free (SPF), specific pathogen resistant (SPR) and specific pathogen tolerant (SPT) as part of the biosecurity strategy for whiteleg shrimp (Penaeus vannamei Boone 1931). Asian Fish Soc 31:112–120

Scarfe AD, Palić D (2020) Chapter 3. Aquaculture biosecurity: practical approach to prevent, control, and eradicate diseases. In: Kibenge FSB, Powell MD (eds) Aquaculture health management, Academic Press, pp 75–116. https://doi.org/10.1016/B978-0-12-813359-0.00003

Sitjà-Bobadilla A, Oidtmann B (2017) 5 - integrated pathogen management strategies in fish farming. In: Jeney G (ed) Fish diseases. Academic press, pp 119–144. https://doi.org/10.1016/B978-0-12-804564-0.00005-3

Snieszko S, Piotrowska W, Kocylowski B, Marek K (1938) Badania bakteriologiczne i serogiczne nad bakteriami posocznicy karpi. Memoires de l'Institut d'Ichtyobiologie et Pisciculture de la Station de Pisciculture Experimentale a Mydlniki de l'Universite Jagiellonienne a Cracovie Nr 38

Stentiford GD, Neil DM, Peeler EJ, Sheilds JD, Small HJ, Flegel TW, Vlak JM, Jones B, Morado F, Moss S, Lotz J (2012) Disease will limit future food supply for the global crustacean fishery and aquaculture sector. J Invertebr Pathol 110(2):141–157. https://doi.org/10.1016/j.jip.2012.03.013

Vijayan KK, Sanil NK (2010) Biosecurity in fisheries and aquaculture in the Indian context. In: Winter school on vistas in marine biotechnology

Vose D (2008) Risk analysis: a quantitative guide, 3rd edn. Wiley, West Sussex, England

World Organisation for Animal Health (OIE) (2006) Aquatic animal health code, 9th edn. OIE, Paris

World Organisation for Animal Health (OIE) (2019) Aquatic animal health code, 22nd edn. OIE, Paris

World Organisation for Animal Health (OIE) (2022) Aquatic animal health code, 25th edn. OIE, Paris

WTO (1994) Agreement on the application of sanitary and phytosanitary measures. In: The results of the Uruguay round of multilateral trade negotiations: the legal texts. General Agreement on Tariffs and Trade (GATT), World Trade Organization, Geneva, pp 69–84. http://www.wto.org/english/docs_e/legal_e/15-sps.pdf

Yanong RP, Erlacher-Reid C (2012) Biosecurity in aquaculture, part 1: an overview. USDA Southern Regional Aquaculture Center, Stoneville, pp 25–59

Yanong RP, Francis-Floyd R, Petty BD (2022) Production medicine and biosecurity in aquaculture. Production medicine and biosecurity in aquaculture—exotic and laboratory animals—MSD veterinary manual. msdvetmanual.com

Chapter 15
Vaccination as an AMR Control/Preventive Measure in Aquaculture

Kim D. Thompson and Thao Mai

Abstract The stressful conditions associated with intensive fish farming increase the fish's susceptibility to infectious diseases. Rising water temperatures due to climate change are also likely to see increased disease outbreaks and the emergence of new pathogens within farming systems. Disease outbreaks result in substantial economic losses to the aquaculture industry globally, in the region of several billion US dollars annually. Antibiotics are commonly used to control bacterial disease outbreaks. Antimicrobial resistance (AMR) has emerged as a global threat to human and animal health. In aquaculture, AMR poses significant challenges due to the widespread use of antimicrobial agents to prevent and treat diseases. There are increasing concerns about their impact on the environment, the development of antimicrobial resistance, and the risk of antibiotic residues in fish products. Vaccination has been shown to be effective in controlling infectious disease outbreaks and is routinely used for some fish species, especially for higher-value species such as Atlantic salmon. Vaccine use has been shown to reduce the use of antibiotics in fish farming. The main advantage of vaccines is that they are used prophylactically to prevent disease outbreaks from occurring, thus removing the need to use antibiotics. However, there is a lack of vaccines for many pathogens, and where vaccines are available, their uptake can be poor for some fish species, e.g. tilapia. Most commercial vaccines are based on killed whole-cell preparations administered to fish by intraperitoneal injection. These can be expensive and logistically difficult to administer in some countries, so farmers prefer to treat disease outbreaks with antibiotics rather than investing in disease prevention. This chapter provides an overview of the applications and benefits of vaccination, explores various vaccine types and their applications, and evaluates the impact of vaccination on AMR control.

Keywords Intensive fish farming · Tilapia

K. D. Thompson (✉) · T. Mai
Moredun Research Institute, Pentlands Science Park, Penicuik, UK
e-mail: kim.thompson@moredun.ac.uk

P. Elumalai, S. Lakshmi (eds.), *Antimicrobial Resistance in Aquaculture and Aquatic Environments*, https://doi.org/10.1007/978-981-97-7320-6_15

15.1 Introduction

Aquaculture has surpassed capture fisheries in the past 10 years as the most rapidly expanding sector in animal production globally. It now produces approximately 94.4 million tonnes of aquatic animals, valued at around 295.7 billion US dollars (FAO 2024). Aquaculture, a vital industry for meeting the rising global demand for seafood, faces significant challenges due to its intensification and expansion. These challenges, if not addressed, threaten the industry's long-term sustainability. A primary concern within aquaculture systems is the spread of diseases, which can lead to devastating effects such as high mortality rates, stunted growth, and poor aquatic production. These disease outbreaks result in significant economic losses for farmers and have adverse environmental impacts. Estimates suggest infectious diseases cause a loss of about 10% in global aquaculture production each year, translating to a loss exceeding 10 billion US dollars (Adams 2019).

When confronted with disease outbreaks, aquaculture operators often turn to antimicrobial agents, including antibiotics, to control and prevent disease spread. However, excessive and inappropriate use of antimicrobials contributes to the emergence and spread of antimicrobial resistance (AMR) in aquaculture systems (Van Boeckel et al. 2015; Phu et al. 2016; Watts et al. 2017). AMR occurs when microorganisms evolve to resist the effects of antimicrobial drugs, making these drugs less effective or completely ineffective. This resistance poses a significant threat to the health and welfare of aquatic animals and, importantly, human health. Resistant pathogens can be transmitted to humans through consuming contaminated seafood, for example, or through the environment, highlighting the urgent need to address AMR in aquaculture (Reverter et al. 2020).

Various strategies are employed for controlling disease within aquaculture farming systems. These include enhancing farm environments, breeding species that are resistant to diseases, maintaining stringent biosecurity measures, utilising probiotics, employing advanced techniques for disease detection, and using vaccines (Bondad-Reantaso et al. 2005, 2023; Henriksson et al. 2018; Austin et al. 2022; Wang et al. 2023).

Vaccination has become one of the key methods for controlling disease in aquaculture (Adams 2019). Vaccination is a sustainable and practical approach to enhance the immune response of aquatic animals against diseases, reducing the need for antimicrobials, especially in salmon aquaculture (O'Neill 2015; Henriksson et al. 2018; Norwegian Ministries 2020). By preventing disease outbreaks through vaccination, the reliance on antibiotics can be significantly diminished, thereby slowing the development and spread of AMR.

Adopting vaccination strategies in aquaculture practices marks a shift towards more responsible and eco-friendly disease management. This approach emphasises proactive health measures instead of treatments once the disease has occurred. Addressing AMR through vaccination not only protects the fish's health but also helps ensure public health and ecological equilibrium in fish farm environments. Adopting vaccination strategies is in line with worldwide initiatives to minimise the use of antimicrobials in food production systems. While they may not be the ultimate solution for all diseases, vaccinations play a significant role in a comprehensive approach to fighting AMR (Henriksson et al. 2018; Wang et al. 2023).

15.2 Vaccination: Principles and Benefits

Vaccines used in aquaculture are designed to combat prevalent diseases and play a crucial role in promoting the health and production efficiency of aquatic species (Ellis 1988; Ma et al. 2019). Their primary function is to expose the fish's immune system to specific pathogens in a controlled manner, using antigens or antigenic components derived from the pathogens. Antigens are molecules that can trigger an immune response when detected by an organism's immune system. They are typically found on the surface of pathogens such as bacteria, viruses, and parasites and are usually proteins or polysaccharides, although they can also be lipids or nucleic acids, depending on the pathogen. Each antigen has specific structures called epitopes that are recognised by immune cells, meaning that the immune response is tailored to target each specific type of pathogen. These molecules are distinct enough from the body's molecules to be recognised as foreign. One of the key features of the immune system is its ability to remember antigens it has previously encountered. When the immune system is exposed to an antigen for the first time, it creates memory cells that remain in the body. If the same antigen is re-encountered, these memory cells recognise it and mount a quicker and stronger response, often neutralising the pathogen before it can cause significant harm. Vaccination mimics an infection, which prompts an immune response (Ellis 1988). This response includes activating immune cells and producing antibodies, proteins designed to recognise and neutralise the pathogen (Munang'andu et al. 2020). Consequently, when the fish encounters the actual pathogen in the future, its immune system is better prepared to fight off the infection quickly and effectively.

When fish are vaccinated, their immune system responds by recognising the components of the vaccine through innate immune cells such as macrophages and dendritic cells. Antigens from the vaccine are processed by antigen-presenting cells and presented to adaptive immune cells, i.e. T cells and B cells, leading to their activation and differentiation (Du et al. 2022). B cells differentiate into plasma cells that secrete pathogen-specific antibodies, primarily immunoglobulin M (IgM), the main antibody isotype produced by fish in response to pathogens (Smith et al. 2019; Salinas et al. 2021). Immunoglobulin isotypes IgT and IgD are also found in fish, but they play different roles in the immune response. IgT is an antibody isotype mainly associated with mucosal immunity, while IgD is present in small amounts and has roles that are not yet fully understood in fish immunity (Salinas 2015; Salinas et al. 2021). Memory cells, formed during vaccination, "remember" the antigens and play a crucial role in the secondary immune response (Du et al. 2022). In mammals, the secondary immune response is characterised by rapid and robust production of antibodies upon re-exposure to a pathogen. This leads to a more effective and quicker pathogen clearance compared to the primary response. Additionally, memory T cells play a critical role in recognising and destroying infected cells, contributing to the overall efficiency of the secondary response. The evidence for a distinct and well-defined secondary immune response in fish following vaccination is less extensive than in mammals. While fish possess adaptive immunity and can generate memory responses, the characteristics of the secondary response differ from those observed in mammals (Bengtén and Wilson 2015). Some studies have reported increased antibody production and specific immune responses upon re-exposure to a pathogen after vaccination in fish, indicating the presence of memory cells (Ma et al. 2019). However, the magnitude and kinetics of the secondary response can vary depending on factors such as the type of vaccine used, the specific pathogen, and the fish species involved.

Vaccination programmes can create a concept known as "herd immunity" (Imtiaz et al. 2023). When a large percentage of the fish population is vaccinated, disease transmission is interrupted, protecting both vaccinated and unvaccinated individuals (Gudding 2014). Unvaccinated animals benefit from reduced exposure to pathogens due to the decreased disease prevalence within the population (Assefa and Abunna 2018). Herd immunity reduces the overall disease burden in the aquaculture system, thereby reducing the need for antimicrobial interventions. Several factors influence the establishment and maintenance of herd immunity in aquaculture. The percentage of vaccinated individuals within the population directly affects herd immunity. High vaccine coverage is necessary to prevent disease transmission effectively. The vaccine's effectiveness in inducing protective immune responses influences the development of herd immunity. Vaccines with high efficacy ensure a higher level of herd immunity within the population (Lipsitch and Siber 2016). The properties of the pathogen, such as its transmission dynamics, virulence, and persistence, can impact the establishment and sustainability of herd immunity. Implementing robust biosecurity measures, such as controlling the movement of animals, preventing contact with wild populations, and ensuring hygienic practices, can enhance herd immunity by reducing the introduction and spread of diseases.

The benefits of vaccination in disease management go beyond reducing the reliance on antimicrobials. Vaccination can also improve animal welfare by reducing the occurrence and severity of diseases and minimising the stress associated with disease outbreaks and treatments. Additionally, it contributes to the sustainability and productivity of the aquaculture industry by minimising production losses, improving the overall health and quality of farmed animals, and enhancing the industry's reputation.

Although developing and administering vaccines incurs initial costs, the long-term economic benefits outweigh the costs associated with disease treatment and the losses from disease outbreaks. Vaccines are especially prevalent in the husbandry of high-value fish like rainbow trout and Atlantic salmon (Adams 2019). Many of these vaccines are multivalent, protecting against several pathogens with just one application. Thorarinsson and Powell (2006) pointed out that the cost of vaccination, unlike antibiotics, can be calculated in advance and factored into the production budget. Vaccines have been shown to increase profitability by reducing fish deaths, improving product quality, and enhancing fish welfare. Vaccines, used prophylactically, also reduce the reliance on antibiotics. For species with lower economic value, like tilapia, the expensive nature and logistical hurdles of vaccines prompt farmers to opt for antibiotics, given the smaller profit margins involved.

15.3 Vaccine Types and Applications in Aquaculture

Duff's work in 1942 marked the start of vaccine development in aquaculture, focusing on a furunculosis vaccine for cutthroat trout (*Oncorhynchus clarkia*) using inactivated *Aeromonas salmonicida*. It was not until the 1980s that the first commercial aquaculture vaccines appeared in the United States of America, targeting diseases such as enteric redmouth in rainbow trout (*Oncorhynchus mykiss*) and vibriosis in farmed Atlantic salmon (*Salmo salar*) (Busch 1982; Tebbit and Goodrich 1982). These vaccines were marketed in Scotland and Scandinavia during the following decade. In the 1980s, only two commercial fish vaccines were available. These initial vaccines, made of formalin-killed cells, were delivered by immersion. The efficacy of the furunculosis vaccine was initially low, prompting the addition of an adjuvant and a switch to injectable delivery (Gudding et al. 1999). The number of commercial vaccines for fish has risen significantly, with at least 50 vaccines currently available for different fish species, mainly based on formalin-killed cell formulations (Du et al. 2022). These are primarily delivered by Merck Sharp and Dohme (Norbio, Intervet, AVL), Elanco (AquaHealth via Novartis Animal Health), and Zoetis (Apothekernes Laboratorium/Alpharma/Pharmaq) for higher valued farmed fish species (Barnes et al. 2022). Other types of vaccines, such as live attenuated, subunit, and DNA, have also been commercialised for use in aquaculture. There is interest in developing RNA and vector vaccines for fish, but these have only been tested in laboratory trials. Each vaccine type has its advantages and considerations, including safety, efficacy, and ease of administration (Adams 2019).

15.3.1 Inactivated Vaccines

Inactivated vaccines, also referred to as killed vaccines, play a crucial role in preventing and controlling viral and bacterial diseases, with many commercially available vaccines formulated with inactivated pathogens (Du et al. 2022). Unlike live attenuated vaccines, which use a weakened form of the pathogen, inactivated vaccines contain pathogens that have been killed or inactivated. This means they cannot replicate or cause disease in vaccinated fish. Creating an inactivated vaccine involves growing the pathogen in a controlled environment and then killing it using physical or chemical methods, such as heat or formalin treatment (Tlaxca et al. 2015). The inactivated pathogen retains its immunogenic properties, reflecting its ability to provoke an immune response. One of the main advantages of inactivated vaccines is their safety. Since the pathogen cannot replicate, there is no risk of its reverting to a virulent form and causing disease in the vaccinated population. Inactivated vaccines typically stimulate a strong humoral immune response, especially if they are adjuvanted, characterised by the production of antibodies. However, their ability to elicit a robust cellular immune response is generally weaker than live vaccines. Inactivated vaccines are often administered with adjuvants to enhance the immune response, which boosts the fish's immune response to the vaccine. The administration of inactivated vaccines in aquaculture usually requires injection, which can be labour-intensive and stressful for the fish (Du et al. 2022). This is one of the limitations of using inactivated vaccines in large-scale aquaculture operations. However, for valuable species or smaller operations, the benefits of vaccination can outweigh these challenges. Another consideration is the duration of immunity provided by inactivated vaccines. The immunity conferred by these vaccines may not last as long as live vaccines, potentially necessitating booster vaccinations to maintain protection (Adams 2019). In terms of storage and handling, inactivated vaccines are often more stable than live vaccines. The preparation of these vaccines is relatively simple and can be adapted for multivalent formulations. Despite these benefits, it is necessary to ensure thorough inactivation of the bacteria, and high vaccination doses may be required to elicit protection, which can result in unwanted reactions (Du et al. 2022).

15.3.2 Live Attenuated Vaccines

Live attenuated vaccines work by introducing a weakened form of the pathogen to the fish, which is incapable of causing the disease but can still provoke an immune response. The development of live attenuated vaccines involves modifying the pathogen to lose its virulence or disease-causing ability while retaining its immunogenic properties, which are necessary to stimulate the immune system (Shoemaker et al. 2009; Ma et al. 2019). This can be achieved through various methods, such as genetic modification or culturing the pathogen under conditions that induce mutations, including antibiotic-induced attenuation, leading to a loss of virulence.

A method using rifampicin, an antibiotic, for chemically inducing mutations has proved effective in creating live bacterial vaccines, especially for Gram-negative bacteria in the United States (Shoemaker et al. 2007; Norqvist et al. 1989; LaFrentz et al. 2008). This technique involves serially growing the bacteria in increasing concentrations of rifampicin and has been used to develop vaccines against various fish bacteria (Ma et al. 2019), e.g. *Vibrio anguillarum* (Norqvist et al. 1989) and *Flavobacterium psychrophilum* (LaFrentz et al. 2008). Additionally, chemical compounds such as acriflavine dye and novobiocin have been trialled for attenuating pathogens like *Streptococcus agalactiae*, *S. iniae*, *Edwardsiella ictaluri*, and *Aeromonas hydrophila* (Ma et al. 2019).

One of the key advantages of live attenuated vaccines is their ability to induce a strong and long-lasting immune response. Because these vaccines closely mimic a natural infection, they often generate a more robust form of immunity compared to inactivated or subunit vaccines (Levine and Sztein 2004). This includes the activation of both the humoral (antibody-mediated) and cellular and mucosal immune responses, and generally, only one immunisation is required (Du et al. 2022). Another benefit is the potential for oral administration. Live attenuated vaccines can be delivered through the water (Adamek et al. 2022; Liu et al. 2022) or incorporated into feed (Klafack et al. 2022), simplifying the vaccination process and reducing stress for the fish. However, there are also risks associated with live attenuated vaccines. There is a small chance that the attenuated pathogen could revert to a virulent form, which could cause disease in the vaccinated or wild fish populations if the vaccine is released into the environment. Moreover, these vaccines require careful handling and storage to maintain efficacy. Additionally, the development of live attenuated vaccines is often more complex and time-consuming than other vaccine types. Each vaccine must be specifically tailored to the pathogen and the fish species, and extensive testing is required to ensure safety and efficacy.

15.3.3 Subunit Vaccines

Subunit vaccines represent a more targeted approach than live attenuated or inactivated vaccines. These vaccines contain specific parts of the pathogen, composed of proteins, peptides, or polysaccharides, rather than the entire organism. Developing subunit vaccines involves identifying and isolating specific antigens of the pathogen, the components that the immune system recognises and responds to. The antigens are then purified and used in the vaccine formulation. Advanced biotechnological methods, such as recombinant DNA technology, are used to produce the antigens, whereby the gene for a pathogen antigen is inserted into a recombinant expression vector and expressed as a gene product, either as a recombinant peptide or protein, using either a prokaryotic or eukaryotic expression system (Hansson et al. 2000). Bacteria can produce antigens efficiently and in significant amounts. However, they lack post-translational modification capabilities, and there is a risk of contamination with endotoxins in the final expressed products. Yeast, on the other hand, is not only able to secrete proteins but can also modify them post-translationally. Mammalian expression systems can also modify expressed products post-translation; however, producing vaccines in this type of system can be expensive (Khan 2013), especially for use in aquaculture.

One of the main advantages of subunit vaccines is their high safety profile. Since they do not contain live organisms, there is no risk of the vaccine causing the disease it is meant to prevent, eliminating any concerns about reversion to a virulent form. However, subunit vaccines often trigger weaker immune responses. They may not be as effective in stimulating a strong cellular immune response compared to live vaccines or whole-cell vaccines, whether killed or live. This limitation stems from fewer immune-stimulating components and the absence of replication or exposure to a full range of antigens that are present in whole-cell vaccines (Ma et al. 2019).

Subunit vaccines generally elicit a humoral immune response, with antibodies produced only to the antigens present in the vaccine. Subunit vaccines are often administered with adjuvants to enhance the immune response (Holten-Andersen et al. 2004).

Another important aspect of subunit vaccines is their specificity. Because they can be tailored to target specific parts of a pathogen, they can be highly effective against certain diseases. However, this specificity also means their development requires a detailed understanding of the pathogen and the host's immune response. The production of subunit vaccines can be more complex and costly than traditional vaccines. The antigen identification, isolation, and purification processes require advanced technology and expertise. This can make subunit vaccines more expensive, which might limit their use in situations where the economic value of the fish species justifies the cost of their use.

Virus-like particles (VLPs) offer significant advantages as subunit vaccine platforms thanks to their ability to mimic the structure of viruses without being capable of replication. The self-assembly feature of viral structural proteins, creating structures similar to an actual virus, has enabled the creation of a new class of subunit vaccines based on VLPs (Dhar and Allnutt 2011). These particles feature densely packed viral surface proteins that accurately represent the shapes of viral epitopes, crucial for triggering robust immune responses from T and B cells (Akahata et al. 2010). With their small size, typically between 20 and 200 nm in diameter, VLPs have been shown to enhance the immune response of mammals to the vaccine. This non-replicative nature makes VLPs safer than live attenuated viruses (Noad and Roy 2003). The baculovirus system offers a rapid, cost-efficient approach for high-yield recombinant protein (e.g. VLPs) production, making it a favoured method for heterologous protein synthesis (Bedekar et al. 2022). Some VLP-based vaccines receive approval for humans from the Food and Drug Administration for hepatitis B and human papillomavirus (Hotez and Bottazzi 2022). No VLP vaccines have been marketed for fish as yet.

15.3.4 DNA Vaccines

DNA vaccines represent cutting-edge technology for aquaculture vaccines. These vaccines employ a fragment of the pathogen's genetic material, typically DNA, to stimulate an immune reaction. In contrast to traditional vaccines that introduce components of the pathogen itself, DNA vaccines instruct host cells to generate specific antigens. The fundamental principle behind DNA vaccines involves incorporating a gene encoding a pathogen's antigen into a plasmid, a small, circular piece of DNA. When this plasmid is introduced into the fish, usually by injection, the cells internalise the DNA, encoding the biological information necessary for manufacturing the antigen. Subsequently, this antigen is presented on the cell surface, prompting the fish's immune system to identify it as foreign and initiate an immune response against it. DNA vaccines have several unique characteristics that activate both innate and adaptive immune responses (Collins et al. 2019). A notable advantage of DNA vaccines lies in their capacity to activate both humoral (antibody-mediated) and cellular immune responses (Pollard and Bijker 2021), providing robust protection against the targeted disease. Evidence of immunity has been shown to last for up to 2 years after DNA vaccination against infectious hematopoietic necrosis virus (IHNV) (Kurath et al. 2006) and up to 17 months following vaccination for viral haemorrhagic septicaemia virus (VHSV) (Sepúlveda and Lorenzen 2016).

The development and production of DNA vaccines can be more straightforward and cost-effective than traditional vaccine methods. Once the target gene is identified, it can be quickly and efficiently produced in substantial quantities. These vaccines offer a significant benefit due to their adaptability and the speed and ease with which they can be modified and manufactured in response to a new pathogen (Pollard and Bijker 2021). Another merit of DNA vaccines is their high level of safety. As they do not contain live or inactivated pathogens, there is no risk of the vaccine causing the disease.

One challenge with DNA vaccines is the delivery method, however. Presently, injection is the primary means of administering these vaccines, which can be labour-intensive and stressful for the fish. Ongoing research is exploring alternative delivery methods, such as oral or immersion-based techniques, which would be more practical in aquaculture (Collins et al. 2019). DNA vaccines generally exhibit favourable stability and storage characteristics, as DNA is relatively robust and often does not require extremely cold conditions for preservation. This simplifies transport and handling compared to certain traditional vaccines. Nonetheless, the regulatory approval process for DNA vaccines can be intricate and stringent, given the relatively novel nature of this technology, particularly in the context of food-producing animals like fish.

Hansen et al. (1991) were the first to demonstrate that a DNA plasmid could express a protein following its injection into the muscle of fish. DNA vaccines have made significant strides in aquaculture, targeting various diseases afflicting fish populations. One of the earliest examples is the vaccine for IHNV, a rhabdovirus that causes severe disease in salmon and trout. This vaccine, notable for its early regulatory approval, has effectively reduced mortality rates and controlled disease spread among these fish species (Kurath et al. 2006). Similarly, a DNA vaccine against VHSV has shown promise in protecting a range of freshwater and marine fish from this pervasive viral threat (Lorenzen et al. 2002). Another rhabdovirus, spring viraemia of carp virus (SVCV), which significantly impacts carp and other cyprinids, has been targeted by DNA vaccine research, showing encouraging results in disease prevention (Embregts et al. 2017). The first DNA vaccine to receive commercial licensing for use in aquaculture was APEX-IHN (Novartis/Elanco) in 2005, designed to protect Atlantic salmon against IHNV in British Columbia, Canada, representing the first DNA vaccine licensed for any veterinary use in animal husbandry production globally (Salonius et al. 2007). In 2017, CLYNAV (Elanco), a polyprotein-encoding DNA vaccine against salmon pancreas disease virus (SPDV) infection in Atlantic salmon, was given marketing authorisation by the European

Commission through the European Medicines Agency for use within the EU (Collins et al. 2019). Collins et al. (2019) have reviewed the progress in DNA vaccine research aimed at combating some of the major viral diseases affecting fish.

15.3.5 RNA Vaccines

RNA vaccines, a relatively new and innovative technology in the field of immunisation, are starting to make their mark in aquaculture research. These vaccines use messenger RNA (mRNA) to instruct cells in the host organism to produce specific antigens of a pathogen, thereby triggering an immune response. Unlike DNA vaccines, RNA vaccines do not need to enter the cell nucleus and work directly in the cytoplasm. Two types of RNA-based vaccines exist: non-replicating mRNA vaccines and self-replicating mRNA vaccines. Non-replicating mRNA (NRM) is characterised by the presence of 5′ and 3′ untranslated regions (UTRs), a 5′ cap structure, and a polyadenylated 3′ tail. Upon entering the cell cytoplasm, the NRM is promptly translated into its corresponding protein. In contrast, self-amplifying (SAM) mRNA possesses the same structural features as NRM but also contains replicase enzymes that facilitate self-replication of the mRNA within the cell, producing multiple mRNA copies. These copies are then translated into the target protein (Bedekar et al. 2022).

The mRNA vaccine is encapsulated into lipid nanoparticles for delivery to protect it from degradation and facilitate its entry into the cells. Once inside, the mRNA is used by the host cell's machinery to produce the pathogen's antigen. The antigen is then displayed on the cell's surface, prompting the fish's immune system to recognise it as foreign and initiate an immune response. One of the key advantages of RNA vaccines is their safety profile. Since the mRNA does not integrate into the host's genome and degrades naturally after a short period, there is no risk of long-term genetic changes in the vaccinated organism. RNA vaccines are known for their rapid development and production capabilities. The mRNA sequences can be designed and synthesised quickly once the genetic information of the pathogen is known. This feature is particularly beneficial in responding to emerging diseases or producing vaccines for various pathogens. The immune response induced by RNA vaccines can be potent and specific, generating both humoral and cellular responses. However, the actual effectiveness and duration of immunity in the context of aquaculture are subjects of ongoing research. RNA vaccines are more effective than traditional plasmid DNA vaccines in inducing antigen-specific cellular immune responses (Leitner et al. 1999).

One of the challenges with RNA vaccines is their delivery and stability. mRNA is inherently unstable and susceptible to degradation, necessitating sophisticated delivery systems and stringent storage conditions, usually at very low temperatures. These requirements can pose logistical challenges, particularly in large-scale or remote aquaculture operations. The administration of RNA vaccines currently relies on injection, like DNA vaccines. Given the novelty of this technology, especially in food-producing animals, regulatory agencies are likely to require extensive data on safety, efficacy, and environmental impact before approving widespread use.

15.3.6 Vector Vaccines

Vector vaccines use a carrier, often a harmless virus or bacterium, to deliver genetic material from a targeted pathogen into the cells of fish (Adams et al. 2008). This method marries the concepts of live attenuated and subunit vaccines, aiming to provoke a strong, specific immune response while ensuring safety. The idea of vector vaccines is to employ a non-pathogenic vector to transport genes that code for specific antigens of a disease-causing pathogen. When the vector enters the fish's cells, these genes are expressed, producing antigens within the host organism (Du et al. 2022). This process allows the fish's immune system to recognise these antigens, thereby building immunity against the disease-causing pathogen. This response typically encompasses both the humoral (antibody-mediated) and cellular aspects of the immune system. The endogenous expression of the antigen, as it occurs within the host's cells, closely mimics a natural infection, potentially offering a more effective form of immunity compared to traditional inactivated or subunit vaccines. Regarding safety, vector vaccines are generally considered safe since the vector is non-pathogenic and the antigen represents only a small, non-infectious part of the pathogen. This minimises the vaccine's risk of causing the disease or reverting to a more virulent form. The immunity conferred by vector vaccines can be both stable and durable, which might reduce the necessity for frequent booster vaccinations. Additionally, these vaccines are versatile, capable of being designed to protect against various pathogens and adaptable to changes in the pathogen, such as mutations. It is important to note that while these vector vaccines represent a significant advancement in aquaculture disease control, their development and application are complex, requiring careful consideration of various factors such as vaccine safety, efficacy, and cost-effectiveness.

The first viral vector was engineered in 1972 using the SV40 virus (Jackson et al. 1972). This technology advanced when the hepatitis B surface antigen gene was inserted into a vaccinia virus, creating the first recombinant viral vector (McCann et al. 2022). Since the vaccinia virus vector was introduced in 1984, multiple viruses have been explored as vaccine vectors (Humphreys and Sebastian 2018). Various viruses, such as adenovirus, adeno-associated virus, retrovirus, cytomegalovirus, Sendai virus, and lentiviruses, have been used as vaccine vectors (Ura et al. 2014). Vaccinia and adenovirus became the most popular choices for their strong immune responses. Viral vector vaccines have undergone human clinical trials for diseases like Zika, influenza, respiratory syncytial virus, HIV, and malaria before being used for COVID-19 vaccines (Sasso et al. 2020). See Du et al. (2022) for examples of bacterial and viral recombinant live vector vaccines currently being investigated for aquaculture.

15.3.7 Autologous Vaccines

Developing commercial vaccines for the aquaculture industry faces significant challenges due to the high costs and extensive time required for development. Given the vast diversity of fish species cultured and their susceptibility to various diseases, creating licensed vaccines for all fish pathogens is not always feasible (Adams 2019). Autologous (or autogenous) vaccines also have an essential role in controlling disease outbreaks in aquaculture, especially where no licenced vaccine exists for a particular disease. For this reason, they are regarded as emergency vaccines with more straightforward routes for regulatory approval. These vaccines are developed by isolating pathogens from infected animals within the same aquaculture population system and offer customised protection against specific pathogens (Yanong 2011). They tend to be formalin-inactivated vaccines, formulated into "personalised" vaccines only for use in the farming system where the pathogen was isolated. Autogenous vaccines offer a quicker and more adaptable solution to combat local strains of rapidly evolving pathogens, unlike licensed commercial vaccines, which may not effectively address the continuous emergence of new serotypes due to genetic variability within pathogen species. These tailor-made vaccines can be developed and modified more swiftly, providing a crucial advantage in managing diseases caused by pathogens with fast-changing antigens, where generic vaccines fall short (Barnes et al. 2022). By employing autologous vaccines when no commercial products are available, aquaculture producers can effectively manage disease outbreaks, minimise economic losses, and reduce the need for antibiotics, contributing to better fish health and welfare. The regulation of these vaccines varies between regions, with governmental agencies overseeing safety and efficacy to ensure their proper use in aquaculture, and their administration is often provided by or under the supervision of a licensed veterinarian. Barnes et al. (2022) provide further information on using autologous vaccines in aquaculture, particularly in low- to middle-income countries.

15.3.8 Adjuvants

Adjuvants are a diverse group of substances added to injection vaccines to enhance and prolong the immune response to the vaccine (Guy 2007). However, because even related compounds can exhibit vastly different immunomodulating activities, recent classifications have emphasised the immunological responses they trigger rather than their chemical nature and physical properties, even though the precise mechanisms of immunostimulation of many of these substances remain unknown (Tafalla et al. 2013). Different types of adjuvants have varying mechanisms of action, and their selection depends on the fish species, target pathogen, and desired immune response. The current classification system distinguishes adjuvants into two main types: Signal 1 and Signal 2 facilitators (Tafalla et al. 2013). They help improve the uptake and presentation of antigens by immune cells, stimulating both the innate and adaptive arms of the immune system. Adjuvants activate various components of the innate immune system, triggering inflammation and recruiting immune cells to the site of vaccination. Signal 1 and 2 facilitators are required to activate specific T and B lymphocytes (Ribeiro and Schijns 2010). Montanides are notable examples of Signal 1 adjuvants supplied by SEPPIC for injectable vaccines. At the same time, beta-glucans, alums, saponins, poly I:C, synthetic oligonucleotides, cytokines, and flagellin are examples of Signal 2 adjuvants, designed to augment the immune response by providing co-stimulatory signals during antigen recognition (Tafalla et al. 2013). These provide additional co-stimulatory signals to immune cells, effectively responding to the vaccine antigens. Adjuvants can prolong antigen exposure by forming depots at the injection site, leading to a sustained immune response. Furthermore, they help generate immunological memory, ensuring a quicker and more effective response upon subsequent exposure to the pathogen. Adjuvants improve vaccine efficacy, reduce the required antigen dose, and promote longer-lasting immunity. There are few mucosal adjuvants for immersion or oral delivery, however. A commercial adjuvant for oral and immersion applications is available from SEPPIC. Delivery systems, such as nanoparticle-based carriers or liposomes, are also being explored to optimise vaccine delivery and enhance immunogenicity, discussed in more detail below. Ongoing research may lead to the development of new and improved adjuvants to further enhance fish vaccine efficacy, especially for mucosal vaccines (Adams 2019).

15.4 Administration Methods

Vaccine administration methods vary depending on the target population's vaccine type, species, and size (Gudding et al. 1999). Standard methods include injection, immersion, bath, and oral delivery (Plant and Lapatra 2011; Adams 2019; Du et al. 2022). Du et al. (2022) discuss the mechanisms of action associated with these various vaccination methods.

15.4.1 Injection Vaccination

Injection vaccination is the most used method to vaccinate fish because it has high efficacy, stimulating a rapid and effective immune response and resulting in strong and long-lasting protection against pathogens (Mitchell 1995). Injection vaccination, either manually or by machine, ensures that each fish receives the precise amount of vaccine needed for optimal protection. One of the main disadvantages is that the process can be labour-intensive. Individually injecting each fish can be time-consuming and requires skilled personnel, leading to higher labour costs (Dunn et al. 1990; Plant and Lapatra 2011). Moreover, the injection process can stress the fish, potentially impacting their growth rates and making them more susceptible to secondary infections if not handled with care. Mishandling during injection can also cause injuries to the fish, leading to tissue damage or infection, further affecting their health and performance. Using a vaccination machine can reduce the stress experienced by fish during the immunisation process and significantly reduce the amount of labour and vaccination time compared to manual vaccination (Liang et al. 2015).

15.4.2 Immersion Vaccination

Immersion vaccination is commonly employed to administer vaccines to many fish simultaneously, allowing significant numbers of fish to be vaccinated quickly and efficiently (Du et al. 2022). The vaccine used for immersion vaccination is typically diluted in the water at an appropriate concentration, and the fish are left to swim in the vaccine solution for a specific period to ensure sufficient exposure and uptake. The fish's skin and mucous membranes then take up the vaccine, triggering an immune response. This route of administration is generally more suitable for smaller-sized fish, such as fry and fingerlings, rather than larger adult fish. This is because smaller fish have a larger surface area relative to their body mass, allowing for better vaccine uptake through their mucosal surfaces (skin, gills, lateral line, and gastrointestinal tract) during immersion, where it stimulates a mucosal immune response in the fish. It is subsequently dispersed to the peripheral immune tissues via blood circulation, eliciting a systemic immune response (Du et al. 2017). Also, the cost of the vaccine can be prohibitive for vaccinating larger fish by immersion. Immersion vaccination is less labour-intensive than injection vaccination, reducing the need for skilled personnel and saving on labour costs. However, lower levels of protection can result from immersion vaccination compared to injection vaccination (Du et al. 2022). Various studies have explored methods to enhance immersion immunity, such as booster vaccinations, immunostimulants/adjuvants, delivery vehicles, physical preconditioning with low-frequency ultrasound, hyperosmotic solutions, and live attenuated and DNA vaccines (Bøgwald and Dalmo 2019). Mucoadhesive polymers, commonly used in drug delivery, are also being developed as vaccine carriers for immersion immunisation (Alawdi and Solanki 2021).

15.4.3 Oral Vaccines

Oral vaccines offer significant advantages for immunising fish. The gut offers several distinct advantages as a route for delivering antigens to fish, including simplified vaccination procedures, reduced stress for the fish, suitability for both small and large fish, and the lack of side effects (Mutoloki et al. 2015). Despite this, a limited number of oral vaccines are used by the aquaculture industry vaccines. For instance, in 2014, only two oral viral vaccines had been commercialised (Dhar et al. 2014). Developing effective oral vaccines can be complex, as ensuring the stability of vaccine antigens during digestion and successful delivery to immune cells pose technical challenges. Degradation of antigens due to gastric secretions can result in poor vaccine potency (Mutoloki et al. 2015). Vaccine formulations may need protective coatings or adjuvants to enhance their stability and efficacy. Ensuring uniform vaccine uptake among all individuals within a population can also be challenging. Designing effective oral vaccines is challenging due to our lack of knowledge of inducing strong protective immunity rather than inducing immune tolerance (Mutoloki et al. 2015). When vaccine antigens are ingested, they interact with immune cells in the digestive system, activating both systemic and mucosal immune responses (Xue et al. 2013; Chen et al. 2014) and triggering the production of specific antibodies. IgT has been associated with mucosal immunity (Munang'andu et al. 2015b; Zhang et al. 2010), although its role in immune protection against specific pathogens remains to be determined. The precise dosing for individual fish can be difficult to control, requiring careful monitoring and adjustment of vaccination protocols. The palatability of the vaccine is another important consideration to ensure the fish will eat the vaccine-containing feed. The stability and storage requirements of oral vaccines must also be the reliability of oral vaccines for aquaculture species, especially through antigen encapsulation (Thompson et al. 2023). For additional insights into oral vaccines in aquaculture, consult the reviews by Mutoloki et al. (2015), Embregts and Forlenza (2016), and Radhakrishnan et al. (2023).

15.4.4 Vaccine Encapsulation

Encapsulation of antigens within nanoparticles can protect them from degradation, increasing their stability and allowing for controlled antigen release. Nanovaccines are a type of vaccine that utilises nanotechnology to improve vaccine efficacy and delivery. The nanoparticles are used as carriers to deliver vaccine antigens, adjuvants, or other immunomodulatory agents that enhance the immune response (Vinay et al. 2018; Kheirollahpour et al. 2020; Pati et al. 2018; Curley and Putnam 2022). Nanoparticle-based vaccine delivery can enhance the specificity and strength of the immune response, leading to improved vaccine performance. These delivery systems offer distinct advantages by encapsulating antigens within nanoparticles or attaching them to the nanoparticle's surface. When encapsulated, antigens are shielded from the acidic conditions of a fish's digestive system, making this method ideal for oral administration in aquaculture. Surface-displayed antigens enable direct interaction with immune system pattern recognition receptors, such as Toll-like receptors on antigen-presenting cells (APCs), facilitating the uptake of antigens by these cells (Curley and Putnam 2022). This process leads to a more potent initiation of innate, humoral, cellular, and mucosal immunity than traditional antigen delivery (Pati et al. 2018). Nanoparticles, by their adjuvant properties, can also boost the immune response to proteins that are normally not very immunogenic, like some recombinant proteins (Fujita and Taguchi 2017). Their large surface area allows for incorporating greater amounts of antigenic material than is possible with standard vaccine forms. Moreover, nanoparticles improve the solubility and permeability of vaccines, resulting in heightened mucosal immunity, fewer adverse effects than injectable vaccines, and the ability to target the delivery to specific mucosal sites (Munang'andu et al. 2015a). Positively charged nanoparticles are particularly effective, as they are more readily taken up by immune cells at the vaccination site, amplifying the immune response (Fröhlich 2012; Fromen et al. 2015; Dacoba et al. 2017). The controlled and extended release of antigens from nanoparticles minimises the necessity for repeated booster doses. Furthermore, nanovaccines can be engineered for sustained release, ensuring continuous antigen exposure and immune activation, which contributes to the vaccine's overall efficacy (Thompson et al. 2023).

Nanoparticle vaccines utilise diverse formulations to boost immune reactions in fish using materials (Thompson et al. 2023). These platforms offer advantages in antigen delivery and immune stimulation without the need for frequent boosters or adjuvants. Inorganic options provide a high surface area for effective antigen adsorption and modification, improving vaccine stability and release. Polymeric versions allow for controlled antigen exposure and enhanced uptake by immune cells due to their biodegradable and biocompatible nature. Lipid-based nanoliposomes and nanoemulsions encapsulate hydrophilic and hydrophobic antigens, protecting them from degradation and enabling targeted delivery. Structures like ISCOMs and carbon nanotubes facilitate efficient antigen presentation and cellular entry. Virus-like particles mimic natural infections to provoke strong immune responses without the risks associated with live pathogens. These nanoparticle vaccines represent a versatile and potent approach to fish vaccination, offering broad-spectrum protection against various pathogens.

15.4.5 Timing of Vaccine Delivery

In aquaculture, the timing of vaccination should be carefully considered to ensure that the animals are at an appropriate age and have developed sufficient immune competence to mount an effective response to the vaccine. Vaccination is usually initiated when the animals reach a suitable age and size, where their immune systems can respond adequately to the vaccine and produce protective antibodies. Pre-grow-out vaccination is commonly practised, ensuring that animals have acquired immunity before being transferred to grow-out facilities or ponds. Seasonal timing of vaccination may be considered to address disease risks associated with specific environmental conditions. Emergency vaccination campaigns may be conducted to contain the spread of disease and protect vulnerable populations in response to disease outbreaks. To maintain long-term immunity, some species may require revaccination at regular intervals (Adams 2019).

15.5 Vaccine Development

The development of vaccines for various fish pathogens is increasingly complex, with a notable absence of commercial vaccines for several key pathogens. Research in this field is advancing through sophisticated techniques such as the molecular sequencing of pathogens, reverse vaccinology, and proteomics (Adams 2019; Jose Priya and Kappalli 2022). These methods enable the precise selection of antigens to create recombinant and DNA vaccines. Additionally, artificial intelligence is being integrated with reverse vaccinology to enhance the prediction of immunogenic epitopes for inclusion in these advanced vaccine formulations. There is a pressing need for innovative approaches to develop affordable, safe, and easy-to-administer vaccines, especially for less valuable species like tilapia (Thompson et al. 2023). This is crucial to encourage farmers to vaccinate their fish, aiming to reduce their dependence on antibiotics. Current research is exploring the potential of nanoparticles in creating immersion and oral vaccines for these species (Thompson et al. 2023). However, the paramount challenge remains to produce these vaccines at a cost low enough to be feasible for widespread use by farmers (Can et al. 2023).

Vaccine development and application vary depending on the target species in aquaculture. Each species may have unique immune characteristics, pathogen profiles, and environmental conditions that influence the choice of vaccine type, administration method, and timing. Therefore, specific vaccination strategies must be tailored to the requirements of each aquaculture species (Adams 2019). Developing vaccines for aquaculture involves several stages, including antigen selection, adjuvant formulation, safety testing, and efficacy evaluation (Cowan et al. 2016). Vaccine safety is paramount to avoid adverse reactions and ensure the well-being of vaccinated individuals. Extensive testing is conducted to assess the safety profile of vaccines, including toxicity studies, immunogenicity assessments, and evaluation of potential side effects. Efficacy evaluation involves assessing the vaccine's ability to induce protective immune responses and its effectiveness in preventing disease.

15.5.1 Reverse Vaccinology

Reverse vaccinology is an innovative approach with the potential to revolutionise vaccine development for aquaculture. Traditionally, vaccine development can be laborious and time-consuming. Reverse vaccinology, however, adopts a pioneering methodology that leverages cutting-edge genomic and bioinformatics tools to analyse the entire genetic makeup of pathogenic microorganisms, allowing scientists to pinpoint specific antigens important in the infection process (Dadar et al. 2016). Once these antigens are identified, they are synthetically produced and used to formulate targeted vaccines, such as subunit, DNA, or RNA vaccines. This approach expedites the vaccine development process compared to traditional approaches and ensures greater precision in immunising fish against the pathogen. By employing bioinformatics to predict immunogenic sequences and identify specific regions for recombinant protein expression, the development timeline for vaccines has been reduced from the conventional 5–10 years to just 1–2 years (Rappuoli 2000).

Artificial intelligence (AI) aids reverse vaccinology for antigen discovery (Sharma et al. 2022). With its ability to rapidly analyse vast amounts of genetic data from various fish species, AI algorithms can identify potential antigens that stimulate robust immune responses. By predicting antigenic regions and epitopes within pathogens, AI facilitates the selection of highly effective and specific vaccine candidates. Moreover, machine learning models have the potential to analyse complex interactions between antigens and the fish immune system, leading to the design of tailored vaccines that offer enhanced protection.

15.5.2 Regulatory Aspects

Implementing effective vaccination strategies requires careful consideration of several factors. This includes selecting the appropriate vaccines for the target species and diseases, optimising vaccination protocols and schedules, and ensuring proper storage and administration of vaccines. Regulatory frameworks and guidelines are essential to ensure the quality, safety, and efficacy of vaccines used in aquaculture (Cowan et al. 2016). Collaborations between industry stakeholders, research institutions, and regulatory bodies are crucial in developing and implementing effective vaccination programmes. Vaccination programmes can be tailored to target specific diseases prevalent in the region, providing a proactive approach to disease prevention.

The safety and efficacy of vaccines used in aquaculture are crucial for their successful implementation (Cowan et al. 2016). Regulatory frameworks are in place to ensure that vaccines meet rigorous standards. Regulatory authorities assess the safety and efficacy of vaccines before approving use in aquaculture. These evaluations include pre-clinical trials, field trials, quality control assessments, and post-marketing surveillance to monitor vaccine performance. Regulatory approval and registration processes vary across countries and regions. Streamlining these processes and ensuring accessibility to approved vaccines are important for facilitating their widespread use in aquaculture.

15.6 Challenges and Limitations

Vaccine development, production, and administration costs can be a significant barrier, particularly for small-scale aquaculture operations (Barnes et al. 2022). Access to affordable vaccines, especially in developing regions, remains a challenge. Collaboration between industry, academia, and regulatory bodies is crucial to address these cost-related challenges and ensure the accessibility of vaccines.

The formulation and stability of vaccines pose challenges in aquaculture. Environmental conditions, such as temperature and salinity, can impact the stability and efficacy of vaccines during storage and transportation. Research efforts are focused on developing more stable formulations and improving vaccine delivery systems to ensure the efficacy of vaccines under diverse environmental conditions (Du et al. 2022).

Aquatic organisms exhibit species-specific variations in immune responses, which influence vaccine efficacy. Genetics, age, physiological state, and environmental factors can affect the immune response of fish to vaccines. Understanding these variations and tailoring vaccine strategies is essential for optimising vaccine efficacy in different aquaculture species.

Achieving high vaccination coverage within aquaculture populations can be challenging, with factors such as logistics, infrastructure, training, and stakeholder engagement influencing vaccine coverage and compliance. Education, awareness campaigns, and efficient vaccination programmes are necessary to promote vaccine uptake and compliance among aquaculture producers.

15.7 Impact of Vaccination on AMR, Future Perspectives, and Recommendations

Vaccination plays a crucial role in reducing the dependence on antimicrobial agents in aquaculture (Barnes et al. 2022). By preventing diseases and minimising the occurrence of outbreaks, vaccines decrease the need for therapeutic antimicrobial treatments. This reduction in antimicrobial use helps mitigate the selection pressure that drives the emergence and spread of AMR. Vaccination contributes to reducing disease prevalence and transmission in aquaculture systems. Vaccinated individuals are less susceptible to infections, decreasing the pathogen load within populations and subsequently limiting the transmission of diseases to non-vaccinated individuals. This reduction in disease transmission indirectly reduces the potential for the development and dissemination of AMR.

Disease outbreaks in aquaculture often lead to the use of antimicrobial agents, including antibiotics, to control infections. This practice can contribute to the development and spread of AMR in aquatic animals and the environment. Implementing effective vaccination programmes can reduce the incidence of diseases, resulting in lower reliance on antibiotics and decreased selection pressure for AMR. Vaccination, therefore, plays a significant role in mitigating the overuse of antibiotics in aquaculture. The effectiveness of antibiotics in treating bacterial infections is gradually diminishing due to the emergence of resistant strains. Vaccination reduces the need for antibiotics, helping to preserve their efficacy for critical cases where no alternative treatment is available. Vaccination contributes to the long-term sustainability of aquaculture operations by minimising the use of antibiotics. It also reduces AMR in aquatic environments. This helps to maintain the ecological balance and minimises the risk of AMR transmission to humans through the food chain.

Continued advancements in vaccine technology hold great potential for enhancing the efficacy, safety, and stability of vaccines in aquaculture. Together with new biotechnology advancements such as reverse vaccinology and proteomic approaches, novel vaccine platforms, such as recombinant proteins, viral-like particles, and nanotechnology-based delivery systems, offer opportunities for more targeted and efficient vaccines. Vaccination should be part of an integrated disease management strategy in aquaculture. Combining vaccination with good husbandry practices, biosecurity measures, and appropriate diagnostics can provide a holistic approach to disease prevention and control, reducing reliance on antimicrobials and minimising the development of AMR. Establishing robust surveillance and monitoring systems is crucial for assessing the effectiveness of vaccination programmes and detecting changes in disease prevalence and pathogen dynamics. Surveillance data can help identify emerging diseases, monitor vaccine efficacy, and inform decision-making processes in aquaculture.

References

Adamek M, Matras M, Rebl A, Stachnik M, Falco A, Bauer J et al (2022) Do not let it get under your skin! Vaccination protects the skin barrier of common carp from disruption caused by cyprinid herpesvirus 3. Front Immunol 13:787021. https://doi.org/10.3389/fimmu.2022.787021

Adams A (2019) Progress, challenges, and opportunities in fish vaccine development. Fish Shellfish Immunol 90:210–214. https://doi.org/10.1016/j.fsi.2019.04.066

Adams A, Aoki T, Berthe C, Grisez L, Karunasagar I et al (2008) Recent technological advancements on aquatic animal health and their contributions toward reducing disease risks—a review. In: Diseases in Asian aquaculture VI. Fish health section. Asian Fisheries Society, Colombo, pp 71–88

Akahata W, Yang ZY, Andersen H, Sun S, Holdaway HA, Kong WP et al (2010) A virus-like particle vaccine for epidemic Chikungunya virus protects nonhuman primates against infection. Nat Med 16(3):334–338. https://doi.org/10.1038/nm.2105

Alawdi S, Solanki AB (2021) Mucoadhesive drug delivery systems: a review of recent developments. J Sci Res Med Biol Sci 2(1):50–64. https://doi.org/10.47631/jsrmbs.v2i1.2

Assefa A, Abunna F (2018) Maintenance of fish health in aquaculture: review of epidemiological approaches for prevention and control of infectious disease of fish. Vet Med Int 2018:1–10. https://doi.org/10.1155/2018/5432497

Austin B, Lawrence AL, Can E, Carboni C, Crockett J, Demirtaş Erol N, Schleder DD, Kayis S, Karacalar U, Kizak V, Kop A, Thompson K, Ruiz CAM, Serdar O, Seyhaneyildiz Can Ş, Watts S, Yücel Gier G (2022) Selected topics in sustainable aquaculture research: current and future focus. Zenodo 1(2):74–125. https://doi.org/10.5281/zenodo.7032804

Barnes AC, Rudenko O, Landos M, Dong HT, Lusiastuti A, Phuoc LH, Delamare-Deboutteville J (2022) Autogenous vaccination in aquaculture: a locally enabled solution towards reduction of the global antimicrobial resistance problem. Rev Aquac 14(2):907–918. https://doi.org/10.1111/raq.12633

Bedekar MK, Kole S, M., M. (2022) Types of vaccines used in aquaculture. In: Makesh M, Rajendran KV (eds) Fish immune system and vaccines. Springer, Singapore. https://doi.org/10.1007/978-981-19-1268-9_3

Bengtén E, Wilson M (2015) Antibody repertoires in fish. Results Probl Cell Differ 57:193–234. https://doi.org/10.1007/978-3-319-20819-0_9

Bøgwald J, Dalmo RA (2019) Review on immersion vaccines for fish: an update 2019. Microorganisms 7(12):627. https://doi.org/10.3390/microorganisms7120627

Bondad-Reantaso MG, Subasinghe RP, Arthur JR, Ogawa K, Chinabut S, Adlard R, Tan Z, Shariff M (2005) Disease and health management in Asian aquaculture. Vet Parasitol 132(3–4):249–272. https://doi.org/10.1016/j.vetpar.2005.07.00

Bondad-Reantaso MG, MacKinnon B, Karunasagar I, Fridman S, Alday-Sanz V, Brun E, Groumellec ML, Li A, Surachetpong W, Karunasagar I, Hao B, Urbani R, Caputo A (2023) Review of alternatives to antibiotic use in aquaculture. Rev Aquac 15(4):1421–1451. https://doi.org/10.1111/raq.12786

Busch RA (1982) Enteric redmouth disease (Yersinia ruckeri). In: Les antigenes des micro-organisms pathogenes des poissons. Collection Fondation Marcel Merieux, pp 201–222

Can E, Austin B, Steinberg CE, Carboni C, Sağlam N, Thompson K, Yiğit M, Seyhaneyildiz Can S, Ergün S (2023) Best practices for fish biosecurity, well-being, and sustainable aquaculture. Sustain Aquatic Res 2(3):221–267. https://doi.org/10.5281/zenodo.10444855

Chen L, Klaric G, Wadsworth S, Jayasinghe S, Kuo TY, Evensen O et al (2014) Augmentation of the antibody response of Atlantic salmon by oral administration of alginate-encapsulated IPNV antigens. PLoS One 9(10):e109337. https://doi.org/10.1371/journal.pone.0109337

Collins C, Lorenzen N, Collet B (2019) DNA vaccination for finfish aquaculture. Fish Shellfish Immunol 85:106–125. https://doi.org/10.1016/j.fsi.2018.07.012

Cowan G, Smith P, Christofilogiannis P (2016) Fish vaccines: the regulatory process and requirements from the laboratory bench to a final commercial product, including field trials. In: Adams A (ed) Fish vaccines (Birkhäuser advances in infectious diseases). Springer, Basel. https://doi.org/10.1007/978-3-0348-0980-1_5

Curley SM, Putnam D (2022) Biological nanoparticles in vaccine development. Front Bioeng Biotechnol 10:867119. https://doi.org/10.3389/fbioe.2022.867119

Dacoba TG, Olivera A, Torres D, Crecente-Campo J, Alonso MJ (2017) Modulating the immune system through nanotechnology. Semin Immunol 34:78–102. https://doi.org/10.1016/j.smim.2017.09.007

Dadar M, Dhama K, Vakharia VN, Hoseinifar SH, Karthik K, Tiwari R, Khandia R, Munjal A, Salgado-Miranda C, Joshi SK (2016) Advances in aquaculture vaccines against fish pathogens: global status and current trends. Rev Fish Sci Aquac 25(3):184–217. https://doi.org/10.1080/23308249.2016.1261277

Dhar A, Allnutt F (2011) Challenges and opportunities in developing oral vaccines against viral diseases of fish. J Marine Sci Res Dev 1(2)

Dhar AK, Manna SK, Allnutt FCT (2014) Viral vaccines for farmed finfish. Virusdisease 25(1):1–17. https://doi.org/10.1007/s13337-013-0186-4

Du Y, Tang X, Sheng X, Xing J, Zhan W (2017) The influence of concentration of inactivated Edwardsiella tarda bacterin and immersion time on antigen uptake and expression of immune-related genes in Japanese flounder (Paralichthys olivaceus). Microb Pathog 103:19–28. https://doi.org/10.1016/j.micpath.2016.12.011

Du Y, Hu X, Miao L, Chen J (2022) Current status and development prospects of aquatic vaccines. Front Immunol 13:1040336. https://doi.org/10.3389/fimmu.2022.1040336

Dunn EJ, Polk A, Scarrett DJ, Olivier G, Lall S, Goosen MFA (1990) Vaccines in aquaculture: the search for an efficient delivery system. Aquac Eng 9(1):23–32. https://doi.org/10.1016/0144-8609(90)90009-O

Ellis AE (1988) General principles of fish vaccination. In: Fish vaccination. Academic Press, New York, pp 1–19

Embregts CW, Forlenza M (2016) Oral vaccination of fish: lessons from humans and veterinary species. Dev Comp Immunol 64:118–137. https://doi.org/10.1016/j.dci.2016.03.024

Embregts CWE, Rigaudeau D, Veselý T, Pokorová D, Lorenzen N, Petit J, Houel A, Dauber M, Schütze H, Boudinot P, Wiegertjes GF, Forlenza M (2017) Intramuscular DNA vaccination of juvenile carp against spring viremia of carp virus induces full protection and establishes a virus-specific B and T cell response. Front Immunol 8:1340. https://doi.org/10.3389/fimmu.2017.01340

FAO (2024) The state of world fisheries and aquaculture 2024 – Blue transformation in action. Rome. https://doi.org/10.4060/cd0683en

Fröhlich E (2012) The role of surface charge in cellular uptake and cytotoxicity of medical nanoparticles. Int J Nanomedicine 7:5577–5591. https://doi.org/10.2147/IJN.S36111

Fromen CA, Robbins GR, Shen TW, Kai MP, Ting JP, DeSimone JM (2015) Controlled analysis of nanoparticle charge on mucosal and systemic antibody responses following pulmonary immunization. Proc Natl Acad Sci USA 112(2):488–493. https://doi.org/10.1073/pnas.1422923112

Fujita Y, Taguchi H (2017) Nanoparticle-based peptide vaccines. In: Skwarczynski M, Toth I (eds) Micro- and nanotechnology in vaccine development. Elsevier, pp 149–170. https://doi.org/10.1016/B978-0-323-39981-4.00008-7

Gudding R (2014) Vaccination as a preventive measure. In: Gudding R, Lillehaug A, Evensen O (eds) Fish vaccination. Wiley, Oxford, UK, pp 12–21

Gudding R, Lillehaug A, Evensen Ø (1999) Recent developments in fish vaccinology. Vet Immunol Immunopathol 72(1-2):203–212. https://doi.org/10.1016/S0165-2427(99)00133-6

Guy B (2007) The perfect mix: recent progress in adjuvant research. Nat Microbiol 5:505–517. https://doi.org/10.1038/nrmicro1681

Hansen E, Fernandes K, Goldspink G, Butterworth P, Umeda PK, Chang K-C (1991) Strong expression of foreign genes following direct injection into fish muscle. FEBS Lett 290:73–76. https://doi.org/10.1016/0014-5793(91)81229-2

Hansson M, Nygren PAK, Stahl S (2000) Design and production of recombinant subunit vaccines. Biotechnol Appl Biochem 32:95–107. https://doi.org/10.1042/ba20000034

Henriksson PJG, Rico A, Troell M, Klinger DH, Buschmann AH, Saksida S, Chadag MV, Zhang., W. (2018) Unpacking factors influencing antimicrobial use in global aquaculture and their implication for management: a review from a systems perspective. Sustain Sci 13:1105–1120. https://doi.org/10.1007/s11625-017-0511-8

Holten-Andersen L, Doherty TM, Korsholm KS, Andersen P (2004) Combination of the cationic surfactant dimethyl dioctadecyl ammonium bromide and synthetic mycobacterial cord factor as an efficient adjuvant for tuberculosis subunit vaccines. Infect Immun 72(3):1608–1617. https://doi.org/10.1128/IAI.72.3.1608-1617.2004

Hotez PJ, Bottazzi ME (2022) Whole inactivated virus and protein-based COVID-19 vaccines. Annu Rev Med 73(1):55–64. https://doi.org/10.1146/annurev-med-042420-113212

Humphreys IR, Sebastian S (2018) Novel viral vectors in infectious diseases. Immunology 153(1):1–9. https://doi.org/10.1111/imm.12829

Imtiaz N, Anwar Z, Waiho K, Shi C, Mu C, Wang C, Qingyang W (2023) A review on aquaculture adaptation for fish treatment from antibiotic to vaccine prophylaxis. Aquac Int 32:2643. https://doi.org/10.1007/s10499-023-01290-6

Jackson DA, Symons RH, Berg P (1972) Biochemical method for inserting new genetic information into DNA of simian virus 40: circular SV40 DNA molecules containing lambda phage genes and the galactose operon of Escherichia coli. Proc Natl Acad Sci USA 69(10):2904–2909. https://doi.org/10.1073/pnas.69.10.2904

Jose Priya TA, Kappalli S (2022) Modern biotechnological strategies for vaccine development in aquaculture—prospects and challenges. Vaccine 40(41):5873–5881. https://doi.org/10.1016/j.vaccine.2022.08.075

Khan KH (2013) Gene expression in mammalian cells and its applications. Adv Pharm Bull 3(2):257–263. https://doi.org/10.5681/apb.2013.042

Kheirollahpour M, Mehrabi M, Dounighi NM, Mohammadi M, Masoudi A (2020) Nanoparticles and vaccine development. Pharm Nanotechnol 8(1):6–21. https://doi.org/10.2174/2211738507666191024162042

Klafack S, Schroeder L, Jin Y, Lenk M, Lee PY, Fuchs W et al (2022) Development of an attenuated vaccine against koi herpesvirus disease (KHVD) suitable for oral administration and immersion. NPJ Vaccines 7:106. https://doi.org/10.1038/s41541-022-00525-6

Kurath G, Garver KA, Corbeil S, Elliott DG, Anderson ED, LaPatra SE (2006) Protective immunity and lack of histopathological damage two years after DNA vaccination against infectious hematopoietic necrosis virus in trout. Vaccine 24:345–354. https://doi.org/10.1016/j.vaccine.2005.07.068

LaFrentz BR, LaPatra SE, Call DR, Cain KD (2008) Isolation of rifampicin-resistant Flavobacterium psychrophilum strains and their potential as live attenuated vaccine candidates. Vaccine 26:5582–5589. https://doi.org/10.1016/j.vaccine.2008.07.083

Leitner WW, Ying H, Restifo NP (1999) DNA and RNA-based vaccines: principles, progress, and prospects. Vaccine 18:765–777. https://doi.org/10.1016/s0264-410x(99)00271-6

Levine MM, Sztein MB (2004) Vaccine development strategies for improving immunization: the role of modern immunology. Nat Immunol 5(5):460–464. https://doi.org/10.1038/ni0504-460

Liang R, Li J, Ruan Y, Zhu S (2015) The design and parameter optimization of automatic positioning mechanism for automatic injection system of fry vaccine. Am Soc Agric Biol Eng Ann Int Meeting 1:1–18. https://doi.org/10.13031/aim.20152188971

Lipsitch M, Siber GR (2016) How can vaccines contribute to solving the antimicrobial resistance problem? mBio 7(3):e00428–e00416. https://doi.org/10.1128/mBio.00428-16

Liu YL, He TT, Jiang XL, Sun SS, Wang LK, Nie P et al (2022) Development of a hyper-adhesive and attenuated Edwardsiella ictaluri strain as a novel immersion vaccine candidate in yellow catfish (Pelteobagrus fulvidraco). Microb Pathog 167:105577. https://doi.org/10.1016/j.micpath.2022.105577

Lorenzen N, Lorenzen E, Einer-Jensen K, LaPatra SE (2002) Immunity induced shortly after DNA vaccination of rainbow trout against rhabdoviruses protects against heterologous virus but not against bacterial pathogens. Dev Comp Immunol 26:173–179. https://doi.org/10.1016/s0145-305x(01)00059-3

Ma J, Bruce TJ, Jones EM, Cain KD (2019) A review of fish vaccine development strategies: conventional methods and modern biotechnological approaches. Microorganisms 7(11):569. https://doi.org/10.3390/microorganisms7110569

McCann N, O'Connor D, Lambe T, Pollard AJ (2022) Viral vector vaccines. Curr Opin Immunol 77:102210. https://doi.org/10.1016/j.coi.2022.102210

Mitchell H (1995) Choosing a furunculosis vaccine: points to consider. Bull Aquac Assoc Can 95:30–37

Munang'andu HM, Mutoloki S, Evensen Ø (2015a) A review of the immunological mechanisms following mucosal vaccination of finfish. Front Immunol. https://doi.org/10.3389/fimmu.2015.00427

Munang'andu HM, Mutoloki S, Evensen O (2015b) Overview of challenges limiting the design of protective mucosal vaccines for finfish. Front Immunol. https://doi.org/10.3389/fimmu.2015.00542

Munang'andu HM, Salinas I, Tafalla C, Dalmo RA (2020) Editorial: vaccines and immunostimulants for finfish. Front Immunol 11:573771. https://doi.org/10.3389/fimmu.2020.573771

Mutoloki S, Munang'andu HM, Evensen Ø (2015) Oral vaccination of fish—antigen preparations, uptake, and immune induction. Front Immunol 6:519. https://doi.org/10.3389/fimmu.2015.00519

Noad R, Roy P (2003) Virus-like particles as immunogens. Trends Microbiol 11:438–444. https://doi.org/10.1016/S0966-842X(03)00208-7

Norqvist A, Hagstrom A, Wolf-Watz HANS (1989) Protection of rainbow trout against vibriosis and furunculosis by the use of attenuated strains of vibrio anguillarum. Appl Environ Microbiol 55:1400–1405. https://doi.org/10.1128/aem.55.6.1400-1405.1989

Norwegian Ministries (2015–2020) Norwegian Government's National Strategy against antimicrobial resistance. Norwegian Ministry of Health and Care Services. https://cdn.who.int/media/

docs/default-source/antimicrobial-resistance/amr-spc-npm/nap-library/national-strategy-against-antibiotic-resistance-2015-2020.pdf?sfvrsn=3c7ded3a_1&download=true
O'Neill J (2015) Antimicrobials in agriculture and the environment: reducing unnecessary use and waste. The review on antimicrobial resistance. AMR-review.org
Pati R, Shevtsov M, Sonawane A (2018) Nanoparticle vaccines against infectious diseases. Front Immunol 9:2224. https://doi.org/10.3389/fimmu.2018.02224
Phu TM, Phuong NT, Dung TT, Hai DM, Son VN, Rico A, Clausen JH, Madsen H, Murray F, Dalsgaard A (2016) An evaluation of fish health-management practices and occupational health hazards associated with Pangasius catfish (Pangasianodon hypophthalmus) aquaculture in the Mekong Delta, Vietnam. Aquac Res 47:2778–2794. https://doi.org/10.1111/are.12728
Plant KP, Lapatra SE (2011) Advances in fish vaccine delivery. Dev Comp Immunol 35(12):1256–1262. https://doi.org/10.1016/j.dci.2011.03.007
Pollard AJ, Bijker EM (2021) A guide to vaccinology: from basic principles to new developments. Nat Rev Immunol 21:83–100. https://doi.org/10.1038/s41577-020-00479-7
Radhakrishnan A, Vaseeharan B, Ramasamy P, Jeyachandran S (2023) Oral vaccination for sustainable disease prevention in aquaculture-an encapsulation approach. Aquac Int J Eur Aquac Soc 31(2):867–891. https://doi.org/10.1007/s10499-022-01004-4
Rappuoli R (2000) Reverse vaccinology. Curr Opin Microbiol 3(5):445–450. https://doi.org/10.1016/s1369-5274(00)00119-3
Reverter M, Sarter S, Caruso D, Avarre JC, Combe M, Pepey E, Pouyaud L, Vega-Heredía S, de Verdal H, Gozlan RE (2020) Aquaculture at the crossroads of global warming and antimicrobial resistance. Nat Commun 11(1):1870. https://doi.org/10.1038/s41467-020-15735-6
Ribeiro CMS, Schijns VEJC (2010) Immunology of vaccine adjuvants. Methods Mol Biol 626:1–14. https://doi.org/10.1007/978-1-60761-585-9_1
Salinas I (2015) The mucosal immune system of teleost fish. Biology 4(3):525–539. https://doi.org/10.3390/biology4030525
Salinas I, Fernández-Montero Á, Ding Y, Sunyer JO (2021) Mucosal immunoglobulins of teleost fish: a decade of advances. Dev Comp Immunol 121:104079. https://doi.org/10.1016/j.dci.2021.104079
Salonius K, Simard N, Harland R, Ulmer JB (2007) The road to licensure of a DNA vaccine. Curr Opin Investig Drugs 8:635–641
Sasso E, D'Alise AM, Zambrano N, Scarselli E, Folgori A, Nicosia A (2020) New viral vectors for infectious diseases and cancer. Semin Immunol 50:101430. https://doi.org/10.1016/j.smim.2020.101430
Sepúlveda D, Lorenzen N (2016) Can VHS virus bypass the protective immunity induced by DNA vaccination in rainbow trout? PLoS One 11:e0153306. https://doi.org/10.1371/journal.pone.0153306
Sharma A, Virmani T, Pathak V, Sharma A, Pathak K, Kumar G, Pathak D (2022) Artificial intelligence-based data-driven strategy to accelerate research, development, and clinical trials of COVID vaccine. Biomed Res Int 2022:7205241. https://doi.org/10.1155/2022/7205241
Shoemaker CA, Klesius PH, Evans JJ (2007) Immunization of eyed channel catfish, Ictalurus punctatus, eggs with monovalent Flavobacterium columnare vaccine and bivalent F. columnare and Edwardsiella ictaluri vaccine. Vaccine 25:1126–1131. https://doi.org/10.1016/j.vaccine.2006.09.055
Shoemaker CA, Klesius PH, Evans JJ, Arias CR (2009) Use of modified live vaccines in aquaculture. J World Aquacult Soc 40(5):573–585
Smith NC, Rise ML, Christian SL (2019) A comparison of the innate and adaptive immune systems in cartilaginous fish, ray-finned fish, and lobe-finned fish. Front Immunol 10:2292. https://doi.org/10.3389/fimmu.2019.02292
Tafalla C, Bøgwald J, Dalmo RA (2013) Adjuvants and immunostimulants in fish vaccines: current knowledge and future perspectives. Fish Shellfish Immunol 35(6):1740–1750. https://doi.org/10.1016/j.fsi.2013.02.029

Tebbit GL, Goodrich TD (1982) Vibriosis and the development of bacterins for its control. In: Les antigenes des micro-organisms pathogenes des poissons. Collection Foundation Marcel Merieux, pp 225–248

Thompson KD, Rodkhum C, Bunnoy A, Thangsunan P, Kitiyodom S, Sukkarun P, Yostawornkul J, Yata T, Pirarat N (2023) Addressing nanovaccine strategies for Tilapia. Vaccines 11(8):1356. https://doi.org/10.3390/vaccines11081356

Thorarinsson R, Powell DB (2006) Effects of disease risk, vaccine efficacy, and market price on the economics of fish vaccination. Aquaculture 256(1–4):42–49

Tlaxca JL, Ellis S, Remmele RL Jr (2015) Live attenuated and inactivated viral vaccine formulation and nasal delivery: potential and challenges. Adv Drug Deliv Rev 93:56–78. https://doi.org/10.1016/j.addr.2014.10.002

Ura T, Okuda K, Shimada M (2014) Developments in viral vector-based vaccines. Vaccine 2(3):624–641. https://doi.org/10.3390/vaccines2030624

Van Boeckel TP, Brower C, Gilbert M, Grenfell BT, Levin SA, Robinson TP, Teillant A, Laxminarayan R (2015) Global trends in antimicrobial use in food animals. Proc Natl Acad Sci USA 112:5649–5654. https://doi.org/10.1073/pnas.1503141112

Vinay TN, Bhat S, Gon Choudhury T, Paria A, Jung M-H, Shivani Kallappa G, Jung S-J (2018) Recent advances in application of nanoparticles in fish vaccine delivery. Rev Fish Sci Aquac 26(1):29–41. https://doi.org/10.1080/23308249.2017.1334625

Wang B, Thompson KD, Wangkahart E, Yamkasem J, Bondad-Reantaso MG, Tattiyapong P, Jian J, Surachetpong W (2023) Strategies to enhance tilapia immunity to improve their health in aquaculture. Rev Aquac 15(S1):41–56. https://doi.org/10.1111/raq.12731

Watts JEM, Schreier HJ, Lanska L, Hale MS (2017) The rising tide of antimicrobial resistance in aquaculture: sources, sinks and solutions. Mar Drugs 15(6):158. https://doi.org/10.3390/md15060158

Xue R, Liu L, Cao G, Xu S, Li J, Zou Y et al (2013) Oral vaccination of BacFish-vp6 against grass carp reovirus evoking antibody response in grass carp. Fish Shellfish Immunol 34(1):348–355. https://doi.org/10.1016/j.fsi.2012.11.024

Yanong RP (2011) Use of vaccines in finfish aquaculture. US Department of Agriculture, Cooperative Extension Service, Washington, DC. University of Florida, IFAS, FL. A. & M. University Cooperative Extension Program, and Boards of County Commissioner

Zhang YA, Salinas I, Li J, Parra D, Bjork S, Xu Z, LaPatra SE, Bartholomew J, Sunyer JO (2010) IgT, a primitive immunoglobulin class specialized in mucosal immunity. Nat Immunol 11(9):827–835. https://doi.org/10.1038/ni.1913

Chapter 16
The Global Approach: Aquaculture and National Action Plans for Antimicrobial Resistance

Darren M. Green and Andrew P. Desbois

Abstract National Action Plans (NAPs) for Antimicrobial Resistance (AMR) underpin the coordinated international response to the global problem of AMR, which includes bacterial antibiotic resistance. Nations prepare their NAPs often under the guidance of the Quadripartite, comprised of the Food and Agriculture Organization (FAO) of the United Nations, the United Nations Environment Programme (UNEP), the World Health Organization (WHO) and the World Organisation for Animal Health (WOAH). NAPs are drafted following the guiding principles of the Global Action Plan (GAP) on AMR that was developed and proposed by the WHO, FAO and WOAH in 2015. Each NAP is prepared to account for the individual needs of the nation, including the scale of the problem, the resources available, cultural considerations and logistical concerns, and they aim to reflect the state of each country's experience with and progress against AMR. NAPs should take a One Health approach to the AMR issue, which is in recognition of the impact of AMR on humans, animals and the environment; aquaculture often finds itself at the nexus of these One Health compartments. This chapter describes what NAPs are and how they help to mitigate AMR and seeks to raise awareness around how they pertain to the aquaculture sector.

Keywords Aquatic animals · Fisheries · Fish farming · Governance · Shrimp farming

16.1 Introduction

Antimicrobials, including antibiotics, are used in aquaculture, and, as with all other uses of these agents, there is a risk of the emergence, selection and enrichment of resistant microorganisms in these aquatic culture systems (Santos and Ramos 2018;

D. M. Green · A. P. Desbois (✉)
Institute of Aquaculture, University of Stirling, Stirling, UK
e-mail: andrew.desbois@stir.ac.uk

P. Elumalai, S. Lakshmi (eds.), *Antimicrobial Resistance in Aquaculture and Aquatic Environments*, https://doi.org/10.1007/978-981-97-7320-6_16

Watts et al. 2017). As such, these environments may be a sink of antimicrobial-resistant microorganisms and the genes that confer resistance (Watts et al. 2017). Many aquaculture systems are relatively open and highly interconnected with the other One Health compartments of humans, animals and the wider environment (Brunton et al. 2019; Watts et al. 2017) (Fig. 16.1). Indeed, aquaculture may play a central role, not only as an environment where antibiotic resistance genes and microorganisms may emerge and be selected, but also by providing effective pathways for their transmission, including for zoonotic species and antimicrobial residues (Brunton et al. 2019) (Fig. 16.1). The potential importance of aquaculture in selecting for antimicrobial resistance (AMR) and affording routes to dissemination may explain the recent increase in attention for characterizing antimicrobial usage and AMR in farmed aquatic animals (e.g. Lulijwa et al. 2020; Schar et al. 2020; Desbois et al. 2021). Hence, it is essential to include aquaculture activities in all initiatives aiming to mitigate or diminish the problems posed by AMR.

The Global Action Plan (GAP) on AMR is the global approach to address the problems posed by AMR (WHO 2015). It provides a governance framework for the coordinated actions towards this challenge for different countries and sectors, including aquaculture industries (WHO 2015). The GAP has provided a platform for the development and implementation of National Action Plans (NAPs) on AMR, with each country taking its share of responsibility for tackling AMR. This chapter describes what NAPs are and how they help to mitigate AMR and seeks to raise awareness for how they pertain to the aquaculture sector. Greater awareness for NAPs will allow aquaculture businesses to plan more effectively in anticipation of

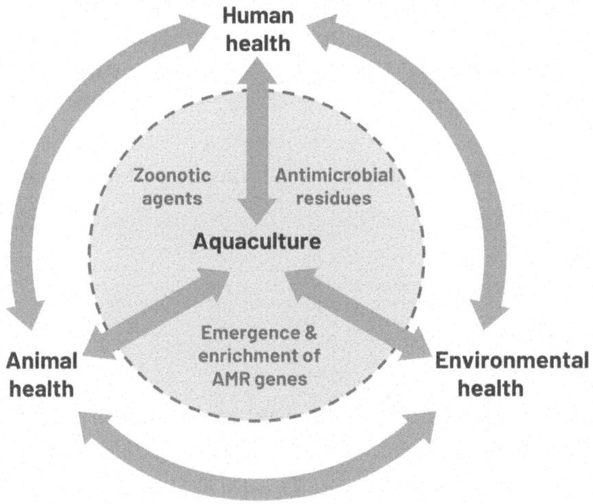

Fig. 16.1 Conceptual diagram showing aquaculture at the heart of the One Health concept, with potentially important roles in connecting human, animal and natural environment compartments. Aquaculture systems are places where antibiotic resistance genes and microorganisms may emerge and be selected, similar to the other compartments, but the often-open nature of the aquaculture systems provides efficient pathways for the transmission of antimicrobial residues and resistant microorganisms, including zoonotic species, to the other compartments

changes in policies that will affect their operations, whilst these documents also provide opportunities for the research community to contribute to meeting the goals described in each strategy.

16.2 What Is the GAP and What Is a NAP?

The GAP is a high-level strategy document intended for national policymakers that has the overall goal of ensuring that we can continue to prevent and treat infectious diseases with safe and effective medicines (WHO 2015). To achieve this, the GAP provides a framework of actions that Member States, the GAP Secretariat and international and national partners need to take, and this includes the requirement to develop bespoke plans at national level (i.e. a NAP) to tackle the AMR problem (WHO 2015). In recognition of the increasing threat posed globally by AMR, in 2014 the WHO, in consultation with the Food and Agriculture Organization (FAO) of the United Nations and World Organisation for Animal Health (then OIE, now WOAH), initiated and led the development of the GAP (WHO 2015). The GAP outlines five key objectives for tackling AMR: to raise awareness of the problem; strengthen research and surveillance; reduce incidence of infections; optimize the use of antimicrobials; and increase investment in interventions (WHO 2015; Fig. 16.2). The GAP built upon National Action Plans to counter AMR already developed by several countries, for example Cyprus (Republic of Cyprus Ministry of Health 2012) and the United Kingdom (UK; Department of Health 2000), to provide a coordinated global framework and to offer advice for preparing such plans.

At its heart, the GAP emphasizes the desire for improved infection prevention whilst maintaining access to antimicrobials when needed (WHO 2015). It foresaw the requirement for whole-of-society engagement through a One Health approach to the AMR challenge, with the need to consider human, animal and environmental compartments and their interactions (WHO 2015). The need to consider activities associated with farmed aquatic species was stated explicitly, thus indicating an awareness for the potential importance of aquaculture in the global AMR problem (WHO 2015). Indeed, a body of evidence has accumulated from around the world that supports a potentially significant role for aquaculture in AMR emergence and transmission (Reverter et al. 2020; Schar et al. 2021). However, fewer established initiatives on AMR pertain specifically to aquaculture, particularly around data collection, such as is required by certain existing efforts coordinated internationally for humans and livestock (Schrijver et al. 2018; Mesa Varona et al. 2020; Mshana et al. 2021). This means new resource or diversion of existing funding is needed to capture new data relevant to the aquaculture sectors in each country.

Relevant to aquaculture, the GAP recognized a need for WOAH to update and maintain the Aquatic Animal Health Code (AAHC) to include specific guidance pertinent to AMR (WOAH 2022) and for WOAH to develop and provide more guidance for antimicrobial susceptibility testing of aquatic bacteria (WHO 2015). These actions can support the GAP objective to optimize the use of antimicrobial

Objective 1: Improve awareness and understanding of antimicrobial resistance through effective communication, education and training

Objective 2: Strengthen the knowledge and evidence base through surveillance and research

Objective 3: Reduce the incidence of infection through effective sanitation, hygiene and infection prevention measures

Objective 4: Optimize the use of antimicrobial medicines in human and animal health

Objective 5: Develop the economic case for sustainable investment that takes account of the needs of all countries, and increase investment in new medicines, diagnostic tools, vaccines and other interventions

Fig. 16.2 The five objectives of the Global Action Plan on Antimicrobial Resistance (World Health Organization 2015)

medicines for aquatic animal health (WHO 2015). Meanwhile, the GAP Secretariat was charged with developing standards and guidance concerning the presence of antimicrobial agents and their residues in the environment, particularly in water, wastewater, food and feed (WHO 2015).

The GAP and the framework contained within prepared a platform for the development and implementation of a bespoke plan by each individual Member State: a NAP. Each Member State can develop their NAP to take account of the country's needs, resources available, present situation with AMR and rate of progress, as well as other contexts. NAPs underpin the coordinated international response to AMR, which includes bacterial antibiotic resistance, and encompass activities associated with farming aquatic animals. The success of the GAP in addressing AMR in the long term is linked intrinsically to the success of the NAPs and progress made nation by nation (WHO 2015; WHO, FAO, and OIE 2019).

Since the GAP was published, many nations have developed or are developing NAPs, and, as of 2023, more than 130 nations have a NAP deposited in the *Library of AMR National Action Plans*, a publicly accessible resource containing NAPs maintained by the WHO (WHO 2023a; Fig. 16.3). In total, some 170 NAPs have been developed, and 38 are under development, with others yet to be approved and lodged in the WHO-hosted repository (WHO 2022a, 2023a). In some cases, these represent second-generation NAPs, for example Tanzania (2017–2022 and 2023–2028 [United Republic of Tanzania 2022]), Ireland (2017–2020 and 2021–2025 [Government of Ireland 2021]) and Indonesia (2017–2019 and 2020–2024 [Coordinating Ministry for Human Development and Cultural Affairs of

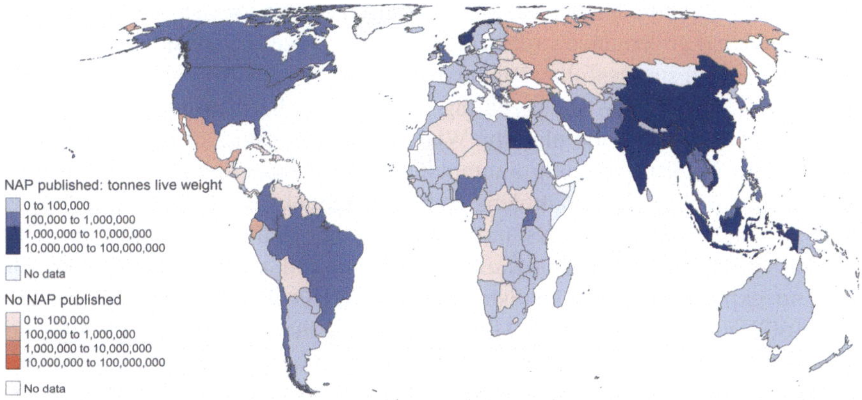

Fig. 16.3 A map showing aquaculture output by country and whether the country has a National Action Plan (NAP) available in the *Library of AMR National Action Plans* on the World Health Organization (WHO) website (WHO 2023a). Countries with a NAP are shown in blue and those without a NAP are red, with darker colours indicating greater aquaculture output. Aquaculture output was calculated as mean production of finfish and crustaceans during 2015–2019 from data available in the *Fisheries and Aquaculture Statistics* database at the Food and Agriculture Organization website (FAO 2023)

the Republic of Indonesia 2021]), and the WHO maintains sequential versions (WHO 2023a). Figure 16.3 shows countries with and without NAPs in the *Library of AMR National Action Plans* and their contribution to total world aquaculture output.

16.3 How Are NAPs Prepared?

The process of NAP preparation is distinct between countries, with a specific government department often leading and coordinating the process, usually in consultation with other relevant departments and broader stakeholders. Approaches to tackling AMR are multifaceted, and wide engagement with stakeholders is important to ensure the NAP and its content will be suited to different sectors in the local context. There are specific needs for each nation, due to differences in composition of sectors and scale of associated problems, for example, balance between human and animal sectors, and the resources available. In some cases, evidence is provided in the NAP for the stakeholder engagement process, such as the way opinions were sought and gathered, who the key informants and stakeholders groups were, how they were represented and how these discussions influenced the document (e.g. Thailand NAP [Ministry of Public Health and Ministry of Agriculture and Cooperatives 2017]; Willemsen et al. 2022). To ensure the most effective measures are taken, and the greatest chance that the NAP goals are implemented effectively, it is important to ensure early involvement in the process of key powers and affected groups (Kickbusch and Gleicher 2012). NAPs also account for each country's individual cultural and governance structures, which are crucial to consider if the strategy is to be realized successfully (Pant et al. 2022).

Support and guidance when preparing a NAP have been provided by several authors and agencies. In particular, the WHO maintains a suite of resources on its website (WHO 2023b), including a template document that countries may use to begin preparing a NAP and an implementation handbook (WHO 2022b), although this latter resource focuses largely on antibiotic usage and AMR in humans. The NAP template encourages consideration of human and animal health, although there is no explicit reference to fish or water (WHO 2023b). Since 2016, the Tripartite of WHO, WOAH and FAO (now enlarged to the Quadripartite through the addition of the United Nations Environment Programme [UNEP]) has provided a manual to support the development of NAPs that aims to assist countries in developing or refining their NAPs (WHO, FAO, and OIE 2016). This manual outlines a process of five steps: establishing a governance for the NAP; situational analysis; planning; initiating implementation; and periodic review of the NAP and progress. The Tripartite guidance recommends countries utilize the Strength, Weakness, Opportunity, Threat (SWOT) model in NAP preparation, and it advocates the application of the SMART method to draw up objectives, such that each objective should be *Specific*, *Measurable*, *Achievable*, *Relevant* and *Time-bound*. The Tripartite SMART "A" is *Achievable*, however an alternative (and the original) interpretation of the "A" is *Assignable*, and both these aspects are important from a NAP governance perspective.

An alternative model for drawing up a NAP is provided by the parallel *AMR Framework for Action* published by the Interagency Coordination Group on AMR (Interagency Coordination Group on AMR 2017), and, for example, the UK used this approach to prepare its NAP for 2019–2024 (HM Government 2019). These two documents provide quite different suggestions for structuring a NAP: The GAP is organized hierarchically around five key objectives (Fig. 16.2); by contrast, the *AMR Framework for Action* is designed as a lattice, where the levers of awareness, surveillance, funding, policy and championing are applied across three content areas: reducing need and unintentional exposure; optimized use of medicines; and innovation (Interagency Coordination Group on AMR 2017). Thus, equity, governance and awareness are intrinsic to each content area rather than separate first-rank headings, and this may explain some of the stylistic diversity in NAPs where different philosophies of writing have been taken (Carelli et al. 2023).

ReAct is an independent network that works to combat antibiotic resistance and is funded by the Swedish International Development Cooperation Agency and Uppsala University (ReAct 2023). The ReAct website maintains resources for NAP development, including a toolbox, with guidance rooted in a basic five-step cycle: engaging stakeholders; assessing the current situation; planning and drafting the NAP; implementing the plan and policies as a published working document; and evaluating progress, which then leads back to the start of the cycle and the development of revised or next-generation plans (ReAct 2023). This cycle adheres closely to that recommended initially by the Tripartite (WHO, FAO,, and OIE 2016), and resources and materials for aquaculture are included on the website, such as example interventions and links to relevant case studies (ReAct 2023). Specific guidance for incorporating aquaculture during the development of a NAP has also been published by Bondad-Reantaso et al. (2020), which includes a six-point list of recommended actions covering from scoping the potential risk through to AMR surveillance, and the authors identified further guidance documents of relevance to aquaculture for NAP creators. Nevertheless, whilst preparing a NAP may be relatively simple, and with a wealth of information to draw upon, ensuring the document is fit for purpose and contains a strategy well suited to the context of an individual country is far more challenging.

16.4 What Is in a NAP?

Each NAP describes the high-level strategy of that nation towards the goal of reducing and addressing problems caused by AMR. In essence, the NAP sets out the path towards the development and implementation of wide-ranging policies across government and provides justification for the introduction of policy interventions expected to facilitate progress towards reduced burden of AMR at national level whilst contributing to international efforts and cooperation on this issue.

NAPs vary considerably in length, breadth and content, which reflects their nature as bespoke documents (Willemsen et al. 2022; Carelli et al. 2023; Fig. 16.4), currently stretching from China's 2022–2025 NAP at 7 pages (Medical Administration and Hospital Authority 2022) to the NAP for Kuwait of 285 pages (Moghnieh et al. 2022). Nevertheless, page count can belie the content, with some NAPs containing extensive appendices and annexes whilst others present detailed plans in a more condensed format. NAPs have been written in various languages, English most commonly (Willemsen et al. 2022), while several NAPs present an English-language version alongside the national language, such as the NAP for Cambodia published in 2019 (Ministry of Health; Ministry of Agriculture, Forestry and Fisheries; Ministry of Environment 2019).

The NAP document begins with narrative description of background information and context, with the forewords and introductions setting the scene at global and national levels, documenting the severity, distribution and impact of the AMR problem and antibiotic use and rationalizing the purpose of the document. Often, the national situation is conveyed through antibiotic usage and AMR prevalence trends. NAP committee and governance structures may be provided, which can include key contacts and organograms, whilst glossaries of terms and lists of authors and stakeholders involved in or consulted during preparation are often also presented (Willemsen et al. 2022). The main part of each NAP is typically a hierarchical narrative tackling in turn the strategic objectives against AMR of the country, potentially following one of the structure proposals outlined in Sect. 16.3, or otherwise.

At the core of every NAP is reducing the need and requirement for using antimicrobials across all sectors and, where antimicrobials are necessary, ensuring that

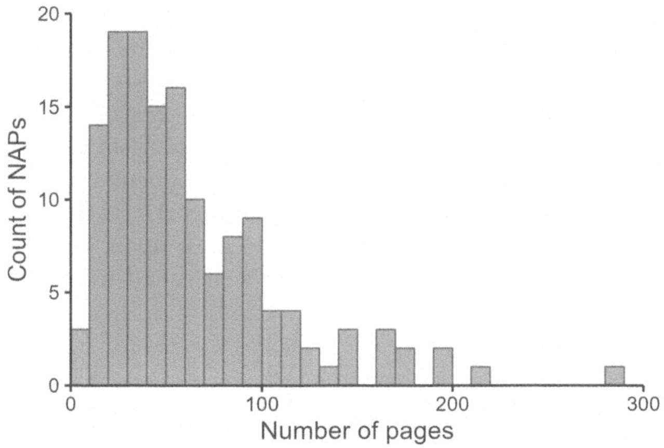

Fig. 16.4 Bar chart showing the distribution of page counts of National Action Plans available in the *Library of AMR National Action Plans* on the World Health Organization (WHO) website (WHO 2023a, b)

their use is responsible and prudent; in common with other sectors, concerns for inappropriate use of antimicrobials in aquaculture have been reported (Cabello 2006). Actions towards this aspiration span the five objectives of the GAP and meaningful progress requires the introduction of a wide range of effective interventions.

The development and preparation of a NAP to target AMR include the collecting together of such interventions: actions and activities to be undertaken by governmental bodies (be it subnational, national or supranational), or intended for other parties, that are posited to be beneficial in reducing the burden of AMR and fulfil one or more of the GAP objectives. Interventions can take various forms such as legislation and regulations, education and awareness building, infrastructure support and development, or market (dis)incentives. One analysis of interventions applied in aquaculture has defined seven types, with legislation and regulations, and voluntary instruments, found to be most common and with the other intervention types grouped together as industry rules and standards; commercial technology and alternatives to antimicrobials; on-farm management practices; learning and awareness-raising activities; and activities with co-benefits (Garza et al. 2022). The interventions in this analysis were categorized according to several axes, including function (restricting or enabling), scope (international to local), implementer, compulsion, and target stakeholder(s) (Garza et al. 2022).

Typical interventions found in NAPs to achieve responsible and prudent use of antibiotics in aquaculture include direct legislation to control the supply of antimicrobials. In 2022, 59% of countries with developed NAPs regulated their prescription and sale for aquatic animals, and 54% regulated prescription and sale of medicated feed for aquatic animals (WHO 2022a). A smaller number of countries

(44%) have national systems in place for monitoring antimicrobial sales and usage in the sector.

Countries may commit to improved awareness and provide education, for example, by insisting aquatic medicine practitioners follow therapeutic guidelines, as found in the Norway NAP published in 2015 (Norwegian Ministries 2015) and overseen by the Norwegian Food Safety Authority and the *VetReg* database (AACTING 2023), or introducing the topic of AMR into the high school curriculum, such as the communication programme for awareness described in the Sudan NAP (Federal Ministry of Health and Ministry of Animal Resources 2017).

Efforts to strengthen the knowledge and evidence base around antimicrobial use and prevalence of resistance may be bolstered by introducing or strengthening national AMR surveillance programmes, including in aquaculture. In Japan, since 1999 the Japanese Veterinary Antimicrobial Resistance Monitoring System (JVARM) has monitored antimicrobial usage and antimicrobial-resistant organisms, including for farmed aquatic animals, and the Japan NAP published in 2016 committed to strengthening this system through wider surveillance, standardization of testing methods across species and linking with the parallel human surveillance system called the Japan Nosocomial Infections Surveillance (JANIS [Government of Japan 2016; JANIS 2023]). Furthermore, the most recent Japan NAP published in 2023 described a new system that is intended to be implemented to understand per-farm antimicrobial usage, though implementation details were not provided (Government of Japan 2023). National surveillance programmes for AMR in live aquatic animals are relatively poorly developed globally, with only 13% of countries reporting systematic collection and reporting of these data at national level out of countries with NAPs available in the WHO database (WHO 2022a).

A notable focus for interventions in several NAPs is zoonotic diseases, reflecting their special importance from a One Health perspective. The NAP for the USA published in 2015 (Federal Task Force on Combating Antibiotic-Resistant Bacteria 2015) committed to enhancement of monitoring of zoonotic pathogens on farm and at slaughter, alongside research into linkages between human and animal commensal microbiota and zoonotic pathogens, while plans for similar research and vaccine development against zoonoses are presented in the Japan NAP published in 2023 (Government of Japan 2023). Some countries consider the farm-to-fork problems of antimicrobial-resistant microorganisms, as well as antimicrobial residues entering the human food chain, such as with the development of guidelines for the detection of food-borne outbreaks and strengthening of food testing for antimicrobial residues as planned in the NAP for Pakistan published in 2017 (Ministry of National Health Services, Regulations, and Coordination 2017).

Actions to reduce burden of infections through improving effective biosecurity may be achieved by a commitment to implement the AAHC, which can strengthen animal health and improve farming practices and animal welfare, leaving animals less susceptible to disease outbreaks (WOAH 2022). Intention to align veterinary provision with both the WOAH terrestrial and aquatic animal health codes is noted, without particular detail, in the Laos NAP (Anonymous 2019), and these international standards have also been adopted for implementation by South Africa's

Department of Agriculture, Forestry and Fisheries (Departments of Health and Agriculture, Forestry and Fisheries for the Republic of South Africa 2018).

Interventions that align with Objective 4 of the GAP—to optimize antimicrobial usage—can relate to improved antibiotic stewardship, such as measures to introduce or revise policies that take account of WHO and WOAH guidance on the use of critically important antibiotics (CIAs) and ensuring the use of antibiotics for growth promotion is prohibited. Progress on banning the use of antibiotics for growth promotion varies: In the EU and UK, application of antibiotics for this purpose has been banned since 2006 (European Parliament and Council 2003); by contrast, the use of antimicrobials as growth promoters in South Africa is regulated by a somewhat dated law that permits over-the-counter purchase by farmers (The Union of South Africa Department of Agriculture 1947). Still, research into use behaviours around antimicrobial prescribing is a priority area noted in the South Africa NAP published in 2018 (Departments of Health and Agriculture, Forestry and Fisheries for the Republic of South Africa 2018), but this is complicated by the need for collaboration across government departments (Rudolph 2022). Meanwhile, CIAs are still used across the world in aquaculture, including in countries with long-standing commitments to eradicating the application of these agents, such as the UK (UK-VARSS 2022).

In addition, NAPs may indicate ways that investment in research of alternatives to antimicrobial medicines will be encouraged, such as new government funding initiatives. Alternatives being pursued actively in aquaculture include vaccination, phage and prophylactic biological products such as probiotics (Bondad-Reantaso et al. 2023). In its NAP published in 2017, Canada committed to developing a fast-track processing for licensing both antimicrobials and alternatives thereto (Public Health Agency of Canada 2017), while the UK 2019–2024 NAP indicated an intent to stimulate research into both vaccines and other alternatives to antimicrobials such as novel therapies and nutrition, though few details were provided for how this vision will be accomplished (HM Government 2019).

NAPs may contain qualitative and quantitative outputs and targets that allow for monitoring and evaluation of the commitments and/or of particular interventions, though many NAPs lack this level of detail. The monitoring and evaluation plan in the NAP may consider success both in terms of implementation and impact on outcomes (see Sect. 16.6). Targets are encouraged but not set universally, as contexts vary, and many possible measures for monitoring and evaluating different actions in the NAPs have been proposed, particularly so these align with the core indicators of the GAP (WHO, FAO, and OIE 2019).

One document structure seen commonly in NAPs is the Tripartite suggestion to divide it into strategic plans (i.e. goals, objectives, priorities and interventions), operation plans (activities, implementation, budgeting and costing) and monitoring and evaluation plans (performance indicators, targets and timelines, data collection and reporting). A good example of a NAP conforming to this suggested structure is the Philippines document where, for example, Key Strategy 2 (strengthen surveillance and laboratory capacity) has three objectives and activities for animal health, for which Activity 2.3.4 is to implement surveillance and monitoring systems for

AMR and AMU in food-producing animals and their products (Inter-Agency Committee on Antimicrobial Resistance 2019). Annex B of the Philippines NAP, composing about a quarter of the whole document, expands on the objectives and planned activities by providing details for the timeline, the intended output of the activity (surveillance and monitoring report in food-producing animals), a responsible agency (Department of Agriculture) and a specific budget allocation (₱3b), which adheres closely to the key principles for formulating SMART objectives.

For a country to write a NAP is an early stage in its response to AMR, and the declarations of intent contained within must be followed up by effective implementation, which is a more demanding and complex task. By way of example, whilst the Pan-Canadian AMR Framework (Public Health Agency of Canada 2017) was published in 2017, the implementation plan document, the Pan-Canadian Action Plan, is still pending (Somanader et al. 2022).

16.5 How Are NAPs Implemented?

Each NAP strategy is implemented at national level typically via various government departments and agencies, and these will have impacts on government and industrial activities, as well as the population more generally. Implementation of interventions may be by legislative instruments or by encouraging sectors to adopt them voluntarily. Aquaculture producers may become aware of new policies in the NAP being implemented through communications with various government agencies or via industry representatives, and they may or may not be aware that the driver for the change is the NAP. Nevertheless, aquaculture stakeholders, particularly those with responsibility for aquatic animal health, can prepare themselves better for the future by becoming familiar with the NAP content relevant to them and planning accordingly for matters that may be expected to impact them.

Whilst formulating a NAP may be relatively straightforward, ensuring that the steps required to succeed in realizing the broad vision are described and then implemented effectively is a major challenge. Which policies and interventions will be implemented and by whom may be apparent from the NAP itself, but the document might instead indicate a broader area for action where the solutions need to be developed first by an agency before being implemented. Coverage in this respect varies widely between NAPs and often it can be difficult to determine exactly how a particular goal will be realized. For example, in many NAPs the agencies responsible and accountable for particular policies are not stated or are merely implied (e.g. the UK NAP [HM Government 2019]), and this lack of detail represents a shortcoming of many NAPs (Patel et al. 2023). Even where specific agencies are earmarked to lead on a particular action, lack of coordination with other relevant parties can see plans falter (Joshi et al. 2021). Nevertheless, NAPs are high-level documents and—rightly or wrongly—implementation plans may not have been intended document content from the outset.

In some cases, implementation plans may be presented in appendices (e.g. the Philippines NAP [Inter-Agency Committee on Antimicrobial Resistance 2019]), found in supplementary documents (e.g. the Australia NAP [Australian Government 2016]) or intended to be developed subsequently, either stated in the plan (e.g. the Canada NAP [Public Health Agency of Canada 2017]) or not, and again this can be unclear. Though NAPs are deposited in the WHO *Library of AMR National Action Plans*, ancillary documents such as implementation plans may not be, as is the case of the first Australia NAP (Australian Government 2016). Furthermore, absence of an intervention from the subsequent version of a NAP does not mean necessarily that the action has ceased: For example, the second USA NAP makes it explicit that it builds on the first NAP, thereby avoiding an exhaustive listing of every national activity in the AMR arena and therefore only comprising lists of "continuing and new actions…[of]…highest priority" (Federal Task Force on Combating Antibiotic-Resistant Bacteria 2020).

The problem of who is responsible and accountable for implementing different measures in a NAP can be even more difficult to determine for policies that apply to aquaculture. The farming of aquatic animals sometimes falls under the oversight of government departments responsible for terrestrial animals, but this is not always the case, and aquaculture can be within the responsibilities of a dedicated and specific body, or be within the scope of other governmental agencies such as the environment brief. Sometimes—and even more problematic—different aspects of aquaculture are overseen by different departments or, worse, are shared across departments, with no single entity having overall responsibility, with devolution and federalism sometimes injecting complexity. Moreover, often it is not always explicit in the NAP whether a policy is intended to apply to aquaculture or not, which again can lead to important goals and policies falling between stools of who has ownership (see Sect. 16.7).

Taken together, this highlights the sometimes-complex nature of implementing policies in the NAPs of different countries related to aquaculture activities, which will impact on the likely success of the plan and the strategic objectives and measures detailed within.

16.6 How Are the NAPs Evaluated?

Approaches to evaluating a NAP vary according to intent and the stage of the NAP cycle addressed, be that design, implementation or assessing performance. Aspects that can be evaluated in the design stage include the developmental process undertaken to produce the NAP, the presentation of the document and its content such as the interventions included, clarity of governance (e.g. SMART objectives) and the formulation of a plan for monitoring and evaluation (e.g. Anderson et al. 2019; Orubu et al. 2020; Chua et al. 2021; Fuller et al. 2022; Willemsen et al. 2022; Cassini and Tacconelli 2023; Charani et al. 2023; Patel et al. 2023). Once a NAP has been published for implementation, aspects that can be evaluated include the effectiveness of its implementation (are NAP commitments carried out and, if so, how well?) and success in terms of progress towards targets and concomitant generation of *outputs*. Further downstream can be considered the impact of the NAP or its individual components against *outcomes* such as the extent to which the output(s) effected a reduction in AMR infections in target species and finally how outcomes support an overarching *goal* such as defined in the GAP: reduced impact of infectious diseases on human and animal health (WHO, FAO, and OIE 2019; Fig. 16.5). To distinguish the two latter levels of focus, there is a difference between a particular narrow outcome being realized and one that contributes substantially to broader goals without unexpected side effects. Moreover, it can be helpful when considering evaluation to associate questions with the different elements of the NAP process (Fig. 16.5).

Evaluation is a process conducted to assess an activity or strategy as systematically and impartially as possible, and it can be performed internally (i.e. within the country itself) or externally by a third party, with the former offering the potential for greater accuracy but the latter providing greater independence and potential objectivity (WHO 2013). The evaluation may be performed exclusively on the NAP document itself; however, given NAPs are prepared largely for a domestic audience, dialogue between evaluators and NAP creators may be required to clarify details in these concise documents and achieve the most meaningful insight (Willemsen et al. 2022; Cassini and Tacconelli 2023). Additionally, where NAPs are written to be incremental beyond an existing—potentially undocumented—baseline, as demonstrated in Sect. 16.5 with sequential NAPs for the USA (Federal Task Force on Combating Antibiotic-Resistant Bacteria 2020), this will complicate evaluation efforts based on NAP documents alone. An overview of the development state, progress, coverage and implementation of NAPs at a global level is derived from the Tripartite AMR Country Self-Assessment Survey (TrACSS), which is performed at national level (WHO 2022a). This internal evaluation is completed collectively across all sectors, with a single response submitted to the WHO. The scope of the TrACSS includes countries yet to approve and publish their NAPs, and questions assess progress made against the five GAP objectives, with a consideration for human and animal sectors, and it includes specific questions addressing the aquatic animal health sector separately from terrestrial livestock.

The design stage of the NAP can be evaluated in terms of the process taken to produce the document itself, and this may be performed to assess aspects such as inclusivity and equity in the consultation and other parts of the approach to formulation such as governance structures (Chua et al. 2021; Özçelik et al. 2022; Willemsen et al. 2022; Patel et al. 2023). A commonly applied lens by which to evaluate the NAPs is to consider the range of interventions contained, including the extent to which these commitments align with the five objectives of the GAP (Ogyu et al. 2020; Chua et al. 2021; van Heuvel et al. 2022; Willemsen et al. 2022). The interventions described in the NAP can be evaluated for their appropriateness to a particular context, as well as whether there has been due consideration and coverage of different One Health compartments, activities and sectors (and species), including aquaculture (e.g. Chua et al. 2021; Munkholm et al. 2021; Caputo et al. 2022).

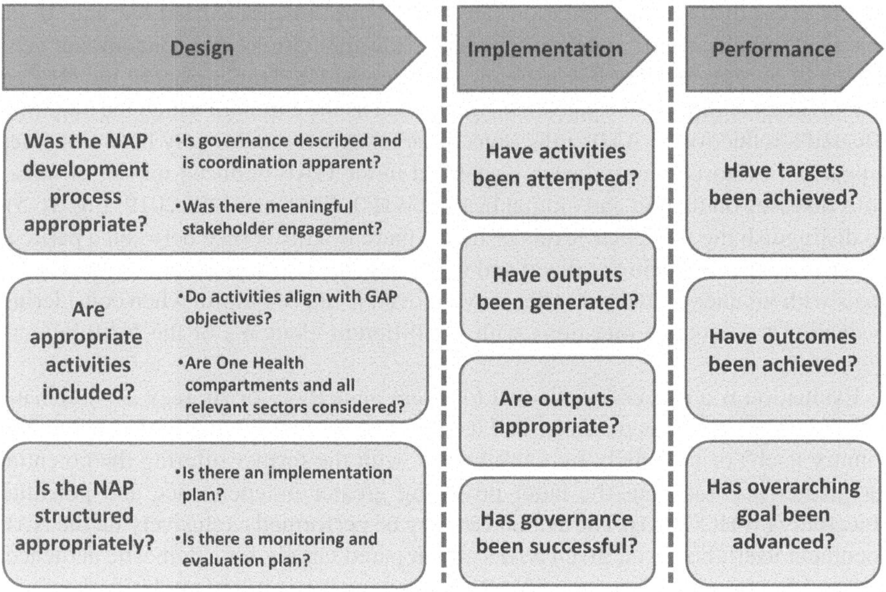

Fig. 16.5 Schematic diagram showing a selection of different ways to evaluate National Action Plans and their development from design to implementation and performance. Example questions that can be posed at each stage are provided

The NAP can be evaluated for clarity of presentation and whether it has incorporated key elements, such as implementation plans and monitoring and evaluation plans, with due consideration shown for how progress will be assessed (Anderson et al. 2019; WHO, FAO, and OIE 2019; Chua et al. 2021; Patel et al. 2023). NAPs may contain details of how success will be evaluated and reported, including the data sources that will be used as evidence, but many lack this information, not only for the national strategy as a whole but also for individual policies and objectives (Patel et al. 2023). When selecting core indicators of success, including appropriate metrics and measures, the GAP monitoring and evaluation plan considered four key criteria for their inclusion: relevance, availability, feasibility and sensitivity (WHO, FAO, and OIE 2019). Much current guidance for NAP preparation encourages greater consideration for monitoring and evaluation (M&E) plans and consideration for aligning data collection and indicators with those in the GAP M&E plan (WHO, FAO, and OIE 2019; Cassini and Tacconelli 2023), as this may enhance the usefulness and relevance of these important documents at global scale.

As many NAPs reach the end of their intended lifetimes, there is a desire to evaluate the success of these strategies, in terms of their implementation and generation of outputs. Effective implementation of a NAP is essential to its likely success, and this can be evaluated by assessing the generation of qualitative and quantitative outputs. Outputs themselves can be evaluated for various aspects including cost-effectiveness and sustainability (Patel et al. 2023). Research on NAP implementation success is growing and will be important for informing the next generation of NAPs, many of which are in preparation and will benefit from lessons learnt from earlier experiences, including failings, from previous NAP efforts and those reported by others (Anderson et al. 2019; Chua et al. 2021; Frumence et al. 2021; Fuller et al. 2022; Willemsen et al. 2022; Patel et al. 2023).

Whether the outputs created lead to a desirable outcome is the following step that can be assessed and evaluated. Realistically measuring *outcomes* requires a longer time frame, perhaps even beyond the lifetime of the NAP. For this reason, it is understandable that most evaluations of NAPs so far have assessed elements of design and now increasingly implementation (e.g. Frumence et al. 2021). The final step in the evaluation process is considering whether progress has been achieved in addressing an overarching *goal*. The AMR issue displays properties distinctive of a wicked problem (Rittel and Webber 1973), which inherently ensures success is difficult to define, although the GAP M&E plan in particular attempts to achieve this (WHO, FAO, and OIE 2019).

Effective evaluation of the entire NAP process is critical for the development of better NAPs that are more likely to lead to effective and successful implementation of policies that meet predefined success criteria and contribute helpfully to achieving the overarching goal of the GAP. When preparing their next NAP, countries can reflect on whether interventions were implemented and outcomes achieved, with many opportunities for learning incorporated into subsequent versions (Anderson et al. 2019; Chua et al. 2021).

16.7 How Well Is Aquaculture Covered in NAPs?

Large variation in attention afforded to aquaculture in different NAPs is to be expected, given that aquaculture represents a very large proportion of farmed animal output for some countries whilst for others aquaculture contributes a tiny proportion of activity and is effectively absent (Fig. 16.3). In the latter scenario in particular, the risks posed by antimicrobial inputs in the animal sector can be expected to be overwhelmingly dominated by livestock. A recent review of NAPs by Caputo et al. (2022) assessed coverage of aquaculture in 95 NAPs and found that, as of 2022, 9 out of the 15 largest aquaculture-producing countries had an aquaculture component integrated specifically in their NAP, with all these countries located in Asia except Norway, which is to be expected since most of the world's aquaculture takes place in Asia (FAO 2023). A further 3 of the top 15 aquaculture-producing countries mentioned aquaculture (Bangladesh, Chile, Thailand), 2 others did not reference aquaculture specifically (Brazil, Egypt), while the final country was yet to submit a NAP (Ecuador). Overall, Caputo et al. (2022) identified 63% of the 95 NAPs acknowledged aquaculture to be an important sector to be included in efforts aimed at containing AMR.

Caputo et al. (2022) noted that the scale of aquaculture industry in a country was not necessarily reflected in the coverage of these activities in the country's

NAP. Even where aquaculture production is significant, overall coverage and policies affecting aquatic animals are still mixed. For example, the France NAP published in 2022 does not refer to "fish" or "aquaculture", and only makes one mention to "water" that is in reference to limiting environmental pollution in terms of antimicrobial residues (Ministère des Solidarités et de la Santé 2022). However, the France NAP is light on references to animals in general, and, by contrast, though the Norway NAP published in 2015 is a much shorter document (36 versus 91 pages), it contains 80 references to "animal", 29 to "fish" and 10 to "water", which is perhaps understandable given Norway's substantial salmon industry (Norwegian Ministries 2015). Across the globe, some countries with small aquaculture industries cover these activities well. For example, Sudan produces a relatively small amount of aquaculture products, estimated at 9850 tonnes in 2020, compared with 32 million head of cattle in the same year (FAO data), but has named the Animal Production Fisheries Administration as the responsible authority for several interventions including "strengthening disease prevention and biosecurity in animals, food and agriculture" and "implementing antimicrobial stewardship", which at least imply aquaculture will be included in these initiatives.

It is not always clear in a specific NAP, certainly to an external audience, whether a particular policy is intended to apply to the sector or not. This may reflect the cultural gulf between local and global audiences, as the intended primary audience of a NAP is the population of the home country and the nuances of language and the full implementation details of these high-level strategy documents may not be apparent or readily available to observers outside the country. Nevertheless, third parties can play an important role in supporting the development and evaluation of the strategy but will need to probe to get a thorough understanding, with dialogue probably necessary to achieve this effectively. Lack of inclusion of aquaculture can also be due to a lack of expertise in NAP creators for this sector, and recent efforts have been made to better support the inclusion of aquaculture into NAPs (e.g. Bondad-Reantaso et al. 2020).

Multiple references to aquaculture might belie an explicit, focused attention on aquaculture beyond general interventions. For example, the Japan NAP published in 2016 (Government of Japan 2016) has high—over 50—mentions of the word "aquaculture", but half of these are in the repeated phrase "livestock production and aquaculture", which may well imply interventions intended for aquaculture, but may not imply interventions specifically so. Conversely, the NAP for Cambodia 2015–2017 (Ministry of Health; Ministry of Agriculture, Forestry and Fisheries; Ministry of Environment 2014) has no explicit mentions of fish or aquaculture; nevertheless interventions are listed that have as the responsible body the Ministry of Agriculture, Forestry and Fisheries (MAFF), such as to establish a national AMR surveillance system integrating human, veterinary and food safety data. In such cases, there is the possibility of interventions being inclusive of aquaculture though not explicit in the text. Aquaculture coverage is made slightly more explicit in the second-generation NAP for Cambodia published in 2019, with several references to fisheries made, including in relation to interventions (Ministry of Health; Ministry of Agriculture, Forestry and Fisheries; Ministry of Environment 2019).

Synonyms complicate the evaluation of aquaculture coverage in NAPs. For example, the UK 2019–2024 NAP (HM Government 2019) contains reference to aquaculture, whereas the word is absent in the Thailand 2017–2021 NAP (Ministry of Public Health and Ministry of Agriculture and Cooperatives 2017). Instead, the Thailand NAP refers to "fisheries", which, in turn, in the UK tends to apply to capture fisheries rather than aquaculture. Both definitions are covered under the FAO definition of "fishery" but have varied usages by country. Only a very small handful of NAPs such as Malaysia refer explicitly to "shrimp", specifically in the phrase "fish and shrimp" (Ministry of Health Malaysia 2017). It is likely that in many NAPs, shrimp are covered implicitly under the hypernym "fish", in the same way that "fish" might well be implicitly covered under "animal". Indeed, the FAO definition of a fish includes "molluscs, crustaceans and any aquatic animal which is harvested"; hence the somewhat tautological term "finfish" is offered for clarification. Moreover, "shrimp" and "prawn" themselves vary in usage across countries using English. The reverse of this is found where a term is used conservatively, excluding a subpart. For example, the Pakistan NAP published in 2017 (Ministry of National Health Services, Regulations, and Coordination 2017) mentions "animals" or "livestock" alongside "poultry", which suggests "animal" is being used in a conservative sense, implying lack of coverage explicitly of fish and aquaculture, with genuine omission of aquaculture rather than implicit inclusion.

There are several caveats to assessing the content of NAP documents in this simplistic textual way. First, it is better to assess the specific interventions that relate to aquaculture, as proof the NAP intends to operate in this area. We may wish also to evaluate the degree to which interventions are SMART as recommended by Tripartite, and discount content or interventions which do not read as credible. Second, NAPs are produced in multiple languages and local versions thereof. To compare world NAPs, analysis will need knowledge of multiple languages, or risk the vagaries of automatic translation. Third, nuances of the culture and politics of the country concerned can impact understanding. Overall, we cannot without finer study assume that a NAP without mention of words such as "aquaculture" or "fish" does not hide a national strategy that incorporates aquaculture implicitly, particularly where this can be considered correct practice according to the FAO itself.

16.8 What Are the Shortcomings of the NAPs and GAP?

Several shortcomings of NAPs have been highlighted above, and this section does not seek to compile a comprehensive list, with key limitations and weaknesses already well described by others (e.g. Caputo et al. 2022; Patel et al. 2023). As high-level strategy documents, NAPs often contain few details of how the plans will be implemented—such as a lack of detail for responsible and accountable parties for each action—and, if necessary, enforced, as well as provisions for how each of the components of the strategies will be evaluated. Often, it is not clear whether further documents are associated with the NAP that do provide the necessary details covering implementation plans.

Details on how activities will be resourced may be missing. A lack or shortage of resources needs to be considered, and transnational approaches to certain objectives may help to achieve national objectives in a more resource-efficient manner. The lack of an accountable or responsible agency for implementation of policies in a particular strategic area, or against an objective or target, obviously increases the risk of implementation failing to happen, or being inefficient or ineffective. Targets and timescales for meeting objectives may also be lacking.

One explanation for the lack of criteria pertinent to monitoring and evaluation in many NAPs could be the delay between publication of the GAP and its own monitoring and evaluation plan, which was not made available publicly until four years later in 2019 (WHO, FAO, and OIE 2019). With an increasing number of NAPs now available to NAP creators, access to the GAP monitoring and evaluation plan and expanding amounts of guidance for this aspect, it is hoped that newer NAPs will address this shortcoming. Moreover, inclusion of the SMART approach to objectives may help to address several shortcomings around implementation.

Some NAPs may have been prepared in the absence of meaningful stakeholder engagement for different sectors to varying extents, and this may have included the aquaculture industry (Willemsen et al. 2022). Engagement of NAP creators with the aquaculture sector appears to be mixed, with the TrACSS responses revealing that 67% of countries with a NAP in the WHO database had indicated active engagement of the aquatic animal health sector in multisector coordination mechanisms (WHO 2022a). Lack of stakeholder involvement can mean that the NAP does not accommodate sector needs appropriately, especially with respect to meeting intended objectives. This can reduce the odds of effective implementation of the strategy across the intended sectors, undermining the likelihood of success for the NAP in whole or part. Whilst lack of stakeholder involvement may have occurred during the preparation of some NAPs, when future NAPs are prepared, the leaders of this process are encouraged to engage all relevant stakeholders in a wide consultation across all aspects of the NAP, and this will likely happen given lessons learnt from earlier iterations or the experiences of neighbouring countries and other partners (Willemsen et al. 2022).

The diversity in NAP presentation leads to the conclusion that some are presented more effectively than others, not only in general but also in relation to aquaculture (e.g. see Sect. 16.7). Moreover, whilst the presentational diversity of NAPs is welcome from the perspective of each nation being able to develop a bespoke plan for their needs, this approach makes comparisons difficult, particularly when aiming to monitor and evaluate progress meaningfully at global level (Willemsen et al. 2022). More detailed guidance is one approach that may help to improve NAPs, especially for aquaculture-relevant content, as this area had received relatively little attention, and an overall lack of guidance specifically for aquaculture is one possible reason why some NAPs appear to lack full consideration of these activities.

Variations in vocabulary and meaning can create difficulties in understanding—certainly to external audiences, though not necessarily restricted to these—with this issue exacerbated by imprecise language (e.g. see Sect. 16.7). There are also terminological differences between NAPs in the words used for the structural components, with objectives, actions, strategies, outputs, outcomes and goals often used variably to mean the same or different things across the documents. One solution to this would be the provision of a glossary, with clarification of key terms. However, these documents are intended primarily for domestic consumption, so understanding of the NAP by this primary audience is really the key requirement.

Future NAPs may pay more consideration to the ability to evaluate the NAP by third party, but not if the cost is likely to be fewer NAPs or NAPs of lesser quality. This may be achieved by encouraging NAP preparers to engage more with the GAP authors, which may help to achieve better outcomes. They may be able to help with more coordination and help to develop mutually beneficial partnerships on a global scale, not only between countries but also for activities, and help to identify gaps in activities that may not have been considered. Opening up to external and internal scrutiny may help with better NAPs, as well as helping to align with a global strategy.

Finally, the NAP approach encouraged by the GAP itself has shortcomings, with the bespoke nature of NAPs making it difficult to monitor and evaluate progress meaningfully at global level, even if this is achievable at national level. Global impacts will always be challenging to assess, and, though harmonization of NAP content and presentation may not be desirable, improved support and guidance may contribute to simplifying the complex nature of GAP evaluation. The GAP and associated documents need to be responsive to new knowledge and provide updated and improved guidance where necessary (Anderson et al. 2019), particularly as more data becomes available for the effectiveness of NAPs and NAP interventions and for the relative importance of different One Health compartments and processes that impact AMR emergence and transmission. Whilst the GAP and resultant NAPs seem to offer the right path, experience should provide a means to improving the next generation of these documents.

16.9 Conclusions and Future Perspectives

Global aquaculture is a user of antimicrobials, including antibiotics, meaning this sector contributes to the global problem of AMR, including the emergence, selection and enrichment of resistant microorganisms (Santos and Ramos 2018; Watts et al. 2017). Moreover, many aquaculture systems are relatively open and highly interconnected with the other One Health compartments, meaning that in various places aquaculture may play a particularly important role in the transmission of resistant microorganisms, resistance genes and antimicrobial residues (Fig. 16.1). Accordingly, aquaculture activities, particularly the farming of aquatic animals, were highlighted rightly by the GAP as requiring consideration, where the GAP is the coordinated global approach to tackling AMR and provided an initial framework for the development of national-level strategy plans in terms of NAPs.

Despite shortcomings, the NAPs constitute our best approach to addressing AMR globally, as these bespoke plans are developed to account for differences between countries and their specific needs, including diversity in culture, approach to governance, resource availability and national scale of the AMR problem and progress against it across different local sectors. In countries where aquaculture activities pose AMR risks, it is essential to include the aquaculture sector in appropriate initiatives to tackle the problem. Still, conversion of intention into implementation of the strategic vision set out in each NAP remains a major challenge. Aquaculture receives a variable degree of attention in NAPs, and it is not always clear whether a specific policy is intended to apply to the sector or not. Even in the absence of explicit inclusion, it is vital that aquaculture recognizes the ambition of the GAP and responds to the challenge of AMR with actions that align to applicable NAPs where possible.

Lessons learnt from previous NAPs and their development—both from reflection within a country and by comparison with other nations—can ensure new NAPs address the shortcomings of former iterations, including elements of presentation (Willemsen et al. 2022). The next generation of NAPs may see improvement in appropriate coverage for aquaculture; however more detailed and nuanced guidance on how to support inclusion of aquaculture in NAPs would be helpful, as will greater awareness amongst the industry globally on how to engage with the NAP development process. Furthermore, greater recognition of the role of the environment in the AMR problem may lead to enhanced attention for aquaculture because aquaculture activities often sit at the nexus of human, terrestrial agriculture and environmental compartments of One Health. Indeed, future NAPs will almost certainly include more strategies relating to the natural environment, particularly since UNEP joined the Tripartite in 2022, forming the Quadripartite partnership. This expanded alliance is coordinating international efforts to combat AMR through a One Health approach, having developed the *One Health Joint Plan of Action (2022–2026)*, which has "curbing the silent pandemic of AMR" as one of the six main action tracks (FAO, UNEP, WHO, and WOAH 2022).

Greater awareness by the aquaculture sector concerning NAPs, their preparation and the content impacting aquaculture operations will help businesses to plan effectively by aligning their activities with NAP objectives and to anticipate likely policy changes that will affect their industry. This heightened awareness will open up opportunities to contribute to and influence the content of new NAPs. Meanwhile, enhanced appreciation for NAPs and their content amongst relevant researchers will allow them to play their part in achieving NAP goals.

The aquaculture sector globally has a responsibility to contribute to fulfilling the goals outlined in the GAP, and this is achieved most effectively by helping to meet the objectives of relevant NAPs.

Box 16.1 Thailand: A Case Study in NAP Evolution and Development
Reflecting the cyclic process of preparing and evaluating NAPs, Thailand demonstrates a country of high aquacultural importance with long-standing AMR strategy. Concern over antibiotic use in aquaculture in Thailand and potential AMR development was recognized for several decades (Saitanu et al. 1994): Holmström et al. (2003) found that 74% of shrimp farmers in Thailand interviewed used antibiotics in pond management, mostly as a prophylactic against disease outbreaks. Reduction in antimicrobial usage over the past two decades was driven particularly by external market pressures and refusals of contaminated imports by the European Union and USA (Vass et al. 2008). Several changes in practices and policies addressed these legitimate concerns to ensure international exports of shrimp from Thailand, including a switch to whiteleg shrimp (*Litopenaeus vannamei*) from the tiger shrimp (*Penaeus monodon*) due to greater availability of specific-pathogen-free (SPF) juveniles, which contributed to reduced reliance on antimicrobials (Henriksson et al. 2018).

Since the 1990s, coordinated policies against AMR have been developed in Thailand (Sumpradit et al. 2017), predating the GAP (WHO 2015). Activities centred on human antibiotic use and emerged from the Thailand Ministry of Public Health (Sumpradit et al. 2017). In 1998, the National AMR Surveillance Centre was established, and since 2005 this has been a WHO Collaborating Centre (CC) for AMR Surveillance for the Southeast Asia region. Currently, there are two CCs in Thailand (the only ones in Southeast Asia as of 2023): the Centre for Antimicrobial Resistance Surveillance and Training and the Centre for Antimicrobial Resistance (AMR) Prevention and Containment. Another pre-GAP initiative was the *Antibiotics Smart Use* programme (ASU; Sumpradit et al. 2012), again focused on the human health sector. A broader One Health approach initiative was the Thailand Antimicrobial Resistance Containment and Prevention Program (AMRCP), which operated in its first phase from 2012 to 2016 (Thamlikitkul et al. 2015). This programme's ten *operational actions* largely mirrored the five objectives of the GAP, including estimating burden (action 1), education (action 6) and regulation of supply (action 5), in particular banning the use of antibiotics as growth promoters in food-animal farming and stronger control of prescription through veterinarians (Thamlikitkul et al. 2015). Still, the AMRCP remained largely driven by the human health sector agenda, although noting veterinary issues.

Sumpradit et al. (2017) relate development of the Thailand National Strategic Plan on Antimicrobial Resistance, the country's first NAP, active from 2017 to 2021. Early steps included a wide stakeholders meeting organized by the Ministry of Public Health in 2014, followed by appointment in 2015 of a Coordinating and Integrating Committee on AMR, led by the same ministry (Ministry of Public Health Ministry of Agriculture and Cooperatives 2017). Completed and endorsed by the Cabinet in 2016, the plan is organized into six strategy areas around surveillance, regulation of antimicrobial distribution, infection prevention and control (IPC) in humans, IPC (i.e. biosecurity) in agriculture animals, public awareness and governance mechanisms. With the Ministry of Agriculture and Cooperatives (MOAC) involved in its drafting, animal health is intrinsic to AMR concerns in Thailand, with progressive joint initiatives between these two ministries including harmonization of surveillance protocols (Sumpradit et al. 2021). MOAC is a large ministry encompassing multiple departments including the Department of Livestock Development and the Department of Fisheries; however, aquaculture and aquatic animal health is more implied than explicit in the narrative of the Thailand NAP, with the word "animal*" appearing 55 times but only 7 appearances for "*fish*" and 1 for "shrimp" (* denotes wildcarding).

Writing is only the start of the NAP process: The biggest task is implementation. Evaluation and monitoring of the NAP are performed by the Thai government and the WHO Country Cooperation Strategy Programme (Sumpradit et al. 2017). An independent mid-term review was carried out in 2020 (WHO 2020) across six areas, of which AMR was one. Sumpradit et al. (2021) reported mixed levels of progress against NAP targets but, pleasingly, a 49% reduction in antimicrobial use in animals from 2017 to 2019, exceeding the 30% target, though this reduction remains unresolved at the level of aquaculture versus livestock (Lekagul et al. 2023). Notable successes include capacity building for microbiology laboratories in both human and animal sectors (specifically including aquaculture), raising of aquaculture AMR awareness and promotion of appropriate antibiotic use, and in investigating antimicrobial use in aquaculture (Sumpradit et al. 2021).

Preparatory work continues ahead of the next NAP. For example, the "AMR Dialogues" project assembled public and other stakeholders through online national and in-person events, to provide an AMR stakeholder map and engagement strategy (Poomchaichote et al. 2021). The current 2017–2021 cycle was recently extended until 2022; thus, the next NAP is expected to cover 2023–2027 (Sumpradit 2022), with discussion in 2022 over governance mechanisms. Dialogue and documentation internal to the new NAP are unlikely to be accessible to an external audience during this phase.

References

AACTING (2023) Systems for quantification of antimicrobial usage. https://aacting.org/. Accessed 5 June 2023

Anderson M, Schulze K, Cassini A, Plachouras D, Mossialos E (2019) A governance framework for development and assessment of national action plans on antimicrobial resistance. Lancet Infect Dis 19:e371–e384. https://doi.org/10.1016/S1473-3099(19)30415-3

Anonymous (2019) National strategic plan on antimicrobial resistance in Lao PDR 2019–2023. https://www.who.int/publications/m/item/lao-pdr-national-strategic-plan-on-antimicrobial-resistance-in-lao-pdr-2019-2023. Accessed 2 June 2023

Australian Government (2016) Implementation plan: Australia's first national antimicrobial resistance strategy 2015–2019. https://www.amr.gov.au/sites/default/files/2022-10/implementation-plan-australia-s-first-national-antimicrobial-resistance-strategy-2015-2019.pdf. Accessed 2 June 2023

Bondad-Reantaso M, Lavilla-Pitogo C, Lopez MML, Hao B (2020) Guidance in development of aquaculture component of a National Action Plan on antimicrobial resistance. Asian Fish Sci 33.S1:119–124. https://doi.org/10.33997/j.afs.2020.33.S1.017

Bondad-Reantaso MG, MacKinnon B, Karunasagar I, Fridman S, Alday-Sanz V, Brun E, Le Groumellec M, Li A, Surachetpong W, Karunasagar I, Hao B, Dall'Occo A, Urbani R, Caputo A (2023) Review of alternatives to antibiotic use in aquaculture. Rev Aquac 15:1421. https://doi.org/10.1111/raq.12786

Brunton LA, Desbois AP, Garza M, Wieland B, Mohan CV, Häsler B, Tam CC, Le PNT, Phuong NT, Van PT, Nguyen-Viet H, Eltholth MM, Pham DK, Duc PP, Linh NT, Rich KM, Mateus ALP, Hoque MA, Ahad A, Khan MNA, Adams A, Guitian J (2019) Identifying hotspots for antibiotic resistance emergence and selection, and elucidating pathways to human exposure: application of a systems-thinking approach to aquaculture systems. Sci Total Environ 687:1344–1356

Cabello FC (2006) Heavy use of prophylactic antibiotics in aquaculture: a growing problem for human and animal health and for the environment. Environ Microbiol 8:1137–1144. https://doi.org/10.1111/j.1462-2920.2006.01054.x

Caputo A, Bondad-Reantaso M, Karunasagar I, Hao B, Gaunt P, Verner-Jeffreys D, Fridman S, Dorado-Garcia A (2022) Antimicrobial resistance in aquaculture: a global analysis of literature and national action plans. Rev Aquac 15:568–578. https://doi.org/10.1111/raq.12741

Carelli DE, Mitsouli ET, Ogne JB, Pierre J (2023) The best laid plans? international governance perspectives in AMR national action plans in Europe. Eur J Public Health 33:682. https://doi.org/10.1093/eurpub/ckad080

Cassini A, Tacconelli E (2023) Potential for improvement in governance and national action plans to overcome antimicrobial resistance. Lancet 23:640–642. https://doi.org/10.1016/S1473-3099(22)00826-X

Charani E, Mendelson M, Pallett SJC, Ahmad R, Mpundu M, Mbamalu O, Bonaconsa C, Nampoothiri V, Singh S, Peiffer-Smadja N, Anton-Vazquez V, Moore LSP, Schouten J, Kostyanev T, Vlahović-Palčevski V, Kofteridis D, Silva Corrêa J, Holmes AH (2023) An analysis of existing national action plans for antimicrobial resistance—gaps and opportunities in strategies optimising antibiotic use in human populations. Lancet Glob Health 11:e466–e474. https://doi.org/10.1016/S2214-109X(23)00019-0

Chua AQ, Verma M, Hsu LY, Legido-Quigley H (2021) An analysis of national action plans on antimicrobial resistance in Southeast Asia using a governance framework approach. Lancet Reg Health West Pac 7:100084. https://doi.org/10.1016/j.lanwpc.2020.100084

Coordinating Ministry for Human Development and Cultural Affairs [of the Republic of Indonesia] (2021) The national action plan on antimicrobial resistance control year 2020–2024. https://www.who.int/publications/m/item/indonesia-second-national-action-plan-on-antimicrobial-resistance-2020-2024. Accessed 14 June 2023

Department of Health (2000) UK antimicrobial resistance strategy and action plan. https://webarchive.nationalarchives.gov.uk/ukgwa/20130107105354/http:/www.dh.gov.uk/prod_consum_dh/groups/dh_digitalassets/@dh/@en/documents/digitalasset/dh_4078448.pdf. Accessed 2 May 2023

Departments of Health and Agriculture, Forestry and Fisheries for the Republic of South Africa (2018) South African antimicrobial resistance national strategy framework. https://www.who.int/publications/m/item/south-africa-south-african-antimicrobial-resistance-national-strategy-framework-a-one-health-approach. Accessed 2 June 2023

Desbois AP, Garza M, Eltholth M, Hegazy YM, Mateus A, Adams A, Little DC, Høg E, Mohan CV, Ali SE, Brunton LA (2021) Systems-thinking approach to identify and assess feasibility of potential interventions to reduce antibiotic use in tilapia farming in Egypt. Aquaculture 540:736735. https://doi.org/10.1016/j.aquaculture.2021.736735

European Parliament and Council (2003) Regulation (EC) No 1831/2003 of the European Parliament and of the council of 22 September 2003 on additives for use in animal nutrition. Official J Eur Union (English version) 46:29–43

FAO (2023) Fisheries and aquaculture statistics. https://www.fao.org/fishery/statistics-query/en/aquaculture/aquaculture_quantity. Accessed 2 May 2023

FAO, UNEP, WHO & WOAH (2022) Global plan of action on one health. Towards a more comprehensive one health, approach to global health threats at the human-animal-environment interface. FAO, UNEP, WHO & WOAH, Rome, Italy. https://doi.org/10.4060/cc2289en

Federal Ministry of Health & Ministry of Animal Resources (2017) National action plan on antimicrobial resistance. https://www.who.int/publications/m/item/sudan-national-action-plan-on-antimicrobial-resistance. Accessed 2 June 2023

Federal Task Force on Combating Antibiotic-Resistant Bacteria (2015) National action plan for combating antibiotic-resistant bacteria. https://www.cdc.gov/drugresistance/us-activities/national-action-plan.html. Accessed 5 June 2023

Federal Task Force on Combating Antibiotic-Resistant Bacteria (2020) National action plan for combating antibiotic-resistant bacteria, 2020–2025. https://aspe.hhs.gov/reports/national-action-plan-combating-antibiotic-resistant-bacteria-2020-2025. Accessed 3 June 2023

Frumence G, Mboera LEG, Sindato C, Katale BZ, Kimera S, Metta E, Durrance-Bagale A, Jung AS, Mshana SE, Clark TG, Rweyemamu M, Legido-Quigley H, Matee MIN (2021) The governance and implementation of the National Action Plan on antimicrobial resistance in Tanzania: a qualitative study. Antibiotics 10:273. https://doi.org/10.3390/antibiotics10030273

Fuller WL, Hamzat OT, Aboderin AO, Gahimbare L, Kapona O, Yahaya AA, Kasambara W, Nikiema J-B, Ilboudo DW, Mpundu MM (2022) National action plan on antimicrobial resistance: an evaluation of implementation in the World Health Organization Africa region. J Public Health Africa 13:2000

Garza M, Mohan CV, Brunton L, Wieland B, Häsler B (2022) Typology of interventions for antimicrobial use and antimicrobial resistance in aquaculture systems in low- and middle-income countries. Int J Antimicrob Agents 59:106495. https://doi.org/10.1016/j.ijantimicag.2021.106495

Government of Ireland (2021) Ireland's second one health national action plan on antimicrobial resistance 2021–2025. https://www.gov.ie/en/publication/d72f1-joint-action-on-antimicrobial-resistance. Accessed 6 June 2023

Government of Japan (2016) National action plan on antimicrobial resistance (AMR). https://www.mhlw.go.jp/file/06-Seisakujouhou-10900000-Kenkoukyoku/0000138942.pdf. Accessed 2 June 2023

Government of Japan (2023) Second national action plan on antimicrobial resistance (AMR). https://www.who.int/publications/m/item/japan-second-national-action-plan-on-antimicrobial-resistance-2023-2027. Accessed 2 June 2023

Henriksson PJG, Rico A, Troell M, Klinger DH, Buschmann AH, Saksida S, Chadag MV, Zhang W (2018) Unpacking factors influencing antimicrobial use in global aquaculture and their implication for management: a review from a systems perspective. Sustain Sci 13:1105–1120. https://doi.org/10.1007/s11625-017-0511-8

HM Government (2019) Tackling antimicrobial resistance 2019–2024—the UK's five-year national action plan. Global and Public Health Group, Emergency Preparedness and Health Protection Policy Directorate, UK. https://www.gov.uk/government/publications/uk-5-year-action-plan-for-antimicrobial-resistance-2019-to-2024. Accessed 30 June 2023

Holmström K, Gräslund S, Wahlström A, Poungshompoo S, Bengtsson B-E, Kautsky N (2003) Antibiotic use in shrimp farming and implications for environmental impacts and human health. Int J Food Sci Technol 38:255–266. https://doi.org/10.1046/j.1365-2621.2003.00671.x

Inter-Agency Committee on Antimicrobial Resistance (2019) Philippine national action plan on antimicrobial resistance 2019–2023. https://pharma.doh.gov.ph/the-philippine-action-plan-to-combat-antimicrobial-resistance-2019-2023/. Accessed 6 June 2023

Interagency Coordination Group on Antimicrobial Resistance (2017) AMR Framework for Action. https://cdn.who.int/media/docs/default-source/antimicrobial-resistance/amr-gcp-tjs/iacg/other-documents/20170818-amr-ffa-v01.pdf?sfvrsn=796a325e_6. Accessed 10 May 2023

JANIS (2023) Japan nosocomial infections surveillance, Ministry of Health, Labour and Welfare, https://janis.mhlw.go.jp/english/about/index.html. Accessed 10 May 2023

Joshi MP, Hafner T, Twesigye G, Ndiaye A, Kiggundu R, Mekonnen N, Kusu N, Berthé S, Lusaya EP, Acho A, Tuala RT, Siddiqua A, Kaboré H, Aidara SS, Guzman J (2021) Strengthening multisectoral coordination on antimicrobial resistance: a landscape analysis of efforts in 11 countries. J Pharm Policy Practice 14:27. https://doi.org/10.1186/s40545-021-00309-8

Kickbusch I, Gleicher D (2012) Governance for health in the 21st century. WHO regional Office for Europe, Copenhagen, Denmark. https://apps.who.int/iris/handle/10665/326429. Accessed 7 June 2023

Lekagul A, Kirivan S, Tansakul N, Krisanaphan C, Srinha J, Laoprasert T, Kaewkhankhaeng W, Tangcharoensathien V (2023) Antimicrobial consumption in food-producing animals in Thailand between 2017 and 2019: the analysis of national importation and production data. PLoS One 18:e0283819. https://doi.org/10.1371/journal.pone.0283819

Lulijwa R, Rupia EJ, Alfaro AC (2020) Antibiotic use in aquaculture, policies and regulation, health and environmental risks: a review of the top 15 major producers. Rev Aquac 12:640–663. https://doi.org/10.1111/raq.12344

Medical Administration and Hospital Authority (2022) Notice on the issuance of the National Action Plan to Curb Microbial Resistance (2022–2025) [translated]. http://www.nhc.gov.cn/yzygj/s7659/202210/2875ad7e2b2e46a2a672240ed9ee750f.shtml. Accessed 14 June 2023

Mesa Varona O, Chaintarli K, Muller-Pebody B, Anjum MF, Eckmanns T, Norström M, Boone I, Tenhagen B-A (2020) Monitoring antimicrobial resistance and drug usage in the human and livestock sector and foodborne antimicrobial resistance in six European countries. Infect Drug Resist 13:957–993. https://doi.org/10.2147/IDR.S237038

Ministère des Solidarités et de la Santé (2022) National strategy for preventing infections and antibiotic resistance. https://sante.gouv.fr/IMG/pdf/national_strategy_for_preventing_infections_and_antibiotic_resistance_2022-2025_pdf. Accessed 3 June 2023

Ministry of Health Malaysia (2017) Malaysian action plan on antimicrobial resistance (MyAP-AMR) 2017–2021. https://www.who.int/publications/m/item/malaysia-malaysian-action-plan-on-antimicrobial-resistance-(myap-amr). Accessed 2 June 2023

Ministry of Health; Ministry of Agriculture, Forestry and Fisheries; Ministry of Environment (2014) National strategy to combat antimicrobial resistance 2015–2017. https://cdn.who.int/media/docs/default-source/antimicrobial-resistance/amr-spc-npm/nap-library/cambodia-strategy-to-combat-amr-3-years-22-july-2014(final).pdf. Accessed 5 June 2023

Ministry of Health; Ministry of Agriculture, Forestry and Fisheries; Ministry of Environment (2019) Multi-sectoral action plan on antimicrobial resistance in Cambodia 2019–2023. https://cdn.who.int/media/docs/default-source/antimicrobial-resistance/amr-spc-npm/nap-library/cambodia-multi-sectoral_actionplan_on_amr_2019-2023.pdf. Accessed 3 June 2023

Ministry of National Health Services, Regulations & Coordination (2017) Antimicrobial resistance national action plan. https://www.nih.org.pk/wp-content/uploads/2018/08/AMR-National-Action-Plan-Pakistan.pdf. Accessed 3 June 2023

Ministry of Public Health and Ministry of Agriculture and Cooperatives (2017) Thailand's national strategic plan on antimicrobial resistance 2017–2021. https://www.who.int/publications/m/item/thailand-national-strategic-plan-on-antimicrobial-resistance-2017-2021. Accessed 6 June 2023

Moghnieh R, Omar AA, Al-Motawaa A (2022) Kuwait national action plan on antimicrobial resistance 2022. https://www.who.int/publications/m/item/kuwait-national-action-plan-on-antimicrobial-resistance. Accessed 14 June 2023

Mshana SE, Sindato C, Matee MI, Mboera LEG (2021) Antimicrobial use and resistance in agriculture and food production systems in Africa: systematic review. Antibiotics 10:976. https://doi.org/10.3390/antibiotics10080976

Munkholm L, Rubin O, Bækkeskov E, Humboldt-Dachroeden S (2021) Attention to the Tripartite's one health measures in national action plans on antimicrobial resistance. J Public Health Policy 42:236–248. https://doi.org/10.1057/s41271-021-00277-y

Norwegian Ministries (2015) National strategy against antibiotic resistance 2015–2020. https://www.regjeringen.no/contentassets/5eaf66ac392143b3b2054aed90b85210/antibiotic-resistance-engelsk-lavopploslig-versjon-for-nett-10-09-15.pdf. Accessed 2 June 2023

Ogyu A, Chan O, Littmann J, Pang HH, Lining X, Liu P, Matsunaga N, Ohmagari N, Fukuda K (2020) National action to combat AMR: a one-health approach to assess policy priorities in action plans. Br Med J Glob Health 5:e002427. https://doi.org/10.1136/bmjgh-2020-002427

Orubu ESF, Sutradhar I, Zaman MH, Wirtz VJ (2020) Benchmarking national action plans on antimicrobial resistance in eight selected LMICs: focus on the veterinary sector strategies. J Glob Health 10:020414. https://doi.org/10.7189/jogh.10.020414

Özçelik EA, Doucet C, Kang H, Levy N, Feldhaus I, Hashiguchi TCO, Lerouge A, Cecchini C (2022) A comparative assessment of action plans on antimicrobial resistance from OECD and G20 countries using natural language processing. Health Policy 126:522–533. https://doi.org/10.1016/j.healthpol.2022.03.011

Pant I, Khosla S, Lama JT, Shanker V, AlKhaldi M, El-Basuoni A, Michel B, Bitar K, Nsofor IM (2022) Decolonising global health evaluation: synthesis from a scoping review. PLOS Global Public Health 2:e0000306. https://doi.org/10.1371/journal.pgph.0000306

Patel J, Harant A, Fernandes G, Mwamelo AJ, Hein W, Dekker D, Sridhar D (2023) Measuring the global response to antimicrobial resistance, 2020–21: a systematic governance analysis of 114 countries. Lancet Infect Dis 23:706–718. https://doi.org/10.1016/S1473-3099(22)00796-4

Poomchaichote T, Osterrieder A, Prapharsavat R, Naemiratch B, Ruangkajorn S, Thirapantu C, Sukrung K, Kiatying-Angsulee N, Sumpradit N, Punnin S, Limmathurotsakul D, Cheah, & P.Y. (2021) "AMR Dialogues": a public engagement initiative to shape policies and solutions on antimicrobial resistance (AMR) in Thailand. Wellcome Open Research 6:188. https://doi.org/10.12688/wellcomeopenres.17066.2

Public Health Agency of Canada (2017) Tackling antimicrobial resistance and antimicrobial use: a Pan-Canadian framework for action. https://www.canada.ca/en/health-canada/services/publications/drugs-health-products/tackling-antimicrobial-resistance-use-pan-canadian-framework-action.html. Accessed 2 June 2023

ReAct (2023) ReAct. https://www.reactgroup.org/. Accessed 17 May 2023

Republic of Cyprus Ministry of Health (2012) Cyprus national strategy for combating antibiotic resistance. https://www.who.int/publications/m/item/cyprus-national-action-plan-on-antimicrobial-resistance. Accessed 26 May 2023

Reverter M, Sarter S, Caruso D, Avarre J-C, Combe M, Pepey E, Pouyaud L, Vega-Heredía S, de Verdal H, Gozlan RE (2020) Aquaculture at the crossroads of global warming and antimicrobial resistance. Nat Commun 11:1870. https://doi.org/10.1038/s41467-020-15735-6

Rittel HWJ, Webber MM (1973) Dilemmas in a general theory of planning. Policy Sci 4:155–169

Rudolph, N. (2022) Farmer access to antibiotics 'revised' in South Africa. Food For Mzansi. https://www.foodformzansi.co.za/farmer-access-to-antibiotics-revised-in-south-africa/. Accessed 2 June 2023

Saitanu K, Amornsin A, Kono F, Tsai C-E (1994) Antibiotic residues in tiger shrimp (*Penaeus monodon*). Asian Fish Sci 7:47–52

Santos L, Ramos F (2018) Antimicrobial resistance in aquaculture: current knowledge and alternatives to tackle the problem. Int J Antimicrob Agents 52:135–143. https://doi.org/10.1016/j.ijantimicag.2018.03.010

Schar D, Klein EY, Laxminarayan R, Gilbert M, Van Boeckel TP (2020) Global trends in antimicrobial use in aquaculture. Sci Rep 10:21878. https://doi.org/10.1038/s41598-020-78849-3

Schar D, Zhao C, Wang Y, Larsson DGJ, Gilbert M, Van Boeckel TP (2021) Twenty-year trends in antimicrobial resistance from aquaculture and fisheries in Asia. Nat Commun 12:5384. https://doi.org/10.1038/s41467-021-25655-8

Schrijver R, Stijntjes M, Rodríguez-Baño J, Tacconelli E, Babu Rajendran N, Voss A (2018) Review of antimicrobial resistance surveillance programmes in livestock and meat in EU with focus on humans. Clin Microbiol Infect 24:577–590. https://doi.org/10.1016/j.cmi.2017.09.013

Somanader DS, Brunskill I, Perrin M, Barkema HW, Hillier S, Hindmarch S, Weese JS, Wright GD, Morris AM (2022) Canada has an opportunity to address antimicrobial resistance through COVID-19 recovery spending. Lancet Reg Health Am 16:100393. https://doi.org/10.1016/j.lana.2022.100393

Sumpradit N (2022) Reflection: mobilizing domestic resources for AMR interventions country perspective: Thailand. Presentation at Bi-Regional (Africa & Asia) Policy AMR Workshop, 12–14 April 2022. https://www.ivi.int/wp-content/uploads/2022/04/Day_2_Nithima_Sumpradit_Mobilizing_Domestic_Resources_for_AMR_Interventions_Reflections_from_Thailand.pdf. Accessed 6 June 2023

Sumpradit N, Chongtrakul P, Anuwong K, Pumtong S, Kongsomboon K, Butdeemee P, Khonglormyati J, Chomyong S, Tongyoung P, Losiriwat S, Seesuk P, Suwanwaree P, Tangcharoensathien V (2012) Antibiotics Smart use: a workable model for promoting the rational use of medicines in Thailand. Bull World Health Organ 90:905–913. https://doi.org/10.2471/BLT.12.105445

Sumpradit N, Wongkongkathep S, Poonpolsup S, Janejai N, Paveenkittiporn W, Boonyarit P, Jaroenpoj S, Kiatying-Angsulee N, Kalpravidh W, Sommanustweechai A, Tangcharoensathien V (2017) New chapter in tackling antimicrobial resistance in Thailand. Br Med J 358:j3415. https://doi.org/10.1136/bmj.j2423

Sumpradit N, Wongkongkathep S, Malathum K, Janejai N, Paveenkittiporn W, Yingyong T, Chuxnum T, Vijitleela A, Boonyarit P, Akaleephan C, Manosuthi W, Thienthong V, Srinha J, Wongsrichai S, Laoprasert T, Athipunyakom P, Kriengchaiyaprug N, Intarukdach K, Numsawad S, Somjetanakul N, Punnin S, Kiatying-Angsulee N (2021) Thailand's national strategic plan on antimicrobial resistance: progress and challenges. Bull World Health Organ 99:661–673. https://doi.org/10.2471/BLT.20.280644

Thamlikitkul V, Rattanaumpawan P, Boonyasiri A, Pumsuwan V, Judaeng T, Tiengrim S, Paveenkittiporn W, Rojanasthien S, Jaroenpoj S, Issaracharnvanich S (2015) Thailand antimicrobial resistance containment and prevention program. J Global Antimicrob Resist 3:290–294

The Union of South Africa Department of Agriculture (1947) Bill to provide for the registration of fertilizers, farm feeds, sterilizing plants and certain remedies; to regulate the importation and sale of fertilizers, farm feeds, seeds and certain remedies, and to provide for matters incidental thereto. Government Gazette. https://www.gov.za/sites/default/files/gcis_document/201505/act-36-1947.pdf. Accessed 9 June 2023

UK-VARSS (2022) Veterinary antibiotic resistance and sales surveillance report (UK-VARSS 2021). Veterinary Medicines Directorate, New Haw, Addlestone, UK

United Republic of Tanzania (2022) The national action plan on antimicrobial resistance 2023–2028. https://www.who.int/publications/m/item/united-republic-of-tanzania-second-national-action-plan-on-antimicrobial-resistance. Accessed 6 June 2023

van Heuvel L, Caini S, Dückers MLA, Paget J (2022) Assessment of the inclusion of vaccination as an intervention to reduce antimicrobial resistance in AMR national action plans: a global review. Glob Health 18:85. https://doi.org/10.1186/s12992-022-00878-6

Vass M, Hruska K, Franek M (2008) Nitrofuran antibiotics: a review on the application, prohibition and residual analysis. Veterinarni Medicina 53:469–500. https://doi.org/10.17221/1979-VETMED

Watts JEM, Schreier HJ, Lanska L, Hale MS (2017) The rising tide of antimicrobial resistance in aquaculture: sources, sinks and solutions. Mar Drugs 15:158. https://doi.org/10.3390/md15060158

WHO (2013) WHO evaluation practice handbook. WHO Press, Geneva, Switzerland

WHO (2015) Global action plan on antimicrobial resistance. WHO Press, Geneva, Switzerland
WHO (2020) Independent mid-term evaluation of the WHO-thailand country cooperation strategy 2017–2021. https://www.who.int/publications/m/item/independent-mid-term-evaluation-of-the-who-thailand-country-cooperation-strategy-2017-2021. Accessed 30 June 2023
WHO (2022a) Tripartite AMR Country Self-assessment Survey—TrACSS (6.0) 2022. https://www.who.int/publications/m/item/tripartite-amr-country-self-assessment-survey%2D%2D-tracss-(6.0)-2022. Accessed 28 June 2023
WHO (2022b) WHO implementation handbook for national action plans on antimicrobial resistance: guidance for the human health sector. WHO Press, Geneva, Switzerland
WHO (2023a) Library of national action plans. https://www.who.int/teams/surveillance-prevention-control-AMR/national-action-plan-monitoring-evaluation/library-of-national-action-plans. Accessed 2 May 2023
WHO (2023b) Supporting countries with national action plan implementation. https://www.who.int/activities/supporting-countries-with-national-action-plan-implementation. Accessed 2 May 2023
WHO, FAO & OIE (2016) Antimicrobial resistance: a manual for developing national action plans. WHO Press, Geneva, Switzerland. https://www.who.int/publications/i/item/antimicrobial-resistance-a-manual-for-developing-national-action-plans. Accessed 5 June 2023
WHO, FAO & OIE (2019) Monitoring and evaluation of the global action plan on antimicrobial resistance: framework and recommended indicators. WHO Press, Geneva, Switzerland
Willemsen A, Reid S, Assefa Y (2022) A review of national action plans on antimicrobial resistance: strengths and weaknesses. Antimicrob Resist Infect Control 11:90. https://doi.org/10.1186/s13756-022-01130-x
WOAH (2022) Aquatic animal health code. https://www.woah.org/en/what-we-do/standards/codes-and-manuals/aquatic-code-online-access/. Accessed 7 June 2023

The manufacturer's authorised representative in the EU is Springer Nature Customer Service Centre GmbH, Europaplatz 3, 69115 Heidelberg, Germany. If you have any concerns regarding our products, please contact ProductSafety@springernature.com

Printed and bound by CPI Group (UK) Ltd, Croydon, CR0 4YY
01/10/2026
02225956-0013